Flexible Metal–Organic Frameworks

Structural Design, Synthesis and Properties

Inorganic Materials Series

Series editors:
Duncan W. Bruce, *University of York, UK*
Dermot O'Hare, *University of Oxford, UK*
Richard I. Walton, *University of Warwick, UK*

Titles in the Series:

1: Pre-combustion Carbon Dioxide Capture Materials
2: Post-combustion Carbon Dioxide Capture Materials
3: Solar Energy Capture Materials
4: Bioinspired Inorganic Materials: Structure and Function
5: Energy Storage and Conversion Materials
6: Two-dimensional Inorganic Nanomaterials for Conductive Polymer Nanocomposites
7: The Chemistry of Inorganic Biomaterials
8: Computer Simulation of Porous Materials: Current Approaches and Future Opportunities
9: Inorganic Thermoelectric Materials: From Fundamental Concepts to Materials Design
10: Biomedical Applications of Inorganic Materials
11: Photocatalysis Using 2D Nanomaterials
12: Mechanical Behaviour of Metal–Organic Framework Materials
13: Flexible Metal–Organic Frameworks: Structural Design, Synthesis and Properties

How to obtain future titles on publication:
A standing order plan is available for this series. A standing order will bring delivery of each new volume immediately on publication.

For further information please contact:
Book Sales Department, Royal Society of Chemistry, Thomas Graham House, Science Park, Milton Road, Cambridge, CB4 0WF, UK
Telephone: +44 (0)1223 420066, Fax: +44 (0)1223 420247
Email: booksales@rsc.org
Visit our website at books.rsc.org

Flexible Metal–Organic Frameworks

Structural Design, Synthesis and Properties

Edited by

Susumu Kitagawa
Kyoto University, Japan
Email: kitagawa@icems.kyoto-u.ac.jp

ROYAL SOCIETY OF CHEMISTRY

Inorganic Materials Series No. 13

Print ISBN: 978-1-83916-452-1
PDF ISBN: 978-1-83916-661-7
EPUB ISBN: 978-1-83916-662-4
Print ISSN: 2472-3819
Electronic ISSN: 2472-3827

A catalogue record for this book is available from the British Library

The Royal Society of Chemistry is a charity, registered in England and Wales, Number 207890, and a company incorporated in England by Royal Charter (Registered No. RC000524), registered office: Burlington House, Piccadilly, London W1J 0BA, UK, Telephone: +44 (0) 20 7437 8656.

For further information see our website at www.rsc.org

Printed in the United Kingdom by CPI Group (UK) Ltd, Croydon, CR0 4YY, UK

Series Preface

Materials chemistry has grown hugely in the past few decades and continues to evolve rapidly. This is especially the case since the subject now finds itself central to many of the major challenges that face global society, such as those concerning energy and the environment; but applications of materials are also found in many everyday situations, with ongoing developments in technology around electronics and the miniaturisation of devices. Although the study of materials is a highly interdisciplinary topic, the role of innovative, fundamental chemistry continues to lie at the heart of the discovery of new materials, the understanding of their atomic-scale structure and the relationship of structure to a useful property.

With the *Inorganic Materials Series*, our aim is to provide a series of themed volumes, each reflecting the diversity of the topic as it continues to develop over the coming years, with many chosen to relate closely to emerging practical applications. The use of the title *Inorganic Materials* is by no means limiting, but instead emphasises the study of the chemistry of all elements in the Periodic Table and how they impact the development of new materials. The chapters in each book will have a pedagogical flavour with a target audience from first-year postgraduate student upwards, but with wide subject coverage ranging from continuous inorganic solids, to molecular and soft matter systems. We will use expert, active researchers as editors and contributors to provide an up-to-date perspective of their field. At the same time our aim is to provide an international perspective, so to reflect the diversity and interdisciplinarity of the now very broad area of the subject.

Duncan W. Bruce, York, UK
Dermot O'Hare, Oxford, UK
Richard I. Walton, Warwick, UK

Inorganic Materials Series No. 13
Flexible Metal–Organic Frameworks: Structural Design, Synthesis and Properties
Edited by Susumu Kitagawa

Published by the Royal Society of Chemistry, www.rsc.org

Preface

1 Porous Materials

Activated carbon, a representative porous material, was used for medical purposes in ancient Egypt (1550 BCE), as described in papyrus; it is still widely used today for water purification and other purposes. In 1753 CE, some three thousand years later, the first zeolites were discovered from natural ores as new porous materials and in the first half of the 20th century, zeolites were successfully synthesised artificially. Activated carbon, zeolites, and clay minerals are now well-known examples of porous materials, which are used in our daily lives and in industry. Activated carbon has an excellent adsorption function due to its various pore sizes, but it is not crystalline because it does not have regular pores. Conversely, zeolites have strongly bound inorganic elements such as Al^{3+} and silicate, giving them stable porous crystals with a regular structure. Guests, such as gas molecules, organic molecules, and ions, are incorporated into these pores. In other words, the stability of the framework is essential for the porous structure.

Thus, existing porous materials have been used for many years as an integral part of human life. If materials with functions such as storage and separation that surpass the functions of activated carbon and zeolites, or materials with completely new porous functions are discovered, and can be produced rationally using low-energy processes, they are expected to bring about revolutionary changes in human life. It is desirable to design nano spaces at will to realise such functions of porous materials, but it has been difficult to do so at the molecular level in zeolites and activated carbon. For this reason, no major scientific progress has been made in porous materials since the discovery of zeolites. In the late 1990s, however, the development of metal complex materials with stable pores based on a completely new concept, called

Inorganic Materials Series No. 13
Flexible Metal–Organic Frameworks: Structural Design, Synthesis and Properties
Edited by Susumu Kitagawa

Published by the Royal Society of Chemistry, www.rsc.org

porous coordination polymers (PCPs) or metal–organic frameworks (MOFs), made it possible to control pore size and shape at the molecular level.

2 Coordination Polymers (CPs) Constructed by Organic Ligands

CPs are infinite network solids prepared using coordination bonds from metal ions and organic or inorganic ligands. These contain structural motifs extremely important in CP design: (1) one-dimensional chains and ladders; (2) two-dimensional sheets such as square and honeycomb lattices; and (3) three-dimensional lattices such as diamond and rock salt. Especially in CPs with organic ligands, the combination of metal cations and ligands is almost unlimited, making it possible to construct extremely diverse network structures. There are about 90 000 structures of a vast number of CPs so far, among which more than 30 000 CPs with porous structures have been synthesised.[1]

In the late 1980s and early 1990s, structural chemistry studies were mainly focused on network structures themselves. CPs were synthesised using Cu(I) ions, which are prone to forming infinite network structures, and attention was paid to polymer-like structures (*e.g.*, similarities to minerals and natural structures). The earliest examples were dinitrile CP $[Cu\{NC(C_4H_8)CN\}_2]NO_3$,[2] tris-nitrile CP, $[Cu\{C(C_6H_4CN)_4\}]$,[3] bipyridine CP $[Cu_2(4,4'\text{-bpy})_3(NO_3)_2](H_2O)_3$,[4] and pyrazine CP, $\{[Cu_2(pyz)_3(MeCN)_2](PF_6)_2(Me_2CO)\}$.[5] It is no exaggeration to state that the chemistry of CP using organic ligands started with Cu(I) ions. Interestingly, the last CP contains organic acetone molecules in the pores of the honeycomb network, an early good example of CP crystals with potentially accessible porous structure. It was not clear at that time whether stable pores could be obtained. Crystals with permanent porosity, which are stable even in the absence of guests, have been long awaited.

3 Porous Coordination Polymers (PCPs) and Metal–Organic Frameworks (MOFs)

In 1997, CPs of $[M_2(4,4'\text{-bpy})_3(NO_3)_4]$ (M = Co, Ni, Zn, 4,4′-bpy = 4,4′-bipyridine) were synthesised.[6] These showed a robust porous framework even after the removal of guests, and reversible gas (nitrogen, oxygen and methane) adsorption at room temperature between 1 and 35 atm. It is significant that the gas adsorption capacity at room temperature and high pressure is realised, and that it functions stably enough even under practical conditions. The following year, $[Zn(bdc)(H_2O)](DMF)$ was synthesised (bdc = benzene-1,4-dicarboxylate),[7] and its pore properties were investigated by low temperature and low-pressure carbon dioxide and nitrogen adsorption measurements. Subsequently, the stability of the crystal structure of $[M_2(4,4'\text{-bpy})_3(NO_3)_4]$ was reconfirmed by the detailed temperature dependence of the guest-free single crystal structure.[8] This clearly supported the presence of CP with robust pores. Since then, the following decades were marked by functional chemistry based on nanosized pores in CPs, resulting in those with various pores (sizes and shapes) and various network backbones. This chemistry has developed rapidly, involving many researchers around the world. Regarding the terminology of MOFs and PCPs, it is recommended[9] that MOFs are defined as 'organometallic frameworks consisting of a coordination network of organic ligands containing potential voids'. This is synonymous with PCPs.

4 Discovery of Flexibility in MOFs

The third development in CPs (Figure 1) is a demonstration of the flexibility and dynamic properties inherent in the MOF structure. Flexible MOFs were predicted just before the year 2000,[10] and shortly thereafter, in early 2000, several flexible MOFs were synthesised: [Cu(BF_4)$_2$(4,4′-bpy)] (ELM1),[11] [Cr(III)(OH)·{bdc}] (MIL-53),[12,13] [Cu$_2$(pzdc)$_2$(dpyg)] (CPL7) (pzdc = pyrazine-2,3-dicarboxylate; dpyg = 1,2-di(4-pyridyl)glycol),[14] [Cu$_2$(dhbc)$_2$(4,4′-bpy)] (CID1) (dhbc = 2,5-dihydroxybenzoate), and [Cu(bdc)(4,4′-bpy)$_2$].[15] These gave sigmoidal adsorption isotherms, which show no adsorption at low pressure but a sudden gate-opening behaviour when sufficient pressure is applied. This type of isotherm does not belong to any of the seven classifications (six-type adsorption isotherms by IUPAC). These exhibit a structural change from crystal to crystal in response to the pressure of adsorbates, a feature that distinguishes them from other porous materials and is reminiscent of the induced fit mechanism of bioenzymes and the cooperative phenomenon of haemoglobin.

CPL7 and CID1 showed a different gate-opening pressure of the sigmoid curve for the adsorption of nitrogen, oxygen, carbon dioxide, and methane at room temperature. The clear recognition of such a weakly interacting low-boiling gas, even in a supercritical state, was unprecedented and a big

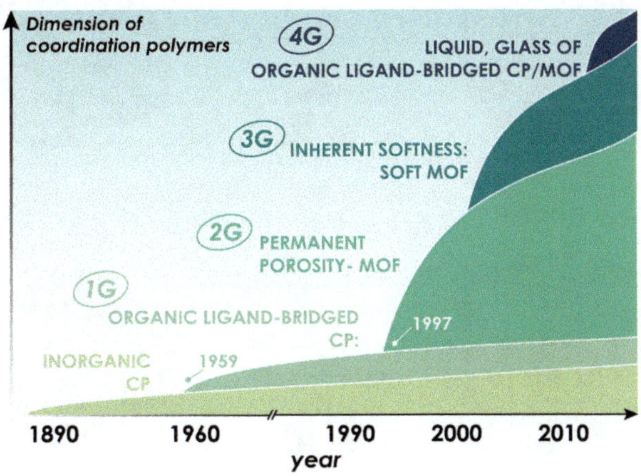

Figure 1 Discontinuous progress in MOF chemistry. Reproduced from ref. 17 with permission from John Wiley & Sons, Copyright 2020.

surprise at the time. In other words, at the beginning of the 21st century, a new material called the flexible MOF was born, with excellent properties that clearly set it apart from existing porous materials, and it was a milestone in MOF science. Later, it was comprehensively viewed as a crystal phase transition involving deformation of the entire crystal structure and local molecular motion within a crystal, caused not only by chemical stimuli (gas adsorption) but also by physical stimuli (temperature, pressure, light, electric field). The flexibility and dynamics of these structures and the multi-stability due to their porosity distinguish them from conventional porous solids such as carbon, zeolites, phosphates and mesoporous silica.

5 Scientific Significance of MOF Development

CPs are categorised into three generations (Figure 2).[10] (1) CPs that collapse by this operation as the first-generation type (1G). (2) Substances that have a stable spatial structure and can be used (called PCP or MOF) as the

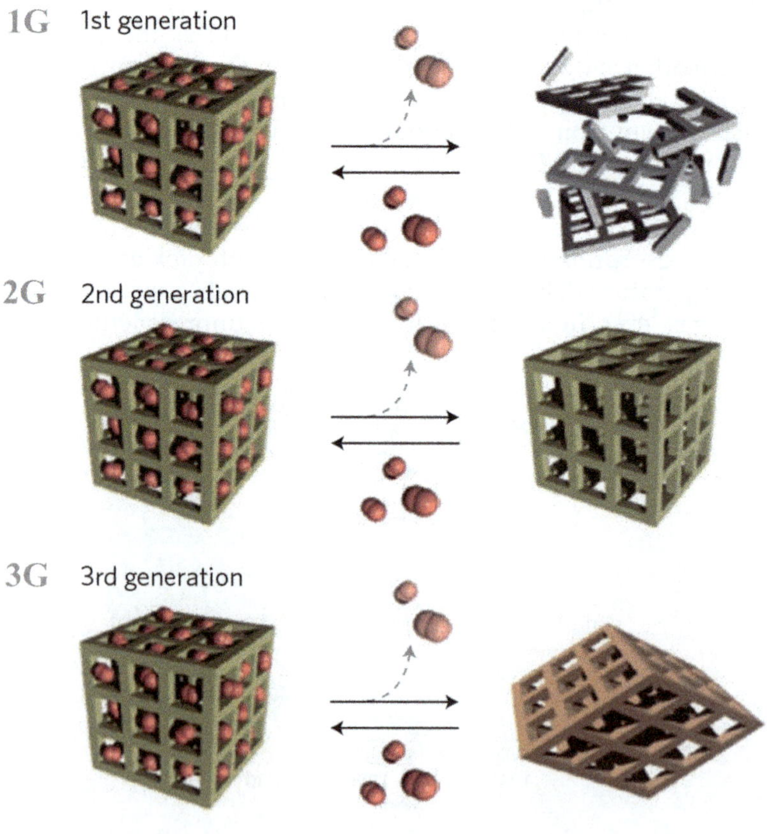

Figure 2 Classification of coordination polymers. Reproduced from ref. 16 with permission from Springer Nature, Copyright 2009.

second-generation type (2G). (3) The flexibility of the metal–organic ligand backbone is a characteristic of the third-generation type (3G), which is expected to have a dynamic structure and function. Later, it became clear that this property encompassed not only MOFs but also a wide range of other organic crystals, and the term soft porous crystal (SPC) emerged as a term to broadly include materials with that attribute.[16]

Three important developments in CP along the timeline of the last few decades have had a major impact on the direction of research trends (Figure 1).[17] The first development is the discovery of organic ligand-bridging structures (first generation, 1G). The application of organic ligands to CPs has expanded chemistry in terms of modularity and design and has given rise to numerous new networks. Single-crystal X-ray analysis was essential to characterise these structures. The first organic ligand-bridged coordination polymers characterised by X-ray crystallography were constructed from the d^{10} metal ion Cu^+ and adiponitrile, as mentioned previously.[2]

The second noteworthy progression was the development of CPs with robust and permanent porosities by gas adsorption, which opened up a new direction for the chemistry of porous materials. In general, CPs are isolated as stable crystals with guest ions and/or solvents filling the pores. To utilise the pores, these guest molecules must be removed (by decompression and heating, *etc.*). Porous materials are characterised by their high density of introduced voids, in contrast to dense solid crystals. Crystal structures of organic molecule-based networks with pores are often disrupted or deformed by the removal of guest ions or molecules, and their porous function cannot be utilised. Regarding the mechanical and thermodynamic stability of the porous structure, it was common knowledge that porous materials using organic materials do not provide a robust porous framework. The demonstration of the strong and permanent porosity of CPs by gas adsorption in the late 1990s was an epoch-making event in CPs and opened a new direction in the chemistry of porous materials. Since then, many such porous crystals have been synthesised and are now also called MOFs and PCPs (2G).

The third development was the demonstration of the intrinsic flexibility and dynamic properties of the MOF structures. In condensed matter systems such as metal oxides, structural phase transitions of crystalline phases responsive to external physical stimuli are frequently observed and have been studied as a treasure trove of diverse solid–state properties such as conductivity, dielectricity, and magnetism. Structural phase transitions have also been well-studied in molecular organic crystals. When compared to external physical stimuli, the phenomenon of incorporation of chemical species into crystals is a kind of chemical reaction that also generates space in the solid for their diffusion and trapping. To incorporate guest molecules into a crystal, the crystal structure itself is broken (amorphous) or a structural phase transition occurs to give rise to a new crystalline phase. There is a competition between the interaction energy with guests and MOF that contributes to the stabilisation of the system and the energy of the structural change that contributes to its destabilisation. For the guests to penetrate the dense structure, and for a phase change to occur and a new crystalline phase to be formed, it is important that a new phase overcomes the energetically negative effect of the structural change, and that the interaction energy complements it and moves in the direction of the stabilisation. Molecular crystals of metal complexes are no exception and crystals in which metal complex-assembled crystals change colour (solvatochromism, vapochromism) on their exposure to moisture or organic (mainly solvent) vapour, were discovered early on.

Solid crystals are conventionally thought of as solid, but are there any that are soft enough to incorporate molecules? Since zeolites have a robust structure with pre-prepared pores, the introduction of guest molecules rarely changes the structure. It is common knowledge that materials with a porous structure have a mechanically and thermodynamically stable pore structure. The author studied the precedent of structural phase transitions induced by external stimuli in many solid-state materials and came up with the idea that MOFs have an intrinsically flexible framework that undergoes structural deformation in response to external stimuli, especially the concentration (pressure) of guest species, spreading cooperatively throughout a crystal. Subsequently, framework deformations of MOFs (flexible MOFs) were discovered due to the incorporation of guests, and it became clear that a phase transition occurs readily in MOF families, which is an outstanding feature of MOFs. This finding broke the common sense that porous materials have rigid pores and caused a paradigm shift in MOFs, where the materials are capable of framework deformation to suit the guest. This property has established a position as a completely new porous material (3G) that is distinct from conventional materials such as activated carbon and zeolites.

6 Scope of This Book

MOFs have opened up new horizons in practically all branches of engineering, physics, chemistry, biology and medicine (Figure 3).[18,19] They are increasingly being used for the storage, separation, and catalytic transformation of

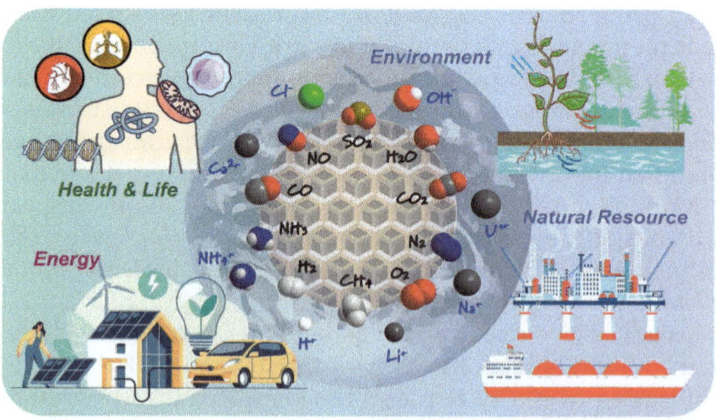

Figure 3 Future prospects for MOFs contribution to social issues.

carbon dioxide, methane, and other gaseous and aromatic molecules, as well as drugs, with thousands of research reports being published every year. There are now many books and review articles on MOFs, but few have been written on flexible MOFs. Their cooperative phenomena lead to multiple chemical or physical changes upon stimulation of the materials, leading to their designation as multifunctional or 'smart' materials. Flexible MOFs are unique materials with structural flexibility not found in other porous materials. These structural features and functions produced by flexible MOFs are being developed for practical use because they are unique and cannot be found in other materials today.

This book provides a basic and accessible overview of the historical background of the chemistry of flexible MOFs, their characteristics, especially their design and synthesis, dynamic structural analysis, functions, theoretical treatment and interpretation of mechanisms, and their applications. The chapters in this book are written by researchers who were involved in flexible MOF chemistry from an early stage and have developed the field. This book also presents many examples of flexible MOFs and summarises the research methods used to study them. The understanding of this chemistry will be of great help to young researchers working on porous materials, as well as to those who are already familiar with conventional porous materials, in creating new materials.

Acknowledgments

I would like to thank Professor Richard Walton, series editor, for his help and support in launching this project in May 2020. I would also like to thank Ms Katie Morrey (Commissioning Editor), Merlin Fox (Associate Publisher), Amina Headley (Editorial Assistant) and the Royal Society of Chemistry. I also thank Professor Shigeyoshi Sakaki, Assistant Professors Ken-ichi Otake and Shotaro Hiraide, Drs Ping Wang, Ziqian Xue, Maryam Nurhuda and Nathan C. Flanders, for their help in my peer review.

References

1. P. Z. Moghadam, A. Li, S. B. Wiggin, A. Tao, A. G. P. Maloney, P. A. Wood, S. C. Ward and D. Fairen-Jimenez, *Chem. Mater.*, 2017, **29**, 2618.
2. Y. Kinoshita, I. Matsubara, T. Higuchi and Y. Saito, *Bull. Chem. Soc. Jpn.*, 1959, **32**, 1221.
3. B. F. Hoskins and R. Robson, *J. Am. Chem. Soc.*, 1989, **111**, 5962.
4. O. M. Yaghi and H. Li, *J. Am. Chem. Soc.*, 1995, **117**, 10401.
5. S. Kitagawa, M. Munakata and T. Tanimura, *Inorg. Chem.*, 1992, **31**, 1714.
6. M. Kondo, T. Yoshitomi, K. Seki, H. Matsuzaka and S. Kitagawa, *Angew. Chem. Int. Ed. Engl.*, 1997, **36**, 1725.
7. H. Li, M. Eddoudi and O. M. Yaghi, *J. Am. Chem. Soc.*, 1998, **120**, 8571.
8. C. Kepert and M. Rosseinsky, *Chem. Commun.*, 1999, 375.

9. S. R. Batten, N. R. Champness, X.-M. Chen, J. Garcia-Martinez, S. Kitagawa, L. Öhrström, M. O'Keeffe, M. P. Suh and J. Reedijk, *Pure Appl. Chem.*, 2013, **85**, 1715.
10. S. Kitagawa and M. Kondo, *Bull. Chem. Soc. Jpn.*, 1998, **71**, 1739.
11. D. Li and K. Kaneko, *Chem. Phys. Lett.*, 2001, **335**, 50.
12. C. Serre, F. Millange, C. Thouvenot, M. Noguès, G. Marsolier, D. Louër and G. Férey, *J. Am. Chem. Soc.*, 2002, **124**, 13519.
13. G. Férey, *Dalton Trans.*, 2016, **45**, 4073.
14. R. Kitaura, K. Fujimoto, S. Noro, M. Kondo and S. Kitagawa, *Angew. Chem., Int. Ed.*, 2002, **41**, 133.
15. R. Kitaura, K. Seki, G. Akiyama and S. Kitagawa, *Angew. Chem., Int. Ed.*, 2003, **42**, 428.
16. S. Horike, S. Shimomura and S. Kitagawa, *Nat. Chem.*, 2009, **1**, 695.
17. S. Horike, S. Nagarkar, T. Ogawa and S. Kitagawa, *Angew. Chem.*, 2020, **59**, 6652.
18. S. Horike and S. Kitagawa, *Nat. Mater.*, 2022, **21**, 983.
19. S. Kitagawa, *Acc. Chem. Res.*, 2017, **50**, 514.

Contents

Inorganic Materials Series No. 13
Flexible Metal–Organic Frameworks: Structural Design, Synthesis and Properties
Edited by Susumu Kitagawa

Published by the Royal Society of Chemistry, www.rsc.org

CHAPTER 1

Structure, Design and Synthesis

AAMOD V. DESAI[a,b], YONG-SHENG WEI[c], SUJIT K. GHOSH*[a,d] AND SATOSHI HORIKE*[c,e,f]

[a]Department of Chemistry, Indian Institute of Science Education and Research (IISER), Pune 411008, India; [b]EaStCHEM School of Chemistry, University of St Andrews, North Haugh, St Andrews KY16 9ST, UK; [c]Institute for Integrated Cell-Material Sciences, Institute for Advanced Study, Kyoto University, Yoshida-Honmachi, Sakyo-ku, Kyoto 606-8501, Japan; [d]Centre for Water Research (CWR), Indian Institute of Science Education and Research (IISER), Pune 411008, India; [e]Department of Chemistry, Graduate School of Science, Kyoto University, Katsura, Nishikyo-ku, Kyoto 615-8510, Japan; [f]Department of Materials Science and Engineering, School of Molecular Science and Engineering, Vidyasirimedhi Institute of Science and Technology, Rayong, 21210, Thailand
*E-mail: sghosh@iiserpune.ac.in, horike.satoshi.3r@kyoto-u.ac.jp

List of Abbreviations

abdc	2-aminobenzene-1,4-dicarboxylate
ad	adipate
aip	5-aminoisophthalate
Amal	allylmaloate
AzDC	4,4′-azobenzene dicarboxylate
azpy	azopyridine
BBCDC	9,9′-([1,1′-biphenyl]-4,4′-diyl)bis(9*H*-carbazole-3,6-dicarboxylate)
1,4-bdc or bdc	1,4-benzene dicarboxylate

Inorganic Materials Series No. 13
Flexible Metal–Organic Frameworks: Structural Design, Synthesis and Properties
Edited by Susumu Kitagawa

Published by the Royal Society of Chemistry, www.rsc.org

bdp	1,4-benzenedipyrazolate
benztb	*N,N,N′,N′*-benzidine tetrabenzoate
BME-bdc	2,5-bis(2-methoxyethoxy)-1,4-benzenedicarboxylate
bpa	1,2-bis(4-pyridyl)acetylene
4,4′-bpdc	4,4′-biphenyl dicarboxylate
bpe	1,2-(bis(4-pyridyl)ethylene)
bpee	bipyridine ethylene
BPnDC	benzophenone-4,4′-dicarboxylate
bpp	1,3-bis(4-pyridyl)-propane
bpt	3,6-bis(4-pyridyl)-1,2,4,5-tetrazine
BPT	biphenyl-3,4′,5-tricarboxylate
bpy	4,4′-bipyridine
1_{st}-Br_2	bistable switch 4-(2,7-dibromo-9*H*-fluoren-9-ylidene)-3-methyl-1,2,3,4-tetrahydrophenanthrene
bza	benzoate
CAU	Christian-Albrechts-Universität zu Kiel
cdc	*trans*-1,4-cyclohexanedicarboxylate
3-CH_3-spy	3-methylstyrylpyridine
CID	coordination polymer with interdigitated structure
dabco	1,4-diazabicyclo[2.2.2]octane
dhbc	2,5-dihydroxybenzoate
dpe	1,2-di(4-pyridyl)ethylene
dpNDI	*N,N′*-di(4-pyridyl)-1,4,5,8-naphthalenetetracarboxdiimide
dps	4,4′-dipyridylsulfide
dpt	3-(2-pyridyl)-5-(4-pyridyl)-1,2,4-triazolate
DUT	Dresden University of Technology
ELM	elastic layer structured metal–organic framework
etz	3,5-diethyl-1,2,4-triazolate
FcDC	1,1′-ferrocenedicarboxylate
FTR	frustrated trigonal-symmetry rotator
fu-bdc	functionalised 1,4-benzene dicarboxylate
fum	fumarate
Hdhbc	2,5-dihydroxybenzoate
ip	isophthalate
L0	bis(pyridyl)dithienylethene
L2	4,4′-((4-(*tert*-butyl)-1,2-phenylene)bis(oxy))dibenzoate
L3	2,3-difluoro-1,4-bis(4-pyridyl)benzene
L4	1,2,4,5-benzenetetracarboxylate
L5	2,2′-(1,3,5,7-tetraoxo-5,7-dihydropyrrolo[3,4-*f*]isoindole-2,6(1*H*,3*H*)-diyl)dipropionate
L6	5-(*tert*-butyl)-*N*1,*N*3-di(pyridin-4-yl)isophthalamide
miba	4-(1*H*-2-methylimidazol-1-yl)benzoate
MIL	Materials Institut Lavoisier
MIP	Materials of the Institute of Porous Materials from Paris
ndc	naphthalene dicarboxylate
3-pia	*N*-(3-pyridyl)isonicotinamide

pcu	primitive cubic
ppt	3-(2-pyrazinyl)-5-(4-pyridyl)-1,2,4-triazolate
PXRD	powder X-ray diffraction
pyrdc	pyridine-2,3-dicarboxylate
pytpy	2,4,6-tris(4-pyridyl)pyridine
pzdc	pyrazine-2,3-dicarboxylate
SCXRD	single-crystal X-ray diffraction
SHF	Sheffield framework
SNU	Seoul National University
2-stp	2-sulfoterephthalate
SU	Stockholm University
tci	tris(2-carboxyethyl)isocyanurate
UTSA	University of Texas at San Antonio
VTTF	2,2′-[1,2-bis(4-benzoic acid)-1,2-ethanediylidene] bis-1,3-benzodithiole

1.1 Introduction

Soft porous crystals have received increasing attention owing to their attractive features as structurally dynamic functional solids.[1] These materials form the 3rd generation of porous coordination polymers (PCPs)/metal–organic frameworks (MOFs) that reversibly transform with at least one of the bi-/multi-stable phases bearing porosity. This transition between different states is driven by external stimuli or by the presence of certain species in the voids. Apart from developing these materials for specific applications,[2,3] there is growing interested in the community to understand the structural factors that cause softness, and thereby potentially predicting or designing novel flexible materials.[4,5] Structural response in MOFs is broadly divided into the following types—breathing, swelling, linker rotation and network movement.[6] These responses are usually observed in their guest adsorption–desorption profiles, with a signature gate-opening adsorption and desorption with hysteresis. Apart from solvent/guest inclusion or release, the transformations can also be triggered by a physical stimulus, such as temperature, light or mechanical pressure.[7] While breathing and swelling modes involve discernible changes in cell volume,[8] some other modes such as linker rotation offer dynamic response without any or negligible change in the overall framework.

Flexibility, in general, can be linked to supramolecular interactions and the stability of the framework in the transformed state. The metal–ligand coordination bond is one of the vital factors that governs the structure and its stability, not only at the local level, but also can impact the overall packing and consequent properties in MOFs. The constituents—metal and linker—also have a role at their individual level in bestowing softness to the system. Likewise, the intraframework interactions influence dynamicity, especially in compounds featuring integrated self-assembly of coordination nets. Guest molecules that occupy the void spaces of MOFs during synthesis, often drive framework flexibility due to strong interactions with the framework backbone. MOFs are typically synthesised at high temperatures and/or pressures from solution. It is seen that using the same building units, varying structures, and topologies are obtained. Pertinent to the discussion of this chapter, in certain examples a flexible structure is the product obtained only under specific reaction conditions.[9] Additionally, the non-structural reagents—guest molecules—can also direct the self-assembly to a particular conformation by functioning as the template.[10] Another route of inclusion of guest molecules is by binding to the metal centre and being a part of the coordination sphere.

There are different approaches to analysing the library of flexible compounds; in this chapter, we view flexibility from the perspective of (a) framework building blocks and (b) topological level. It is worth noting, that softness in a multi-component system is essentially a combination of different factors working simultaneously, and a strict distinction is inappropriate. However, based on a wider pool of compounds certain trends are drawn and a dominating factor in every case can be identified. The examples listed in the subsequent sections are not exhaustive but mainly aim to demonstrate a flavour of different structural aspects from the literature on flexible MOFs.

1.2 Flexibility From the Viewpoint of Structural Components

In general, flexibility in frameworks and especially with non-interpenetrated structures, is typically ascribed to either the building units and/or guest molecules. The release/inclusion of guests in such systems often leads to different structural responses, such as sliding, swelling or pore contraction (Figure 1.1). The broader context of structural softness can be narrowed down to the choice and character of the building blocks. It is expected that the use of certain metal cations with variable coordination numbers in conjunction with linkers having suitable donor groups offer a dynamic coordination node, and lead to the formation of non-rigid frameworks. In certain cases, the presence of a hemilabile coordination bond may also render local flexibility.[11,12] Ligands with flexible bonds are known to impart flexibility, but not necessarily lead to formation of flexible MOFs.[13] Another subset involves ligands with functional groups at their core that respond dramatically to an external stimulus. This motion can in turn alter the coordination environment and impart structural dynamicity.

Guest molecules also play a major part in the flexibility of MOFs. Owing to accessible binding sites within the pore walls, with ligand or metal, the host–guest interaction can be stronger than the coordination sphere rigidity to afford a breathing behaviour. However, guest molecules that bind to the framework component may also lead to a flexible response upon their release. This chapter describes several examples of flexibility that are seen as an outcome of either the ligands, metals or guest molecules.

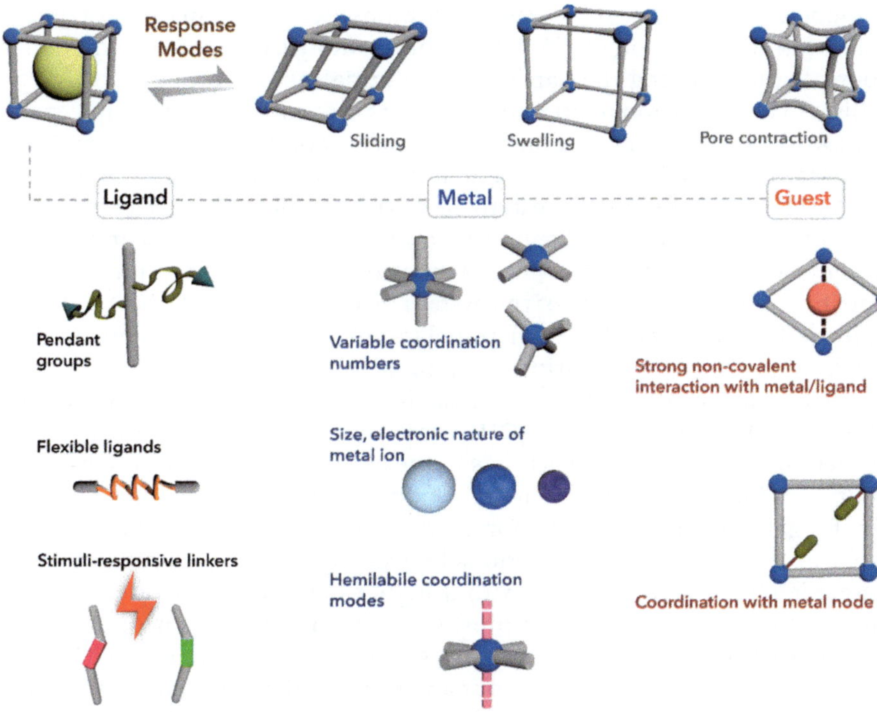

Figure 1.1 Schematic illustration of different modes of flexibility presented from the viewpoint of the building blocks in a MOF.

1.3 Ligands

1.3.1 Effect of Secondary Functional Groups

The ligand in MOFs is typically functionalised for application-specific demands. However, the addition of substituents on the organic linker is seen to have a profound effect on the dynamic nature of flexible MOFs. This was demonstrated in a comprehensive study of MIL-88: $[Fe_3O(L)_3(H_2O)_2(OH)]$ (where L = 1,4-bdc (88B), 4,4′-bpdc (88D)), which represents one of the prominent series of flexible MOFs, whose breathing behaviour is driven by host–guest interactions (described in Section 1.4).[14] By functionalising the linkers in two MOFs of the series—MIL-88B and MIL-88D—with a range of groups, such as Cl, Br, CH_3, OH, NH_2, NO_2, CF_3 and F, the swelling properties were studied. In addition, the number of functional groups per ligand varied between 1, 2 and 4. The swelling properties for the MOF were seen to have a direct correlation to the size, number and nature of the secondary functional group. For instance, the amplitude of swelling was reported to be lower as the number of the CH_3 groups per linker were increased from two to four, in both MIL-88B and MIL-88D. This also led to the formation of permanent porosity in certain cases. Additionally, changes in the host–guest and intraframework interactions brought about by the nature of functional groups led to lowering the barrier for the initial diffusion of guests, which facilitated the inclusion of non-polar solvents that could otherwise not cause framework swelling in the non-functionalised compounds. Another MOF that breathes reversibly (discussed in Section 1.4) is MIL-53: [M^{III}(OH)(bdc)]; MIL-53(Al)-NH_2 represents an extension of the parent MOF with a secondary pendant group.[15] Although ligand functionalisation is usually seen to influence guest binding properties, it was reported that the amine group mainly contributed to the flexibility of the system by interacting with the $AlO_4(OH)_2$ chains.[16] The presence of the functional group led to the formation of a distinct large pore (lp) phase during methane adsorption at very high pressures, that is absent for MIL-53(Al).[17] On these lines, the breathing behaviours for MIL-53(Al)-OH and MIL-53(Al)-$(OH)_2$ were reported by testing their ability for methane adsorption (298 K).[18] As in the case with the amine, the hydroxy group also interacted with the metal chain and noticeable differences were seen for gate-opening pressures for the two MOFs. This was attributed to the extent of hydrogen bond interactions between the hydroxy groups with the inorganic chain, with the MOF with a higher density of functional groups opening the pore at higher pressures. As a control, when the hydroxy group was replaced with methoxy *i.e.*, MIL-53(Al)-OCH_3, it was found that MOF no longer exhibited breathing behaviour and presented a type-I adsorption profile with no hysteresis, corresponding to a rigid structure. This reinforced the role of hydrogen bonding involving the substituent group regarding the flexibility of the framework.

Co(bdp) represents another well-studied MOF in the literature.[19] The structure of the pristine phase comprises Co^{2+} in tetrahedral coordination to N-atoms of the deprotonated bdp^{2-} ligands. The nitrogen adsorption isotherms at low temperatures (77 and 87 K) for the desolvated compound showed uncommon multistep filling, which was ascribed to a combination of adsorbate–adsorbent interactions and structural transformations. The gate-opening behaviour along with a strong hysteresis was also seen in the case of hydrogen adsorption, that exhibited strong temperature and pressure dependence. In a subsequent study involving a combination of *in situ* experiments, it was revealed that the desolvated structure underwent a stepwise structural reorganisation during pore opening from the closed form.[20] From a Co^{2+} centre in a square-planar geometry for the guest-free phase, the opening of the pore involved a transition to a tetrahedral conformation. This was also accompanied by the reorientation of the benzene rings in the ligand along with a change in dihedral angles of the coordinated pyrazolate rings.

This characteristic behaviour was further examined by adding substituents to the benzene ring in a series of MOFs: Co(F-bdp), Co(*p*-F_2-bdp), Co(*o*-F_2-bdp), Co(bdp-d_4) and Co(*p*-Me_2-bdp).[21] All the compounds exhibited a stepwise N_2 adsorption, with subtle changes, except for Co(*p*-Me_2-bdp) that had a discernible profile. Using methane as a probe adsorbate, larger differences in pore opening pressures were noted (Figure 1.2). Structural

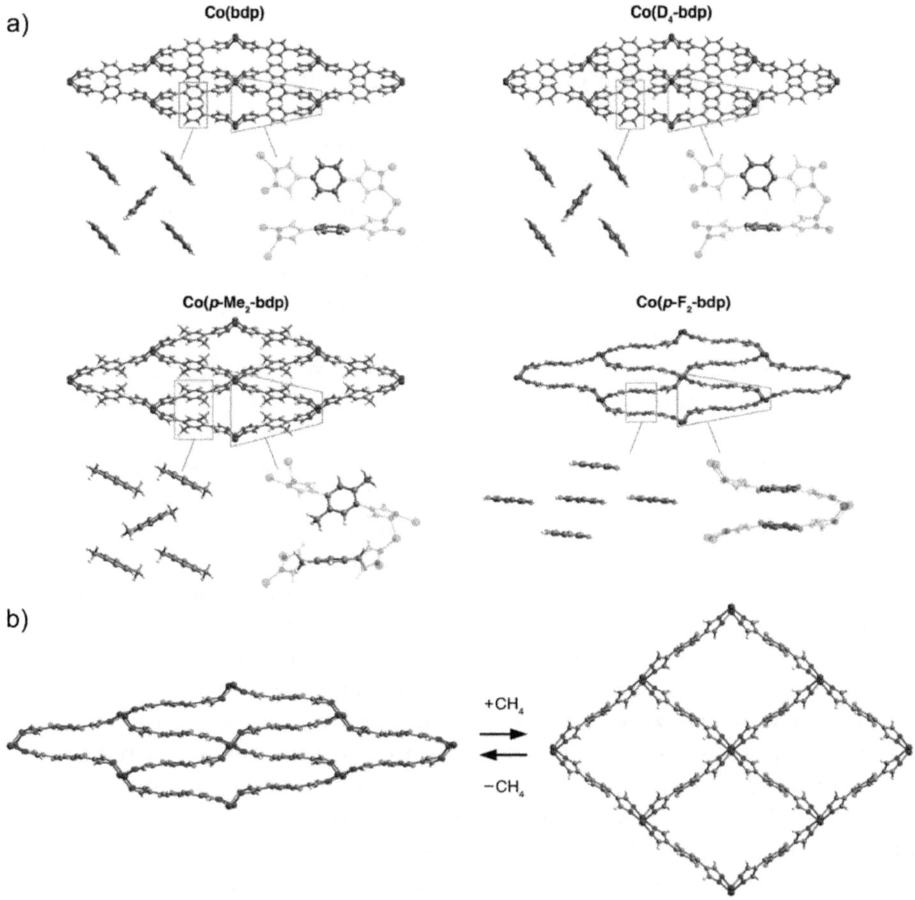

Figure 1.2 (a) Packing view for the activated phases and corresponding zoomed version showing differences in the arrangement of the functionalised linkers. (b) Structural changes in Co(p-F_2-bdp) upon methane adsorption (20 bar) and desorption. Reproduced from ref. 21 with permission from American Chemical Society, Copyright 2016.

characterisation suggested the presence of edge-to-face π–π interactions in the guest-free phase stabilised the compound (Figure 1.2(a)). It was seen that the methyl substituent led to greater stability of the closed pore, while fluoro substituents weakened the non-covalent interaction. Pillar-layered MOFs such as $[Zn_2(bdc)_2(dabco)]_n$, were reported to exhibit some degree of breathing behaviour.[22,23] The addition of a functional group to the ligand, such as amino or alkoxy, was seen to enhance the degree of breathing.[24,25] Also, carrying out post-synthetic modification on an $-NH_2$ functional MOF, with linear alkyl anhydrides $(O[CO(CH_2)_nCH_3]_2$, $n = 0$ to 5), led to modulation of the breathing behaviour.[24] It was found that the nature of the organic substituent stabilised either the large or narrow pore form. Subsequently, in an extensive study on Zn(II)-based MOFs *viz.* $[Zn_2(fu\text{-}bdc)_2(dabco)]_n$, the dangling groups covalently bound to the phenyl ring were systematically varied.[26] It was observed that the addition of long substituents led to a greater degree of breathing in such systems. This behaviour was reasoned to the interaction of the dangling groups with the framework backbone, and with the incoming guests. Instead of using aliphatic substituents, the effect of using aromatic substituents was demonstrated in modulating the flexibility at the local level, and consequently that of the overall structure in two isostructural 3D MOFs: $[Zn_2(bdc\text{-}R)_2(bpy)](solvent)]_n$ (where R = NO_2 or H).[27] Owing to the presence of nitro groups in the dangling side chain, strong π–π interactions with the ligand core were observed. For the other MOF, no such interactions were possible, and the pendant groups were arranged in a way that led to the expansion of the void volume (~34%) and instead had CH–π interactions with the central phenyl ring. The activated phase for both compounds obtained by heating under vacuum reversibly regenerated the pristine structure when exposed to DMF. It was found that activation of the nitro-containing MOF by supercritical CO_2 drying led to the slightly different guest-free structure. The product obtained from both activation methods had the interlayers locked by the strong interaction between the nitroaromatic group and central phenyl ring. The differences in the structure based on the method of activation were found to be due to the stress caused by the pressure of liquid CO_2, which was reflected in a change in the Zn-paddlewheel unit, with one Zn–O clipped after desolvation. This result reinforced the impact of pressure and temperature in soft porous crystals.[1] The same compound was also explored for its ability to screen xylene isomers as the guests participate in π–π interactions with the functional components.[28] It was observed that *o*-xylene (oX) was adsorbed preferentially from a mixture of isomers, as it had the most cooperative interaction with the compound. Although secondary functional sites offer a wider range of possibilities for tuning flexibility, it was also noted that in certain structures, changing the atomic composition of the ligand alters structural softness.[29] Two isostructural MOFs—$[Fe(ppt)_2]$ and $[Fe(dpt)_2]$—were examined for their adsorption properties. $[Fe(ppt)_2]$ exhibited flexibility, while $[Fe(dpt)_2]$ retained a rigid framework. The differences were suggested to arise from the varying coordination ability of the linker's constitution, with Hppt having $pK_b = 0.65$, while Hdpt had $pK_b = 8.8$. Consequently, rigidity and flexibility were seen having a direct correlation with the softness of the coordination bond.

1.3.2 Stimuli-responsive Ligands

Apart from substitution of the organic linker, another type of post-synthetic modification arises from functional sites within the core of the ligands. One of the more studied functions in MOFs is that of photodimerisation reactions that lead to the rearrangement of bonds in a coordination polymer/MOF and subsequently alter the pore chemistries. In this regard a flexible Mn(II)-based MOF—[Mn(bdc)(dpe)]—was reported to exhibit guest-dependent gate-opening behaviour.[30] The metal centre (Mn(II)) had a distorted coordination environment in the pristine phase, while the guest-free phase led to shrinkage of the narrow pore. The distance between the olefinic groups was 4.02 Å, which satisfied Schimdt's criterion for photodimerisation (<4.2 Å). Unlike the use of secondary functional groups, the orientation of the pore wall was deemed to distinguish between the two gases that were used in the study, CO_2 and C_2H_2. The pore opening from the desolvated phase involved an end-on interaction for CO_2, whereas for C_2H_2 the CH–π interaction required pore expansion and undergo gate-opening adsorption. Subsequently, a photoinduced [2 + 2] cycloaddition reaction was carried out (Figure 1.3). The resulting framework after light irradiation showed a distinct colour change on account of loss of conjugation, and a significantly different orientation of the phenylene rings was obtained. Not only was the total adsorption reduced, but the selectivity for the two gases was absent. This resulted from the merging of narrow pore pockets into a single channel pore and unfavourable configuration of the phenylene rings for interaction with the guest.

The idea of cross-photodimerisation has also been investigated in a few other MOFs, for instance, the [2 + 2] cycloaddition reaction in a 3D, Zn(II)-based MOF: $[Zn_2(H_2O)_2(Amal)_2(bpe)]$.[31] The pristine MOF crystallised in *ins* topology, which upon irradiation underwent photoreaction resulting in the formation of a cyclobutane ring and the resulting structure had a new topology. This was seen as a broader strategy to prepare unsymmetrical ligands based on an unsaturated backbone. It is worth noting that this approach of photochemical [2 + 2] reactions using functional ligands in CPs and MOFs has been increasingly applied and found suitable for a range of structural transformations and applications.[32–34]

In another report, an uncommon example of a material showing photoinduced non-linear contraction (PINC) in a 2D Zn(II)-MOF—$[Zn_2(bdc)_2(3\text{-}CH_3\text{-}spy)_2]\cdot H_2O$—was noted.[35] The parent compound showed anisotropic positive thermal expansion (PTE). The photochemical reaction led

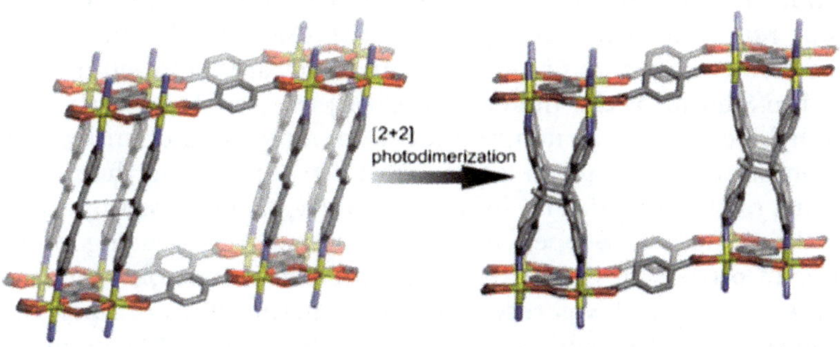

Figure 1.3 Change in shape of the pore in [Mn(bdc)(dpe)] as a result of a photodimerisation reaction. Reproduced from ref. 30 with permission from American Chemical Society, Copyright 2016.

to the formation of a 3D structure, by virtue of cyclobutane rings between the interdigitated layers. Interestingly, when the reaction was carried out by varying the wavelength of irradiation, it was found that the lattice parameter along *a*-axis and unit cell volume shrunk most at $\lambda = 385$ nm. It was found that apart from the reaction in the linker, the Zn-paddle wheel unit also underwent distortion.

A common theme with photochemical reactions is the large stress and strain inflicted on the polymeric structure. To be able to accommodate the structural changes accompanying photoinduced reactions, an approach of using linkers with active groups in flexible structure was proposed.[36] The dithienylethene (DTE) moieties are known for the photoresponsiveness, and it was included as the linker core in the Zn(II)-based interpenetrated MOF $[Zn_4(bdc)_4(L_0)_2{\cdot}4DMF{\cdot}H_2O]_n$. The DTE could undergo a reversible transition from open to closed form upon irradiation of UV radiation in several solvents, and also in the solid-state. The flexible nature of the MOF was ascribed to enable an effective reaction, as the changes in the bonds caused significant structural changes. Similarly, the photoisomerisation of azo bonds from *trans* to *cis* upon UV irradiation offers an attractive pathway to bestow local flexibility in MOFs. Such a study was reported in a Zn(II)-based, interpenetrated MOF bearing potentially two linkers with photoactive sites, $Zn(AzDC)(BPE)_{0.5}$.[37] A dynamic photoinduced isomerisation was observed, which caused remarkable differences in guest (CO_2) uptake, with release of CO_2 being observed when the compound was exposed to radiation during adsorption. This was caused due to the increase in surface energy caused by *trans–cis* isomerisation and consequent weakening of intermolecular interactions with the adsorbate. These results further reinforce the efficacy of soft porous crystals for photo responsiveness.

1.3.3 Ligands Based on Flexible Bonds

Ligands based on flexible bonds, either as donor groups or part of the core, are attractive candidates to bestow structural dynamicity. An example of a guest-induced structural transformation in a MOF with soft coordination bonds was reported in the Co(II)-based MOF-Co(VTTF).[38] The desolvated phase led to the formation of a non-porous structure that included a bond between the S-atom of the linker with Co(II) centres. During the adsorption of ethylene gas, the structure underwent a two-step opening with the interaction between the π-rich guest with Co(II) displacing the bond with the S-atoms of the ligand. The inherent lability of the coordination bond allowed the structure to open up completely with increasing pressures and led to a Co(II)–C_2H_4 interaction at all sites. As a control, when adsorption with ethane (C_2H_6) was recorded, the structure did not open even at higher pressures. The hemilabile bond-assisted, guest-dependent structural transformation was termed as a gate-locking/unlocking mechanism. Among early reports of complete structural characterisation by single-crystal-to-single-crystal studies in the domain of flexible MOFs, a reversible transformation in a 2D MOF—{[Cu(pyrdc)(bpp)]·(5H_2O)}$_n$—having a highly flexible pillar (bpp) was observed.[39] The pristine phase comprised five-coordinate Cu(II) bonded to four independent ligands. However, when the compound was heated, the colour of the compound changed to dark blue, and the guest water was lost leading to the shirking of the pores. Structural information suggested the breaking of a coordination bond at one Cu-centre with the N-atom of the pyridyl group (Figure 1.4). This transition was accompanied by a large change in the orientation and bonding direction of the flexible pillar ligand. Upon resolvation, the original phase was recovered, and the guest-free phase exhibited gate-opening adsorption towards CO_2 (195 K) and vapours of polar solvents, *e.g.*, methanol and ethanol (298 K). The same group also presented such behaviour in a 3D MOF—{[Ln(tci)(H_2O)]·4H_2O}$_n$ (where Ln = Ce or Gd)—that showed a dynamic and reversible structural behaviour upon partial loss of water molecules.[40] The resulting phase was stable and had a change in its pore shape that exhibited the typical gate-opening behaviour towards water adsorption at 298 K. The flexibility and stability of the partially evacuated compound after heating was ascribed to the flexible –CH_2–CH_2– linkages in the ligand.

Another case of a dynamic MOF is {[$Zn_4O(L2)_3(DMF)_2$]·xG}$_n$ (G = guest) – referred to as DynaMOF-100 – in which the linker contains flexible ether linkers.[41] The pristine phase of the compound underwent significant

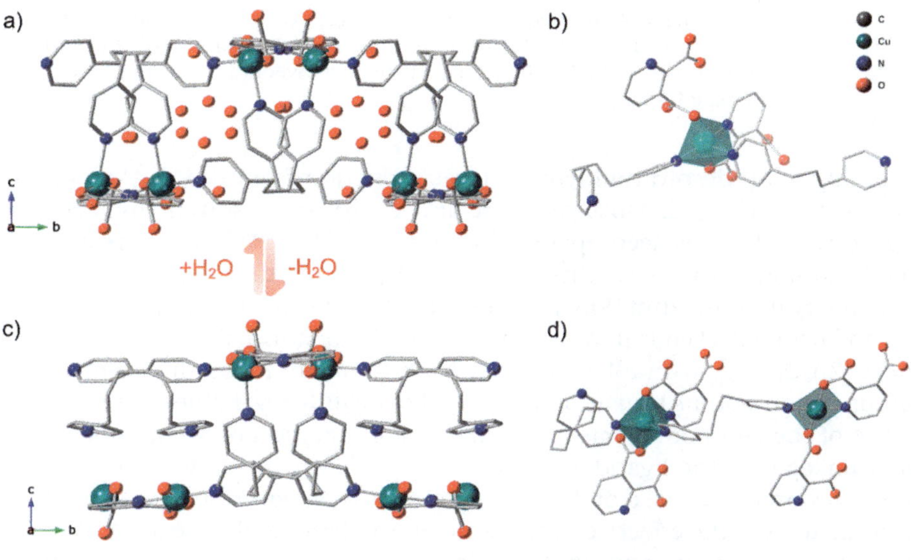

Figure 1.4 Reversible structural changes in {[Cu(pyrdc)(bpp)]·(5H_2O)}n from (a) pristine to (c) desolvated. Corresponding coordination environments for the metal cation are shown alongside (b) pristine and (d) desolvated. Reproduced from ref. 39 with permission from American Chemical Society, Copyright 2005. (H-Atoms omitted for clarity, colours: N, blue; C, grey; O, red; Cu, cyan.)

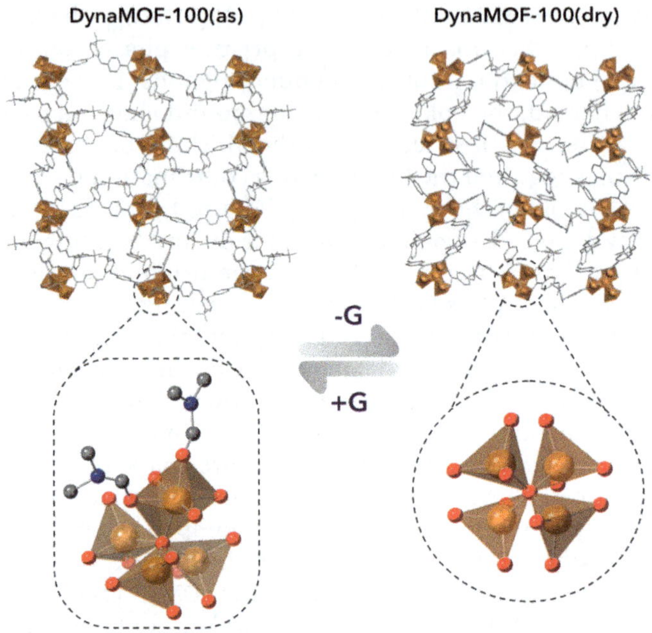

Figure 1.5 Single-crystal-to-single-crystal transformation in DynaMOF-100 upon desolvation, with coordination environment for Zn(II)-centres shown alongside (H-atoms and free guest molecules are omitted for clarity, colours for zoomed figure: Zn, orange; O, red; N, blue; C, grey). Reproduced from ref. 41, https://doi.org/10.1038/srep05761, under the terms of the CC BY 4.0 license http://creativecommons.org/licenses/by-nc-sa/4.0/.

squeezing upon desolvation (void volume reduction from ~3441 Å^3 to ~916 Å^3), resulting from the loss of guest and coordinated solvent molecules (Figure 1.5). This was accompanied by slippage of the 2D layers to result in an overall non-porous structure. Additionally, the pristine structure having a secondary building unit (SBU)—$\{Zn_4O(O_2C^-)_6(DMF)_2\}$—with mixed tetrahedral and octahedral coordinating Zn(II) cations transformed to a non-porous phase $[Zn_4O(L2)_3]_n$ with all Zn(II) centres in tetrahedral coordination. This porous to non-porous transition was found to be fully reversible. The flexible nature of the compound was ascribed to a combination of accommodative ether linkages in the ligand and the stability of Zn(II) in varying coordination environments. Based on low-temperature CO_2 adsorption (195 K) on the desolvated MOF, the effective pore diameter was estimated to be 5.1 Å. When single-component adsorption for xylene isomers (kinetic diameter ~0.6 nm) was performed at 298 K, it was found that only *p*-xylene could open the framework, while its isomers were unable to enter the pores. A re-solvated phase of *p*-xylene in presence of DMF was obtained, which suggested the pore opening was a combination of guest fitting and favourable π–π bonding interactions with the linker, facilitated by the flexible ether bonds.

DynaMOF-100 was also able to distinguish the industrially relevant C8-mixture (styrene and ethyl benzene) by accommodating styrene and remaining in the closed pore form for ethyl benzene.[42] Like the previous study, styrene was found to have multiple non-covalent interactions with the MOF, that stabilised the included guest molecules. Another set of 2D Zn(II)-based flexible MOFs was reported, that contracted upon desolvation and expanded for vapours of various solvents, chiefly being able to separate C8-isomers.[43] Two MOFs with different pillar linkers—$[Zn_2(aip)_2(pillar)]\cdot 2DMF$ (pillar = bpy or bpe)—had roughly the same breathing behaviour, except the MOF with bpe showed higher dynamic character due to the presence of flexible core in the linker. When the activated MOFs were exposed to different guests—small organic solvents and C8 hydrocarbons—it was noted that the extent of pore expansion had a correlation with the size of the adsorbate. Similar to the ether links described above, the flexible sp^2-hybridised N-nodes in DUT-13[44]—$Zn_4O(benztb)_{1.5}$—can drive the linker from a staggered conformation in the open-pore form to an eclipsed orientation in the closed-pore phase.[45] This breathing behaviour was triggered by the adsorption of the guests and was found to be fully reversible.

1.3.4 MOFs Based on Aliphatic Ligands

An early example of a flexible MOF came using an aliphatic linker in a 3D MOF: $\{[La_2(ad)_3(H_2O)_4]\cdot 6H_2O\}_n$.[46] It was suggested that the presence of an aliphatic linker could absorb any structural distortions and prevent collapse upon removal of the guest solvent. This was realised when the MOF was heated at 115 °C, which led to complete removal of free water and partial loss of coordinated water. The addition and removal of water from the system was reported to be a reversible process, that highlighted the role of an aliphatic backbone. Another example of such flexibility induced by bulky guests, *e.g.*, xylenes, was reported in a system, structurally related to MIL-53.[47] The Al-MOF—$[Al(OH)(cdc)]\cdot H_2O$ (referred as CAU-13)—comprised of the linkers in a 1 : 1 ratio of *a*,*a*- and *e*,*e*-conformations that exhibited slight breathing in response to water and CO_2.[48] In both the cases, the primary mode of interaction was with the bridging μ_2-OH groups in the MOF. However, although π–π interactions with framework pore walls were absent, when vapour adsorption for xylenes was carried out, the structure was able to fully open with all the linkers in *e*,*e*-configuration (Figure 1.6). The effective cell volume increase in response to xylene inclusion was ~25%. Such behaviour was also observed for the Ga(III)-analogue,[49] however the In(III) and Cr(III)-MOFs did not exhibit breathing.[50,51] Although generally non-aromatic linkers are expected to bestow higher flexibility, it was found that using fumaric acid as the linker in a MIL-53(Al) analogue led to formation of an almost robust structure.[52] However, using a fully saturated ligand—adipic acid (H_2ADP)—gave rise to a flexible variant MIL-53-ADP,[53] which displayed breathing behaviour upon hydration and removal of water. The linker was seen to change its conformation and shorten its length, from 5.79 and 5.90 Å to 5.43 and 5.70 Å in the hydrated forms. Also, the framework expanded by ~10% with the inclusion

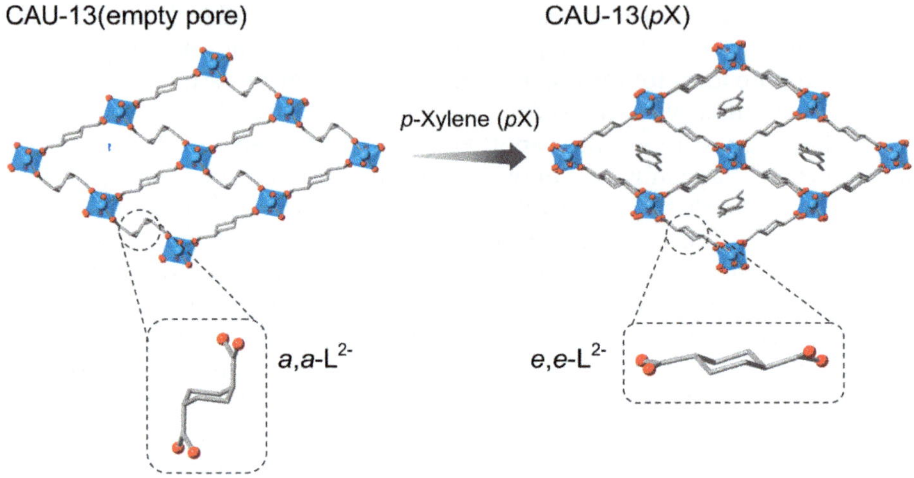

Figure 1.6 Change in structure and orientation of the ligand from evacuated phase of CAU-13 (left) to *p*-xylene included (right).[47] (H-Atoms omitted for clarity, colours: Al, light blue; C, grey; O, red.)

of water. The flexibility directed by the adaptability of the backbone further reiterated the utility of aliphatic linkers.

Zr(IV)-Based MOFs have received increasing attention, owing to the thermal and chemical stability bestowed by the strong coordination bonds in such systems. Consequently, these compounds are commonly seen to be topologically rigid. In an interesting study, the role of linker backbone and substituents in a series of isostructural MOFs based on 10-connected Zr_6 SBU $[Zr_6(\mu_3\text{-}O)_4(\mu_3\text{-}OH)_4]$ and C4-ligands was examined.[54] The selection of ligands—fumaric acid (F), succinic acid (S) and malic acid (M)—comprised variation in degrees of freedom and in the case of malic acid an additional hydroxy substituent. The resulting MOFs—MIP-203-F, MIP-203-S and MIP-203-M, respectively, —exhibited structural flexibility with varying degrees. Upon desolvation, all the compounds showed an opening of the pore when exposed to different polar solvents. Similarly, when the compounds were tested for gas adsorption, differences were noted in terms of pore opening. MIP-203-F exhibited a swelling response, while MIP-203-S, having the highest degree of freedom, was found to have linker rotation to accommodate guests. MIP-203-M having a secondary functional group exhibited the highest selectivity for CO_2/N_2 separation.

1.4 Metal Centres

1.4.1 Influence of Coordination Character of Metal Centre

The flexibility of coordination bonds in MOFs also provides scope for breathing behaviour, and strongly depends upon the nature and size of the cation. Such coordination softness-induced transformation was observed in a

Cd(II)-based MOF built from a mixed N- and O-donor ligand, miba.[55] The pristine MOF showed six-coordinate Cd(II), binding with four O-atoms of the carboxylates and two N-atoms from the imidazolyl units. Upon desolvation, the linker rotated, and the coordination of the metal nodes changed to five-coordinate, hexahedral. The closed-pore structure obtained after desolvation could be opened after the application of high-pressure CO_2 gas. DFT calculations revealed that the coordination bonds for Cd(II) are flexible, which allows linker rotation during desolvation, to convert the system into a low-energy, non-porous state. Similar flexibility, driven by the dynamic nature of the coordination bonds, was noted in a doubly interpenetrated, pillar-layered framework: [$Zn_2(BPnDC)_2(bpy)$] (referred to as SNU-9).[56] The synthesis of the MOF and its stepwise gas adsorption behaviour was previously studied,[57] when it was observed that the cell volume shrunk by ~28% upon loss of guest molecules. However, unlike typical flexibility reported in interpenetrated MOFs, the breathing behaviour in SNU-9 was attributed to the change in the coordination environment of Zn(II). The coordination for one of the Zn(II) centres changed from distorted octahedral in the pristine phase to distorted square-pyramidal (Figure 1.7). This was also reflected in reorientation of one of the $BPnDC^{2-}$ ligands. It is worth noting, that the concept of soft SBU was previously introduced using another Zn(II)-based interpenetrated, pillar-layered MOF: {[$Zn_2(bdc)_2(L3)$]·2.5DMF·0.5H_2O}$_n$.[58] The pristine structure with a regular paddle-wheel coordination environment (square pyramidal geometry) underwent transformation to a tetrahedrally coordinated Zn(II)-based structure upon heating. This was accompanied by a drastic reduction of the cell volume of ~28%. Also, it was reported that this transition was gradual with the coexistance of multiple phases at the intermediate temperatures.

Generally, Zn(II) is expected to exhibit dynamic coordination environments for most MOFs because of its stability in different coordination geometries. In an interesting finding, such flexibility was seen in a Bi(III)-MOF constructed using a tripodal ligand: [Bi(BPT)·2MeOH] (referred to as SU-100).[59] The Bi(III) ions having coordination number seven, exhibited an inorganic building unit of Bi_2O_{12} dimers, with two O-atoms bridging the Bi(III) cations. When the compound was exposed to different polar solvents such as methanol, DMF and DEF, the structure expanded with up to 10% change in unit cell volume corresponding to the inclusion of DEF. A mechanistic investigation suggested the flexibility of the Bi_2O_{12} unit in modulating the framework breathing. The Bi–O–Bi bond angles were observed to change from 121.1(5)° in the as-synthesised phase to 110.5(3)° in the DEF-included structure, and the bridging carboxylates were seen to move apart from each other during expansion.

In a broader context, the choice of the metal ion has an immense impact on the flexible nature of MOFs. Among the more studied dynamic MOFs in the literature, DUT-49—[$Cu_2(BBCDC)$][60]—which is composed of a tetratopic ligand, has shown structural contraction during gas adsorption.[61] It was found that the compound showed an abrupt decrease in gas uptake (CH_4 adsorption at 111 K) at a certain pressure, forming a metastable phase.

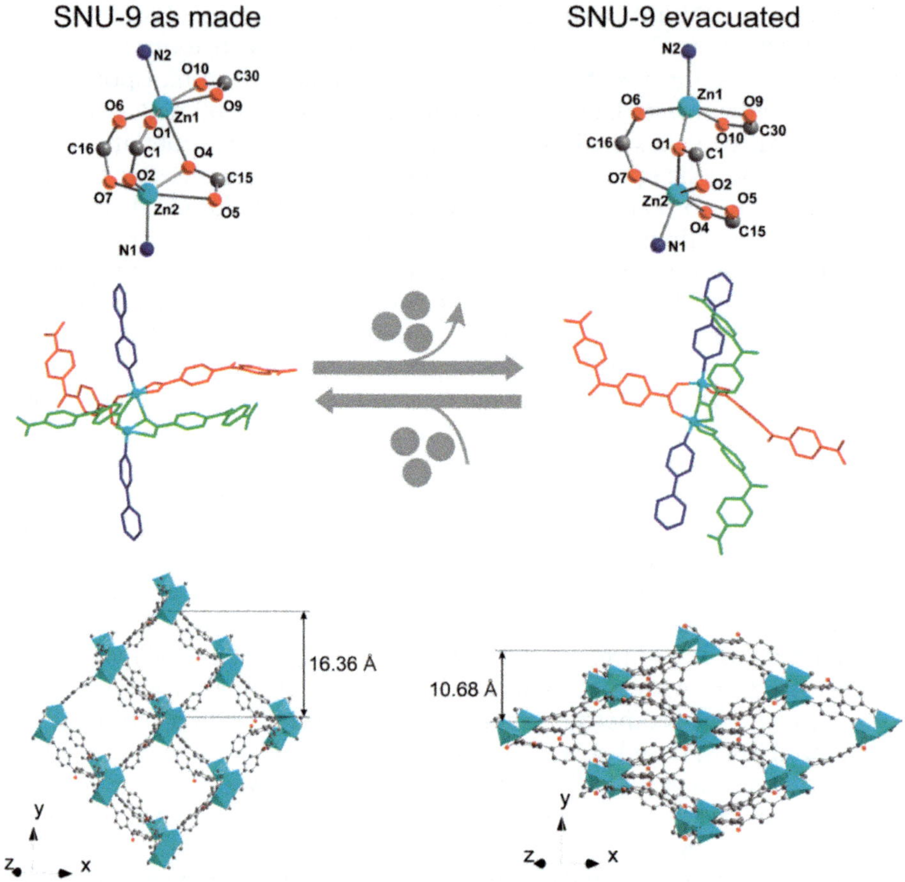

Figure 1.7 Different facets of SNU-9: coordination environment (top), the orientation of the ligands (middle) and structural packing (bottom), in the as-synthesised phase (left) and desolvated phase (right). Reproduced from ref. 56 with permission from American Chemical Society, Copyright 2014.

Detailed *in situ* studies revealed that filling of the mesopores led to a sudden buckling of the ligand leading to the conversion of the open pore form to the closed pore form, termed as negative gas adsorption (NGA). Structurally, this step was accompanied by rotation of the linker along the biphenyl backbone. DUT-49 sustained the NGA profiles for adsorptions at different temperatures and even for other hydrocarbons such as *n*-butane. Also, this behaviour was observed for high-pressure CO_2 adsorption in DUT-49, and also with other systems constructed with different carbazole connecting groups: phenyl (DUT-48), triphenyl (DUT-50) and naphthyl (DUT-46).[62] Subsequently, the role of metal ions in directing the flexibility of frameworks was examined by the same group.[63] By post-synthetically exchanging metal ions in DUT-49(Cu) with Mn(II), Fe(II), Ni(II), Zn(II), Co(II) and Cd(II), a series of

MOFs was prepared. It was found that the Cu(II)-MOF was the most flexible and had the highest porosity, while the Ni(II)-analogue exhibited moderate flexibility. MOFs with other metal ions were unstable upon activation and did not retain porous properties. These trends were ascribed to the role of M–M bonds in the structure, with Cu (2.634 Å) and Ni (2.720 Å) having short distances. Although rarely found, Mn–Mn and Cd–Cd paddle-wheel moieties were observed in the structures although, they were inadequate to sustain stability during the activation procedure.

A study on similar lines was reported by varying the metal ions in a series of 2D MOF *viz.* MCID-1 $[M(ip)(bpy)]_n$ (M = Co, Ni, Cu and Zn) and their response towards guest (water) molecule inclusion was monitored.[64] The structure formed hydrophobic pores upon desolvation for all the M-ions, which then opened up for guest inclusion in the gaps of the 2D layers. However, significant differences were noted for water uptake, which was correlated to the distortion of the coordination environment. CuCID-1 displayed an early onset (P/P_0 = 0.33) for guest inclusion, which was attributed to the Jahn–Teller effect maintaining coordination distortion even after desolvation of the parent phase. ZnCID-1 had an overall similar uptake, with a much higher gate pressure (P/P_0 = 0.6), while NiCID-1 and CoCID-1 did not have any considerable guest inclusion over the entire pressure range. This observation was consistent in mixed-metal MOFs, wherein the presence of Cu(II) or Zn(II) allowed the compounds to open up for water adsorption. This influence of the metal-ion character was also observed in the changing structural dynamics of UTSA-300—$[Zn(dps)_2(SiF_6)]_n$—by replacing Zn(II) with Cu(II) (denoted as NCU-100; UTSA-300-Cu).[65] The structure comprising octahedrally coordinated metal ions exhibited a difference in the M–F bond length (Cu–F: 2.27 Å *vs.* Zn–F: 2.09 Å), owing to the Jahn–Teller distortion. The parent Zn-MOF underwent interlayer space shrinkage but had a narrow pore structure (3.5 × 3.9 × 4.1 Å^3 and 4.3 × 4.3 × 4.1 Å^3) that opened during guest inclusion. However, the Cu-MOF maintained an open pore structure (3.6 × 4.3 × 4.2 Å^3 and 4.7 × 4.5 × 4.2 Å^3), despite a similar interlayer shrinkage. These subtle structural differences were seen to have a profound effect on the application, with the Cu-MOF showing higher uptake of C_2H_2 even at lower pressures.

MIL-53 represents an important type of flexible MOF that exhibits guest-induced breathing behaviour. However, the breathing phenomenon is also strongly dependent upon the choice of the metal centre, as observed in MIL-53(Fe).[67] Likewise other metal ions such as In(III),[68] Ga(III)[66] and Sc(III),[69,70] are seen to have a significant impact on the pore dynamics. The varying evolution of the structures and corresponding thermal stabilities were suggested to be an outcome of the difference in ionic radius of the metal cation in octahedral coordination, and their electronic configuration (Figure 1.8).[66] The implications of these differences in breathing behaviour are discussed in more detail in Section 1.4. Such differences in rigidity *vs.* flexibility based on metal ions were also reported for linkers with bulky backbones—MIL-53-FcDC—where the Ga-MOF showed breathing behaviour towards heavy guests such as iodine and pyrazine, while the Al- and In-analogues did not

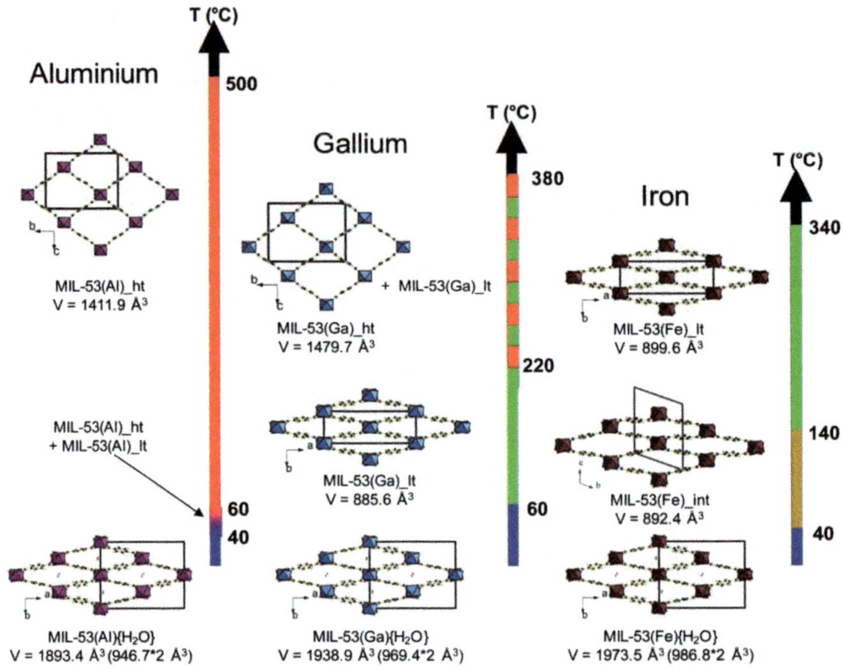

Figure 1.8 Temperature-dependent change in structures for MIL-53 for different metal cations. Reproduced from ref. 66 with permission from the Royal Society of Chemistry.

exhibit breathing.[9,71,72] In addition to exploring various metal cations, mixed-metal and mixed valence approaches have also been tested for MIL-53 structures, including linkers with secondary functional groups such as amino ($-NH_2$).[73–77]

As described in previous section, MOFs having ligands with pendant alkyl chains are seen to present a highly dynamic response. For instance, it was seen that the Zn(II)-compound—$[Zn_2(BME\text{-}bdc)_2(dabco)]_n$—exhibited multiple phases during CO_2 adsorption, with a discernible intermediate phase between the narrow pores in a desolvated state and the open pore form after CO_2 inclusion.[78] Apart from the non-covalent interactions between the pendant groups with the linker backbone, the distortion of the coordination environment was suggested to drive this flexibility. The study on this system was further expanded by preparing a series of isostructural MOFs—$[M_2(BME\text{-}bdc)_2(dabco)]$—with varying metal centres (M = Zn(II), Cu(II), Ni(II) and Co(II)).[79] During desolvation, the extent of pore contraction was seen to alter with the metal node: Co(II) exhibiting the maximum, while Cu(II) with a more rigid coordination environment demonstrated the least (Figure 1.9). These differences were ascribed to the electronic effects of the metal ions in destabilising the square pyramidal geometry of the large pore phase. Again, during CO_2 adsorption, a similar pattern for Zn(II) and Cu(II)-MOFs

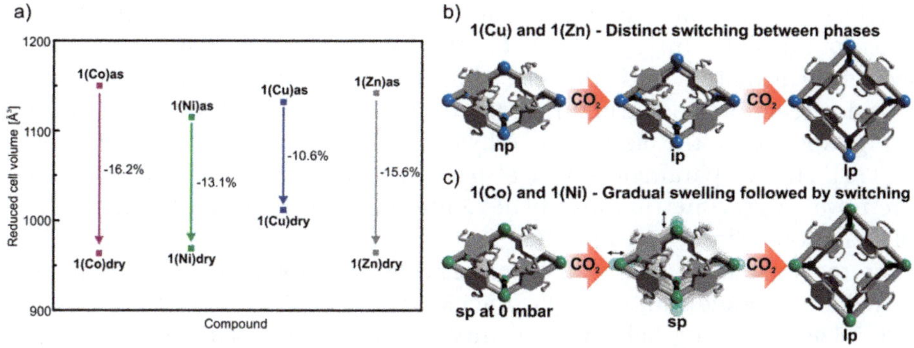

Figure 1.9 (a) Change in cell volume from as-synthesised phase to desolvated for MOFs [M_2(BME-bdc)$_2$(dabco)] with different M(II)-cations. Schematic illustration of structural evolution during CO_2 adsorption for (b) Cu(II) and Zn(II)-MOFs and (c) Co(II) and Ni(II)-analogs. Reproduced from ref. 79 with permission from American Chemical Society, Copyright 2018.

was observed with distinct gate pressures and identification of three states: narrow pore, intermediate and large pore. However, for Ni(II) and Co(II), there was gradual inclusion of the gas before complete conversion to large pore phases at higher pressures. Another study compared the same set of metal cations in DUT-8: [Ni_2(2,6-ndc)$_2$(dabco)].[80] DUT-8(Ni) was earlier seen to exhibit breathing behaviour, and the Co(II)-analog was found to exhibit a similar nature. Interestingly, the Zn(II)-MOF transitioned into a non-porous phase and did not open up. DUT-8(Cu) was seen to maintain an open pore structure upon evacuation.

Apart from transition metals, lanthanides are also seen to impart different features. Owing to higher coordination numbers, typically MOFs constructed from such metal ions have dense packing. This facet was explored by preparing a series of rare-earth (RE) metal ion-based MOFs from a tripodal ligand with acylamide moieties.[81] Compounds prepared from Y(III), Ce(III), Nd(III), Eu(III), Tb(III), Dy(III), Ho(III), and Tm(III) yielded rigid porous structures, with the metal-ions having a coordination number of eight. However, the compound obtained with La(III) exhibited structural differences, with the metal ion having a coordination number of nine, and flexibility upon desolvation. This was also reflected in the adsorption for the La-MOF that exhibited gate-opening with a large hysteresis for the low-temperature (195 K) CO_2 isotherm.

1.4.2 Redox-active Metal Nodes

Another type of flexibility in MOFs is seen by triggering redox reactions at the metal centre. Among the early studies aimed at using MOFs for charge storage, a partially fluorine-doped MIL-53—[Fe(OH)$_{0.8}$(F)$_{0.2}$(bdc)]·H_2O—was reported as a candidate for reversible insertion of Li^+ ions.[82] The MOF was cycled against lithium metal as a counter electrode between 1.5 and 3.5 V. It was found that a mixed-valence Fe^{3+}/Fe^{2+} state could be achieved reversibly during lithiation with the insertion of ~0.6 lithium ions. Apart from the redox reaction, the structural evolution suggested swelling of the MOF during exposure to the electrolyte: ethylene carbonate and dimethyl carbonate (1 : 1). The cell parameters were also found to alter during Li-ion insertion or removal. A gravimetric capacity of 75 mA h g^{-1} was reported for the material at a low discharge rate of C/40.

The response of including a redox-active guest molecule (hydroquinone = H_2Q) in a vanadium-based flexible MOF—MIL-47(V)—was later examined.[83] The breathing behaviour of this compound is discussed further in Section 1.5. The as-prepared MOF, $[V^{III}(OH)(bdc)]$ underwent oxidative loss of guest species upon heating in air to have the metal centre oxidised to +4, resulting in the overall framework of [V(O)(bdc)].[84] The redox-active guest was included in the activated MOF either by heating under ambient air or treatment under anhydrous conditions (Figure 1.10). When heated in air, the guest underwent oxidation to release a proton and electron that reduced the metal centre and converted the bridging oxygen to μ-OH. The resulting *p*-benzoquinone formed a charge-transfer complex with excess hydroquinone, which was loaded in the pores of the MOF. The change in the oxidation state of the metal in MIL-47(V) to +3 is seen to make the system more flexible in developing a strategy to activate the MOF.[85] However, heating activated MIL-47(V) with hydroquinone under inert conditions led to a drastically different outcome. Hydroquinone is known to thermally degrade above 250 °C,

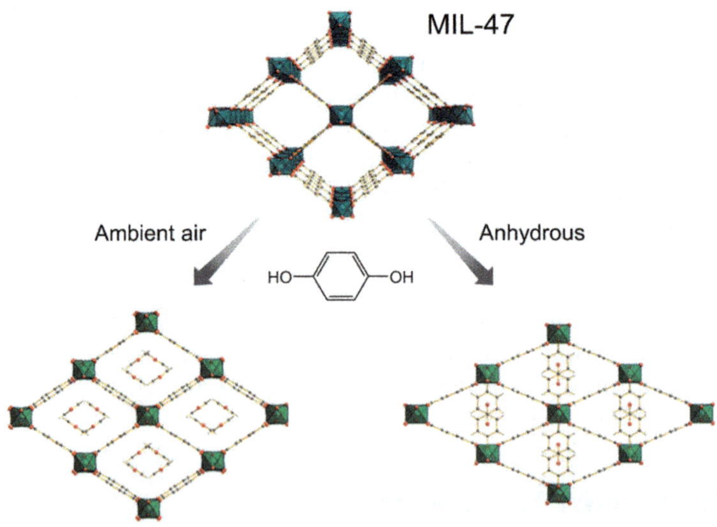

Figure 1.10 Change in the structure of MIL-47 when heated with hydroquinone either under anhydrous conditions (right) or ambient air (left). Reproduced from ref. 83 with permission from American Chemical Society, Copyright 2015.

and led to stepwise conversion to *p*-benzoquinone with *p*-semiquinone radical and *p*-semiquinone radical anion as intermediates. It was suggested that the formation of the radical led to the reduction of V(IV) and protonation of the bridging O-atom. Subsequently, formation of the radical anion would further protonate the μ_2-OH and the radical anion stabilised the inorganic chain, with the loss of a water molecule.

1.5 Guest Molecules

1.5.1 Breathing Behaviour

In an early report in this regard, guest-induced structural transformations were observed in a 2D Co(II)-based MOF: $\{[Co(NCS)_2(3\text{-pia})_2]\cdot 2EtOH\cdot 11H_2O\}_n$.[86] The pristine structure upon heating transformed to an amorphous phase, that exhibited formation of guest-included crystalline phases upon exposure to Me_2CO, THF or water to form $\{[Co(NCS)_2(3\text{-pia})_2]\cdot 4Me_2CO\}_n$, $\{[Co(NCS)_2(3\text{-pia})_2]\cdot 4THF\}_n$ or $\{[Co(NCS)_2(3\text{-pia})_2](H_2O)_2\}_n$. Both acetone and THF included phases upon heating underwent a transition to the crystalline phase, accompanied by a naked eye colour change from pink to violet. As an important finding, this structure could revert to the THF or acetone-included phase reversibly. It was also suggested that the presence of the amide groups in the ligand stabilised the guest-free phases by hydrogen bonding between the layers, and those interactions directed the sliding of the 2D sheets during structural transformations. Among other early compounds in the domain of flexible MOFs, MIL-47 (V)[84] and MIL-53 (Cr)[87,88] were regarded as important findings. Both MOFs are constructed from bdc^{2-} ligands MIL-47: [VO(bdc)]; MIL-53ht: [Cr(OH)(bdc)] and exhibited expansion of the pore upon desolvation. MIL-47 could revert to the as-synthesised phase upon treatment with trimethylbenzene and 2-methyl-1-propanol. In the case of MIL-53, once the unreacted linker was removed from the structure, a reversible transition between the closed and open pores was observed. However, unlike MIL-47 which led to the formation of a hydrophobic pore, the closed-pore form in MIL-53 included water. This was ascribed to the presence of hydroxyl groups in the metal chains that could interact with the water molecule. In a following study of the hydration behaviour in the aluminium analogue of MIL-53(Cr), it was found that the presence of hydrogen bonding interactions between the guest water and frameworks O-atoms were responsible for the breathing behaviour and stability of the narrow pore configuration.[89] These differences in the pore chemistry for the two MOFs were further observed when gas adsorption studies were performed using CO_2 and CH_4.[90] While the CH_4 adsorption isotherm for MIL-53 (Al or Cr) showed a characteristic profile for a nanoporous solid, the response to CO_2 was significantly different. The isotherm had a step at ~6 bar, which was seen for both MIL-53(Cr) and MIL-53(Al). However, when MIL-47 was tested for CO_2 adsorption, it did not show any steps during adsorption and the total amount adsorbed above 10 bar was similar for both sets of MOFs. A more detailed study of the adsorption of

CO_2 by MIL-47(V) was later reported.[91] This difference in behaviour was reasoned to be CO_2 having a permanent quadrupole, that led to MIL-53 undergoing shrinkage of the pore in the first instance and opening subsequently, like the observation for water. A set of investigations were later reported that reinforced the active role of the hydroxy (–OH) group during adsorption of CO_2. Specifically, it was found that both O and H-atoms are involved in strong non-covalent interactions with the guest CO_2, which leads to the breathing behaviour for MIL-53.[92–95]

The understanding was further expanded by studying the adsorption of C1–C4 hydrocarbons in MIL-53(Cr).[96] As reported earlier, no closing of the pore was observed for CH_4. However, in the case of other hydrocarbons, the concurrent presence of the large pore and narrow pore forms was observed, and the extent of breathing was found to depend upon the size of the adsorbate. In terms of energetics, it was found that an initial enthalpy of −20 kJ mol^{-1} was required to initiate the transition from large-to-narrow pores. Interestingly, when this study was expanded for other alkanes (C5–C9),[97] it was found that for all hydrocarbons there was a discernible sub-step, suggesting narrow pore formation as an intermediate. As expected, the total adsorbed amounts lowered with the increasing size of the adsorbate. However, when a different metal centre (MIL-53(Al)) was tried, it was found that the position of the step differentiated noticeably from MIL-53(Cr). Apart from hydrocarbons, this system (MIL-53(Cr)) was also explored for adsorption of polar solvent vapours, *e.g.*, water, methanol and ethanol.[98] It was found that all the three guests squeezed the large pore, but for the alcohols the pore closing was partial. Also, in the case of water, the MOF did not open at any pressure step after closing, whereas for the alcohols the transition to large pore form happened after P/P_0 ~0.2. For all the three guests, the mechanism was proposed to have hydrogen bond interactions between protons of the hydroxy group and O-atoms of the guest, and between protons of the guest and O-atoms of the hydroxy of the inorganic chain. Also, adsorption for ethanol was reported to be most energetic as with a longer alkyl chain, it could simultaneously interact with two metal-chains of the structures. In another study on MIL-53(Al), it was observed that during vapour adsorption of C8-alkyl aromatic compounds (xylene isomers and ethylbenzene), the MOF exhibited contraction of the pore at lower pressures.[99] Following corresponding gate-opening pressures, a separation between the guest molecules was seen along with reopening of the pore. The differences in uptake at higher pressures was correlated to the differential packing of the guests, accompanied with interactions between the methyl groups and carboxylate of the framework. Several groups subsequently reported the synthesis and breathing behaviour of MIL-53(Al) MOFs by addition of secondary functional groups to the ligand, and also by the testing post-synthetic modification in such MOFs.[100–104] The influence of linker rotation and the effect of host–guest interactions with pendant secondary groups towards guest inclusion of liquids was later probed in a series of MIL-53(Al)-X MOFs (X = H, CH_3, NH_2, NO_2 and OH) using NMR spectroscopy.[105] The impact on linker rotation upon the flexibility of MOFs is discussed in more details in Section 1.6. While interaction with guest and linker rotation are key drivers for structural changes in MIL-53 compounds, it was found that mechanical pressure could also be used to control the pore motions.[106] It was proposed to use external pressure to constrain the MOF to a narrow pore (np) form and use that phase for gas adsorption and separation. After adsorption the pressure would be released, that drives the MOF back to the large pore form. This was suggested as a better control for enhancing the regeneration process of the adsorbent.

MIL-53(Fe) is similar to the Cr and Al versions but exhibits different flexibility.[67] The as-synthesised structure consists of water molecules in its pores. Upon desolvation, unlike the Al(III) and Cr(III)-MOFs that undergo pore opening, MIL-53(Fe) exhibits a reversible pore contraction with an intermediate dehydrated structure that has very narrow pores. This difference was ascribed to differences in the M–O–M inorganic chains and the orientation of the ligand. Interestingly, MIL-53(Ga) exhibited a breathing behaviour that lies between the Al and Fe analogues.[66] While the structure opened up completely upon dehydration like Al-MOF, the intermediate step followed the Fe-MOF where a very narrow pore structure was formed. Unlike the Fe-MOF, the intermediate structure had high thermal stability (60–220 °C), before opening the pores completely. MIL-53(Sc) exhibited the formation of a closed pore form upon desolvation, which showed the smallest volume among all analogs of the MIL-53 series.[107] This was attributed to greater flexibility owing to the larger ionic radius for Sc(III). Interestingly, hydration at room temperature to the desolvated structure gave rise to an intermediate that had water molecules included in alternate channels (Figure 1.11).

These structural differences with different metal cations are seen to have a significant impact on the breathing behaviour. Although MIL-53(Fe) does not have a permanent porosity in its evacuated state, it was observed that treatment with a range of liquids gave rise to pore opening to the large pore forms with the inclusion of guest molecules.[108] This process was seen to be rapid and the swelling of the structure was seen to be influenced by the interactions or lack of it with the hydroxy groups in the inorganic chains. Further investigation on this MOF was carried out for its ability to adsorb light alkanes (C1–C4) at 303 K. Unlike Al or Cr-MOFs, MIL-53(Fe) exhibited complex stepwise adsorption for all the hydrocarbons.[109] The structure opened from the very narrow pore (vnp) form to an intermediate np. The steps in the adsorption profiles varied with the size of the adsorbate, and the pore opening process was attributed to a combination of host–guest interactions and the size of the pore for corresponding guest inclusion. However, MIL-53(Sc) exhibited different behaviour for CO_2 adsorption (196 K) which had a two-step process leading to the complete opening of the pore with the presence of a closed pore intermediate from the vnp in the evacuated state.[107] As described in Section 1.3 the presence of secondary functional groups has a significant impact on the breathing behaviour of flexible MOFs. When the linker was functionalised with the $-NO_2$ group, the compound directly opened to the lp form, and the step was at a much lower pressure compared

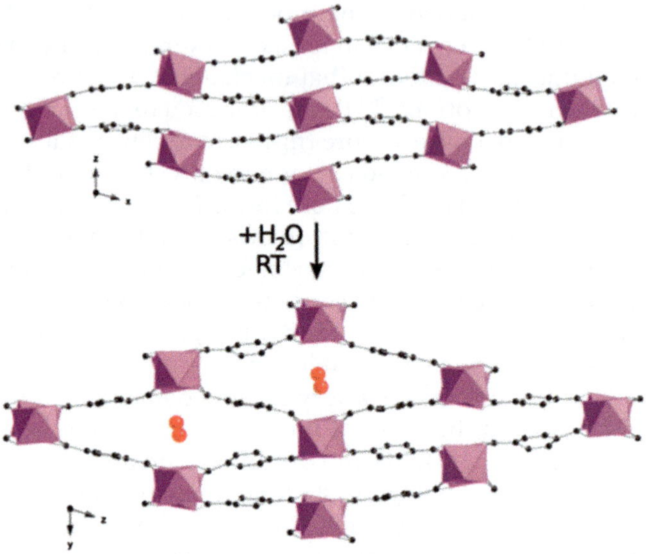

Figure 1.11 Change in the structure of MIL-53(Sc) from closed pore (cp) form (top) to intermediate phase (int) (bottom) upon hydration at room temperature. Reproduced from ref. 107 with permission from the Royal Society of Chemistry.

to the non-functionalised MOF. This was ascribed to the steric hindrance caused by the bulky group. Likewise, it was noted that when the linker in MIL-53(Fe) was functionalised with groups of varying characters, such as –Cl, –Br, $-CF_3$, $-CH_3$, $-NH_2$, –OH, $-CO_2H$, the breathing patterns with respect to the non-functionalised MOF were altered.[110] Although in all the cases the compounds retained their flexibility, the breathing nature was complex and was dependent upon the host–guest interactions as more sites were available for guest interactions. In addition, factors like steric hindrance caused by the pendant groups and intraframework interactions were seen to influence the ability of the compound to breathe.

Another set of MOFs—MIL-88—was found to present a very high degree of breathing. Initially reported using fumaric acid as ligand, the MOF was prepared as a second step after the formation of a trimeric SBU.[111] The resulting MOF—$[Fe_3O(CH_3OH)_3(fum)_3\cdot(CH_3CO_2)\cdot(CH_3OH)_{4.5}]$ (referred to as MIL-88A)—underwent structure shrinkage when heated and desolvated. However, a very large swelling behaviour was observed with ~85% change in cell volume from the fully evacuated state and upon hydration.[112] The guest inclusion was also seen for other polar molecules, *e.g.*, methanol, ethanol and butanol, and the extent of swelling was found to be inversely correlated to the size of the guest. More importantly, this swelling was fully reversible over multiple cycles. The swelling was proposed as a combination of rotation of the metal trimeric SBU and the free rotation of the linker based on single bonds. This system was further expanded into a series of isoreticular MOFs MIL88B (1,4-bdc), MIL-88C (2,6-ndc) and MIL-88D (4,4′-bpdc), that were prepared using a similar synthetic protocol as described initially.[113] Interestingly, all the compounds of this series exhibited large breathing, and the extent of swelling was also found to vary with the nature of the ligand.[114] A detailed study using a number of organic solvents, *e.g.*, pyridine, DMF, DEF, DMC, DMSO, toluene, hexane, lutidine, water and alcohols, suggested that in principle the host–guest interactions drove the swelling behaviour. Broadly, these interactions could be classified into the affinity of polar liquids with the SBU and weak non-covalent (van der Waals, π–π, CH–π) interactions for all solvents with the linkers. Also, for these solvents, the swelling was found to be reversible in all the MOFs.

Similar to the MIL-series MOFs, solvent-dependent breathing was also seen in a In(III)-based MOF: $(Me_2NH_2)[In(abdc)_2]$ (referred to as SHF-61).[115] The compound had a two-fold interpenetrated 3D structure that comprised of uncoordinated cations and occluded solvent molecules.[116] The adsorption properties for the solvent evacuated phase were found to be strongly dependent upon the solvent used for pre-activation treatment. When treated with $CHCl_3$, the structure adopted an open pore form that remained unaltered during adsorption. However, when the pristine phase was heated directly, a continuous breathing phenomenon corresponding to the desolvation of the solvents (DMF and water) was noted. This continuous structural change was found to be fully reversible and was linked to the strong interaction of the solvent molecules with the framework.

1.5.2 Coordinated Solvent or Template Effect

The synthesis of MOFs typically leads to inclusion of water or other solvents in the framework, either as coordinated molecules or free guests within the pore. In certain cases, use of bulky solvents functions as a templating agent that leads to formation of porous architectures. Among the different structures, 2D MOFs are seen to offer higher degree of flexibility in such cases during guest removal. Flexibility of the linker and/or variable coordinating tendency of the metal ions are further seen to facilitate softness in such systems. This was observed in the first instance of gate-opening adsorption in MOFs.[117] The synthesised pre-ELM-11 phase $\{[Cu(BF_4)_2(bpy)(H_2O)_2](bpy)\}_n$, exhibited 2D $Cu(bpy)_2$ layers that were bound by non-covalent interactions. When the compound was activated at different temperatures (348–423 K), differences in N_2 (77 K), Ar (77 K) and CO_2 (273 K) adsorption profiles were seen which could be linked to partial or complete loss of coordinated water molecule. In a following study, it was revealed that the adsorption of the fully dehydrated phase proceeded by clathrate formation of the adsorbates within the layers of the compound.[118,119] This was validated by observations of change in volume of the sample upon CO_2 gas inclusion. Subsequently, several groups tried to elucidate the detailed understanding about the flexible nature of ELM-11.[120,121]

This study was followed by a series of compounds that could possess dynamic behaviour, for instance in a Cu(II)-based 2D MOF: $\{[Cu_2(tci)(OH)(H_2O)_3]1.5H_2O\}_n$.[122] Upon guest removal—both free and coordinated water molecules—the 2D sheets in the structure were seen to slide and contract the pores. The desolvated phase had Cu(II) ion in square-planar geometry (deep blue colour) from an octahedral arrangement (greenish blue) in the pristine phase. This presence of a coordinatively unsaturated site (CUS) in the activated phase was seen to aid the adsorption of water molecules and revert to the pristine structure. The same group had also observed such structural changes in another set of 2D MOFs—$\{[Ce(tci)(H_2O)_2]\cdot 2H_2O\}_n$ and $\{[Pr(tci)(H_2O)_2]\cdot 2H_2O\}_n$—based on a ligand with flexible aliphatic arms.[123] The compounds having central metal ion in a distorted square-antiprismatic geometry with coordinated water, underwent conversion to a 3D phase upon heating accompanied with loss of bound and free water. This phase change was observed to be reversible and seen to occur for both the compounds. Such sliding of 2D MOFs with coordinated water is also seen for several other MOFs as well.[124] Similar flexibility in a 3D Al(III)-MOF having coordinated water $[Al_2(OH)_2(H_2O)_2(L4)]$ (referred to as MIL-118A) was subsequently observed.[125] The compound upon heating lost the coordinated water molecules and transformed to $[Al_2(OH)_2(L4)]$ (MIL-118B). This transition involved the non-bonded carboxylate in the pristine phase participating in coordination to the metal centre instead. Upon hydration of MIL-118B, it was observed that water could now enter the pores and the inorganic layer adjusted to form a rectangular pore shape. The rehydrated phase (MIL-118C) could reversibly transform back to MIL-118B upon dehydration (Figure 1.12).

Likewise, other polar solvents used in the reaction medium often bind to the metal centres, and some compounds exhibit flexible responses upon desolvation. This was seen in a Zn(II)-based 3D MOF $[Zn_2L5_2(DMF)_2]_n$, that

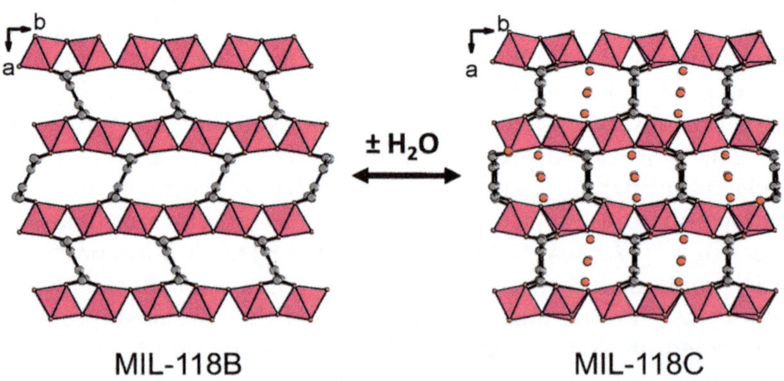

Figure 1.12 Change in structure from desolvated phase (MIL-118B) upon water adsorption to MIL-118C. Reproduced from ref. 125 with permission from American Chemical Society, Copyright 2009.

could squeeze its pores upon loss of coordinated DMF molecules.[126] While the guest-free structure could reversibly return to its pristine form after exposure to DMF, the desolvated phase when left in air adopted a different phase, owing to the coordination of water molecule. Low-temperature CO_2 and CH_4 adsorption exhibited gate-opening behaviour, and due to the presence of CUS, the compound also showed uptake for polar adsorbate such as tetrahydrofuran and dioxane. The same group later observed flexibility in Cu(II)-based MOF constructed from neutral N-donor ligands and using bulky *o*-xylene as a pore template: $\{[Cu(L6)_2(NO_3)_2]\cdot(oX)\cdot(DMF)\}_n$.[127] The pristine phase with a 1D coordination polymer structure, underwent conversion to a non-porous 2D MOF upon desolvation. This transition was accompanied by the bending of the ligand in the presence of Cu(II) allowing structural softness. As a side note, typically for MOFs constructed from neutral N-donor ligands, the anions used in the reaction function as the charge-balancing species in such frameworks.[128,129] Depending upon their coordinating ability, the ions either bind to the metal centres or are present as uncoordinated species in the pores.[130] Not only do these direct functional characters of these system but also tap into the possibility of framework flexibility as the metal–ligand coordination bond is relatively softer compared to the conventionally used carboxylates or azolates.[131–133] In the above example, when NO_3^- anions were exchanged with strongly coordinating SCN^- ions, the equatorial bond lengths for Cu(II) changed, which was accompanied with a noticeable colour change from blue to green and change in structural packing. Previously, a broader case for such MOFs to function as anion sensors was reported. The parent MOF based on Cu(I)—$\{[Cu(pytpy)](NO_3)(CH_3OH)\}_n$—could readily exchange the guest methanol with water and the resulting phase was able to distinguish a range of anions, *e.g.*, F^-, Cl^-, Br^-, N_3^-, SCN^-, CO_3^{2-} and I^-, based on the change in colours.[134] While most of the exchanged products appeared to remain like the water-included structure, subtle changes brought about by the guest (anion) were observed. It is worth noting this subset of MOFs (ionic-MOFs) has evolved considerably since they were initially proposed and are finding avenues in multiple domains, especially on account of their structural softness.

Pillared-layer structures are witnessing increasing attention in the domain of MOFs, as they tend to offer a higher degree of opportunity to tune and modulate structural features in a controlled manner. In an early report in this regard, expansion or contraction of the pore was observed, depending upon the pillar used in the preparation of two similar 3D MOFs: $\{[Cd(pzdc)(azpy)]\cdot 2H_2O\}_n$ and $\{[Cd(pzdc)(bpee)]\cdot 1.5H_2O\}_n$.[135] In both the structures water molecules occupied the pores and the protruded carboxylate moiety of the linker had hydrogen-bonding interactions. Upon desolvation, the MOF with ethylene linker exhibited shrinking owing to favourable H-bond interactions with the carboxylate group. However, the MOF with an azo-bonded linker expanded as proximity with carboxylates was unfavoured. It is worth noting that in certain cases the loss of uncoordinated guest molecules from the framework may lead to irreversible phase transitions that are either crystalline or amorphous.[136]

1.6 Flexibility Design from View of Network Topologies

1.6.1 Interpenetration

Highly porous coordination framework always tends to maximise their packing efficiency, in which the empty spaces are usually occupied by molecules used in the synthesis (*e.g.*, starting materials, solvents, counter ions, *etc.*), or additional coordination networks. These nets are not connected by coordination bonds but cannot be split without any bond breakage and thus termed as 'interpenetration'.[137] The interpenetrated networks can be identified as *n*-fold ($n \geq 2$) frameworks according to the number of the entangled networks in overall packing structures, such as from two to eight, and even up to 54-fold interpenetrated structures.[138,139]

1.6.2 Structure and Property

Generally, the increase in the degree of the interpenetration will reduce void space of frameworks, but can enhance structural stability and flexibility, the latter of which can be shown as displacement between the interpenetrated networks.[6] This dynamic structural transformation behaviour usually stems from the distortion of the single net as well as the sliding action between multiple networks, which always serves as the key factors to regulate and control the property/functionality of the flexible interpenetrated frameworks.[1,140]

1.6.2.1 Gas Sorption

As early as 2002, a two-fold interpenetrated **pcu** network of $[Cu_2(bdc)_2(bpy)]$ was constructed by paddle-wheel $Cu_2(RCOO)_4(py)_2$ cluster and mixed linear linkers of terephthalate (bdc^{2-}) and 4,4′-bipyridine (bpy) (Figure 1.13).[141,168,172] All of the neighbouring layer structures of $[Cu_2(bpy)_2]$ are bridged by the bpy linkers to form a single **pcu** net of the 3D pillared-layer type framework (Figure 1.13(a)). This MOF was demonstrated to present dynamic gate-opening behaviours upon gas (N_2, CH_4) and vapour (methanol) sorption, as evidenced by PXRD (Figure 1.13(b)). SCXRD confirmed that the guest removal can allow the layers to be shifted to each other and the pillared ligand (bpy) will be tilted.[143] The distortion of the pillars and slippage of the layers is topologically accompanied by the displacement of interpenetrated networks regarding each other, finally forming a nonporous closed phase. The dry structure can expand to the guest-included open phase accordingly with the above-mentioned gate-type dynamic sorption behaviours.

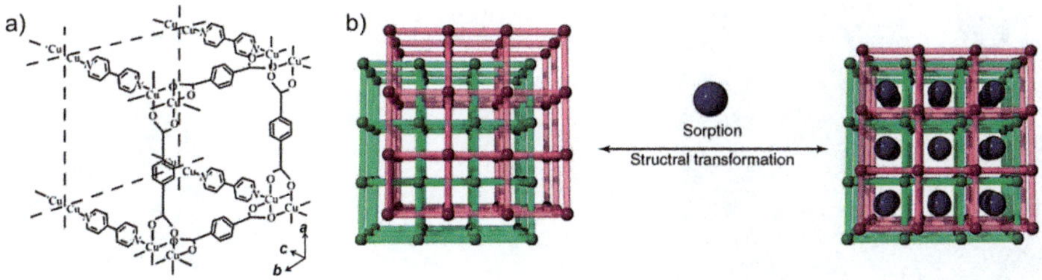

Figure 1.13 (a) The single framework of plausible structures for $[Cu_2(bdc)_2(bpy)]$. Reproduced from ref. 141 with permission from the Royal Society of Chemistry. (b) The proposed displacement between the two-fold interpenetrated networks upon of guest sorption. Reproduced from ref. 142 with permission from the author.

1.6.2.2 Guest Recognition

The displacement of interpenetrated coordination frameworks can be controlled by counter anions with steric hindrance/electric attraction to regulate the pore size for sorption behaviours. For instance, a counter anion-controlled dynamic structure and sorption property was observed based on two-fold interpenetrated framework of $[Ni(bpe)_2(N(CN)_2)]\cdot N(CN)_2\cdot 5H_2O$ (Figure 1.14).[144] The single crystal diffraction analysis showed that its as-synthesised form has two types of channels. The smaller rectangular one (2.81 × 0.61 Å^2) contains the counter anions of $N(CN)_2$, while the larger hexagonal one (6.50 × 4.74 Å^2) accommodates many H_2O molecules. After exchanging the bent $N(CN)_2$ by linear N_3 molecules, the difference in pore size increases because the smaller N_3 allows the two-fold entangled structures to get much closer to each other, as confirmed by PXRD. As a result, CO_2 adsorption isotherms (195 K) showed a 25% increase in uptake after the anion exchange.

1.6.2.3 Sensor

Most of the luminescent MOF sensors can detect a single target molecule in a mixture,[145] while cannot differentiate multiple targets from one another simultaneously (Figure 1.15(a)).[142] For example, a two-fold, interpenetrated, naphthalenediimide-based MOF of $[Zn_2(bdc)_2(dpNDI)]\cdot 4DMF$ was synthesised, showing structural typical dynamics based on the dislocation of two entangled networks upon guest removal/inclusion. This flexible interpenetrated MOF can accommodate a class of aromatic volatile organic compounds (VOCs) with an intense turn-on emission over the entire visible region (benzonitrile: 421 nm, benzene: 439 nm, toluene: 476 nm, *o*-xylene: 496 nm, *m*-xylene: 503 nm, *p*-xylene: 518 nm, and anisole: 592 nm), offering changed luminescent colours detectable by the naked eye (Figure 1.15(b) and 1.15(c)). The unique chemoresponsive, multicolour luminescence should be attributed to the different naphthalenediimide–aromatic guest interaction by induced-fit structural transformation within the interpenetrated framework.

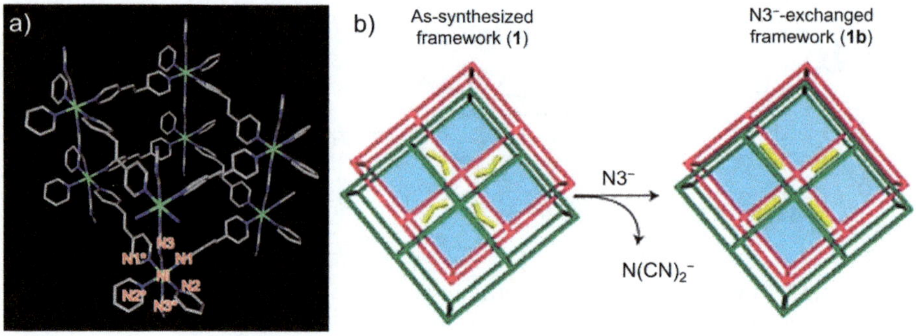

Figure 1.14 (a) The building unit of [Ni(bpe)$_2$(N(CN)$_2$)]. (b) Robust α-polonium-type two-fold interpenetrated porous networks with multiple functionalities. Reproduced from ref. 144 with permission from Springer Nature, Copyright 2007.

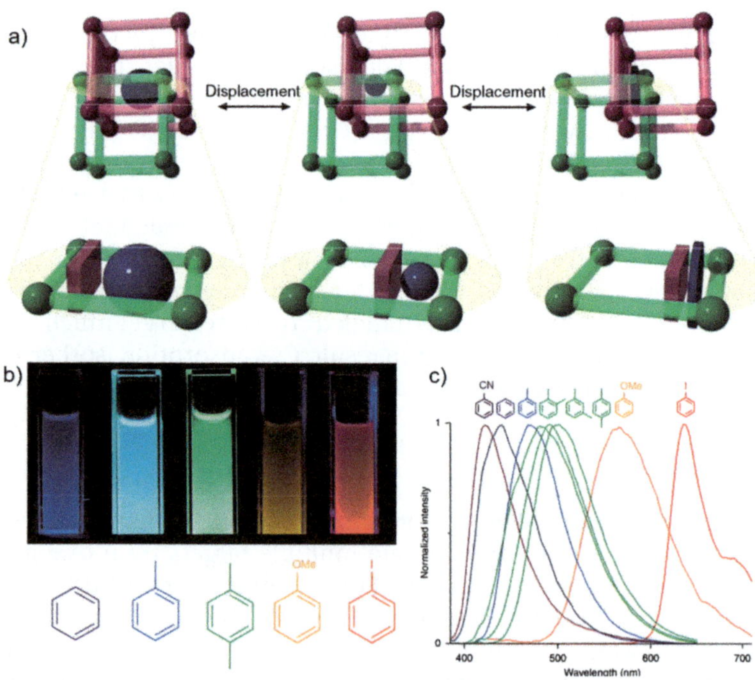

Figure 1.15 (a) The framework entanglements can provide flexibility by altering the spaces in response to different target guest molecules. Chemically non-interconnected frameworks exhibit dynamic movements to effectively detect molecules while maximizing the host–guest interactions. (b) Resulting luminescence of the crystal powders of [Zn$_2$(bdc)$_2$(dpNDI)], suspended in the corresponding VOC liquid with excitation at 365 nm. (c) The height-normalised luminescent spectra of VOCs@[Zn$_2$(bdc)$_2$(dpNDI)] by excitation at 370 nm. Reproduced with from ref. 142 with permission of the author.

1.6.3 Controlling Interpenetration Behaviour

Control over the degree of interpenetration to construct the CPs with desired structures and functionalities remains a great challenge because of a number of factors including temperature, concentration, solvent, time, pH and modulator.[138,146] Here we will focus on summarising recent advances in regulating the degree of interpenetration regarding flexible frameworks and the influence on the resulting properties.

1.6.3.1 Ligand Design

One of the most successful design approaches to control interpenetration behaviours on flexible MOFs is to use metal ions/clusters with different organic ligands to build interpenetrated networks. A two-fold interpenetrated **pcu** network of $[Zn_2(bdc)_2(bpy)]$ (void = 27.7%) was carefully explored for its guest-induced transformation.[147] After guest removal, its guest-free structure retained the original framework connectivity but finally transformed into a closed phase (void = 16.7%), which should be ascribed to the significantly distorted Zn_2 clusters and bent bpy ligands. Additionally, without any steric hindrance of guests, this coordination network slid to a stable position, resulting in a higher symmetry. By virtue of the reversible guest-induced flexibility, the closed phase can expand to an open state and show a hysteretic sorption behaviour upon gases (N_2, H_2, and CO_2).

To further understand the effect of ligand length on flexibility of framework, by using the well-defined $Zn_2(RCOO)_4(py)_2$ cluster with different linear ligands, a series of two-fold interpenetrated mixed-ligand MOFs were constructed: $[Zn_2(bdc)_2(bpy)]$, $[Zn_2(fm)_2(bpe)]$, $[Zn_2(fm)_2(bpa)]$ and $[Zn_2(fm)_2(bpy)]$.[149] The interpenetration here can provide flexibility to pillared-layer type frameworks whose tunable porous properties and sizes can be achieved and demonstrated by the dynamic adsorption behaviours of ethylene and ethane. By using the longer ligands,[150,151] a highly flexible three-fold interpenetrated $[Zn_2$(4,4′-biphenyldicarboxylate)$_2$(4,4′-bis(4-pyridyl)biphenyl)] was synthesised (Figure 1.16(a)).[148] This MOF exhibits six distinct phases and four types of structural transformation in response to various stimuli, which correspond to many observed dynamic structural transformations, including breathing, structural isomerism, shape memory effect, and change in the level of interpenetration (Figure 1.16(b)). Noted that structural flexibility of the interpenetrated frameworks can be finely tuned by the weak interactions (cooperation of local and global motility) from modified ligand substitutions.[27,152]

1.6.3.2 Number of Interpenetrations with Different Flexibility

Although interpenetration has been considered an effective strategy to introduce framework flexibility but increasing the number of interpenetrations is not always a good choice, because it will reduce the necessary space for dynamic behaviours. Significantly different framework flexibilities between

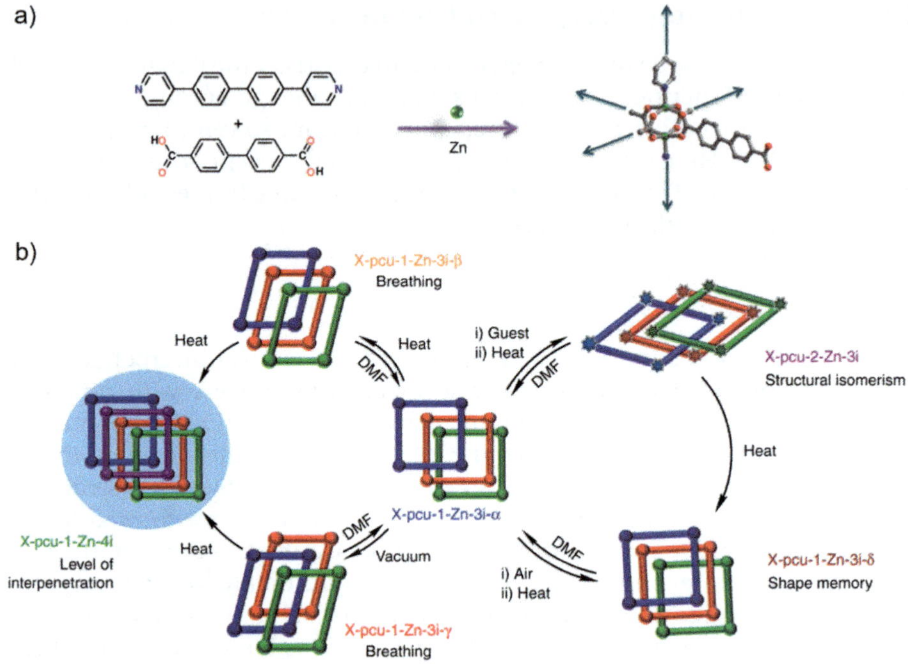

Figure 1.16 (a) The structures of 4,4′-bis(4-pyridyl)biphenyl and 4,4′-biphenyldicarboxylic acid, the pillaring of 2D square network of $[Zn_2(4,4'\text{-biphenyldicarboxylate})_2]$ by 4,4′-bis(4-pyridyl)biphenyl linker gives the **pcu** topology network. (b) Its multi-dynamic nature can be well exemplified by the six structural phases. Reproduced from ref. 148, https://doi.org/10.1038/s41467-018-05503-y, under the terms of the CC BY 4.0 license, http://creativecommons.org/licenses/by/4.0/.

two-fold and three-fold interpenetrated isomers of pillared-layer type $[Zn_2(btdc)_2(bpy)]$ were observed.[153] CO_2 sorption supports that the three-fold interpenetrated one exhibits a type-I isotherm (total uptake of 55 $cm^3\ g^{-1}$ at 195 K), while the two-fold interpenetrated isomer shows a gate-opening-type behaviour (up to 201 $cm^3\ g^{-1}$). Parallel PXRD–adsorption measurements confirmed the structural rigidness of the three-fold interpenetrated isomer and revealed the structural transition from a square to a rhombus-shaped pore for the two-fold interpenetrated one. Obviously, the higher porosity plays an important role in the framework distortion. It should be noted that, besides the interpenetration number, interpenetration direction can also significantly influence framework dynamism.[154]

1.6.3.3 Partial Interpenetration in MOFs

When compared with the various types of fully interpenetrated frameworks, partial interpenetration is another exciting development to study structural flexibility. A tunable change of the interpenetration degrees was carried out

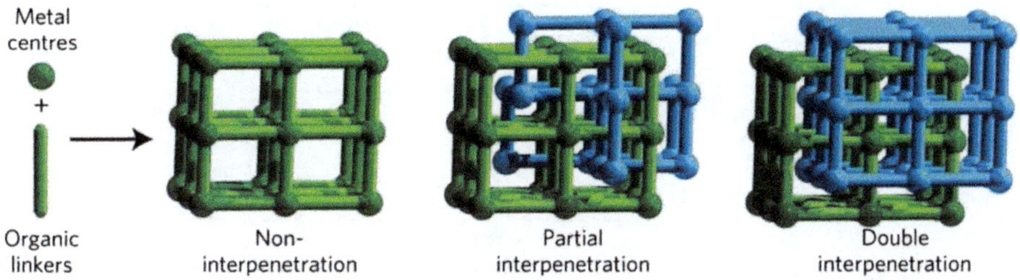

Figure 1.17 The partial interpenetration of NOTT-202 constructed by one dominant framework (green) and the secondary partially formed network (cyan). Reproduced from ref. 156 with permission from Springer Nature, Copyright 2012.

by solvent-mediated grinding/stirring on a **pcu** network of $[Zn_4O(dpdc)_3]$ (MUF-9).[155] By employing different solvents and/or heating times, MUF-9 can be constructed with non-, partial, and two-fold interpenetration. Besides, the similar partial interpenetration was also observed in another two-fold **dia**-type MOF (dia = diamondoid), $(Me_2NH_2)_{1.75}[In(bptc)]_{1.75}\cdot 12DMF\cdot 10H_2O$ (NOTT-202) (Figure 1.17).[156] Interestingly, this MOF consists of two independent networks, one of which is partially formed with a crystallographic occupancy of 0.75. After desolvation, the interpenetrated framework showed sliding behaviours, resulting in a large expansion of pore sizes from 5.0 × 5.0 to 9.0 × 9.0 Å^2 as confirmed by *in situ* SCXRD. This partially interpenetrated framework showed three-step CO_2 adsorption (195 K) behaviour, which should be attributed to the reversible structural sliding transitions during the guest removal/induction. Note that higher temperature synthesis can yield fully interpenetrated rigid **dia** frameworks with a dramatically decreased porosity of <4%.

1.6.3.4 *Crystal Size*

The flexibility of interpenetrated structures relies on not only the chemical compositions and local/long-range connectivity but also crystal size.[157] For example, the downsizing effect was observed upon the crystals of a flexible porous interpenetrated framework of $[Cu_2(bdc)_2(bpy)]$ (void = 35%) (Figure 1.18).[143] A series of crystals with different sizes (methanol as a guest) were synthesised. After guest removal at room temperature, the crystals with 300 × 300 × 30 nm^3 or larger size will shrink to a narrow-pore phase (void = 20%). Surprisingly, smaller nanocrystals can completely (50 × 50 × 20 to 60 × 60 × 20 nm^3) or partially (110 × 110 × 23 to 160 × 160 × 25 nm^3) retain the original open form as observed in the as-synthesised crystals. Their size-dependent flexibility will be different after heating at 200 °C, that is, all samples with different sizes can shrink. Further, methanol sorption (303 K) for the nanocrystals (50 × 50 × 20 nm^3) exhibited a gate-opening isotherm upon the closed-state sample, but an ordinary type-I adsorption behaviour for the open-state sample. By comparison, this similar size-dependent structural flexibility was also experimentally demonstrated in an analogous MOF of $[Cu_2(bdc)_2(bpe)]$. This shape

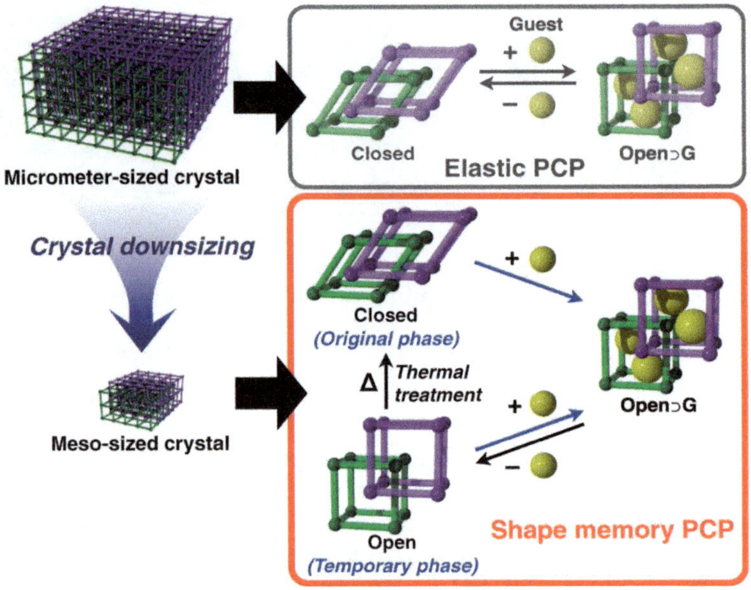

Figure 1.18 Schematic illustration of the shape-memory effect in porous networks by crystal downsizing, which can effectively control the structural flexibility. Reproduced from ref. 143 with permission from The American Association for the Advancement of Science.

memory effect of size-dependent flexibility has been also observed in other flexible frameworks,[140,158–160] and should be attributed to the different degrees of defects in the crystal, which greatly accelerate the phase transition. Obviously, they are unique platform frameworks to study macro- and micro-sized structures, which possess important guiding significance for the function/property modulation of MOFs.

1.6.3.5 External Stimuli

Flexible coordination frameworks can undergo structural transformation only upon external stimuli. Among the various physical forces,[161,162] light irradiation can provide high energy to achieve the structural transformation. To realise light-controlled flexible coordination networks, employing photodynamic compounds as the skeleton of the bridging ligands is one of the most effective strategies. Based on these considerations, an approach through design and synthesis of a two-fold interpenetrated diarylethene-containing pillared-layer framework of $[Zn_4(bdc)_4(L_0)_2{\cdot}(G)]$ was developed.[36] Ligand L_0 serves as the pillar between a non-photochromic layer network of [Zn(bdc)], which accommodates changes in linker geometry after photoisomerisation without framework collapse confirmed by SCXRD. In addition, linker photoisomerisation was also well explored by spectroscopic studies and reversibly photo-modulated CO_2 sorption measurements, showing the possibility to tune the flexibility of the interpenetrated frameworks.[163]

1.6.4 Interdigitation

Interdigitation refers to polythreading of the digited motifs with finite length (usually coordinated ligand moieties) that can project into the accessible voids of adjacent layers in a face-to-face manner. Interdigitation is distinctly different from interpenetration because it did not have any structural penetration.[164]

1.6.4.1 Structure and Property

Up to now, a variety of 1D/2D interdigitated CPs have been successfully designed and constructed. Many of them exhibit high flexibility governed by the regulable weak interactions between interdigitated structures upon guest inclusion and removal.

1.6.4.2 1D Interdigitated CPs

In 2002, a representative 1D interdigitated CP of [Rh^{II}_2(benzoate)$_4$(pyrazine)] was constructed by linking the paddle-wheel Rh_2(benzoate)$_4$ cluster with the bridging ligand of pyrazine in a linear geometry (Figure 1.19(a)).[166] Tight stacking between the benzene and pyrazine rings along the eventual sliding chains enables the high-density packing to reduce the void space of skeletons without any solvents. While this CP was observed to adsorb N_2 with a type I adsorption isotherm (BET surface: *ca.* 350 $m^2\ g^{-1}$) (Figure 1.19(b)). SCXRD analysis showed that, to accommodate the N_2 molecular, the linear chains undergo a significant shift along the *c*-axis (from $C2/c$ to $P\bar{1}$, void = 17.4%) (Figure 1.19(c)).[167] This indicates the underlying dynamic nature of 1D interdigitated CPs, in which the guest inclusion enables the chains to rearrange with each other to become porous structures. Based on this consideration, this kind of CPs have been intensely investigated for sorption of different small molecules, including CO_2, N_2O, NO, CH_4, SO_2, *etc.*[165,168–171]

1.6.4.3 2D Interdigitated CPs

When compared with the 1D frameworks, 2D interdigitated CPs have been investigated more extensively for structure design and property evaluation. A series of 2D interdigitated frameworks based on linear bipyridyl and carboxylate ligands for sorption properties have been illustrated.[172,173] For example, a layer CP of [Cu_2(dhbc)$_2$(bpy)] has an interdigitation by π–π stacking interactions between benzene rings of the dhbc, which serves as the pillars between adjacent layers (Figure 1.20(a)). This π–π stacking offers moderate interaction to stabilise the structure, and its sliding motion gives an open framework when guests are adsorbed. This structural transformation will be

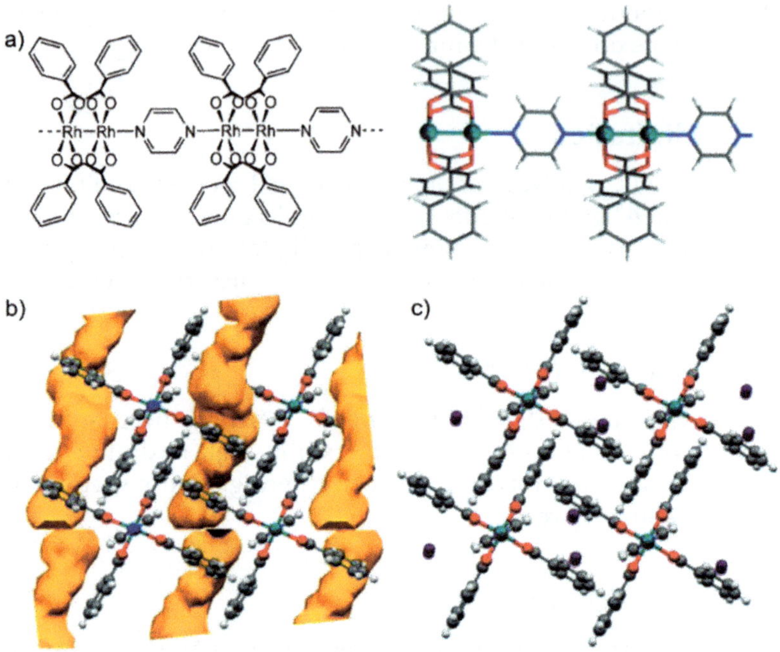

Figure 1.19 (a) Representation of the linear infinite-chain structure of $[Rh^{II}{}_2(benzoate)_4(pyrazine)]$ and (b) guest-free structure with discontinuous void space. (c) Packing structure with inclusion of N_2 molecules. Colour code: dark green, Rh; light grey, C; white, H; red, O; blue, N; purple, N_2. Reproduced from ref. 165 with permission from Elsevier, Copyright 2021.

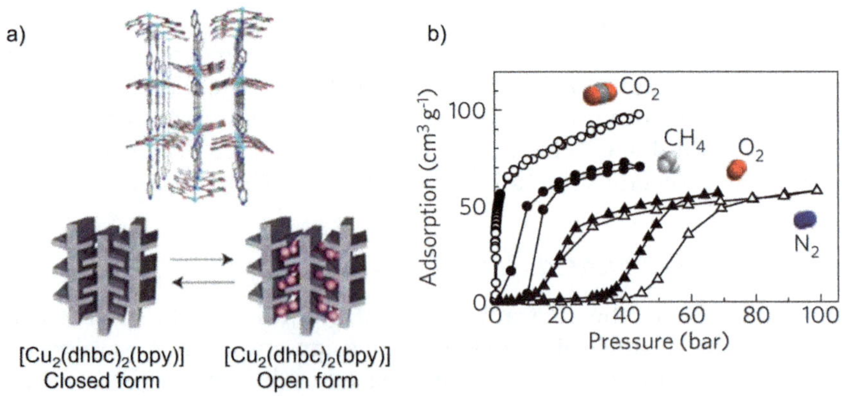

Figure 1.20 (a) The crystal structure of 2D interdigitated framework of $[Cu_2(dhbc)_2(bpy)]$ with the reversible transformation between closed and open structures induced by guest sorption. (b) Gas adsorption isotherms of $[Cu_2(dhbc)_2(bpy)]$ at 298 K, in which all gases are adsorbed at different gate-opening pressures. Reproduced from ref. 1 with permission from Springer Nature, Copyright 2009.

accompanied by the shrinkage of layer distance after guest release. The compound exhibits hysteretic adsorption isotherms for CH_4, CO_2, N_2 and O_2 at 298 K (Figure 1.20(b)).

1.6.5 Controlling Interdigitation Behaviour

1.6.5.1 Ligand Design

One of the most efficient approaches for the design and construction of interdigitated networks is to vary organic ligands to control interdigitation between layered structures to tune the pore sizes and shapes.[174,175] For example, based on the typical interdigitated CP of [Cu_2(dhbc)$_2$(bpy)], V-shaped dicarboxylate of ip was employed to replace the dhbc ligand to construct a similar interdigitated structure of [Zn_2(ip)$_2$(bpy)$_2$] (CID-1).[176] The Zn(II) centre is linked by an isophthalate ligand to create 1D double-chains, which are connected by bpy linkers to give a 2D layer. The layers are interdigitated together to form 3D flexible frameworks with linear non-intersecting arrays of cavities with narrow connecting windows, which respond for its highly selective guest sorption behaviours. Following the guest-responsive flexibility of CID-1, there have already been extensive investigations of ligand design for the isoreticular interdigitated frameworks. Then longer ligands were employed in the construction of a series of similar flexible structures, including V-shaped dicarboxylate ligands for [Zn(ndc)(bpy)] (CID-3) and [Cd(bpndc)(bpy)] (CID-4),[177–179] and dipyridyl linkers for [Zn(ip)(bpb)] (CID-21), [Zn(ip)(bpt)] (CID-22) and [Zn(ip)(bpa)] (CID-23).[180] When compared with the CID-1 (12.2% of void), the replacement of substituent groups of the dicarboxylate/dipyridyl linkers can effectively increase their porosities and modulate the pore sizes and shapes (25.8% for CID-3, 29.4% for CID-4, 26.8% for CID-21 and CID-22, and 27.4% for CID-23). Interestingly, these isoreticular interdigitated frameworks showed highly size-dependent selective gas adsorption properties accompanied by reversible structure transformations.

1.6.5.2 Functional Groups on Ligands

The structural flexibility of the interdigitated networks can be finely tuned by the interchain interactions from varying ligand substitutions.[168,169,181–183] For the 1D interdigitated CP of [Cu_2(benzoate)$_4$(pyrazine)], the *para*-substituent X groups (X = Cl, Br, I and OCH_3) on the benzoate (BA) ligands were utilised to construct a series of interdigitated CPs: [Cu^{II}_2(*p*-XBA)$_4$(pyrazine)].[184] Interestingly, although all of them were non-porous, gate-opening gas sorption behaviour for CO_2 (195 K) and N_2 (77 K) was observed with structural phase transitions. SCXRD analysis confirmed that the magnitude of interchain van der Waals interactions, including halogen···π, π···π, and C–H···π contacts should be associated with the gate-opening sorption pressure and hysteresis behaviours. Importantly, coupled with Hirshfeld surface analysis to evaluate the interactions for the type and magnitude, the structural flexibility would decrease as the larger halogen groups were included. However, the structural transformation and phase transition of the 1D interdigitated CPs can be also altered by the substitutions of the pyrazine. For example, owing to the steric hindrance of alkyl (X) groups from pyrazine derivatives (X-pyz), 1D chains of [Rh_2(bza)$_4$(X-pyz)] (X = 2-methyl, 2,3-dimethyl, or 2-ethyl-3-methyl) were able to switch from the zig-zag to straight upon CO_2 sorption.[168,169]

Different from fine regulation of the flexibility on 1D CPs, judicious choice of substituent groups turns on the flexibility of 2D interdigitated CPs.[181,182] For example, a series of four open frameworks of [Fe(ip)(bpy)] (FeCID-1), [Fe(3,5-pyridinedicarboxylate)(bpy)] (FeCID-2), [Fe(5-NO_2-isophthalate)(bpy)] (FeCID-5), and [Fe(5-MeO-isophthalate)(bpy)] (FeCID-6) were reported. CO_2 sorption (195 K) revealed that all these CIDs were isoreticular networks, while only FeCID-5 exhibited structural flexibility, which should be attributed to the different influences of electronic properties or shapes of substituent groups on the overall packing layered structures.[181]

1.6.5.3 Metal Ions

Metal centres in isostructural flexible interdigitated CPs may exhibit different coordination environments, which can affect the flexibility of whole frameworks.[64] For example, four isostructural CID-1s (M = Co, Ni, Cu, and Zn) were synthesised, where coordination geometries around the metal ions of Co and NiCID-1 were symmetric ideal octahedron with uniform coordination bond lengths (2.0–2.2 Å). However, CuCID-1⊃H_2O and ZnCID-1⊃H_2O showed distorted octahedral geometry (Jahn–Teller effect) revealed by one longer bond length (2.6 Å) between the metal centre and one oxygen atom compared to other coordination bonds. Interestingly, this series of interdigitated frameworks show different dynamic water-adsorption behaviours. The onset pressures (P/P_0) for CuCID-1 and ZnCID-1 were approximately 0.60 and 0.33, while CoCID-1 and NiCID-1 showed almost no water uptake below P/P_0 = 0.9. Crystal structure of CuCID-1⊃H_2O suggested that water molecules in channels form hydrogen bonds with oxygen atoms of the carboxylate ligand, which seemingly leads to the coordination distortion, thus resulting in the gate-opening structural transition for water sorption.

1.6.5.4 Crystal Size

Crystal downsizing can also control the structural transformation of flexible interdigitated CPs.[158] For bulk CID-1 crystals guest removal at 130 °C under vacuum resulted in slight framework shrinkage as demonstrated by the shifting of PXRD peak higher 2θ values. When its crystal size was downsized to the nanoscale (30 × 100 × 500 nm^3), the guest-removal-induced structural shrinkage would reduce significantly, possibly owing to the change in size or surface structure (covered with surfactant) of the crystals. SCXRD and synchrotron PXRD analyses exhibited that bulk- and nano-CID-1 underwent

3.2% and 2.1% reductions in their unit-cell volume after guest removal, respectively.[159]

The structural flexibility of interdigitated frameworks can be also controlled by downsizing bulk crystals to a thin film at the nanoscale, endowed with dynamic gate-opening behaviour. A 2D Hofmann-type CP of Fe(pyridine)$_2$[Pt(CN)$_4$] thin film on the Au/Cr/Si substrate *via* a layer-by-layer (LbL) growth was fabricated.[160] The layered structure consists of $[Pt(CN)_4]^{2-}$ and cyanide-bridged Fe(II) ions coordinated by pyridine molecules, which formed π–π stacking with the adjacent pyridine rings to form a 3D interdigitated framework. Although this CP in bulk showed no uptake of guest molecules because of the densely packed layers, when fabricated as thin film with downsizing to 16 nm thick, the nanoscale thin film exhibited significant gate-opening behaviours for the vapours of water, methanol, ethanol and acetonitrile (Figure 1.21(a)). *In situ* PXRD studies revealed that the vertical interlayer space was expanded by 1.6% after ethanol adsorption to allow ethanol molecules to diffuse (Figure 1.21(b)). The dynamic structures in thin films should contribute to the lowering of potential barrier height in the gate-opening-type guest uptake, which resulted in specific response for guests and sorption kinetics that differs from the bulk crystals.[185]

1.6.5.5 *Solvent-induced Exfoliation*

Apparently, inclusion of guest molecules into 2D interdigitated CPs will significantly affect or weaken the interlayer interdigitation between adjacent layers, which have been widely used to result in the breakage of bulk layered crystals to form nanosheets, usually called 'solvent-induced exfoliation'.[186,187] For example, water molecules were introduced into the interlayer pockets of bulk crystals of [CuCl(dipyrimidinedisulfide)]·nH$_2$O to produce ultrathin nanosheets (<2 nm in thickness).[186] Besides, the spontaneous exfoliation of

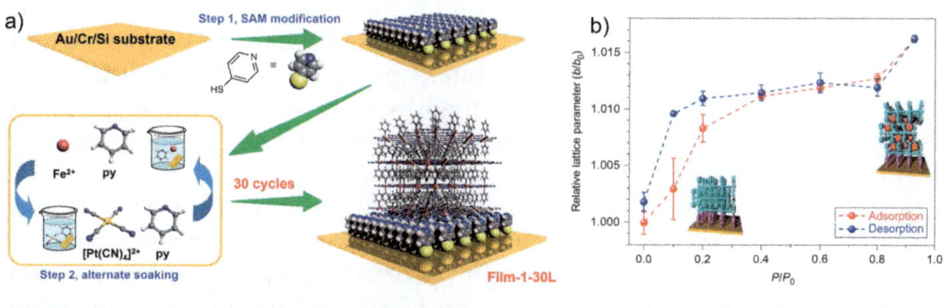

Figure 1.21 (a) Schematic illustration of the nanoscale fabrication of Fe(pyridine)$_2$[Pt(CN)$_4$] thin film. Fe, red; Pt, orange; C, grey: N, blue; H, pink. (b) Relative lattice constant (b/b_0) calculated from the peak positions of 020 and 040 plotted against the relative pressure (P/P_0) of ethanol vapour. Reproduced from ref. 160 with permission from Springer Nature, Copyright 2016.

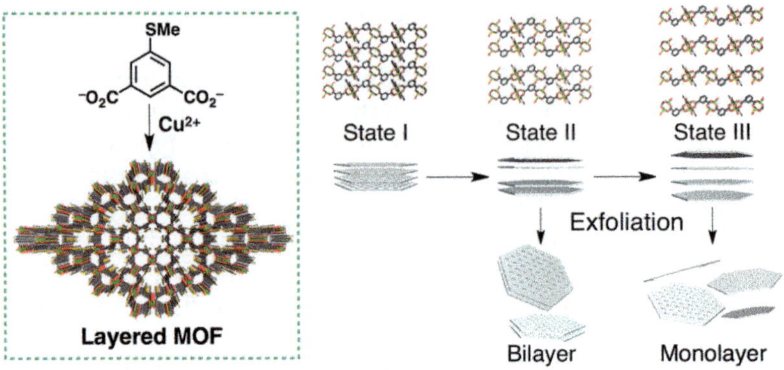

Figure 1.22 Synthesis of the layered Cu-MOF and single-crystal X-ray structures of the as-synthesised (state I), the bilayer-expanded (state II), and the monolayer-expanded (state III) crystals in the reversible expansion. Reproduced from ref. 187 with permission from American Chemical Society, Copyright 2019.

a bulk-layered Cu(II)-MOF by use of tetrahydrofuran (THF) with significant interlayer expansion was demonstrated, and then further delamination of the expanded bulk-layered framework into the bilayer and even monolayer nanosheets in 2-methyltetrahydrofuran (MeTHF) and dioxane, respectively (Figure 1.22).[187]

1.6.6 Ligand Rotation and Vibrations

1.6.6.1 Local Dynamics of Ligands in MOFs

Due to the high porosity and structural flexibility around the coordination bonds of MOFs, the ligands in the structure exhibit movements such as rotation and vibration. For example, the rotation of the benzene ring of the ligand upon adsorption and desorption of the guest has been found from X-ray structural analysis (Figure 1.23).[188] $Zn_4O(NTB)_2$ has a 3D structure and the benzene ring of the NTB ligand rotates upon adsorption and desorption of the guest. Since the structure obtained by X-ray structure analysis is an average structure, it is not possible to quantitatively determine the frequency at which the ligands rotate.

In 2006, a report was published in which the protons of the MOF ligand were deuterated and the rotational frequency was quantitatively observed by solid-state ^2H NMR spectroscopy.[189] [CdNa(2-stp)$(pyz)_{0.5}(H_2O)$] is a 3D structure in which the layered structure is bridged by the pillar pyrazine (Figure 1.24(a)). Solid-state ^2H NMR spectroscopy was performed on [CdNa(2-stp)$(pyz\text{-}d_4)_{0.5}(H_2O)$], in which the four protons of pyrazine were deuterated, in the temperature range of 193 to 293 K. The characteristic Pake-doublet patterns were observed (Figure 1.24(b)). The pattern fitting was performed by simulating the rotational motion of pyrazine around the *C*2

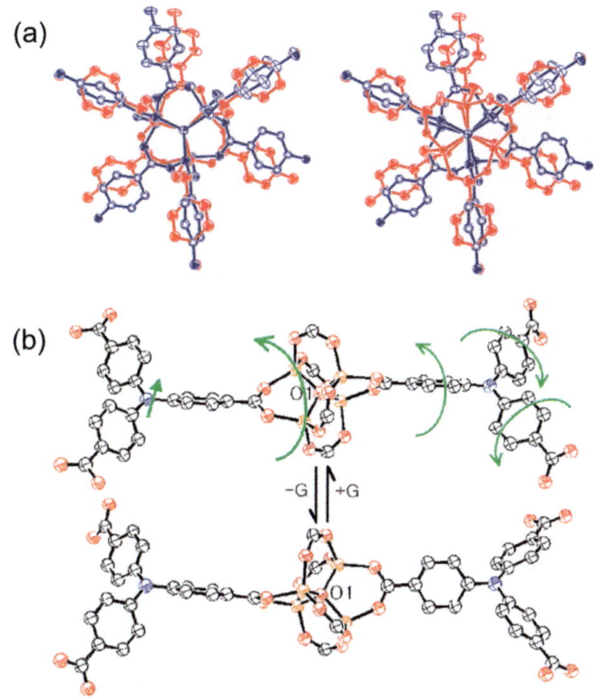

Figure 1.23 (a) Crystal structures between guest-containing state (red) and guest-free state (blue) of $Zn_4O(NTB)_2$. (b) Rearrangements of the framework components upon guest removal and adsorption. Colour scheme: Zn, yellow; O, red; N, blue; C, grey. Reproduced from ref. 188 with permission from American Chemical Society, Copyright 2005.

axis (Figure 1.24(c)). Each fitting showed good agreement, and the rotational frequency was found to vary in the range of 2×10^4 Hz (193 K) to 1×10^5 Hz (293 K). From the Arrhenius plot, the activation energy of rotation was found to be as low as 7.7 kJ mol^{-1}. Considering that the typical energy of an O–H···O hydrogen bond is about 20 kJ mol^{-1}, the energy barrier of the ligand rotation is quite small. In this work, similar experiments were carried out for $Zn_2(1,4\text{-ndc-}d_6)_2(dabco)$. This is a 3D pillared-layer type structure composed of a Zn^{2+} dinuclear paddle-wheel cluster. The rotation frequency of the ligand was 3×10^6 Hz (203 K) to 5×10^7 Hz (223 K). The activation energy of rotation was 53 kJ mol^{-1}. When benzene was introduced as a guest into this MOF, the rotation was found to stop.

A similar investigation was carried out for $[Zn_4O(bdc)]$ (Figure 1.25(a), MOF-5). Cubic $Zn_4O(bdc\text{-}d_4)$ was prepared by use of deuterated 1,4-benzenedicarboxylate (bdc-d_4), and solid-state ^2H NMR measurements were carried out in the range of 300 to 435 K (Figure 1.25(a)).[190] The motion of bdc is mainly a two-site π-flip of the benzene ring, and its activation energy was 47 kJ mol^{-1}. The high activation energy of rotation of the carboxylates compared with

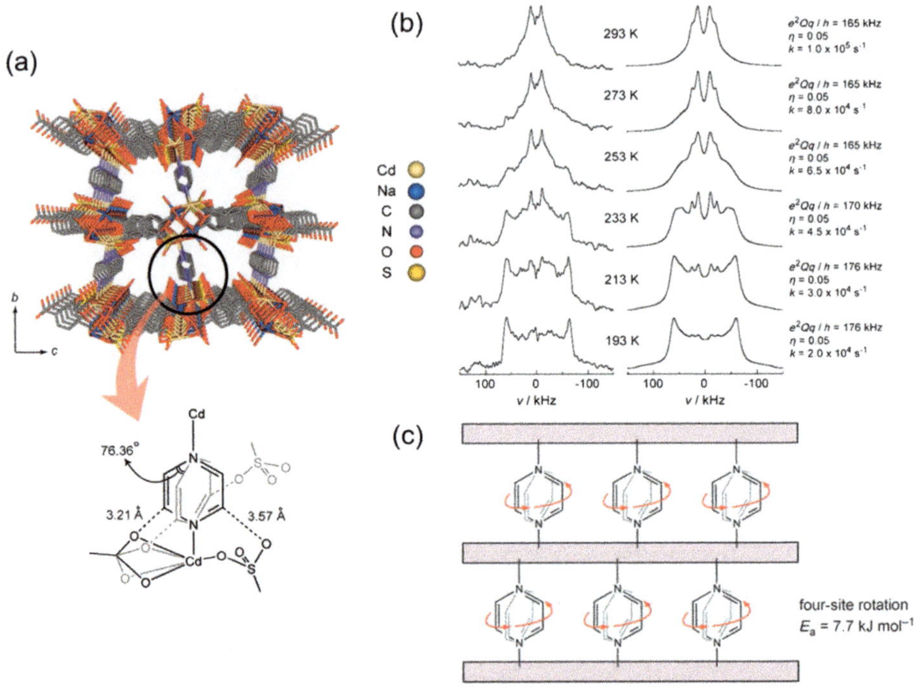

Figure 1.24 (a) Crystal structure and (b) experimental (left) and simulated (right) ^{2}H NMR spectra of [CdNa(2-stp)(pyz-d$_4$)$_{0.5}$(H$_2$O)] at 193–293 K. The jump rates of the pyrazine rings (k), and the nuclear quadrupole coupling constants (e^2Qq/h) and asymmetry parameters (h) are indicated. (c) Illustration of the four-site rotation of the pyrazine rings in [CdNa(2-stp)(pyz)$_{0.5}$(H$_2$O)]. The activation energy (E_a) is indicated. Reproduced from ref. 189 with permission from John Wiley & Sons, Copyright 2006.

N-donor ligands found in [CdNa(2-stp)(pyz)$_{0.5}$(H$_2$O)] is due to their electronic conjugation. Spectroscopy suggested that the rotation of the ligands incorporated into MOFs could be manifested in a variety of ways and could be controlled by external stimuli. Solid-state ^{2}H NMR measurements of [V(OH)(bdc-d$_4$)] (MIL-47) and [Cr(OH)(bdc-d$_4$)] (MIL-53), which consist of paramagnetic metals, have also been reported (Figure 1.25(b)).[191] Although the Pake doublet patterns are distorted by the paramagnetic metal ions, pattern simulation is successful. [Cr(OH)(bdc-d$_4$)] has a flexible structure with two phases, a narrow pore, and a large pore, while [V(OH)(bdc-d$_4$)] has a rigid structure and one phase exists. Both compounds show π-flipping of benzene rings of bdc-d$_4$ ligands, and the activation energies are 45 kJ mol^{-1} for [V(OH)(bdc-d$_4$)] and 41 kJ mol^{-1} for [Cr(OH)(bdc-d$_4$)]. The activation energies of both ligands are comparable to that of Zn$_4$O(bdc-d$_4$) (47 kJ mol^{-1}), but the rotational frequencies are more than an order of magnitude lower. This difference could reflect the constrained local environment for the phenylene ring in a 1D

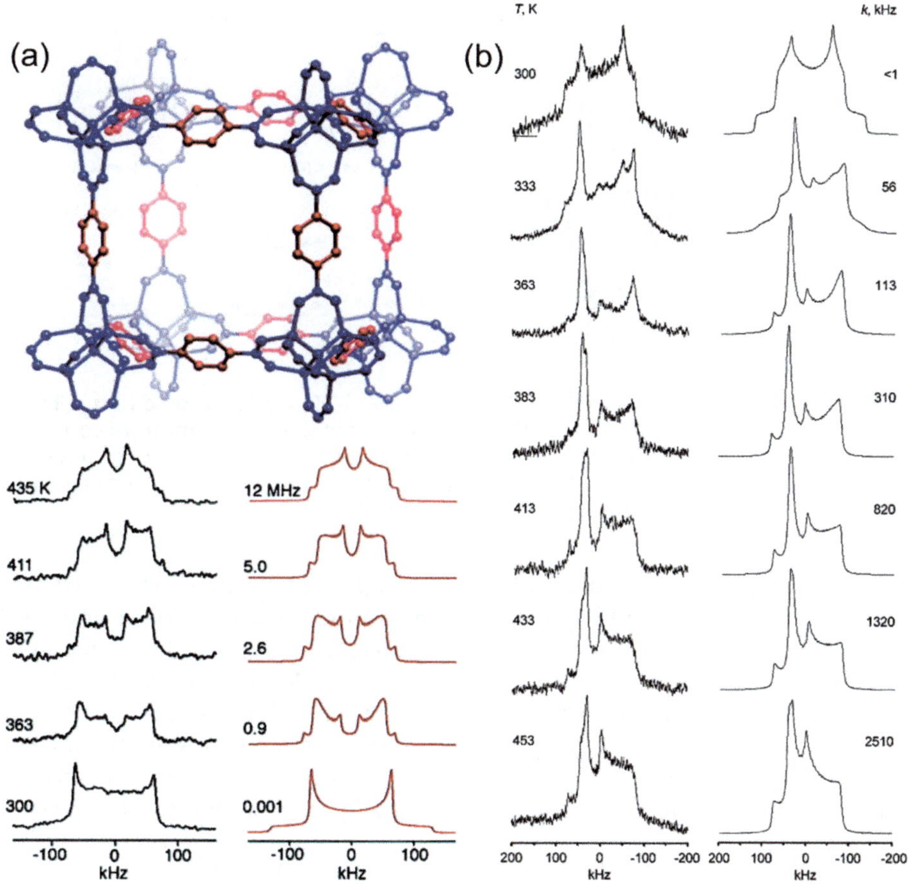

Figure 1.25 (a) Crystal structure of [Zn_4O(bdc)] and selected experimental (left) and calculated (right) quadrupolar echo solid state ^2H NMR of [Zn_4O(bdc-d_4)]. Reproduced from ref. 190 with permission from American Chemical Society, Copyright 2008. (b) Experimental (left) and simulated (right) ^2H NMR temperature dependence of the spectra line shape of the aromatic rings in [Cr(OH)(bdc-d_4)]. k is the π-flip rate constant. Reproduced from ref. 191 with permission from John Wiley & Sons, Copyright 2010.

pore system when compared to a 3D one for [Zn_4O(bdc-4)]. Ring rotation in [$Zr_6O_4(OH)_4$(bdc-4)$_6$] (UiO-66) is also studied by the same procedure.[192]

Preservation of high molecular rotation dynamics, even at low temperatures, is a major challenge. Zn-FTR and Zr-FTR are constructed from bicyclo[1.1.1]pentane-1,3-dicarboxylic acid with Zn^{2+} and Zr^{4+} ions, respectively (Figure 1.26(a)).[193] In both of structures, the carboxylate groups, anchored on metal clusters, serve as the axle while allowing the bicyclic unit to freely rotate. The planes of carboxylate groups are perpendicular to another one (crossed arrangement), while those of Zr-FTR stay in a common plane

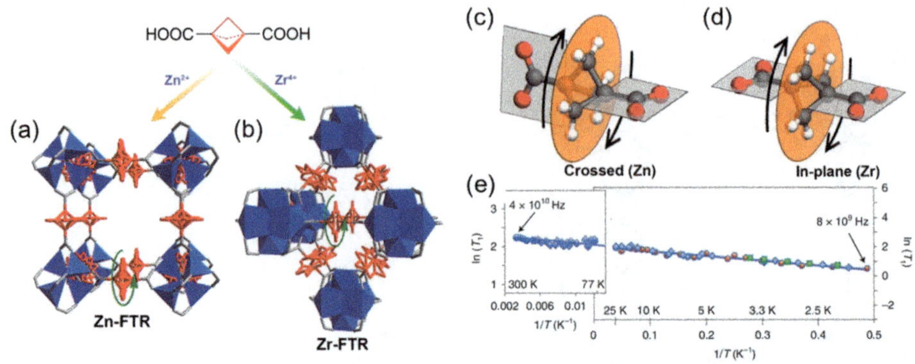

Figure 1.26 Construction of (a) Zn-FTR and (b) Zr-FTR. (c) The crossed (Zn-FTR) and (d) in-plane (Zr-FTR) arrangements of the rotor minimum conformations. (e) ln ^1H T_1 relaxation times with the reciprocal of temperatures from 1.5 to 300 K. Reproduced from ref. 193 with permission from Springer Nature, Copyright 2020.

(in-plane arrangement) (Figure 1.26(b)). Through the potential energy (MP2) scans, the researchers found that the crossed model shows a lower energy barrier than the in-plane conformation (Figures 1.26(c) and (d)). Solid-state NMR analysis gave a low rotation barrier for Zn-FTR to be 6.2 kcal mol^{-1}, with rotors showing the high-frequency motion of 10^{10} Hz (Figure 1.26(e)), even at a temperature below 2 K. The symmetry mismatch between the bicyclopentane rotator (three-fold trigonal) and the carboxylate struts (four-fold orthogonal) in Zn-FTR is the key factor for the hyper fast dynamics at cryogenic temperatures.

1.6.6.2 Controlling Ligand Rotation by Crystal Structure Design

Even with the same crystal structure, changing the crystallite size can change the molecules' packing state and the ligands' rotational motion. [Zn(ip)(bpy)] (CID-1, Figure 1.27(a)) has a 2D layered interdigitation structure.[194] The average size of the crystals prepared by the conventional solvothermal synthesis is 5 μm × 20 μm. When anionic surfactant (AOT, dioctyl sulfosuccinate sodium salt) is used during crystal formation, nanocrystals of [Zn(ip)(bpy)] (NCID-1) can be synthesised. Due to the flexible structure, the cell parameters change upon crystal downsizing (Figure 1.15(a)). Deuterated 4,4-bipyridine (bpy-d_8) was used to prepare samples of two [Zn(ip)(bpy-d_8)] having different crystal sizes, and solid-state ^2H NMR spectra were recorded (Figure 1.27(b)). For the bulk crystals, the rotational frequency was about 4 MHz at 298 K. For the nanocrystals, it was more than 10 MHz at 298 K.

Attempts have been made to control the rotational frequency of the ligands continuously.[195] One approach is to use a ligand-based solid solution. [Zn(5-nitroisophthalate)$_x$(5-methoxyisophthalate)$_{1-x}$(bpy-d_8)], a flexible structure with interdigitated 2D layers can be prepared (Figure 1.28(a)). Crystal cell

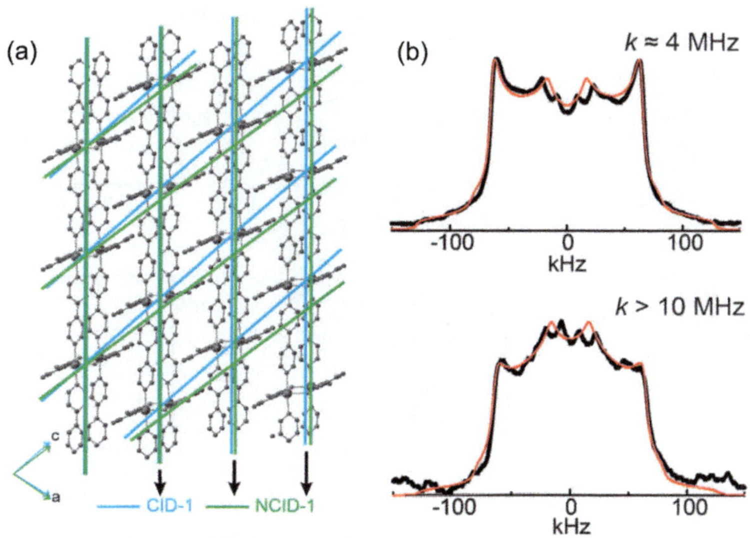

Figure 1.27 (a) Single crystal structure of [Zn(ip)(bpy)] (CID-1) and schematic description of the changes of cell parameters between [Zn(ip)(bpy)] and downsized [Zn(ip)(bpy)] (NCID-1). Blue and green grids represent the periodicity of the relative positions of isophthalate in the 2D layers from the direction of the *b* axis in CID-1 and NCID-1, respectively. (b) The ^{2}H NMR spectra of (a) CID-1 and (b) NCID-1 made from bpy-d_8 at 298 K with schematic illustrations of the frameworks. *k* is the motion rate constant. Black lines show experimental spectra and red lines are simulated spectra. Blue and green hexagons of bpy have higher mobility than the grey parts of bpy-d_8. Reproduced from ref. 194 with permission from the Royal Society of Chemistry.

parameters and bpy-d_8 ligand rotation depend on x (Figure 1.28(b)) and bpy-d_8 ligand rotation varies depending on x. The four-flip/two-flip rotational frequency exhibited by bpy-d_8 varies continuously in the range of 10^3–10^8 Hz at 298 K depending on the value of x. This is because the cell parameters of crystal structures change gradually depending on x. The cell volume also changes continuously depending on x, and the rotation barrier of bpy-d_8 also changes continuously.

Fe(pyrazine)[Pt-(CN)$_4$] is classified as a Hofmann clathrate, a 3D structure consisting of a 2D layer of [FePt(CN)$_4$] bridged by pyrazine, which is known to exhibit Fe^{2+}-derived spin crossover behaviour (Figure 1.29(a)).[196] The transition temperatures are 285 K (heating) and 309 K (cooling). The connection between the rotational motion of pyrazine and the spin crossover has been studied by quasi-elastic neutron scattering (QENS) and solid-state ^{2}H NMR spectroscopy. In the high-spin state at 300 K, the rotational frequency of the pyrazine rings is more than 10^8 Hz, while in the low-spin state at 260 K, it decreases significantly to 3×10^5 Hz (Figure 1.29(b)). The change in the Fe–N distance from 2.22 to 1.96 Å upon spin crossover results in a larger rotational

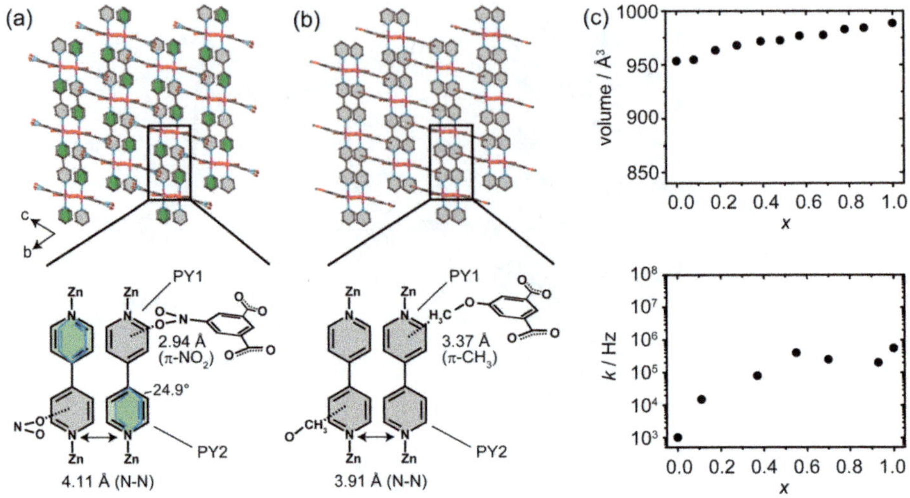

Figure 1.28 2D layer structures of (a) [Zn(5-nitroisophthalate)(bpy)] and (b) [Zn(5-methoxyisophthalate)(bpy)] at 293 K. Local structural information around the ligands are also shown. Green and grey in the pyridyl ring rotors represent the rotating and static moieties, respectively. (c) Cell volumes (upper) and flip frequencies for four-site flip rotation (lower) as a function of the ratio x for Zn(5-nitroisophthalate)$_x$(5-methoxyisophthalate)$_{1-x}$(bpy-d$_8$) with guests at 298 K. Reproduced from ref. 195 with permission from American Chemical Society, Copyright 2015.

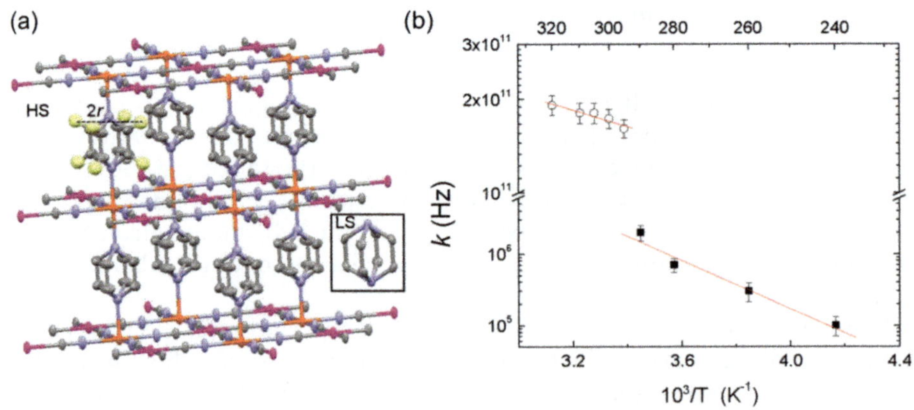

Figure 1.29 (a) Crystal structure of Fe(pyrazine)[Pt-(CN)$_4$] in the high spin state. Fe (orange), Pt (pink), N (blue), C (grey), H (yellow). Inset: detail of the thermal ellipsoids of the pyrazine carbon atoms in the low spin state. (b) Arrhenius plot of the jumping rate k for four-fold jumps in high-spin (circles, from QENS) and in low-spin (squares, from ^{2}H NMR spectroscopy) states. Reproduced from ref. 196 with permission from American Chemical Society, Copyright 2012.

barrier. Interestingly, the rotating pyrazine rings can be stopped after guest inclusion at room temperature.[197] Upon the sorption of six-/five-membered aromatic molecules, the framework expands to a larger cell volume with stabilisation of the high-spin state, while in the case of CS_2 sorption, restriction of network freedom by framework shrinkage results in stabilisation of the low-spin state. In other words, the spin crossover can be well controlled by changing the degree of dynamic motion of pyrazine.

As an example of basic molecular switches and machines, the dynamics of mechanically interlocked molecules (MIMs) have been studied in solution, but in this medium, the molecules are randomly dispersed, and their motions are incoherent. As a strategy to achieve a higher level of molecular organisation, several MIM-based MOFs have been prepared, such as these that contain a mechanically interlocked molecule as the pillaring strut between two periodic Zn^{2+}-carboxylate layers (Figure 1.30(a)).[198] The MIM linker is a [2]rotaxane with a 24-crown-6 (24C6) macrocycle and an aniline-based axle with terminal pyridine donor groups. The single-crystal X-ray structures of MOFs UWDM-2 (1,4-diazophenyl-dicarboxylate) and UWDM-3 (1,4-biphenyl-dicarboxylate) show that both frameworks are large enough to contain the free volume required for rotation of the interlocked 24C6 macrocycle, but the frameworks are interpenetrated (UWDM-2 is three-fold, and UWDM-3 is two-fold). For UWDM-3 the 24C6 rings of the pillaring MIM are positioned directly inside the square openings of neighbouring Zn^{2+} dicarboxylate layers. Variable temperature (VT) ^{2}H SSNMR spectroscopy (Figure 1.30(b)) demonstrated that the 24C6 macrocycles in UWDM-2 and UWDM-3 only undergo restricted motions related to ring flexibility or partial rotation but are incapable of undergoing free rotation. Variable-temperature studies showed that upon activation of UWDM-3 by removing the solvent, a phase change occurs. The new β-phase of UWDM-3 retained crystallinity, and ^{2}H SSNMR spectroscopy demonstrated that the 24C6 macrocyclic ring of the pillared MIM strut is now free enough to undergo full rotation. The phase change is reversible; the β version of the MOF can be reverted to the original α state by re-solvation.

Incorporation of light-driven molecular rotors (photo-switchable molecules) into solid holds great promise to fabricate responsive materials. Researchers synthesised porous switchable frameworks (PSFs) by a Yamamoto cross-coupling between the tetra-*p*-bromo-phenylmethane ($TPM\text{-}Br_4$) and overcrowded alkene ($1_{st}\text{-}Br_2$, usually used as chiroptical switch) (Figure 1.31(a)).[199] Solid-state NMR, Raman and diffuse-reflectance UV-vis spectroscopies confirmed that the switchable rotor embedded in the porous framework maintains the quantitative photo-function and undergoes reversible wavelength-dependent isomerisation upon irradiation by light or heating (Figure 1.31(b)). As a result, they demonstrated photo-modulation for the gas adsorption capacity by striking reduction of the pore volume between irradiated and pristine states, specifically, 20% uptake change of N_2 (77 K) and CO_2 (195 K) at the relative pressure (P/P_0) of 0.6 bar (Figure 1.31(c)–(e)).

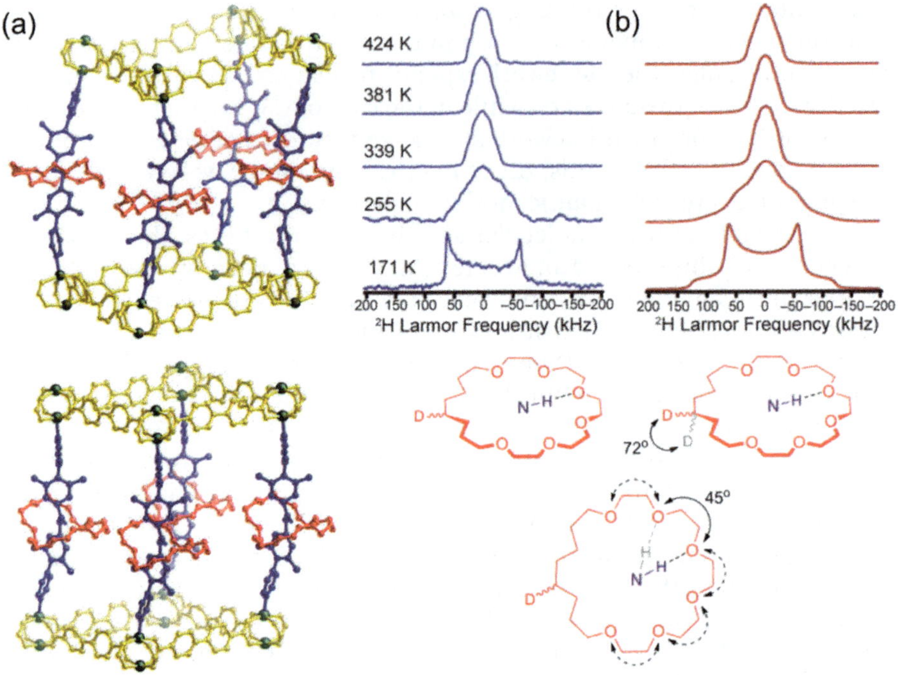

Figure 1.30 (a) Single-crystal X-ray structures of UWDM-2 (top) and UWDM-3 (bottom), showing a portion of one of the interpenetrated lattice frameworks. Blue = aniline axle, red = 24C6 wheel, yellow = carboxylate linkers, green = Zn^{2+} atoms. (b) Experimental (left) and simulated (right) ^{2}H solid-state NMR patterns for β-UWDM-3. Illustrations (below) of the motions of the 24C6 macrocyclic ring relative to the framework axle: (upper left) the slow motion limit where no motion is occurring on the NMR time scale; (upper right) the fast motion limit for jumps between two sites 75° apart; (lower left) partial rotation of the ring over 225° in 45° steps; and (lower right) partial rotation of the ring over 225° in 45° steps combined with jumps over the alkyl portion of the ring resulting in full rotation. Reproduced from ref. 198 with permission from American Chemical Society, Copyright 2014.

1.6.6.3 *Influences of Gas Diffusion to the Ligand Rotation*

Motions such as ligand rotation depend on temperatures and affect gas adsorption behaviour. [Cu(etz)] is a 3D structure that exhibits a crystal structure similar to that of NbO[200] and has pore openings with ethyl groups protruding from them. At low temperatures, such as 77 K, the ethyl groups have stopped moving and block the pores, so no gas is introduced. At 195 K, however, the ethyl groups are mobile, and gas is introduced as diffusion is enhanced. As a result, a reversal of the temperature dependence of gas adsorption, called 'kinetically controlled flexibility', is observed.

The change in ligand motion and pore size with temperature has a significant effect on gas sorption (Figure 1.32).[201] For the material CID-Me, as the

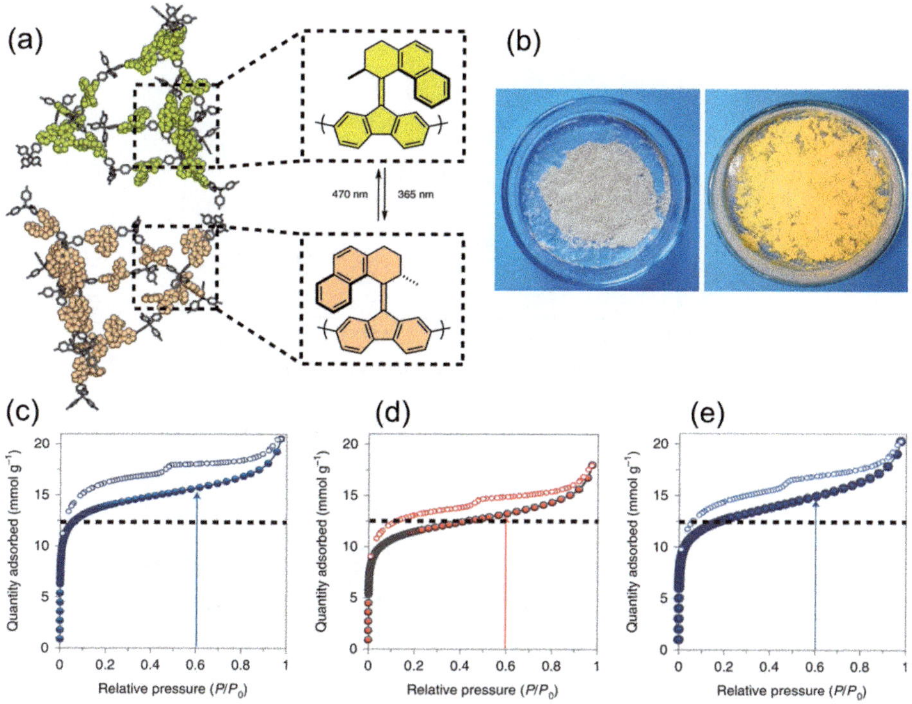

Figure 1.31 (a) Schematic representation of light-induced structural changes in the PSFs on isomerisation from the stable to the metastable isomer. (b) Pictures of the PSF before (left) and after (right) irradiation at 365 nm for 30 min upon isomerisation. N_2 adsorption isotherms at 77 K of the PSF for (c) the pristine material, (d) after irradiation (365 nm) for 54 h and (e) heating. Reproduced from ref. 199 with permission from Springer Nature, Copyright 2020.

temperature increased, the rotors exhibited rotational modes; such rotations dynamically expanded the size of the windows, leading to CO_2 adsorption. The ^2H solid-state NMR and *in situ* ^{13}C MASNMR spectra revealed that [Zn(5-methylisophthalate)(bpy-d_8)] shows both rotation motion of bpy ligands and wobbling motion of 5-methylisophthalate ligands (Figure 1.32(a)). The rotational frequencies of the rotors ($k \sim 10^{-6}$ s) and correlation times of adsorbed CO_2 ($\tau \sim 10^{-8}$ s) were elucidated *via* solid-state NMR studies, which suggest that the slow rotation of the ligand rotors sterically restricts CO_2 diffusion in the pores. This restriction results in an unusually slow CO_2 mobility close to that in the solid state ($\tau \geq 10^{-8}$ s). Once adsorbed at room temperature, CO_2 is robustly stored in the structure under vacuum at 195–233 K because of the steric hindrance of the ligand rotors. This mechanism was also applied to the storage of CH_4 (Figure 1.32(b)).

Although the effective pore size of [Zn(2-methylimidazolate)$_2$] (ZIF-8) is suitable for propylene/propane separation, the imidazolate linker shows some degree of framework flexibility, which severely deteriorates the separation

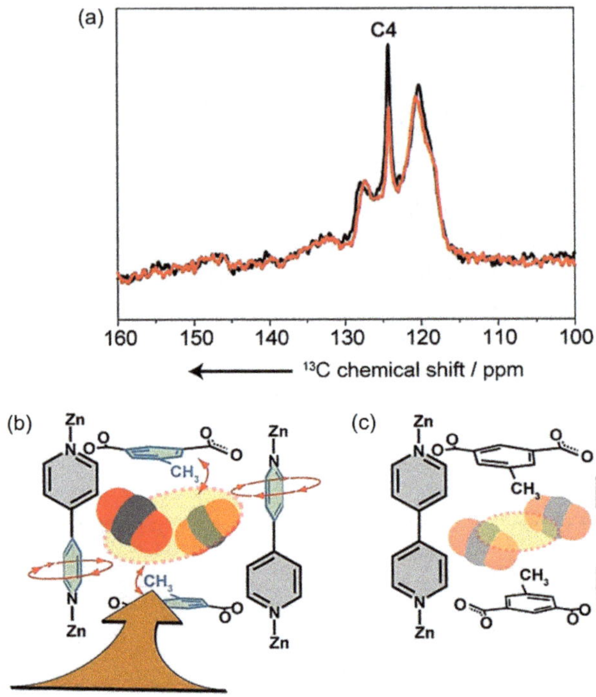

Figure 1.32 (a) ^{13}C MAS-NMR spectrum of CID-Me with CO_2 loading at 304 K before (black line) and after (red line) the sample was exposed to vacuum at 195 K for 16 h. The sharp C4 peak is attributed to free mobile CO_2 in the empty space of the NMR tube and between the particles. Schematic of (b) CO_2 adsorption through the rotational ligand rotors at room temperature and (c) CO_2 storage controlled by steric hindrance of the immobile rotors at 195 K. Reproduced from ref. 201 with permission from John Wiley & Sons, Copyright 2018.

factor of the separation, especially under high pressure and high-temperature conditions.[202] Therefore, inhibition of linker rotation is feasible for providing sharp molecular sieving and improving the separation factor (Figure 1.33). ZIF-8 with space group $I\bar{4}3m$ (denoted as ZIF_$I\bar{4}3m$) undergoes structural transitions within the electric field (500 V mm^{-1}) (Figure 1.33(a)). The symmetry is reduced to the monoclinic space group *Cm* (ZIF_*Cm*), and the symmetry switches further to *R*3*m* (ZIF_ *R*3*m*) at higher electric fields (Figure 1.33(b)).[203] ZIF_*Cm* bears a stiffened linker, exhibits a rigid pore size of 3.6 Å, and molecular sieve effects were observed for various gas pairs, including propylene/propane. In the recent work reported by the same research group,[204] a direct current was utilised during membrane growth. The applied current induced the assembly of ZIF precursors into innately distorted ZIF_8-*Cm*, providing a ZIF-8 membrane with a stiffer framework (Figure 1.33(c)). The resulting membrane with enhanced molecular sieving ability showed a propylene/propane separation factor of up to 300. The hardening of the lattice

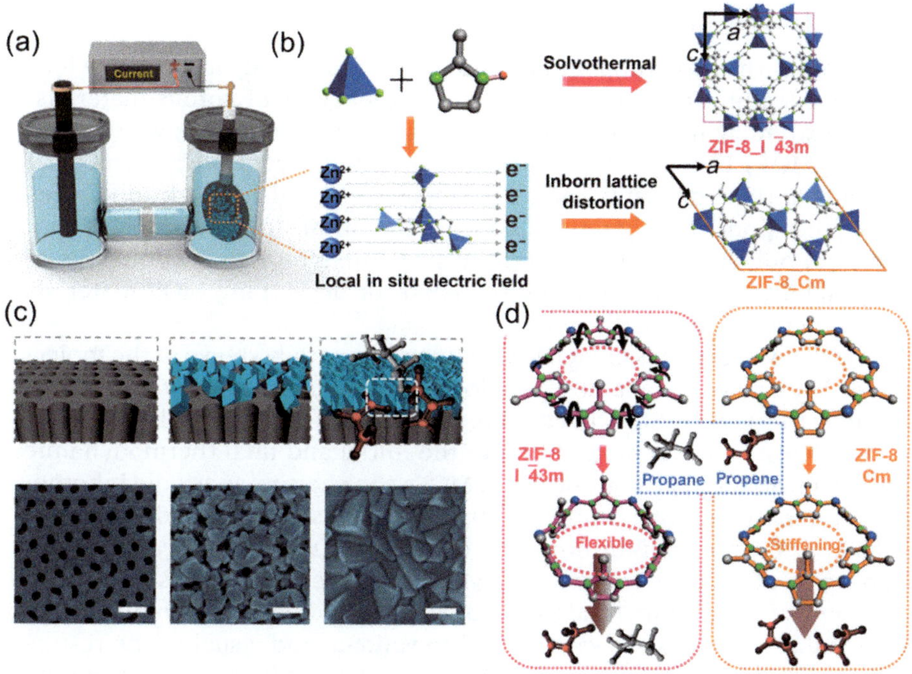

Figure 1.33 (a) Membrane growth in the electrochemical cell. (b) Schematic illustration of electrochemical membrane growth compared with solvothermal growth. (c) Schematic illustration during ZIF-8 layer formation. (d) Difference between ZIF-8_*I*$\bar{4}$3*m* and ZIF-8_*Cm* for propylene/propane separation. Reproduced from ref. 204, https://doi.org/10.1126/sciadv.aau1, under the terms of the CC BY 4.0 license, https://creativecommons.org/licenses/by/4.0/.

and the thickening of the ZIF-8 membrane (up to 200 nm) also limited the permeability of propylene to only 1.74×10^{-8} mol m^{-2} Pa^{-1} s^{-1}.

Flip-flop molecular motion is a dynamic perturbation controlled by thermal energy in bent molecules such as phenothiazine.[205] [Cu(OPTz)] is a Cu-based MOF with a butterfly-type ligand comprising isophthalic acid and phenothiazine-5,5-dioxide (OPTz) moieties (OPTz-ipa), which encodes an unprecedented mechanism of diffusion regulation by kinetic gate functionality enabling efficient gas separation and storage. [Cu(OPTz)] shows a temperature-responsive adsorption behaviour where the maximum uptake is observed at more than 100 K above the boiling point of the adsorbate. The adsorbate molecules are differentiated by each gate-admission temperature, facilitating kinetics-based gas separations of oxygen/argon and ethylene/ethane with high selectivities of ~350 and ~75 at high temperatures, which was previously unattainable by conventional static adsorption-based separation. The long-lasting physical encapsulation of ethylene at ambient conditions is demonstrated, owing to strongly impeded diffusion in distinctive nanochannels.

1.7 Summary and Outlook

Undoubtedly, the structural variation and pronounced dynamic behaviour are unique features of MOFs (compared to other types of porous materials) and are extremely important for properties. As mentioned above, the flexibility of MOFs can be designed/modified through controlling structural components (ligand, metal and guest molecules) and network topologies (interpenetration, interdigitation and ligand rotation, and vibrations). These factors not only help to understand the structure–property relationships of flexible MOFs but also is a new dimension for developing MOF materials with excellent properties.

The structural changes of MOFs can be experimentally shown at the molecular level by virtue of available crystallographic techniques. Crystallographic techniques, especially SCXRD, can directly observe the structural changes of flexible MOFs. However, in addition to the initial and final thermodynamic equilibrium states in crystalline form, MOFs can show nonperiodic inhomogeneous and/or extremely complex superstructures, especially during thermodynamically nonequilibrium states of guest diffusion, chemical reactions and physical stimuli. To study these dynamic behaviours, various characterisation and simulation methods should be used, and better combined. Computational simulations are very helpful to validate and visualise the results obtained by chemical/physical measurements. With a better understanding of the relationships between dynamic behaviour, molecular building blocks and topologies, and external stimuli, one may rationally design and construct new flexible MOFs and control their dynamics to achieve the desired properties and functions.

References

1. S. Horike, S. Shimomura and S. Kitagawa, *Nat. Chem.*, 2009, **1**, 695.
2. Z. Chang, D.-H. Yang, J. Xu, T.-L. Hu and X.-H. Bu, *Adv. Mater.*, 2015, **27**, 5432.
3. N. Behera, J. Duan, W. Jin and S. Kitagawa, *EnergyChem*, 2021, **3**, 100067.
4. S. K. Elsaidi, M. H. Mohamed, D. Banerjee and P. K. Thallapally, *Coord. Chem. Rev.*, 2018, **358**, 125.
5. S. B. Peh, A. Karmakar and D. Zhao, *Trends Chem.*, 2020, **2**, 199.
6. A. Schneemann, V. Bon, I. Schwedler, I. Senkovska, S. Kaskel and R. A. Fischer, *Chem. Soc. Rev.*, 2014, **43**, 6062.
7. F.-X. Coudert, *Chem. Mater.*, 2015, **27**, 1905.
8. G. Férey and C. Serre, *Chem. Soc. Rev.*, 2009, **38**, 1380.
9. J. Benecke, E. S. Grape, A. Fuß, S. Wöhlbrandt, T. A. Engesser, A. K. Inge, N. Stock and H. Reinsch, *Inorg. Chem.*, 2020, **59**, 9969.
10. Z. Zhang and M. J. Zaworotko, *Chem. Soc. Rev.*, 2014, **43**, 5444.
11. R. E. Morris and L. Brammer, *Chem. Soc. Rev.*, 2017, **46**, 5444.
12. L. N. McHugh, M. J. McPherson, L. J. McCormick, S. A. Morris, P. S. Wheatley, S. J. Teat, D. McKay, D. M. Dawson, C. E. F. Sansome, S. E.

Ashbrook, C. A. Stone, M. W. Smith and R. E. Morris, *Nat. Chem.*, 2018, **10**, 1096.

13. Z.-J. Lin, J. Lü, M. Hong and R. Cao, *Chem. Soc. Rev.*, 2014, **43**, 5867.
14. P. Horcajada, F. Salles, S. Wuttke, T. Devic, D. Heurtaux, G. Maurin, A. Vimont, M. Daturi, O. David, E. Magnier, N. Stock, Y. Filinchuk, D. Popov, C. Riekel, G. Férey and C. Serre, *J. Am. Chem. Soc.*, 2011, **133**, 17839.
15. S. Couck, J. F. M. Denayer, G. V. Baron, T. Rémy, J. Gascon and F. Kapteijn, *J. Am. Chem. Soc.*, 2009, **131**, 6326.
16. E. Stavitski, E. A. Pidko, S. Couck, T. Remy, E. J. M. Hensen, B. M. Weckhuysen, J. Denayer, J. Gascon and F. Kapteijn, *Langmuir*, 2011, **27**, 3970.
17. L. Bolinois, T. Kundu, X. Wang, Y. Wang, Z. Hu, K. Koh and D. Zhao, *Chem. Commun.*, 2017, **53**, 8118.
18. T. Kundu, B. B. Shah, L. Bolinois and D. Zhao, *Chem. Mater.*, 2019, **31**, 2842.
19. H. J. Choi, M. Dincă and J. R. Long, *J. Am. Chem. Soc.*, 2008, **130**, 7848.
20. F. Salles, G. Maurin, C. Serre, P. L. Llewellyn, C. Knöfel, H. J. Choi, Y. Filinchuk, L. Oliviero, A. Vimont, J. R. Long and G. Férey, *J. Am. Chem. Soc.*, 2010, **132**, 13782.
21. M. K. Taylor, T. Runčevski, J. Oktawiec, M. I. Gonzalez, R. L. Siegelman, J. A. Mason, J. Ye, C. M. Brown and J. R. Long, *J. Am. Chem. Soc.*, 2016, **138**, 15019.
22. D. N. Dybtsev, H. Chun and K. Kim, *Angew. Chem., Int. Ed.*, 2004, **43**, 5033.
23. K. Uemura, Y. Yamasaki, Y. Komagawa, K. Tanaka and H. Kita, *Angew. Chem., Int. Ed.*, 2007, **46**, 6662.
24. Z. Wang and S. M. Cohen, *J. Am. Chem. Soc.*, 2009, **131**, 16675.
25. S. Henke, R. Schmid, J.-D. Grunwaldt and R. A. Fischer, *Chem. – Eur. J.*, 2010, **16**, 14296.
26. S. Henke, A. Schneemann, A. Wütscher and R. A. Fischer, *J. Am. Chem. Soc.*, 2012, **134**, 9464.
27. P. Wang, K.-i. Otake, N. Hosono and S. Kitagawa, *Angew. Chem.*, 2021, **60**, 7030.
28. P. Wang, T. Kajiwara, K.-i. Otake, M.-S. Yao, H. Ashitani, Y. Kubota and S. Kitagawa, *ACS Appl. Mater. Interfaces*, 2021, **13**, 52144.
29. Y. Ma, R. Matsuda, H. Sato, Y. Hijikata, L. Li, S. Kusaka, M. Foo, F. Xue, G. Akiyama, R. Yuan and S. Kitagawa, *J. Am. Chem. Soc.*, 2015, **137**, 15825.
30. M. L. Foo, R. Matsuda, Y. Hijikata, R. Krishna, H. Sato, S. Horike, A. Hori, J. Duan, Y. Sato, Y. Kubota, M. Takata and S. Kitagawa, *J. Am. Chem. Soc.*, 2016, **138**, 3022.
31. A. D. Volodin, A. A. Korlyukov, E. N. Zorina-Tikhonova, A. S. Chistyakov, A. A. Sidorov, I. L. Eremenko and A. V. Vologzhanina, *Chem. Commun.*, 2018, **54**, 13861.
32. M. Nagarathinam and J. J. Vittal, *Macromol. Rapid Commun.*, 2006, **27**, 1091.
33. G. K. Kole and J. J. Vittal, *Chem. Soc. Rev.*, 2013, **42**, 1755.

34. B. B. Rath and J. J. Vittal, *Acc. Chem. Res.*, 2022, **55**, 1445.
35. Y.-X. Shi, H.-H. Chen, W.-H. Zhang, G. S. Day, J.-P. Lang and H.-C. Zhou, *Chem. – Eur. J.*, 2019, **25**, 8543.
36. Y. Zheng, H. Sato, P. Wu, H. J. Jeon, R. Matsuda and S. Kitagawa, *Nat. Commun.*, 2017, **8**, 100.
37. R. Lyndon, K. Konstas, B. P. Ladewig, P. D. Southon, C. J. Kepert and M. R. Hill, *Angew. Chem., Int. Ed.*, 2013, **52**, 3695.
38. S. Sen, N. Hosono, J.-J. Zheng, S. Kusaka, R. Matsuda, S. Sakaki and S. Kitagawa, *J. Am. Chem. Soc.*, 2017, **139**, 18313.
39. T. K. Maji, G. Mostafa, R. Matsuda and S. Kitagawa, *J. Am. Chem. Soc.*, 2005, **127**, 17152.
40. S. K. Ghosh, S. Bureekaew and S. Kitagawa, *Angew. Chem., Int. Ed.*, 2008, **47**, 3403.
41. S. Mukherjee, B. Joarder, B. Manna, A. V. Desai, A. K. Chaudhari and S. K. Ghosh, *Sci. Rep.*, 2014, **4**, 5761.
42. S. Mukherjee, B. Joarder, A. V. Desai, B. Manna, R. Krishna and S. K. Ghosh, *Inorg. Chem.*, 2015, **54**, 4403.
43. J. Lee, Y. Kim, Y. Son, H. Kim, Y. N. Choi, D. D'Alessandro, P. C. Rao and M. Yoon, *Chem. – Eur. J.*, 2021, **27**, 14851.
44. R. Grünker, I. Senkovska, R. Biedermann, N. Klein, M. R. Lohe, P. Müller and S. Kaskel, *Chem. Commun.*, 2011, **47**, 490.
45. B. Felsner, V. Bon, J. D. Evans, F. Schwotzer, R. Grünker, I. Senkovska and S. Kaskel, *Chem. – Eur. J.*, 2021, **27**, 9708.
46. V. Kiritsis, A. Michaelides, S. Skoulika, S. Golhen and L. Ouahab, *Inorg. Chem.*, 1998, **37**, 3407.
47. F. Niekiel, J. Lannoeye, H. Reinsch, A. S. Munn, A. Heerwig, I. Zizak, S. Kaskel, R. I. Walton, D. de Vos, P. Llewellyn, A. Lieb, G. Maurin and N. Stock, *Inorg. Chem.*, 2014, **53**, 4610.
48. F. Niekiel, M. Ackermann, P. Guerrier, A. Rothkirch and N. Stock, *Inorg. Chem.*, 2013, **52**, 8699.
49. H. Reinsch and D. De Vos, *Microporous Mesoporous Mater.*, 2014, **200**, 311.
50. L. Wang, T. Song, C. Li, J. Xia, S. Wang, L. Wang and J. Xu, *J. Solid State Chem.*, 2012, **190**, 208.
51. I. H. Kim, X. Wang and A. J. Jacobson, *Solid State Sci.*, 2010, **12**, 76.
52. E. Alvarez, N. Guillou, C. Martineau, B. Bueken, B. Van de Voorde, C. Le Guillouzer, P. Fabry, F. Nouar, F. Taulelle, D. de Vos, J.-S. Chang, K. H. Cho, N. Ramsahye, T. Devic, M. Daturi, G. Maurin and C. Serre, *Angew. Chem., Int. Ed.*, 2015, **54**, 3664.
53. H. Reinsch, R. S. Pillai, R. Siegel, J. Senker, A. Lieb, G. Maurin and N. Stock, *Dalton Trans.*, 2016, **45**, 4179.
54. S. Wang, N. Xhaferaj, M. Wahiduzzaman, K. Oyekan, X. Li, K. Wei, B. Zheng, A. Tissot, J. Marrot, W. Shepard, C. Martineau-Corcos, Y. Filinchuk, K. Tan, G. Maurin and C. Serre, *J. Am. Chem. Soc.*, 2019, **141**, 17207.
55. H. Yang, F. Guo, P. Lama, W.-Y. Gao, H. Wu, L. J. Barbour, W. Zhou, J. Zhang, B. Aguila and S. Ma, *ACS Cent. Sci.*, 2018, **4**, 1194.

56. V. Bon, I. Senkovska, D. Wallacher, D. M. Többens, I. Zizak, R. Feyerherm, U. Mueller and S. Kaskel, *Inorg. Chem.*, 2014, **53**, 1513.
57. H. J. Park and M. P. Suh, *Chem. Commun.*, 2010, **46**, 610.
58. J. Seo, C. Bonneau, R. Matsuda, M. Takata and S. Kitagawa, *J. Am. Chem. Soc.*, 2011, **133**, 9005.
59. E. S. Grape, H. Xu, O. Cheung, M. Calmels, J. Zhao, C. Dejoie, D. M. Proserpio, X. Zou and A. K. Inge, *Cryst. Growth Des.*, 2020, **20**, 320.
60. U. Stoeck, S. Krause, V. Bon, I. Senkovska and S. Kaskel, *Chem. Commun.*, 2012, **48**, 10841.
61. S. Krause, V. Bon, I. Senkovska, U. Stoeck, D. Wallacher, D. M. Többens, S. Zander, R. S. Pillai, G. Maurin, F.-X. Coudert and S. Kaskel, *Nature*, 2016, **532**, 348.
62. V. Bon, S. Krause, I. Senkovska, N. Grimm, D. Wallacher, D. M. Többens and S. Kaskel, *Angew. Chem.*, 2021, **60**, 11735.
63. B. Garai, V. Bon, S. Krause, F. Schwotzer, M. Gerlach, I. Senkovska and S. Kaskel, *Chem. Mater.*, 2020, **32**, 889.
64. Y. Kamakura, A. Hikawa, H. Yoshikawa, W. Kosaka, H. Miyasaka and D. Tanaka, *Chem. Commun.*, 2020, **56**, 9106.
65. J. Wang, Y. Zhang, P. Zhang, J. Hu, R.-B. Lin, Q. Deng, Z. Zeng, H. Xing, S. Deng and B. Chen, *J. Am. Chem. Soc.*, 2020, **142**, 9744.
66. C. Volkringer, T. Loiseau, N. Guillou, G. Férey, E. Elkaïm and A. Vimont, *Dalton Trans.*, 2009, 2241.
67. F. Millange, N. Guillou, R. I. Walton, J.-M. Grenèche, I. Margiolaki and G. Férey, *Chem. Commun.*, 2008, 4732.
68. E. V. Anokhina, M. Vougo-Zanda, X. Wang and A. J. Jacobson, *J. Am. Chem. Soc.*, 2005, **127**, 15000.
69. J. P. S. Mowat, S. R. Miller, A. M. Z. Slawin, V. R. Seymour, S. E. Ashbrook and P. A. Wright, *Microporous Mesoporous Mater.*, 2011, **142**, 322.
70. M. Vougo-Zanda, J. Huang, E. Anokhina, X. Wang and A. J. Jacobson, *Inorg. Chem.*, 2008, **47**, 11535.
71. J. Benecke, S. Mangelsen, T. A. Engesser, T. Weyrich, J. Junge, N. Stock and H. Reinsch, *Dalton Trans.*, 2019, **48**, 16737.
72. J. Benecke, A. Fuß, T. A. Engesser, N. Stock and H. Reinsch, *Eur. J. Inorg. Chem.*, 2021, **2021**, 713.
73. F. Nouar, T. Devic, H. Chevreau, N. Guillou, E. Gibson, G. Clet, M. Daturi, A. Vimont, J. M. Grenèche, M. I. Breeze, R. I. Walton, P. L. Llewellyn and C. Serre, *Chem. Commun.*, 2012, **48**, 10237.
74. M. I. Breeze, G. Clet, B. C. Campo, A. Vimont, M. Daturi, J.-M. Grenèche, A. J. Dent, F. Millange and R. I. Walton, *Inorg. Chem.*, 2013, **52**, 8171.
75. J. Bitzer, S.-L. Heck and W. Kleist, *Microporous Mesoporous Mater.*, 2020, **308**, 110329.
76. J. Bitzer, M. Teubnerová and W. Kleist, *Chem. – Eur. J.*, 2021, **27**, 1724.
77. C. M. Rice, Z. H. Davis, D. McKay, G. P. M. Bignami, R. G. Chitac, D. M. Dawson, R. E. Morris and S. E. Ashbrook, *Phys. Chem. Chem. Phys.*, 2020, **22**, 14514.
78. S. Henke, D. C. F. Wieland, M. Meilikhov, M. Paulus, C. Sternemann, K. Yusenko and R. A. Fischer, *CrystEngComm*, 2011, **13**, 6399.

79. A. Schneemann, P. Vervoorts, I. Hante, M. Tu, S. Wannapaiboon, C. Sternemann, M. Paulus, D. C. F. Wieland, S. Henke and R. A. Fischer, *Chem. Mater.*, 2018, **30**, 1667.
80. N. Klein, H. C. Hoffmann, A. Cadiau, J. Getzschmann, M. R. Lohe, S. Paasch, T. Heydenreich, K. Adil, I. Senkovska, E. Brunner and S. Kaskel, *J. Mater. Chem.*, 2012, **22**, 10303.
81. J. Duan, M. Higuchi, M. L. Foo, S. Horike, K. P. Rao and S. Kitagawa, *Inorg. Chem.*, 2013, **52**, 8244.
82. G. Férey, F. Millange, M. Morcrette, C. Serre, M.-L. Doublet, J.-M. Grenèche and J.-M. Tarascon, *Angew. Chem., Int. Ed.*, 2007, **46**, 3259.
83. W. Kaveevivitchai, X. Wang, L. Liu and A. J. Jacobson, *Inorg. Chem.*, 2015, **54**, 1822.
84. K. Barthelet, J. Marrot, D. Riou and G. Férey, *Angew. Chem., Int. Ed.*, 2002, **41**, 281.
85. H. Leclerc, T. Devic, S. Devautour-Vinot, P. Bazin, N. Audebrand, G. Férey, M. Daturi, A. Vimont and G. Clet, *J. Phys. Chem. C*, 2011, **115**, 19828.
86. K. Uemura, S. Kitagawa, M. Kondo, K. Fukui, R. Kitaura, H.-C. Chang and T. Mizutani, *Chem. – Eur. J.*, 2002, **8**, 3586.
87. C. Serre, F. Millange, C. Thouvenot, M. Noguès, G. Marsolier, D. Louër and G. Férey, *J. Am. Chem. Soc.*, 2002, **124**, 13519.
88. F. Millange, C. Serre and G. Férey, *Chem. Commun.*, 2002, 822.
89. T. Loiseau, C. Serre, C. Huguenard, G. Fink, F. Taulelle, M. Henry, T. Bataille and G. Férey, *Chem. – Eur. J.*, 2004, **10**, 1373.
90. S. Bourrelly, P. L. Llewellyn, C. Serre, F. Millange, T. Loiseau and G. Férey, *J. Am. Chem. Soc.*, 2005, **127**, 13519.
91. P. L. Llewellyn, S. Bourrelly, C. Vagner, N. Heymans, H. Leclerc, A. Ghoufi, P. Bazin, A. Vimont, M. Daturi, T. Devic, C. Serre, G. D. Weireld and G. Maurin, *J. Phys. Chem. C*, 2013, **117**, 962.
92. A. Vimont, A. Travert, P. Bazin, J.-C. Lavalley, M. Daturi, C. Serre, G. Férey, S. Bourrelly and P. L. Llewellyn, *Chem. Commun.*, 2007, 3291.
93. C. Serre, S. Bourrelly, A. Vimont, N. A. Ramsahye, G. Maurin, P. L. Llewellyn, M. Daturi, Y. Filinchuk, O. Leynaud, P. Barnes and G. Férey, *Adv. Mater.*, 2007, **19**, 2246.
94. S. Bourrelly, C. Serre, A. Vimont, N. A. Ramsahye, G. Maurin, M. Daturi, Y. Filinchuk, G. Férey and P. L. Llewellyn, *From Zeolites to Porous MOF Materials - The 40th Anniversary of International Zeolite Conference*, 2007, 170, 1008–1014.
95. F. Salles, A. Ghoufi, G. Maurin, R. G. Bell, C. Mellot-Draznieks and G. Férey, *Angew. Chem., Int. Ed.*, 2008, **47**, 8487.
96. P. L. Llewellyn, G. Maurin, T. Devic, S. Loera-Serna, N. Rosenbach, C. Serre, S. Bourrelly, P. Horcajada, Y. Filinchuk and G. Férey, *J. Am. Chem. Soc.*, 2008, **130**, 12808.
97. T. K. Trung, P. Trens, N. Tanchoux, S. Bourrelly, P. L. Llewellyn, S. Loera-Serna, C. Serre, T. Loiseau, F. Fajula and G. Férey, *J. Am. Chem. Soc.*, 2008, **130**, 16926.

98. S. Bourrelly, B. Moulin, A. Rivera, G. Maurin, S. Devautour-Vinot, C. Serre, T. Devic, P. Horcajada, A. Vimont, G. Clet, M. Daturi, J.-C. Lavalley, S. Loera-Serna, R. Denoyel, P. L. Llewellyn and G. Férey, *J. Am. Chem. Soc.*, 2010, **132**, 9488.
99. V. Finsy, C. E. A. Kirschhock, G. Vedts, M. Maes, L. Alaerts, D. E. De Vos, G. V. Baron and J. F. M. Denayer, *Chem. – Eur. J.*, 2009, **15**, 7724.
100. S. Biswas, T. Ahnfeldt and N. Stock, *Inorg. Chem.*, 2011, **50**, 9518.
101. A. Comotti, S. Bracco, P. Sozzani, S. Horike, R. Matsuda, J. Chen, M. Takata, Y. Kubota and S. Kitagawa, *J. Am. Chem. Soc.*, 2008, **130**, 13664.
102. J. Gascon, U. Aktay, M. D. Hernandez-Alonso, G. P. M. van Klink and F. Kapteijn, *J. Catal.*, 2009, **261**, 75.
103. T. Ahnfeldt, D. Gunzelmann, T. Loiseau, D. Hirsemann, J. Senker, G. Férey and N. Stock, *Inorg. Chem.*, 2009, **48**, 3057.
104. C. Volkringer, T. Loiseau, N. Guillou, G. Férey, M. Haouas, F. Taulelle, E. Elkaim and N. Stock, *Inorg. Chem.*, 2010, **49**, 9852.
105. J. Tang, Y. Chu, S. Li, J. Xu, W. Xiong, Q. Wang and F. Deng, *Chem. – Eur. J.*, 2021, **27**, 14711.
106. N. Chanut, A. Ghoufi, M.-V. Coulet, S. Bourrelly, B. Kuchta, G. Maurin and P. L. Llewellyn, *Nat. Commun.*, 2020, **11**, 1216.
107. J. P. S. Mowat, V. R. Seymour, J. M. Griffin, S. P. Thompson, A. M. Z. Slawin, D. Fairen-Jimenez, T. Düren, S. E. Ashbrook and P. A. Wright, *Dalton Trans.*, 2012, **41**, 3937.
108. F. Millange, C. Serre, N. Guillou, G. Férey and R. I. Walton, *Angew. Chem., Int. Ed.*, 2008, **47**, 4100.
109. P. L. Llewellyn, P. Horcajada, G. Maurin, T. Devic, N. Rosenbach, S. Bourrelly, C. Serre, D. Vincent, S. Loera-Serna, Y. Filinchuk and G. Férey, *J. Am. Chem. Soc.*, 2009, **131**, 13002.
110. T. Devic, P. Horcajada, C. Serre, F. Salles, G. Maurin, B. Moulin, D. Heurtaux, G. Clet, A. Vimont, J.-M. Grenèche, B. L. Ouay, F. Moreau, E. Magnier, Y. Filinchuk, J. Marrot, J.-C. Lavalley, M. Daturi and G. Férey, *J. Am. Chem. Soc.*, 2010, **132**, 1127.
111. C. Serre, F. Millange, S. Surblé and G. Férey, *Angew. Chem., Int. Ed.*, 2004, **43**, 6285.
112. C. Mellot-Draznieks, C. Serre, S. Surblé, N. Audebrand and G. Férey, *J. Am. Chem. Soc.*, 2005, **127**, 16273.
113. S. Surblé, C. Serre, C. Mellot-Draznieks, F. Millange and G. Férey, *Chem. Commun.*, 2006, 284.
114. C. Serre, C. Mellot-Draznieks, S. Surblé, N. Audebrand, Y. Filinchuk and G. Férey, *Science*, 2007, **315**, 1828.
115. E. J. Carrington, C. A. McAnally, A. J. Fletcher, S. P. Thompson, M. Warren and L. Brammer, *Nat. Chem.*, 2017, **9**, 882.
116. B. Yuan, D. Ma, X. Wang, Z. Li, Y. Li, H. Liu and D. He, *Chem. Commun.*, 2012, **48**, 1135.
117. D. Li and K. Kaneko, *Chem. Phys. Lett.*, 2001, **335**, 50.

118. A. Kondo, H. Noguchi, S. Ohnishi, H. Kajiro, A. Tohdoh, Y. Hattori, W.-C. Xu, H. Tanaka, H. Kanoh and K. Kaneko, *Nano Lett.*, 2006, **6**, 2581.
119. H. Kanoh, A. Kondo, H. Noguchi, H. Kajiro, A. Tohdoh, Y. Hattori, W.-C. Xu, M. Inoue, T. Sugiura, K. Morita, H. Tanaka, T. Ohba and K. Kaneko, *J. Colloid Interface Sci.*, 2009, **334**, 1.
120. H. Tanaka, S. Hiraide, A. Kondo and M. T. Miyahara, *J. Phys. Chem. C*, 2015, **119**, 11533.
121. A. Kultaeva, V. Bon, M. S. Weiss, A. Pöppl and S. Kaskel, *Inorg. Chem.*, 2018, **57**, 11920.
122. S. K. Ghosh, W. Kaneko, D. Kiriya, M. Ohba and S. Kitagawa, *Angew. Chem., Int. Ed.*, 2008, **47**, 8843.
123. S. K. Ghosh, J.-P. Zhang and S. Kitagawa, *Angew. Chem., Int. Ed.*, 2007, **46**, 7965.
124. S. S. Nagarkar and S. K. Ghosh, *J. Chem. Sci.*, 2015, **127**, 627.
125. C. Volkringer, T. Loiseau, N. Guillou, G. Férey, M. Haouas, F. Taulelle, N. Audebrand, I. Margiolaki, D. Popov, M. Burghammer and C. Riekel, *Cryst. Growth Des.*, 2009, **9**, 2927.
126. B. Joarder, S. Mukherjee, A. K. Chaudhari, A. V. Desai, B. Manna and S. K. Ghosh, *Chem. – Eur. J.*, 2014, **20**, 15303.
127. A. Karmakar, A. V. Desai, B. Manna, B. Joarder and S. K. Ghosh, *Chem. – Eur. J.*, 2015, **21**, 7071.
128. B. Manna, A. V. Desai and S. K. Ghosh, *Dalton Trans.*, 2016, **45**, 4060.
129. A. V. Desai, S. Sharma, S. Let and S. K. Ghosh, *Coord. Chem. Rev.*, 2019, **395**, 146.
130. A. Karmakar, A. V. Desai and S. K. Ghosh, *Coord. Chem. Rev.*, 2016, **307**, 313.
131. K. S. Min and M. P. Suh, *J. Am. Chem. Soc.*, 2000, **122**, 6834.
132. O.-S. Jung, Y. J. Kim, Y.-A. Lee, J. K. Park and H. K. Chae, *J. Am. Chem. Soc.*, 2000, **122**, 9921.
133. S.-i. Noro, R. Kitaura, M. Kondo, S. Kitagawa, T. Ishii, H. Matsuzaka and M. Yamashita, *J. Am. Chem. Soc.*, 2002, **124**, 2568.
134. Y.-Q. Chen, G.-R. Li, Z. Chang, Y.-K. Qu, Y.-H. Zhang and X.-H. Bu, *Chem. Sci.*, 2013, **4**, 3678.
135. T. K. Maji, K. Uemura, H.-C. Chang, R. Matsuda and S. Kitagawa, *Angew. Chem., Int. Ed.*, 2004, **43**, 3269.
136. B. Joarder, A. K. Chaudhari, S. S. Nagarkar, B. Manna and S. K. Ghosh, *Chem. – Eur. J.*, 2013, **19**, 11178.
137. H.-L. Jiang, T. A. Makal and H.-C. Zhou, *Coord. Chem. Rev.*, 2013, **257**, 2232.
138. M. Gupta and J. J. Vittal, *Coord. Chem. Rev.*, 2021, **435**, 213789.
139. H. Wu, J. Yang, Z.-M. Su, S. R. Batten and J.-F. Ma, *J. Am. Chem. Soc.*, 2011, **133**, 11406.
140. J.-P. Zhang, H.-L. Zhou, D.-D. Zhou, P.-Q. Liao and X.-M. Chen, *Natl. Sci. Rev.*, 2017, **5**, 907.
141. K. Seki, *Phys. Chem. Chem. Phys.*, 2002, **4**, 1968.

142. Y. Takashima, V. M. Martínez, S. Furukawa, M. Kondo, S. Shimomura, H. Uehara, M. Nakahama, K. Sugimoto and S. Kitagawa, *Nat. Commun.*, 2011, **2**, 168.
143. Y. Sakata, S. Furukawa, M. Kondo, K. Hirai, N. Horike, Y. Takashima, H. Uehara, N. Louvain, M. Meilikhov, T. Tsuruoka, S. Isoda, W. Kosaka, O. Sakata and S. Kitagawa, *Science*, 2013, **339**, 193.
144. T. K. Maji, R. Matsuda and S. Kitagawa, *Nat. Mater.*, 2007, **6**, 142.
145. Y. Cui, Y. Yue, G. Qian and B. Chen, *Chem. Rev.*, 2012, **112**, 1126.
146. G. Verma, S. Butikofer, S. Kumar and S. Ma, *Top. Curr. Chem.*, 2019, **378**, 4.
147. B. Chen, C. Liang, J. Yang, D. S. Contreras, Y. L. Clancy, E. B. Lobkovsky, O. M. Yaghi and S. Dai, *Angew. Chem., Int. Ed.*, 2006, **45**, 1390.
148. M. Shivanna, Q.-Y. Yang, A. Bajpai, E. Patyk-Kazmierczak and M. J. Zaworotko, *Nat. Commun.*, 2018, **9**, 3080.
149. K. Kishida, S. Horike, Y. Watanabe, M. Tahara, Y. Inubushi and S. Kitagawa, *Asian J. Chem.*, 2014, **9**, 1643.
150. H. Aggarwal, R. K. Das, P. M. Bhatt and L. J. Barbour, *Chem. Sci.*, 2015, **6**, 4986.
151. H. Aggarwal, P. M. Bhatt, C. X. Bezuidenhout and L. J. Barbour, *J. Am. Chem. Soc.*, 2014, **136**, 3776.
152. Y. Gu, J.-J. Zheng, K.-i. Otake, K. Sugimoto, N. Hosono, S. Sakaki, F. Li and S. Kitagawa, *Angew. Chem.*, 2020, **59**, 15517.
153. S. Bureekaew, H. Sato, R. Matsuda, Y. Kubota, R. Hirose, J. Kim, K. Kato, M. Takata and S. Kitagawa, *Angew. Chem., Int. Ed.*, 2010, **49**, 7660.
154. C.-T. He, P.-Q. Liao, D.-D. Zhou, B.-Y. Wang, W.-X. Zhang, J.-P. Zhang and X.-M. Chen, *Chem. Sci.*, 2014, **5**, 4755.
155. A. Ferguson, L. Liu, S. J. Tapperwijn, D. Perl, F.-X. Coudert, S. Van Cleuvenbergen, T. Verbiest, M. A. van der Veen and S. G. Telfer, *Nat. Chem.*, 2016, **8**, 250.
156. S. Yang, X. Lin, W. Lewis, M. Suyetin, E. Bichoutskaia, J. E. Parker, C. C. Tang, D. R. Allan, P. J. Rizkallah, P. Hubberstey, N. R. Champness, K. M. Thomas, A. J. Blake and M. Schröder, *Nat. Mater.*, 2012, **11**, 710.
157. S. Ehrling, H. Miura, I. Senkovska and S. Kaskel, *Trends Chem.*, 2021, **3**, 291.
158. D. Tanaka, A. Henke, K. Albrecht, M. Moeller, K. Nakagawa, S. Kitagawa and J. Groll, *Nat. Chem.*, 2010, **2**, 410.
159. Y. Hijikata, S. Horike, D. Tanaka, J. Groll, M. Mizuno, J. Kim, M. Takata and S. Kitagawa, *Chem. Commun.*, 2011, **47**, 7632.
160. S. Sakaida, K. Otsubo, O. Sakata, C. Song, A. Fujiwara, M. Takata and H. Kitagawa, *Nat. Chem.*, 2016, **8**, 377.
161. S. Krause, N. Hosono and S. Kitagawa, *Angew. Chem.*, 2020, **59**, 15325.
162. S. S. Nagarkar, A. V. Desai and S. K. Ghosh, *Asian J. Chem.*, 2014, **9**, 2358.
163. E. A. Dolgopolova, A. M. Rice, C. R. Martin and N. B. Shustova, *Chem. Soc. Rev.*, 2018, **47**, 4710.
164. G. Chakraborty, I.-H. Park, R. Medishetty and J. J. Vittal, *Chem. Rev.*, 2021, **121**, 3751.

165. M. Lippi and M. Cametti, *Coord. Chem. Rev.*, 2021, **430**, 213661.
166. T. Satoshi, H. Taku, N. Ei-ichi, M. Katsura and M. Wasuke, *Chem. Lett.*, 2002, **31**, 1208.
167. C. Kachi-Terajima, T. Akatsuka, M.-a. Kohbara and S. Takamizawa, *Asian J. Chem.*, 2007, **2**, 40.
168. S. Takamizawa and M.-a. Kohbara, *Dalton Trans.*, 2007, 3640.
169. S. Takamizawa, E.-i. Nataka, T. Akatsuka, R. Miyake, Y. Kakizaki, H. Takeuchi, G. Maruta and S. Takeda, *J. Am. Chem. Soc.*, 2010, **132**, 3783.
170. Y. Takasaki and S. Takamizawa, *J. Am. Chem. Soc.*, 2014, **136**, 6806.
171. S. Takamizawa, *Angew. Chem., Int. Ed.*, 2015, **54**, 7033.
172. R. Kitaura, K. Seki, G. Akiyama and S. Kitagawa, *Angew. Chem., Int. Ed.*, 2003, **42**, 428.
173. Y. Inubushi, S. Horike, T. Fukushima, G. Akiyama, R. Matsuda and S. Kitagawa, *Chem. Commun.*, 2010, **46**, 9229.
174. V. Rubio-Giménez, G. Escorcia-Ariza, C. Bartual-Murgui, C. Sternemann, M. Galbiati, J. Castells-Gil, J. A. Real, S. Tatay and C. Martí-Gastaldo, *Chem. Mater.*, 2019, **31**, 7277.
175. T. Haraguchi, K. Otsubo, O. Sakata, A. Fujiwara and H. Kitagawa, *J. Am. Chem. Soc.*, 2016, **138**, 16787.
176. S. Horike, D. Tanaka, K. Nakagawa and S. Kitagawa, *Chem. Commun.*, 2007, 3395.
177. K. Nakagawa, D. Tanaka, S. Horike, S. Shimomura, M. Higuchi and S. Kitagawa, *Chem. Commun.*, 2010, **46**, 4258.
178. D. Tanaka, K. Nakagawa, M. Higuchi, S. Horike, Y. Kubota, T. C. Kobayashi, M. Takata and S. Kitagawa, *Angew. Chem., Int. Ed.*, 2008, **47**, 3914.
179. S. Lee, S. Jeong, J. Seong, J. Lim, A. Sharma, S. Won, D. Moon, S. B. Baek and M. S. Lah, *Mater. Chem. Front.*, 2021, **5**, 3621.
180. Y. Hijikata, S. Horike, M. Sugimoto, H. Sato, R. Matsuda and S. Kitagawa, *Chem. – Eur. J.*, 2011, **17**, 5138.
181. T. Fukushima, S. Horike, Y. Inubushi, K. Nakagawa, Y. Kubota, M. Takata and S. Kitagawa, *Angew. Chem., Int. Ed.*, 2010, **49**, 4820.
182. S. Horike, M. Sugimoto, K. Kongpatpanich, Y. Hijikata, M. Inukai, D. Umeyama, S. Kitao, M. Seto and S. Kitagawa, *J. Mater. Chem. A*, 2013, **1**, 3675.
183. R. Turo-Cortés, C. Bartual-Murgui, J. Castells-Gil, M. C. Muñoz, C. Martí-Gastaldo and J. A. Real, *Chem. Sci.*, 2020, **11**, 11224.
184. K. Takahashi, N. Hoshino, T. Takeda, S.-i. Noro, T. Nakamura, S. Takeda and T. Akutagawa, *Inorg. Chem.*, 2015, **54**, 9423.
185. Y. Xie, R.-B. Lin and B. Chen, *Adv. Sci.*, 2022, **9**, 2104234.
186. A. Gallego, C. Hermosa, O. Castillo, I. Berlanga, C. J. Gómez-García, E. Mateo-Martí, J. I. Martínez, F. Flores, C. Gómez-Navarro, J. Gómez-Herrero, S. Delgado and F. Zamora, *Adv. Mater.*, 2013, **25**, 2141.
187. V. K.-M. Au, K. Nakayashiki, H. Huang, S. Suginome, H. Sato and T. Aida, *J. Am. Chem. Soc.*, 2019, **141**, 53.
188. E. Y. Lee, S. Y. Jang and M. P. Suh, *J. Am. Chem. Soc.*, 2005, **127**, 6374.

189. S. Horike, R. Matsuda, D. Tanaka, S. Matsubara, M. Mizuno, K. Endo and S. Kitagawa, *Angew. Chem., Int. Ed.*, 2006, **45**, 7226.
190. S. L. Gould, D. Tranchemontagne, O. M. Yaghi and M. A. Garcia-Garibay, *J. Am. Chem. Soc.*, 2008, **130**, 3246.
191. D. I. Kolokolov, H. Jobic, A. G. Stepanov, V. Guillerm, T. Devic, C. Serre and G. Ferey, *Angew. Chem., Int. Ed.*, 2010, **49**, 4791.
192. S. Devautour-Vinot, G. Maurin, C. Serre, P. Horcajada, D. P. da Cunha, V. Guillerm, E. de S. Costa, F. Taulelle and C. Martineau, *Chem. Mater.*, 2012, **24**, 2168.
193. J. Perego, S. Bracco, M. Negroni, C. X. Bezuidenhout, G. Prando, P. Carretta, A. Comotti and P. Sozzani, *Nat. Chem.*, 2020, 845.
194. Y. Hijikata, S. Horike, D. Tanaka, J. Groll, M. Mizuno, J. Kim, M. Takata and S. Kitagawa, *Chem. Commun.*, 2011, **47**, 7632.
195. M. Inukai, T. Fukushima, Y. Hijikata, N. Ogiwara, S. Horike and S. Kitagawa, *J. Am. Chem. Soc.*, 2015, **137**, 12183.
196. J. A. Rodriguez-Velamazan, M. A. Gonzalez, J. A. Real, M. Castro, M. C. Munoz, A. B. Gaspar, R. Ohtani, M. Ohba, K. Yoneda, Y. Hijikata, N. Yanai, M. Mizuno, H. Ando and S. Kitagawa, *J. Am. Chem. Soc.*, 2012, **134**, 5083.
197. M. Ohba, K. Yoneda, G. Agustí, M. C. Muñoz, A. B. Gaspar, J. A. Real, M. Yamasaki, H. Ando, Y. Nakao, S. Sakaki and S. Kitagawa, *Angew. Chem., Int. Ed.*, 2009, **48**, 4767.
198. K. Zhu, V. N. Vukotic, C. A. O'Keefe, R. W. Schurko and S. J. Loeb, *J. Am. Chem. Soc.*, 2014, **136**, 7403.
199. F. Castiglioni, W. Danowski, J. Perego, F. K. Leung, P. Sozzani, S. Bracco, S. J. Wezenberg, A. Comotti and B. L. Feringa, *Nat. Chem.*, 2020, **12**, 595.
200. J. P. Zhang and X. M. Chen, *J. Am. Chem. Soc.*, 2008, **130**, 6010.
201. M. Inukai, M. Tamura, S. Horike, M. Higuchi, S. Kitagawa and K. Nakamura, *Angew. Chem., Int. Ed.*, 2018, **57**, 8687.
202. H. T. Kwon, H.-K. Jeong, A. S. Lee, H. S. An, T. Lee, E. Jang, J. S. Lee and J. Choi, *Chem. Commun.*, 2016, **52**, 11669.
203. A. Knebel, B. Geppert, K. Volgmann, D. I. Kolokolov, A. G. Stepanov, J. Twiefel, P. Heitjans, D. Volkmer and J. Caro, *Science*, 2017, **358**, 347.
204. S. Zhou, Y. Wei, L. Li, Y. Duan, Q. Hou, L. Zhang, L.-X. Ding, J. Xue, H. Wang and J. Caro, *Sci. Adv.*, 2018, **4**, eaau1393.
205. C. Gu, N. Hosono, J. J. Zheng, Y. Sato, S. Kusaka, S. Sakaki and S. Kitagawa, *Science*, 2019, **363**, 387.

CHAPTER 2

Flexible Properties: Adsorptive Storage and Separation

XUE-WEN ZHANG[a], ZI-MING YE[a] AND JIE-PENG ZHANG[a*]

[a]MOE Key Laboratory of Bioinorganic and Synthetic Chemistry, GBRCE for Functional Molecular Engineering, School of Chemistry, IGCME, Sun Yat-Sen University, Guangzhou 510275, China
*E-mail: zhangjp7@mail.sysu.edu.cn

2.1 Introduction

Porous coordination polymers (PCPs) or metal–organic frameworks (MOFs) have been extensively studied for more than 20 years.[1–16] As organic–inorganic hybrids, MOFs show extremely rich structural diversity, and they can be designed and modified to meet the requirements of various applications. Being built by molecular units and reversible coordination bonds, MOFs can facilely crystallise, allowing their structures to be straightforwardly and precisely determined, which is fundamental for understanding the structure–property relationship. Besides the reversible and weakly directional coordination bonds and even weaker supramolecular interactions within/between the coordination networks, numerous rotatable units are also connected by single coordination/covalent bonds in the MOF structures. Therefore, MOFs are much more flexible than conventional porous materials.[17–23] Note that, the flexibility of MOFs refers to the ability of reversible structural transformations without changing the chemical composition of the coordination networks. Some MOFs can undergo reversible or irreversible structural transformations involving composition changes of the coordination

Inorganic Materials Series No. 13
Flexible Metal–Organic Frameworks: Structural Design, Synthesis and Properties
Edited by Susumu Kitagawa

Published by the Royal Society of Chemistry, www.rsc.org

networks (*e.g.*, redox and ion/ligand exchange), which should be classified as reactivity rather than flexibility.

The ability of reversible structural transformation relies on the presence of multiple metastable states on the thermodynamic potential profile. External stimuli that induce the structural transformation of a porous framework can be either chemical or physical, which change or keep the whole host–guest system's chemical composition, respectively. Most examples involve chemical stimuli, *i.e.*, the change of guest loading, during which the change of host–guest interactions alters the potential profile of the whole host–guest system.[24] However, physical stimuli such as a change of temperature, pressure, and electric field can directly alter the thermodynamic potential profile of the host framework.

The flexibility of a MOF is typically illustrated by the change of its crystal structure, during which the structural transformation is highly cooperative in the whole crystallite. Crystal-to-crystal structural transformations can be conveniently identified by powder X-ray diffraction (PXRD) and accurately elucidated by single-crystal X-ray diffraction (SCXRD). Many other diffraction techniques are also powerful. As discussed in Chapter 1, various types of crystal-to-crystal or single-crystal-to-single-crystal structure transformations such as shrinkage, expansion, distortion, network displacement, and network reconstitution have been reported.[17–21] Note that, MOFs are not necessarily crystalline or structurally ordered.[25–27] Besides crystal-to-crystal transformations, MOFs can also undergo complete or partial crystal-to-amorphous transformations.[26–29] During crystal-to-crystal transformations, the intermediate structures can also temporally lose crystallinity.[30] In the context of flexibility, other crystalline molecular materials sustained by hydrogen bonds and weaker interactions or covalent bonds, share the same characteristics with MOFs, although they may have very different chemical or physical properties.[31]

While the framework flexibility of MOFs is mainly determined by the energy differences between multiple metastable states, the energy barrier (E_a) of the structural transformation process is also important but has received less attention.[22,23,32–34] As the typical type of flexibility, reversible guest-induced structural transformations involve negligibly low E_a and an overwhelmingly high E_a for the two counter directions between the metastable and stable states of the host. Therefore, the metastable state can be reached only by guest adsorption, and the host automatically returns to the stable state after guest desorption. By increasing E_a to a suitable level, the automatic transformation from the metastable state to the stable state of the host–guest system can be suppressed until a sufficiently large additional stimulus is applied, giving special flexibility behaviours, such as shape-memory pores and negative changes of pressure/adsorption.[35–37] If the two counter transformation directions both possess negligibly low E_a and/or the potential profile possesses a wide valley, the host framework can keep switching between different structures without the need of changing the guest loading. To show such structural dynamism, *i.e.*, transient structural transformations

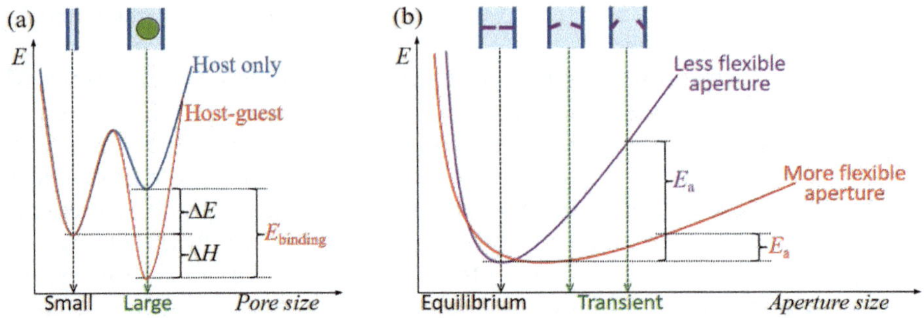

Figure 2.1 Thermodynamics and kinetics of two typical types of framework flexibilities. (a) Thermodynamically controlled, guest-induced structural transformations between equilibrium states (ΔE and E_{binding} refer to energy difference between two metastable states and host–guest binding energy, respectively). (b) Kinetically controlled structural dynamism among non-equilibrium states (represented by gating apertures with different flexibilities). Reproduced from ref. 33 with permission from Elsevier, Copyright 2021.

at the non-equilibrium state, a flexible MOF needs just one equilibrium/crystal structure. When compared with thermodynamically controlled flexibility (TCF), kinetically controlled flexibility (KCF) is more susceptible to physical parameters such as temperature. Without changing the host structure at the equilibrium state, many adsorbents can adsorb guest molecules larger than their pore apertures. The diffusion of oversized guest molecules requires the transient opening/closing of small pore apertures, during which the pore apertures and pore cavities behave like the door/gate and room, respectively. Therefore, the transient opening/closing of pore apertures can be defined as the gating flexibility (Figure 2.1).

Flexibility, as one of the most special features of MOFs, can give rise to not only interesting phenomena involving structural transformability and/or dynamism but also useful properties and functions, which are the main topic of this chapter.

2.2 Adsorption and Desorption

Guest adsorption is the most straightforward and basic property of porous materials, which is related to various applications such as storage, separation/enrichment/purification, catalysis, and sensing. Besides the structures of the porous material and the guest molecule, the guest adsorption amount or guest uptake is also dependent on temperature, pressure and time. According to basic physicochemical principles, the equilibrium guest uptake increases as the pressure increases and/or the temperature decreases until reaching the saturated value determined by the pore volume.

2.2.1 Basic Characters of Adsorption Isotherms

Before reaching saturation, most of the adsorption occurs on the pore surface, meaning that the adsorption amount mainly depends on the surface area of the adsorbent. At even lower concentrations or pressures, the guest molecules are just located at the strongest adsorption sites on the pore surface. Besides special chemical characteristics, such as coordinatively unsaturated metal centres or open metal sites (OMSs), and hydrogen-bonding donors/acceptors, the size and shape of the pores are also critical for the host–guest interactions. When their sizes/shapes are matched, the pore surface and guest molecule can form multipoint interactions to give strong adsorption, which is common for ultramicroporous and microporous adsorbents. Since X–H (X for any atom) covalent bond lengths determined by X-ray diffraction are somewhat shorter than the real ones, they may need to be corrected when the pore sizes are extremely small. For example, the aperture size of SOD-[Zn(mim)$_2$] (MAF-4/ZIF-8, Hmim = 2-methylimidazole)[38,39] are generally calculated as 3.4 Å,[40] but can be calculated as 3.2 Å[41] using the default and corrected C–H bond lengths, respectively. The equilibrium distances between the atoms of the pore surface and guest molecule are slightly shorter than the sum of the corresponding van der Waals radii. Nevertheless, when the pore is obviously smaller than the guest, in the context of effective sizes taking the van der Waals radii into consideration, adsorption cannot occur because of the Pauli repulsion.[42]

The pore size and shape of a flexible adsorbent can change to fit the guest loaded, being similar with the induced-fit action of enzymes. On one hand, the adaptive pores of MOFs can achieve better host–guest recognition than for conventional porous materials. On the other hand, this means that *in situ* measurement of the guest-loaded crystal structure is important for determining precise host–guest interactions. Nevertheless, after adsorption–desorption cycles, especially those involving structural transformations, the crystal structure is more and more difficult to determine due to crystal fragmentation and/or a decrease of crystallinity.[18–20,22] In many cases, the flexibility of MOFs was omitted when the changes of PXRD patterns and/or unit-cell parameters are small. Nevertheless, the crystal structures of MOFs always change more or less under chemical/physical stimuli, and the small changes of PXRD patterns and/or unit-cell parameters may correlate to large changes of the crystal structures and/or properties. For example, as a temperature-controlled framework flexibility, thermal expansion can play an important role in changing the pore size/shape of a flexible MOF. When the crystal structure of the MOF and the gas adsorption property are measured at different temperatures, some examples may show mismatching sizes/shapes between the pore and guest.

2.2.2 Adsorbent Flexibility Indicated by Special Isotherm Shapes

The equilibrium adsorption behaviour of a host–guest system is usually characterised by isotherms that generally exhibit one of the six typical shapes (Type-I to Type-VI) classified by the International Union of Pure and Applied Chemistry (IUPAC). The simplest and most common shape is described by the Langmuir model (Type-I), in which the isotherm slope monotonically decreases as the pressure increases. The isotherm of a flexible adsorbent shows abnormal slope/shape because of the guest-induced structural transformations. For a pore-expanding transformation from a small-pore (sp) state to a large-pore (lp) state, the sudden increase in pore volume gives rise to an additional isotherm step. For a pore-opening transformation from a non-porous state to a porous state, obvious guest adsorption starts above a certain threshold pressure. The assignment of an adsorbent to whether a sp state or a non-porous state may depend on not only the adsorbent structure but also the specific size of the guest. The pore-expanding/pore-opening pressure, *i.e.*, the isotherm inflection point, is thermodynamically dependent on the pore volumes, the host–guest interactions, and the free energy difference of the host at the two states.[24] In practice, an abnormal isotherm shape is usually indicative of the structural transformation of the host. It is more sensitive and objective than the common structural identification technique, PXRD, especially when the structural transformation is small. Although rigid adsorbents can also show special isotherm shapes with an abrupt increase of guest uptake *via* capillary condensation, they require large pores, weak host–guest interactions, and strong guest–guest interactions.

The N_2 adsorption isotherm of MAF-4 is a typical example for identifying adsorbent flexibility. As the first MOF with a regular zeolite structure, MAF-4 is highly stable and is considered to be rigid.[4,38] The N_2 adsorption isotherm of MAF-4 showed an additional step, which was first ascribed to guest rearrangement instead of host flexibility.[39] However, the first-step (340 cm^3 (STP) g^{-1}) and second-step uptakes (420 cm^3 (STP) g^{-1}) are close to and much larger than the theoretical value empirically calculated from its crystal structure (346 cm^3 (STP) g^{-1}), respectively. Variable-temperature SCXRD performed in an open-flow N_2 cryostat showed a sudden and small expansion of unit-cell volume by 0.7% from 113 K to 108 K.[43] More importantly, the number of N_2 molecules in the obtained SCXRD structures are completely consistent with those of the isotherm steps. Further, significant tilting of the imidazolate ligands were observed, which opens the originally non-porous 4-rings for the insertion of N_2 molecules, meaning that the structural change is much larger than that indicated by the unit-cell parameters. SOD-[Zn(mtz)$_2$] (MAF-7, Hmtz = 3-methyl-1,2,4-triazole),[41] an analogue of MAF-4, showed the same N_2-induced pore expansion behaviour, but the structural transformation occurs at a higher temperature (same pressure), because the uncoordinated triazolate nitrogen atoms can provide stronger host–guest interactions (Figure 2.2).

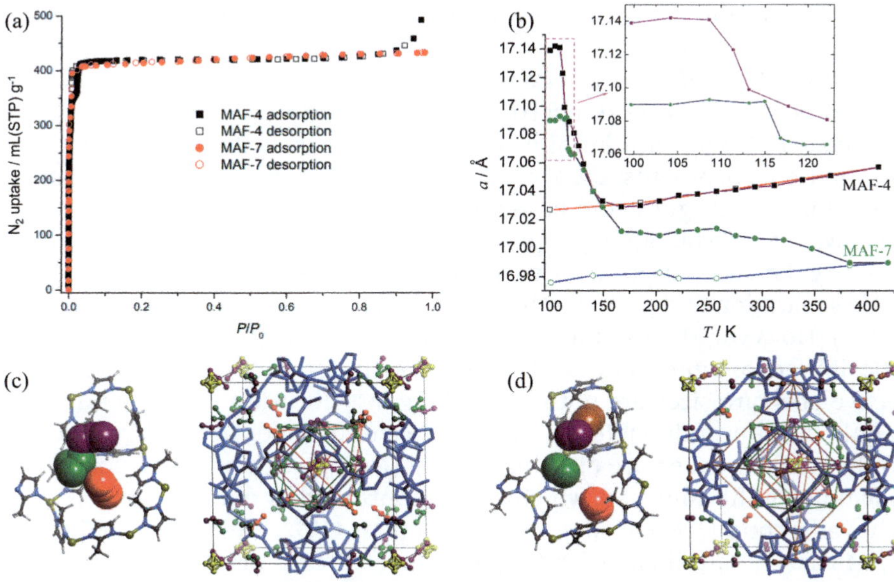

Figure 2.2 N_2 adsorption behaviours of MAF-4 and MAF-7. (a) N_2 adsorption and desorption isotherms at 77 K. (b) SCXRD unit-cell parameters in the whole heating (open symbols) and cooling (solid symbols) cycle (started from the as-synthesised sample containing guest). (c) and (d) Asymmetric (left) and packing (right, thin lines highlight the symmetry relationships of the guest molecules) structures of N_2 saturated MAF-4 measured at 113 K (c) and 109 K (d). Reproduced from ref. 43 with permission from the Royal Society of Chemistry.

The H_2O adsorption isotherm of MAF-7 exhibits a sudden jump from 20 mg g^{-1} at $P/P_0 = 0.24$ to 380 mg g^{-1} at $P/P_0 = 0.37$,[41] which is similar with those of flexible adsorbents showing pore-expanding actions. However, the H_2O uptake just before jumping is much lower than that calculated from its crystal structure. Unlike the linear N_2 molecules, the bent H_2O molecule is not likely able to insert into the open 4-rings of the SOD network. By doping hydrophobic mim^- ligands into MAF-7, the initial H_2O adsorption can be eliminated, and the isotherm inflection point gradually increases. Finally, MAF-4 shows negligible water uptake at $P/P_0 < 0.9$. Therefore, the initial H_2O uptake of MAF-7 results from its relatively high hydrophilicity compared with MAF-4, while the isotherm inflection points can be attributed to capillary condensation of H_2O rather than framework flexibility.

2.2.3 Adsorbent Flexibility Indicated by Slow Adsorption

When guest diffusion is slow, the adsorption kinetics becomes important for the host–guest system. The adsorption amount in an isobar may abnormally decrease as the temperature decreases because the adsorption process can reach equilibrium with more difficulty at lower temperatures. Slow adsorption–desorption is generally disadvantageous for practical applications but can be useful for kinetic separation. In the ideal cases, guest molecules larger than the pore aperture are completely excluded, giving rise to molecular sieving with infinite adsorption selectivity. In classic theories based on rigid adsorbents, the kinetic effect becomes notable when the pore size is similar to the guest and/or there is a strong host–guest interaction.[44,45] Typically, the crystallographic pore/aperture sizes of zeolites/MOFs and the molecular sizes of guests are compared for predicting/explaining their kinetic separation and particularly molecular sieving behaviours (Tables 2.1 and 2.2). However, the usefulness of this method needs to be reconsidered for flexible adsorbents. For flexible MOFs showing structural dynamism, the effective aperture sizes are not static. The guest diffusion rate depends on not only the conventional (equilibrium) sizes of the aperture and guest but also the transient aperture opening/closing (gating) actions of the aperture in the non-equilibrium state which can be controlled by many factors.

The adsorption behaviours of [Cu(detz)] (MAF-2, Hdetz = 3,5-diethyl-1,2,4-triazole) are illustrative for gating apertures.[32] MAF-2 is a three-dimensional (3D) coordination network with **nbo-a** topology, which defines a 3D **bcu** pore system consisting of large cavities (9 Å) and small apertures (1.5 Å) surrounded by the pendant ethyl groups of the triazolate ligands. Considering the rotation/swing motion of ethyl groups, the effective diameter of apertures can increase to 3.6 Å. MAF-2 completely excludes N_2 at 77 K, but can readily adsorb it at 195 K, indicating that the adsorption processes are kinetically controlled, and the motions of ethyl groups are too small to allow N_2 diffusion at 77 K. SCXRD showed that after N_2 adsorption, there is negligible structural change for both the copper triazolate backbone and the ethyl groups, confirming the occurrence of gating flexibility, in which the host involves just one thermodynamically stable state and transforms among non-equilibrium metastable states (Figure 2.3).

MAF-2 can also adsorb larger molecules including methanol (MeOH), ethanol (EtOH), acetonitrile (MeCN) and benzene (Bz). SCXRD showed that the host framework shrinks, expands, or distorts slightly after loading MeOH, EtOH, and MeCN, but the apertures keep closed in all cases. In the crystal structures, MeOH/EtOH/MeCN molecules exhibit as hydrogen-bonding hexamers and show weak interactions with the pore surface. Therefore, their isotherm inflection points result from capillary condensation of the solvent molecules rather than the deformation of the framework. In contrast, the two-step Bz adsorption isotherm is characteristic for a pore-expanding action, and PXRD confirmed that the host framework expanded to the most open state. Obviously, the flexibility of ethyl groups is not sufficient to allow diffusion of these organic molecules, especially for Bz, meaning that the copper triazolate backbone also participates in the gating actions. Nevertheless, MAF-2 can adsorb neither the larger cyclohexane molecules nor the smaller H_2O molecules. The rich adsorption behaviours of MAF-2 show that gating apertures are useful for controlling guest diffusion/adsorption. Besides the

Table 2.1 Key parameters of typical low-boiling point gases.

	Formula	Boiling point[44]/K	Kinetic diameter[44]/Å	van der Waals size[a,b]/Å	Dipole moment[44]/$\times 10^{18}$ esu cm	Quadruple moment[44,219–222]/$\times 10^{26}$ esu cm^2	Polarisability[44]/$\times 10^{-25}$ cm^3
Hydrogen	H_2	20	2.89	2.40 × 2.40 × 3.14	0	0.63	8.042
Nitrogen	N_2	77	3.64–3.80	3.10 × 3.10 × 4.20	0	−1.5	17.403
Carbon monoxide	CO	82	3.69	3.40 × 3.40 × 4.36	0.1098	−2.5	19.5
Argon	Ar	87	3.542	3.76	0	0	16.411
Oxygen	O_2	90	3.467	3.04 × 3.04 × 4.27	0	−0.32	15.812
Methane	CH_4	112	3.758	3.67 × 4.19 × 4.19 or 3.95 × 4.00 × 4.19	0	0	25.93
Krypton	Kr	120	4.0	4.04	0	0	24.844
Xenon	Xe	165	4.3	4.32	0	0	40.44
Carbon dioxide	CO_2	195	3.3	3.40 × 3.40 × 5.39	0	−4.3	29.11
Ethyne	C_2H_2	188	3.3	3.40 × 3.40 × 5.76	0	6.3	33.3–39.3
Ethylene	C_2H_4	169	4.163	3.40 × 4.26 × 4.88	0	N.A.	42.52
Ethane	C_2H_6	184	4.443	3.86 × 4.17 × 4.93	0	N.A.	44.3–44.7
Propene	C_3H_6	225	4.678	3.88 × 4.17 × 6.48	0.366	N.A.	62.6
Propane	C_3H_8	231	4.3–5.188	4.18 × 4.56 × 6.75	0.084	N.A.	62.9–63.7

[a]Structures of polyatomic molecules were optimised by the Dmol^3 module, in Materials studio 5.5, using generalised gradient approximation (GGA) with the Perdew–Burke–Ernzerhof (PBE) functional and the double numerical plus d-functions (DNP) basis set.
[b]van der Waals radii from ref. 223.

Table 2.2 Key parameters of C_{4+} hydrocarbons.

	Formula	Boiling point[44,224]/K	Kinetic diameter/Å	van der Waals size[a,b]/Å	Dipole moment[44,225]/$\times 10^{18}$ esu cm	Polarisability[44,225]/ $\times 10^{-25}$ cm^3	RON[226,227]
Butadiene	C_4H_6	269	4.24[228]	3.40 × 5.25 × 7.97 *cis*-Conformation 3.58 × 5.39 × 7.61	0	86.4	N.A.
1-Butene	C_4H_8	267	4.83[228]	5.24 × 5.48 × 7.87 *cis*-Conformation 4.18 × 5.49 × 7.53	0.359–0.438	79.7–85.2	N.A.
2-Butene	C_4H_8	274	4.61/4.49[228]	4.17 × 5.23 × 7.44 *cis*-Conformation 4.17 × 5.52 × 6.89	0	84.9	N.A.
Iso-butene	C_4H_8	266	4.84[228]	4.17 × 5.77 × 6.72	0.55	80	N.A.
Butane	C_4H_{10}	273	4.687[229]	4.18 × 4.57 × 8.04 *cis*-Conformation 5.07 × 5.48 × 7.57	0.05	82	94
Iso-butane	C_4H_{10}	261	5.278[229]	5.08 × 6.16 × 6.74	0.132	81.4–82.9	102
Pentane	C_5H_{12}	309	4.5[44]	4.16 × 4.57 × 9.30	0	99.9	62
Iso-pentane	C_5H_{12}	301	5.0[44]	5.08 × 6.18 × 8.03	0	N.A.	92
Neo-pentane	C_5H_{12}	283	6.2–6.464[44]	6.05 × 6.15 × 6.73	0.13	102.0	86
Benzene	C_6H_6	353	5.35–5.85[179]	3.40 × 6.71 × 7.38	0	100–107.4	102.7
Cyclohexene	C_6H_{10}	356	5.8[230]	5.06 × 6.74 × 7.40	N.A.	N.A.	N.A.
Cyclohexane	C_6H_{12}	354	6.0–6.182[179]	5.07 × 6.73 × 7.40	0	108.7–110	82.5
Hexane	C_6H_{14}	341	4.3[224]	4.17 × 4.60 × 10.60	N.A.	N.A.	24.8
2-Methylpentane	C_6H_{14}	333	5.0[224]	5.08 × 6.18 × 9.31	0.1	N.A.	73.4
3-Methylpentane	C_6H_{14}	336	5.0[224]	5.16 × 6.20 × 9.25	N.A.	N.A.	74.5
2,2-Dimethylbutane	C_6H_{14}	323	6.2[224]	5.48 × 6.73 × 8.02	N.A.	N.A.	91.8
2,3-Dimethylbutane	C_6H_{14}	331	5.6[224]	5.57 × 6.33 × 8.01	N.A.	N.A.	101.7

Styrene	C_8H_8	418.3	6.8 [224]	3.40 × 6.71 × 9.76	0.13	137	112
Ethylbenzene	C_8H_{10}	409.3	5.8 [224]	5.18 × 6.70 × 9.58	0.59	142	
				Minimum 3.94 × 6.70 × 9.47			
p-Xylene	C_8H_{10}	411.5	5.8 [224]	3.94 × 6.70 × 9.26	0.1	137	
m-Xylene	C_8H_{10}	412	6.8 [224]	3.94 × 7.36 × 9.06	0.37	142	
o-Xylene	C_8H_{10}	417.5	6.8 [224]	4.17 × 7.50 × 7.69	0.64	141	
				Minimum 3.94 × 7.36 × 7.69			

[a] Structures of polyatomic molecules were optimised by the Dmol^3 module, in Materials studio 5.5, using generalised gradient approximation (GGA) with the Perdew–Burke–Ernzerhof (PBE) functional and the double numerical plus d-functions (DNP) basis set.
[b] van der Waals radii from ref. 223.

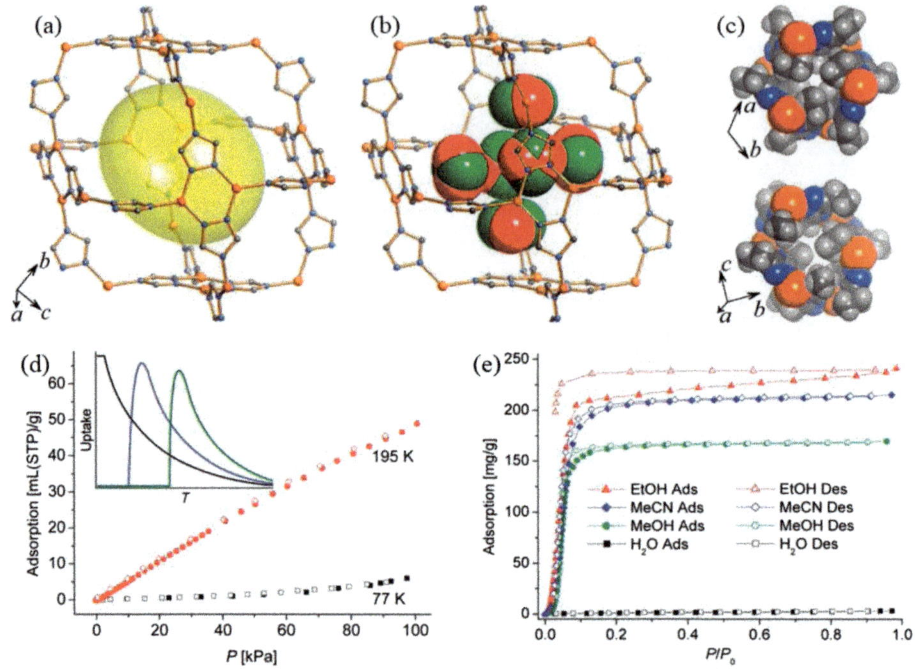

Figure 2.3 Adsorption behaviours of MAF-2. (a)–(c) Crystal structures of the copper triazolate backbones (ethyl groups are omitted for clarity) before ((a) semi-transparent yellow ellipsoid: pore cavity) and after ((b) two-fold disordered N_2 molecules are represented in red and green) N_2 adsorption, as well as two types of ethyl decorated apertures connecting adjacent pore cavities (c). (d) N_2 adsorption and desorption isotherms (inset: adsorption isobars for guests with different effective sizes). (e) Vapour adsorption and desorption isotherms at 298 K. Reproduced from ref. 32 with permission from American Chemical Society, Copyright 2008.

size and flexibility (E_a) of the aperture, the gating process can be also controlled by temperature, the size of guest, and the host–guest interaction at the gating moment. In addition, gating actions can occur during guest-induced structural transformations (Figure 2.3).

Gating pore/adsorption should be common for microporous/ultramicroporous materials. Besides the flexible side groups behaving like door panels, many other types of structures can serve as the gating aperture. The counter ions in zeolites and MOFs may serve as a cork to control guest diffusion.[46–49] More generally, gating behaviours can result from the temporary distortion of the coordination network without independent moving parts. Certainly, flexibility at different parts can cooperate to serve as gating apertures. When the pore apertures are small, slow adsorption and inversed temperature-dependence of adsorption amount can be generally assigned to the gating actions. Some MOFs with relatively large pore apertures may show kinetically controlled gating adsorption behaviours, in which the defects on the crystal surfaces may serve as the gating apertures.[50–53]

2.2.4 Evaluation of Adsorption Amount and Working Capacity

The guest adsorption amount is generally expressed based on the mass/weight (g^{-1}) of the adsorbent which can be directly and reliably measured. However, the volume of the adsorbent (cm^{-3}) is usually more important for practical applications. Actually, when compared with the commonly used gravimetric porosity parameters (uptake, pore volume, surface areas, *etc.*), the volumetric values correlate better with the host–guest structures. Because the adsorbent volume is difficult to measure, the volumetric uptake (cm^{-3}) is usually calculated from the gravimetric uptake (g^{-1}) and the crystallographic density of the empty adsorbent ($g\ cm^{-3}$). Since defects and impurities can lead to large deviations of the sample density, gas adsorption is a simple and useful method for qualitative identification of the sample quality. A typical approach is to compare the pore volume measured by low-temperature gas isotherms and that calculated from the crystal structure. The measured pore volume is also used to convert the measured uptake, *i.e.*, the excess adsorption on the pore surface, to the total uptake that also includes the guests in other positions of the pore, which is important for high-pressure measurements/applications. For flexible adsorbents possessing variable pore volumes, the measurements and calculations are much more complicated. In practical applications, the tap densities of MOFs are generally less than half of the crystallographic values,[54] and the packing efficiency can only reach 70–80% even without shaping (use crystals directly).[55] When filling into the device, flexible MOFs usually stay in their high-density states (with low volume and small pore), which can effectively increase the packing efficiency. However, the deformation of flexible MOFs may lead to the fragmentation of the crystals and pellets.[56] Although fragmentation will stop at certain crystal/pellet sizes rather than continue to framework decomposition/collapse, and generally has no influence on the total adsorption amount,[56,57] gas diffusion in the device may change significantly.[58]

Storage, separation, and other adsorption-based applications require not only adsorption but also desorption. To achieve high working capacities, adsorbents should show large uptake differences between the target adsorption and desorption conditions. Besides decreasing pressure and/or increasing temperature, guest desorption can be also accomplished by other means. For example, steam is widely used in practical separation applications employing highly stable adsorbents, which not only replaces the adsorbed gas but also facilitates heat transfer. The special guest responses of flexible adsorbents are particularly useful for attaining high working capacities. The equilibrium adsorption amount of flexible MOFs may be also dependent on mechanical pressure, electric field, and other external physical stimuli, which can be used to facilitate guest desorption.[59,60] Besides, by introducing photosensitive components in MOFs, the guest uptake can be altered by light irradiation.[61–64] Nevertheless, the role of framework flexibility, especially pore-opening and gating behaviours, are still not comprehensively understood for adsorptive storage and separation.

2.3 Flexible MOFs for Gas Storage and Delivery

While molecules of liquids and solids can be adsorbed, they have sufficiently high densities to be stored and delivered directly in conventional containers. Porous materials may be used as carriers for drug molecules, but the reutilisation of carriers is generally not considered. Having very low densities, gases may need to be stored (delivered in the cylindrical form) under high-pressure and/or low-temperature conditions to meet practical demands.

Gas storage was one of the earliest considered functions for MOF materials. However, the storage capacity may be misunderstood as the gas uptake at an arbitrarily selected condition. Considering practical requirements, gases are generally stored at or near ambient temperatures (*ca.* −40–60 °C) and moderately high pressures. Conventional gas cylinders work below 15 MPa, which can provide a volumetric gas storage density of *ca.* 150 *v*/*v*, being too low for fuels. To be practically useful, the storage pressure using porous materials should be lower than those of conventional gas cylinders. More importantly, not only adsorption but also desorption should be considered. For example, hysteresis between the adsorption and desorption isotherms is generally disadvantageous for storage and delivery. The delivery capacity, storage capacity, or usable storage capacity (USC), *i.e.*, the working capacity for storage and delivery, is the difference in adsorption amounts between the storage/charge temperature/pressure and the discharge temperature/pressure. To allow the gas stored in the container to automatically flow out, the discharge pressure cannot be lower than the atmospheric pressure. However, generally, a higher value is required to attain sufficiently large flow rates.

According to the Langmuir model, adsorption in conventional porous materials is more effective at lower pressures, meaning that working capacity is significantly lower than the adsorption amount at the charge condition. An increase of adsorption affinity decreases the ratio of deliverable uptake, meaning that there is an optimal value for maximising the working capacity (Figure 2.4).[65] For example, based on the Langmuir model, the optimum adsorption enthalpy has been predicted to be 15.1 kJ mol^{-1} for hydrogen (H_2) storage at 298 K and 1.5–30 bar, in which the working capacity is just 63.5% of the total capacity.[66]

Pore volume, pore size and adsorption enthalpy correlate with each other. Generally, a small pore size gives rise to a large adsorption enthalpy and a small pore volume, and *vice versa*. In other words, an optimal adsorption enthalpy requires an optimal pore size/volume, which cannot be unlimitedly large. For example, grand canonical Monte Carlo (GCMC) simulations showed that the optimal pore size for methane (CH_4) storage at 298 K and

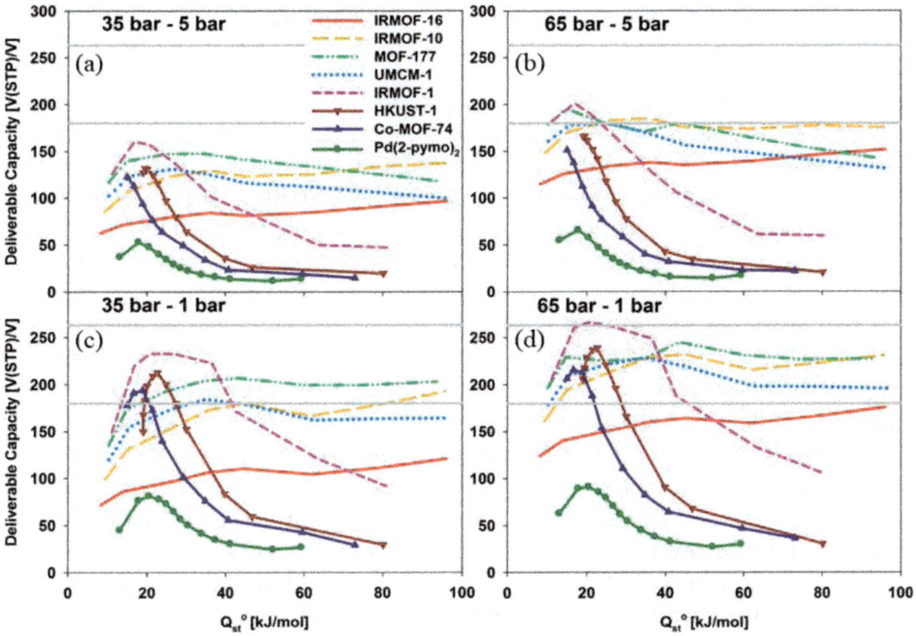

Figure 2.4 Relationships between working capacity at (a) 5–35 bar, (b) 5–65 bar, (c) 1–35 bar and (d) 1–65 bar, and isosteric heat of adsorption at low loadings (Q_{st}). Reproduced from ref. 65 with permission from American Chemical Society, Copyright 2014.

5–65 bar is ~11 Å, which can give a working capacity of ~200 cm^3(STP) cm^{-3}, being ~70% of the total adsorption.[67] Furthermore, it should be noted that the pore surface cannot be completely uniform and there is always heterogeneity (various adsorption sites with different adsorption affinities). Since guest molecules prefer the strongest adsorption sites according to thermodynamic law, the host–guest interaction and adsorption enthalpy generally decrease as the uptake increases, which increases the declining trend of the isotherm slope and decreases the usable storage ratio as compared with the ideal case with uniform pore surface.[66]

The guest–guest interaction also contributes to the adsorption enthalpy, which may lead to the increase of adsorption enthalpy and isotherm slope as the adsorption amount increases.[68] However, this effect can only be used for guests with sufficiently high boiling points, such as CH_4 and H_2O.[41,69–71]

The structural transformation of a flexible MOF during the change of guest loading can give rise to abnormal isotherm shapes/slopes. The most notable one is the S-shaped isotherm associated with a pore-opening transformation, which is obviously beneficial in increasing the working capacity.[72,73]

2.3.1 Flexible MOFs for Adsorptive Storage of H_2

H_2 possesses the highest gravimetric energy density but comparatively low volumetric energy density among all chemical fuels. With an extremely low boiling point of 20 K, it is difficult to liquefy or physically adsorb H_2. The highest H_2 physical adsorption enthalpy reported so far is *ca.* 10 kJ mol^{-1},[74] which is lower than the optimal value of 15.1 kJ mol^{-1}.[66] The highest H_2 adsorption amount at room temperature reported so far is just 10 g L^{-1} at 80 bar,[75] which is much lower than the requirement of 30 g L^{-1} for practical applications.[76]

Because H_2 molecules interact very weakly with each other and with physical adsorbents, H_2 adsorption is difficult to induce any obvious structural transformation in porous materials (or show abnormal isotherm shapes),[77] except at ultra-low temperatures near its boiling point. Special H_2 adsorption behaviours of flexible MOFs can be identified by careful analysis and comparison of the adsorption data. For example, $[M_3(vtz)_6]$ (MAF-123-M, M = Zn, Mn, or Cd, Hvtz = 1,2,3-triazole) is a family of ultramicroporous MOFs possessing pocket-like pores. By varying the metal ion from Zn(II) to Cd(II) you can gradually increase the entrance diameter of the pocket-like pores from 2.2 to 2.6 Å, with an interval of just 0.2 Å. N_2 adsorption isotherms and SCRXD structures showed gradually increased framework flexibility and accessibility of the pocket-like pores. At 77 K and 1.2 atm, the Zn, Mn and Cd analogues show H_2 uptakes of 16.6, 27.6 and 26.7 mg cm^{-3} or 5.7, 10.8 and 8.7 mmol g^{-1}, respectively. Taking 0.1–1.2 atm as the charge–discharge pressure, their working capacities are 11.6, 20.7 and 13.6 mg cm^{-3} or 4.0, 8.2 and 4.4 mmol g^{-1}, which are 70%, 75% and 51% of the total uptakes, respectively. Obviously, the overwhelmingly high H_2 adsorption and storage capacities of the Mn analogue with a medium pore size should be attributed to not only the lowest atomic weight of Mn but also the abnormal isotherm shape. Indeed, the H_2 adsorption enthalpy of the Mn analogue abnormally rises from 5.8 kJ mol^{-1} at zero loading to 7.0 kJ mol^{-1} at 7.1 mmol g^{-1}, indicating the gradual structural transformation of the host. GCMC and density functional theory (DFT) simulations showed that the H_2 binding inside the pore–surface pocket is stronger than the outside, but the narrow pocket entrance hinders H_2 to enter. For the Zn analogue with the smallest entrance, the energy barrier is too high, so H_2 always stays outside the pockets. For the Cd analogue with the largest entrance, the energy barrier is negligibly low, so H_2 can enter the pockets immediately. For the Mn analogue with a suitably high energy barrier, H_2 can enter the pockets to show a high adsorption enthalpy only at high pressures. SCXRD showed that N_2 molecules can enter the pocket-like pores, which induces obvious framework distortion and crystal symmetry decrease. Note that non-stoichiometric adsorption can sometimes lead to structural transformations involving very complicated symmetries and even non-periodic structures, which are difficult to study by crystallography. Since the framework deformation caused by H_2 adsorption should be much smaller than that of N_2, the H_2 isotherms of MAF-123-Mn cannot show obvious inflection points (Figure 2.5).[78]

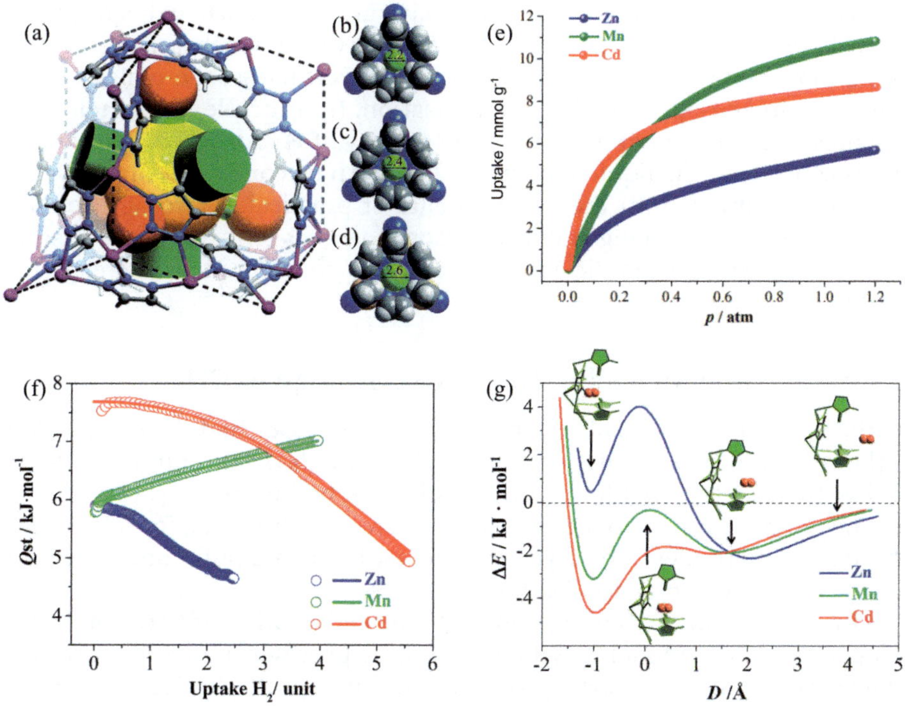

Figure 2.5 H_2 adsorption behaviours of MAF-123-M. (a) Framework and pore structure. (b)–(d) Structures of the pore-surface pockets of MAF-123-Zn (b), MAF-123-Mn (c) and MAF-123-Cd (d). (e) H_2 adsorption isotherms at 77 K. (f) The coverage dependent H_2 adsorption enthalpies (unit: formula unit [$M_3(vtz)_6$]). (g) DFT-derived potential energy surface of a H_2 molecule inserting into the pore-surface pockets (rigid setting, ΔE is binding energy, D is the distance between the pocket entrance and the molecular centroid of H_2). Reproduced from ref. 78 with permission from the Royal Society of Chemistry.

With extremely small molecular sizes, H_2 can diffuse very fast in physical adsorbents, which prevents obvious adsorption–desorption hysteresis at normal temperatures. As a rare example, significant adsorption–desorption hysteresis was observed for [$Ni_2(bpy)_3(NO_3)_4$] (bpy = 4,4′-bipyridine), which possesses pore apertures much smaller than the kinetic diameter of H_2 (Table 2.1). Apparently, gating actions are required for both the adsorption and desorption processes to allow diffusion of the oversized guest molecules. But to show large hysteresis, the gating actions should be more difficult in the desorption process, which was ascribed to the larger steric hindrance effect at higher guest loadings. The reduced aperture flexibility could be attributed to the presence of guest molecules in adjacent pores particularly at high guest loadings. Actually, aperture gating is the transient structural transformations not only limited at the aperture. Distortion of the overall porous framework can also contribute to these actions.

Recently, specially-made cylinders working at 70–80 MPa have been applied for hydrogen fuel cell cars,[79] in which the density of H_2 reaches 39–42 kg m^{-3}, being more than half of its liquid density. Nevertheless, the density of H_2 cannot be increased by further pressurisation. In principle, under relatively low pressures, the apparent volumetric density of H_2 adsorbed/absorbed in solids can exceed the liquid density. So far, it can be concluded that physical adsorption cannot fulfil the requirements of practical H_2 storage applications. Besides introducing chemical adsorption sites to increase the adsorption affinity, framework flexibility should be considered to increase the working capacity.

2.3.2 Flexible MOFs for Adsorptive Storage of CH_4

As the main component of natural gas, CH_4 possesses high heat value, low carbon dioxide (CO_2) emission, and high research octane number (RON), and is an attractive fuel for vehicles. Since the boiling point (112 K) is not too low, large quantities of CH_4 can be delivered in a liquefied form for ocean shipping. Nevertheless, the cryogenic fuel tanks are too bulky and expensive to be used for vehicular applications.[67]

With an appropriate boiling point, physical adsorption of CH_4 at ambient temperatures and moderately high pressures provides a good opportunity to achieve the storage capacity of ultra-high pressure cylinders working at 250 bar, *i.e.*, 263 cm^3 (STP) cm^{-3}.[55] However, adsorbents showing high CH_4 uptakes at moderately high pressures (<250 bar) generally also show high uptakes at 5 bar (the typical working pressure for internal combustion engines),[73] meaning that the working capacities are much lower than the uptakes. So far, the record CH_4 uptakes (for all adsorbents including both rigid and flexible MOFs) at 35, 65 and 80 bar are 235, 267 and 273 cm^3 (STP) cm^{-3}, while the working capacities at 5–35, 5–65 and 5–80 bar are lower than 180, 200 and 220 cm^3(STP) cm^{-3}, respectively.[80,81]

CH_4 adsorption can readily induce the deformation of flexible MOFs at room temperature and moderately high pressures.[57,82,83] Note that the framework density and pore volume of the adsorbent are critical parameters for calculations of the total uptake and working capacity. Since these parameters change with the gas pressure/uptake, and they are difficult to determine, the crystallographic density of the expanded phase and the measured maximum pore volume are generally used, which can somewhat underestimate the working capacity.

The CH_4-induced pore-opening behaviour was first observed for [Cu(dhbc)$_2$(bpy)] (Hdhbc = 2,5-dihydroxybenzoic acid), a mutually interdigitated 2D framework capable of reversible transformation between the non-porous state and the porous state with small 1D channels.[83] At room temperature, CO_2, CH_4, O_2 and N_2 adsorption isotherms showed inflection points at 0.4, 5, 35 and 50 bar, respectively, corresponding well with the order of the boiling points of the gases. After that, the isotherm slopes were near zero, and the saturated (excess) CH_4 uptake was 78 cm^3 (STP) g^{-1}.

The usefulness of pore-opening action for maximising the working capacity for CH_4 storage was first demonstrated by using [Co(bdp)] (H_2bdp = 1,4-benzenedi(4′-pyrazolyl)),[57] which exhibits multistep expansions from the non-porous (collapsed) state to the fully open (expanded) state with a large pore volume of 1.02 $cm^3\ g^{-1}$. To accurately convert the excess CH_4 uptakes to the total uptakes, the pore volume change was regarded as being proportional with the excess uptake. The total CH_4 isotherm at 298 K showed a sharp jump from 11 cm^3 (STP) cm^{-3} at 16 bar to 125 cm^3 (STP) cm^{-3} at 20 bar, then increased to ~158 and ~200 cm^3 (STP) cm^{-3} at 35 and 65 bar, respectively. During desorption, a notable hysteresis appeared, and the hysteresis loop was closed at 7 bar and the CH_4 uptake at 5 bar was negligible (<0.2 mmol g^{-1}). While the total uptakes at 35 and 65 bar were not very high, the working capacities of 155/197 cm^3 (STP) cm^{-3} at 5–35 and 5–65 bar were remarkable, which were 98% and 99% of corresponding total uptakes, respectively. Further, the reversible structural transformations between the collapsed and expanded states were confirmed by *in situ* PXRD (Figure 2.6).

CH_4-induced pore-opening/pore-closing usually show large adsorption-desorption hysteresis. For example, the CH_4 adsorption isotherm of [NiL_2] (X-dia-1-Ni, HL = 4-(4-pyridyl)-biphenyl-4-carboxylic acid) showed a sharp

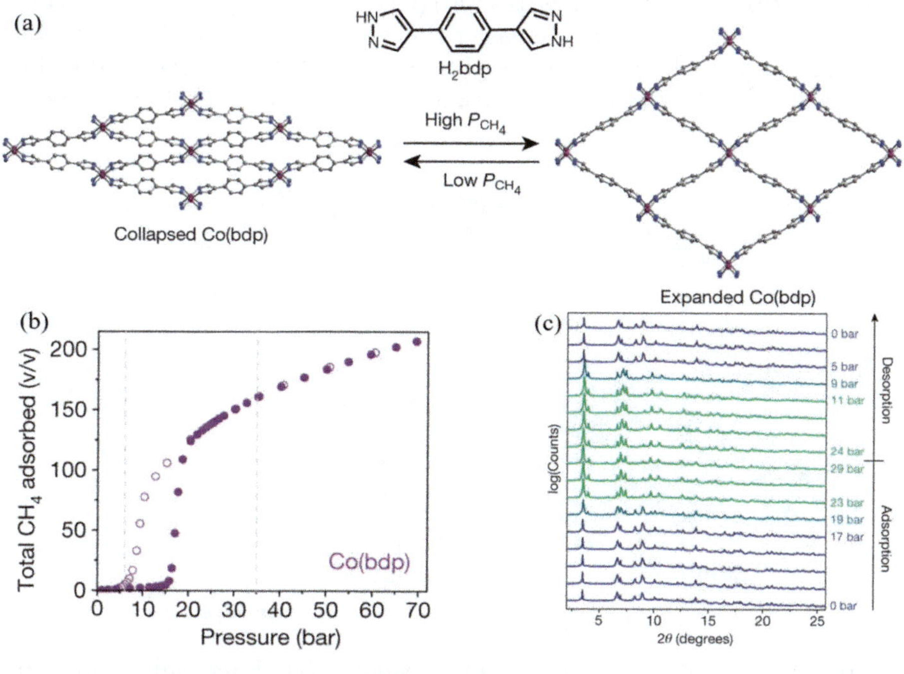

Figure 2.6 CH_4 adsorption behaviours of [Co(bdp)]. (a) Structural transformations between the collapsed and expanded states. (b) Total CH_4 adsorption and desorption isotherms at 298 K. (c) PXRD patterns in CH_4 at 298 K. Reproduced from ref. 57 with permission from Springer Nature, Copyright 2015.

step at ~20 bar and total uptakes of 150 and 189 cm^3 (STP) cm^{-3} at room temperature and 35 and 65 bar, respectively.[82] However, the adsorption–desorption hysteresis was large and the closed-loop pressure was lower than 5 bar. The working capacities at 5–35 and 5–65 bar were 110 and 149 cm^3 (STP) cm^{-3}, being 73% and 79% of the total uptakes, respectively, which were higher than those of most rigid adsorbents but lower than those of [Co(bdp)]. The pore-opening and pore-closing pressures may be shifted to higher values by applying external mechanical pressures, which has been demonstrated by using [Co(bdp)].[57] The framework flexibility and pore-opening/pore-closing pressure of MOFs can also be rationally controlled by other means, such as ligand modification and crystal size/morphology.[23]

Isostructural [Al(OH)(bdc-OH)] (MIL-53(Al)-OH, H_2bdc-OH = 2-hydroxy-1,4-benzenedicarboxylic acid) and [Al(OH)(bdc-$(OH)_2$)] (MIL-53(Al)-$(OH)_2$, H_2bdc-$(OH)_2$ = 2,5-dihydroxy-1,4-benzenedicarboxylic acid) show significantly different pore-opening CH_4 adsorption behaviours and storage performances.[84] When adsorbing CH_4, both MIL-53(Al)-OH and MIL-53(Al)-$(OH)_2$ transformed from the very narrow pore (vnp) state to the lp state, in which the initial transformation pressures in the adsorption isotherms were higher than 15 and 46 bar, and the reverse transformations in the desorption isotherms were completed below 4.1 and 5.8 bar, respectively. The much higher transformation pressures of MIL-53(Al)-$(OH)_2$ were attributed to the additional –OH groups, which increased the intra-framework attraction and impeded framework expansion (Figure 2.7). Because the CH_4 adsorption of MIL-53(Al)-$(OH)_2$ was far from saturation at 65 bar, its 5–65 bar working

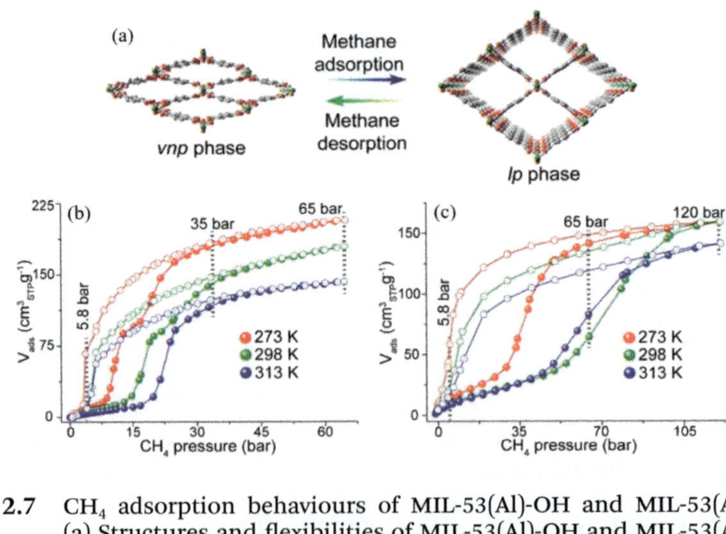

Figure 2.7 CH_4 adsorption behaviours of MIL-53(Al)-OH and MIL-53(Al)-$(OH)_2$. (a) Structures and flexibilities of MIL-53(Al)-OH and MIL-53(Al)-$(OH)_2$. (b) CH_4 adsorption and desorption isotherms of MIL-53(Al)-OH. (c) CH_4 adsorption and desorption isotherms of MIL-53(Al)-$(OH)_2$. Reproduced from ref. 84 with permission from American Chemical Society, Copyright 2019.

capacities (71 cm^3 (STP) cm^{-3}) was much lower than that of MIL-53(Al)-OH (164 cm^3 (STP) cm^{-3}).

Controlling the crystal size and flexibility of an analogous MOF can tune the pore-opening behaviour of CH_4 adsorption.[85] Increasing the DMF–H_2O (DMF = *N*,*N*-dimethylformamide) ratio in the synthetic solvent afforded [Al(OH)(bdc-NH_2)] (MIL-53(Al)-NH_2, H_2bdc-NH_2 = 2-amino-1,4-benzenedicarboxylic acid) with smaller sizes and easier transformation from the vnp state to the lp state. Although the CH_4 uptake at 5.8 bar increased, the uptake at 65 bar increased much more. As the DMF–H_2O ratio gradually increased from 0 : 5 to 2 : 3, the working capacity at 5.8–65 bar gradually increased from 62 to 81 mg g^{-1}. Note that smaller crystals generally exhibit higher energy barriers for the structural transformation and give rise to higher transition pressures, which can be attributed to the presence of less defects per a crystal.[86] The contrary phenomenon of MIL-53(Al)-NH_2 for CH_4 adsorption indicates that introduction of DMF in the synthetic solvent not only reduced the crystal size but also generated higher defect concentration.

So far, the highest CH_4 storage working capacities of flexible MOFs are still not higher than those of rigid adsorbents, which can be attributed to the relatively low pore volumes of the MOFs giving limited CH_4 uptakes at high pressures. Nevertheless, flexible MOFs can offer much higher USC (near 100%), which should be a promising strategy for achieving the target working capacity. Besides a suitably large pore volume in the fully open state, multistep opening of the pore might be useful for optimisation of the host–guest and guest–guest interactions to solve the trade-off issue between pore size and adsorption enthalpy.

2.3.3 Flexible MOFs for Adsorptive Storage of Other Gases

Many gases and vapours have been used for studying the porosity and adsorption behaviours of porous materials, but most of them do not require adsorptive storage. For example, N_2, O_2, Ar and other common gases having low boiling points are difficult to compress, but they can be separated directly from air for large-scale uses. CO_2, NH_3, SO_2 and light hydrocarbons with relatively high boiling points are easy to compress and liquefy.

The adsorption of vapours or easily liquefied gases, such as H_2O, NH_3 and halohydrocarbons, is useful for heat pump applications, which can be regarded as a special kind of adsorptive storage. Nevertheless, their strong guest–guest interactions can lead to capillary condensation and abrupt increase of adsorption, giving high working capacities without the need of flexible adsorbents.

Adsorptive storage may be useful for pure acetylene (C_2H_2), which is an important raw material for manufacturing chemical products and electronic materials. Due to the low explosion limit (2 bar),[76] C_2H_2 is generally stored in the dissolved form in acetone. The existence of acetone vapour is disadvantageous for applications requiring high-purity C_2H_2. For practical gas cylinders, a small amount of a pure C_2H_2 can be delivered, because the working pressure is 1–2 bar and even 1.0–1.5 bar.[87]

With a relatively high boiling point among gases and rich intermolecular binding functionalities, C_2H_2 can be readily adsorbed at ambient conditions and induce significant framework distortions.[88] The C_2H_2 storage capacities were usually evaluated based on the uptake just at one condition (usually room temperature and 1 bar) rather than two conditions (considering both charging and discharging),[89] probably because adsorption isotherms above 1 bar are difficult to measure. On the other hand, considering a working pressure range at 1.0–1.5 bar, most adsorbents cannot offer large working capacity and/or large usable ratio, because the isotherms show very small slopes at this range. For flexible adsorbents, the isotherm inflections point usually locate far below 1.0 bar.

An abnormal isotherm shape with an inflection point near ambient conditions can be used to improve the C_2H_2 storage working capacity.[87] Based on the low-temperature isotherms and the calculated adsorption enthalpy, the isotherm at 298 K measured up to *ca.* 1.0 bar was extrapolated to 1.5 bar. By virtue of the slightly S-shaped C_2H_2 isotherms of MAF-2, a working capacity or USC of 20 cm^3 (STP) cm^{-3} at 1.0–1.5 bar was obtained, which was 40 times higher than that of a gas cylinder. A much higher working capacity was demonstrated by MAF-123-Cd showing more obvious S-shaped C_2H_2 isotherms (Figure 2.8).[78] The electrostatic fields of the pore-surface pockets (electropositive at entrances and electronegative inside pockets) can distinguish the opposite quadrupole moments of CO_2 and C_2H_2 and exclusively adsorb C_2H_2 (Table 2.1). At 298 K, the C_2H_2 isotherm was almost linear, and the uptake was just 2.23 mmol g^{-1} at 1.0 bar. At 273 K, the isotherm showed an obvious S-shape and a much higher uptake of 6.34 mmol g^{-1} at 1.0 bar, indicating a rapid increase of uptake above 1 bar at 298 K. Based on the extrapolated isotherm at 1.0–1.5 bar, MAF-123-Cd showed a USC of 1.3 mmol g^{-1} or *ca.* 1.99 mmol cm^{-3}, which was 98 times higher than that of a gas cylinder. Importantly, the deliverable C_2H_2 was 37% of the adsorption amount, which was not only much higher than those of strong C_2H_2 adsorbents (<20%), but also higher than those of gas cylinders and weak C_2H_2 adsorbents (*ca.* 33%).

Precise control of the pore-opening pressure is obviously useful to improve the C_2H_2 storage working capacity, which may be achieved by using the solid-solution structure.[90] $[Zn_2(bdc)_2(bpy)]$ (Zn-CAT) is a classic flexible framework showing typical guest-induced structural transformations (Figure 2.9).[91] C_2H_2 adsorption and desorption isotherms of Zn-CAT measured at 273 K showed typical pore-opening and pore-closing actions at 44 and 19 kPa, respectively. Isostructural $[Zn_2(bdc\text{-}NH_2)_2(bpy)]$ (Zn-CAT-$(NH_2)_{100}$) and $[Zn_2(bdc\text{-}NO_2)_2(bpy)]$ (Zn-CAT-$(NO_2)_{100}$, $H_2bdc\text{-}NO_2$ = 2-nitro-1,4-benzenedicarboxylic acid) showed similar adsorption behaviours, but the pore-opening/closing pressures shifted to >100/100 kPa and 1.0/0.9 kPa at 273 K, respectively, which can be attributed to the decrease and increase of the host–guest interactions, respectively. By feeding mixed $H_2bdc/H_2bdc\text{-}NH_2$ or $H_2bdc/H_2bdc\text{-}NO_2$ ligands, a series of solid-solution frameworks were synthesised to gradually tune the pore-opening/closing pressures (Figure 2.9). Notably, Zn-CAT-$(NH_2)_{60}$/Zn-CAT-$(NH_2)_{100}$

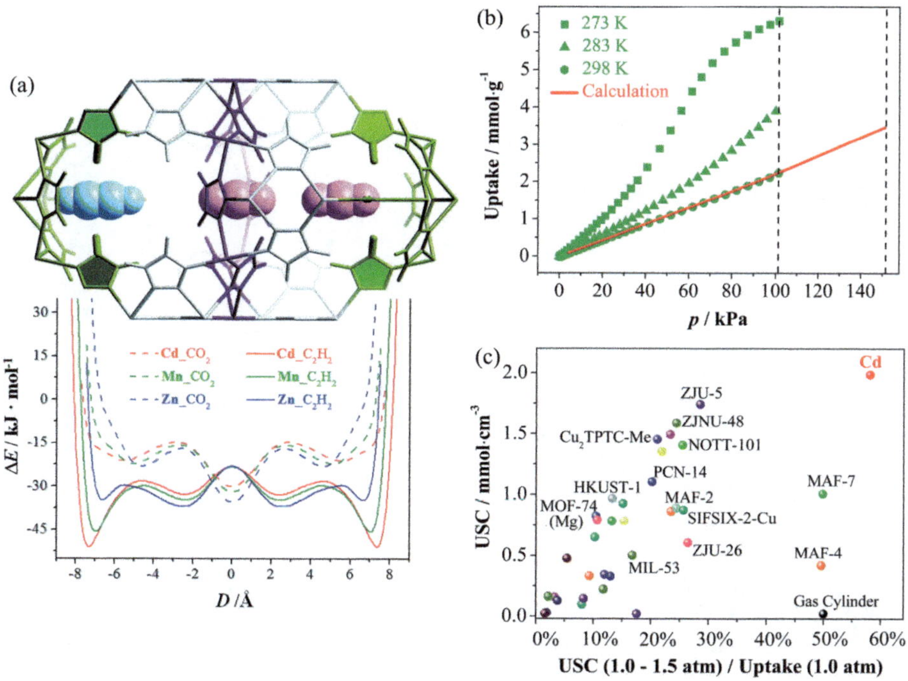

Figure 2.8 C_2H_2 adsorption and storage behaviours of MAF-123-Cd. (a) DFT-derived potential energy surface of a CO_2/C_2H_2 molecule inserting into the pore-surface pockets (rigid setting D is the distance between the centres of the channel and the gas molecule, purple molecule: CO_2, blue molecule: C_2H_2). (b) C_2H_2 adsorption isotherms. (c) Comparison of the usable storage capacities (USCs) and utilisation ratios for C_2H_2 storage parameters of representative MOFs. Reproduced from ref. 78 with permission from the Royal Society of Chemistry.

achieved extremely high C_2H_2 usage capacities of 2.95/4.02 mmol cm^{-3} under charging conditions of 273/243 K and 150/150 kPa and discharging conditions of at 310/298 K and 100/100 kPa, respectively.

Flexible MOFs may be attractive for storage of highly toxic gases such as AsH_3, BF_3, and PH_3, which are also electronic gases used in the semiconductor industry. To avoid leakage, they are stored in vacuum actuated cylinders, which require high tightness and provide very limited working capacity. At room temperature and 1.0 atm, the adsorption amounts of these gases in porous materials can be much higher than that of the gas cylinders, and very high usable ratio for storage and delivery can be obtained by employing adsorbent flexibility with suitable isotherm shapes.

2.4 Adsorptive Separation

By virtue of the diverse pore structures and adsorption sites, MOFs are very promising for selective recognition and separation of similar molecules. Many separation strategies have been reported, which all rely on thermodynamic

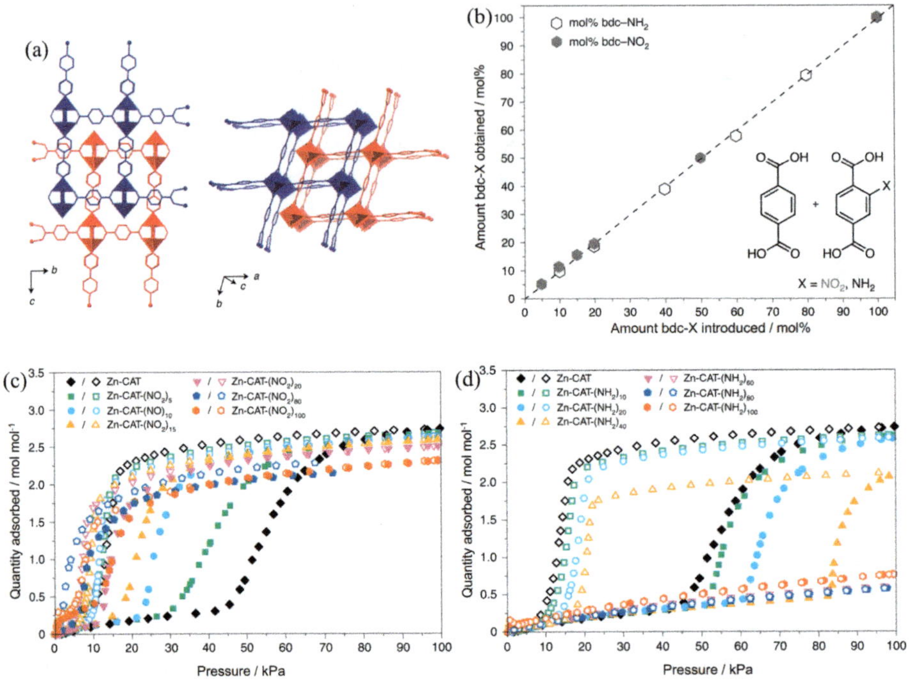

Figure 2.9 C_2H_2 adsorption and storage behaviours of Zn-CAT analogues. (a) Crystal structures of Zn-CAT. (b) Relationships of molar percentage of bdc-NH_2/bdc-NO_2 linker in the obtained material and in the reaction. (c) C_2H_2 adsorption isotherms for Zn-CAT and analogues functionalised by –NH_2. (d) C_2H_2 adsorption isotherms for Zn-CAT and analogues functionalised by –NO_2. Reproduced from ref. 90 with permission from Springer Nature, Copyright 2022.

(different binding at the equilibrium state) and/or kinetic (different diffusion rates) mechanisms. Molecular sieving is a subset of kinetic separation with ideal/infinite selectivity, in which guest molecules larger than the pore apertures are completely excluded. Note that the term sieving is widely used for other common separation mechanisms, which usually means just separation. In this chapter, sieving is used for the ideal kinetic separation based on the size exclusive mechanism. Basically, adsorptive separation by using flexible MOFs is also based on thermodynamic and/or kinetic mechanisms, but the structural transformability of the adsorbents gives rise to many unique properties.

Mixture adsorption is not the simple mathematical addition of single-component adsorptions. Nevertheless, single-component adsorption is easy to measure, and is the basis for studying the separation performance. The differences in single-component adsorption behaviours, usually the isotherms, are widely used for predicting adsorption selectivity. The simplest approaches use the ratio of the uptakes at relevant pressures, which usually underestimate the selectivity, because the competition of different guests is not considered. The ideal adsorption solution theory (IAST) considers the competitive adsorption. However, it requires accurate fitting of isotherms based on appropriate models, which are usually difficult to verify. Further, it is not suitable for many cases, such as adsorbates with significantly different adsorbing abilities, molecular sieves, and flexible adsorbents. It should be pointed out that, low adsorption amounts are generally associated with large relative errors, meaning that high adsorption selectivities, especially those calculated from the single-component data, should be treated with caution. For host–guest systems involving slow diffusions, the kinetic selectivity can be predicted by the ratio of the diffusion constants obtained by fitting the single-component adsorption kinetics (based on appropriate models), in which competitive adsorption is also not considered.

The mixture adsorption/separation behaviour is difficult to study but definitely important for both practical applications and academic research. Among various methods, column breakthrough experiments, being similar to many industrial separation processes, have been increasingly used in recent years. In principle, the adsorption amount, adsorption selectivity, purity, adsorption kinetics, and other useful information can be obtained from properly measured and calculated breakthrough curves. Note that the outlet flow rates rather than the concentrations are required to calculate the adsorption amounts and selectivities.

The isotherm inflection points associated with the pore-opening actions usually differ a lot for different guests.[24] Therefore, in a selected pressure range the adsorption isotherms of different guests may be the same as those for molecular sieving, which has been considered as promising for adsorptive separation. However, in principle, there are many problems for adsorptive separation by using the pore-opening transformation. Thermodynamically, multiple guests can cooperate to open the pore,[92] indicating that they can be adsorbed together. Besides, pore-opening actions are generally cooperative in the whole crystallite, meaning that the opened and unoccupied pores can adsorb other smaller guests (Figure 2.10). To avoid co-adsorption,

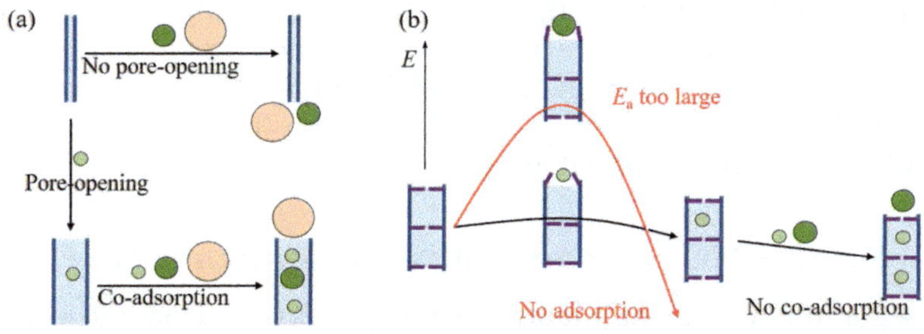

Figure 2.10 Adsorption/separation behaviours of pore-opening and gating actions. (a) Typical structure responses and adsorption behaviours of the pore-opening action toward size-different guests. (b) Structure responses and molecular sieving behaviours of the gating action (with suitable E_a toward size-different guests) Reproduced from ref. 33 with permission from Elsevier, Copyright 2021.

the opened pores should be smaller than the non-target guests and/or the pore-opening action should occur individually rather than in the common cooperative/periodic manner. Therefore, quantitative mixture adsorption/separation experiments are necessary for proving molecular sieving. Further, the inability of adsorption below the pore-opening pressure is disadvantageous for enrichment and purification applications, which exhibits as the leakage flow in the breakthrough curves. Some pore-opening adsorbents showed sharp separation breakthrough curves, but the reasons for avoiding leakage were usually unclear.

As KCF, structural dynamism or gating action of pore aperture mainly influences the adsorption kinetics rather than the isotherm shape. This kind of flexibility is generally considered to be disadvantageous for adsorptive separation, because the effective aperture size is uncertain and larger than the expected one. However, precise design and control of aperture size are everlasting great challenges. Because of the irregular shapes, aperture/guest sizes are usually subjective and difficult to measure and describe (Tables 2.1 and 2.2), meaning that the conventional size comparison approach is not as simple as expected. However, in some cases, usually at very low temperatures, sufficiently large guest molecules diffuse so slowly and can be considered as being completely excluded. In addition, the gating action of each aperture is usually independent, which is useful to avoid co-adsorption (Figure 2.10). Note that in the molecular/quantum world, the atom/molecule/pore shapes and sizes are arbitrary, and all the hosts and guests can be considered as flexible. In this context, all kinetic separation (including molecular sieving) systems involving relatively small pore apertures could be dominated by gating processes. The conventional aperture/molecular sizes, such as crystallographic distances, van der Waals radii, and kinetic diameters, should be considered as important rather than sufficient/necessary parameters for kinetic separation and molecular sieving.[33]

Adsorptive separation is useful for gases, vapours and liquids. Both selectivity and adsorption amount are important criteria of the separation performance, but they generally show the trade-off relationship. Usually, guests having very different boiling points can be separated easily with high adsorption selectivity, so that the adsorption amount is more important. For guests with similar boiling points, separation is difficult, and selectivity is the main target. The following discussions mainly focus on the role of flexibility for adsorptive separation of similar guests.

2.4.1 Flexible MOFs for Adsorptive Separation of H_2–D_2

H_2–D_2 separation is of great importance for applications from nuclear fusion reactors to industrial and scientific research. When compared with H, the heavier isotope D with the lower zero-point energy gives stronger bonding and supramolecular interactions, as well as slower kinetics for the involved reactions, which is also valid for H_2–D_2 adsorption. Thermodynamic separation mechanism mainly uses OMSs to bind the H_2 and D_2 differently, which may give D_2–H_2 selectivity as high as *ca.* 20.[93] To attain sufficiently high adsorption amounts, H_2–D_2 separation is generally carried out at ultra-low temperatures. According to the de Broglie's theory of matter waves, the lighter H_2 shows a larger effective size than D_2, which is more obvious at low temperature. Therefore, H_2 diffuses slower than D_2 when the pore size is comparable to the conventional size of H_2–D_2 (Table 2.1), during which the gating flexibility plays a critical role.

In 2013, the gating effect was studied for H_2–D_2 separation.[94] [$Zn_5Cl_4(bbta)_3$] (MFU-4, H_2bbta = benzobistriazole) possesses two types of cages with the effective diameters of 3.88 and 11.94 Å, which are connected by the small apertures defined by four chloride ligands locating at a square configuration. The maximal diameter of a sphere that can pass the aperture is 2.52 Å. Single-component isotherms showed high and negligible H_2–D_2 uptakes at 70 K and 50 K, respectively, indicating that there is an energy barrier for H_2–D_2 diffusion through the apertures. DFT simulations for H_2 diffusion through the aperture were performed, in which three orientations of the guest molecule were considered. The results showed that the lowest energy barrier occurred when the guest was perpendicular to the aperture direction and simultaneously parallel to the edge of the aperture, rather than being parallel to the aperture direction, which may be attributed to the recognition between the quadruple moments of the aperture and guest. The thermal-desorption spectroscopy (TDS) experiment in 1 : 1 D_2–H_2 at 10 mbar and 60 K for 15 min showed the highest selectivity of 7.5, which decreased significantly as the temperature/contact time increased (Figure 2.11).

In 2019, the temperature dependence of gating behaviours of small apertures with different sizes was observed for H_2–D_2 adsorption and separation.[95] [$Cu(hfipbb)(H_2hfipbb)_{0.5}$] (FMOFCu, H_2hfipbb = 4,4′-(hexafluoroisopropylidene)-bis(benzoic acid)) possesses three kinds of cavities (A, B and C) with sizes of 9.4 × 9.2, 5.6 × 4.2 and 8.8 × 4.7 $Å^2$. Cavities A and B are connected by a small aperture with a diameter of 3.6 Å, while cavities B and C are connected by an even smaller aperture with a diameter of 2.5 Å. Single-component H_2–D_2 isotherms measured at 25–120 K showed that the H_2–D_2 uptake at 1 bar abnormally increased as the temperature increased from 25 K to 77 K. The increasing rate is particularly large at 40–77 K, indicating that the guest diffusion was mainly controlled by the large and the small apertures at relatively low and high temperatures, respectively. TDS experiments were carried out in a 1 : 1 H_2–D_2 mixture at 10 mbar and 25–87 K. At 25, 40 and 77 K, the D_2 uptake was determined as 0.1, 0.37 and 0.14 mmol g^{-1}. The D_2–H_2 selectivity was determined as 14, 6 and 2, respectively. Unlike the single-component D_2 adsorption at 1 bar, the D_2 uptake in the 1 : 1 H_2–D_2 mixture at 10 mbar decreased from 40 to 77 K, implying that the gating process was controlled by not only temperature but also adsorbate pressure (Figure 2.12).

At extremely low temperatures, H_2 adsorption can induce the structural transformation of flexible MOFs. For example, [Al(OH)(bdc)] (MIL-53, H_2bdc = 1,4-benzenedicarboxylic acid) can vary its pore size at different H_2–D_2

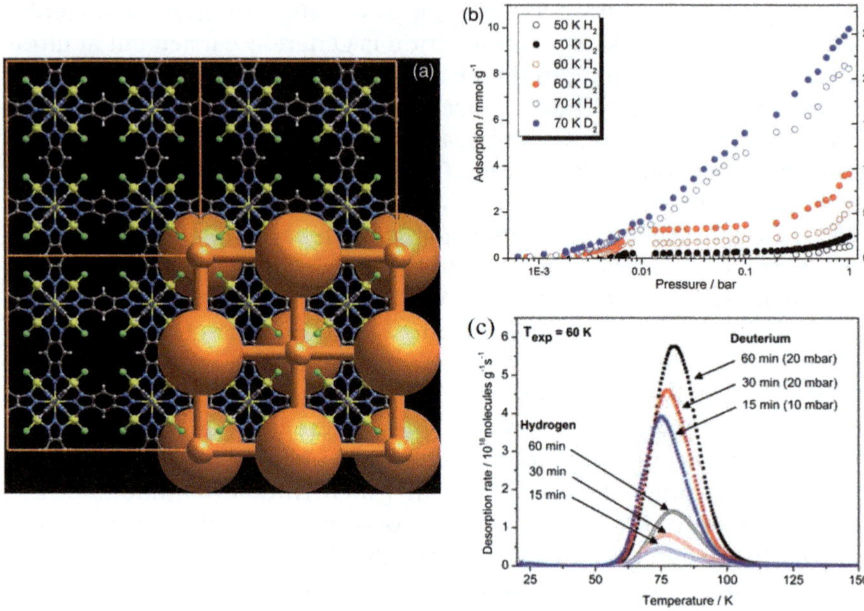

Figure 2.11 H_2–D_2 adsorption–separation properties of MFU-4. (a) Framework and pore structures. (b) Adsorption isotherms. (c) TDS measurements for a 1:1 H_2–D_2 mixture. Reproduced from ref. 94 with permission from John Wiley & Sons, Copyright 2013.

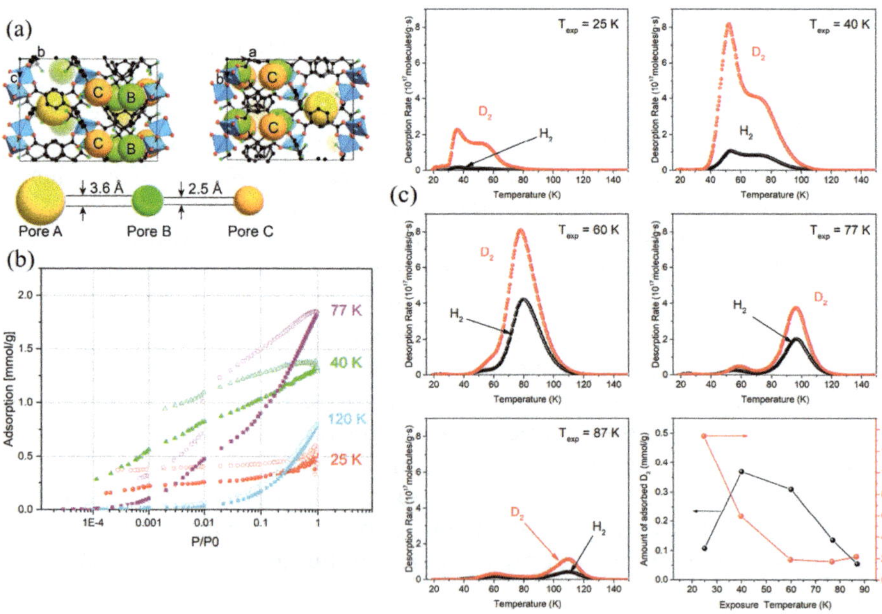

Figure 2.12 H_2–D_2 adsorption–separation properties of FMOFCu. (a) Framework and pore structures. (b) H_2 adsorption and desorption isotherms. (c) TDS measurements for a 1:1 H_2–D_2 mixture at 25–87 K. Reproduced from ref. 95 with permission from American Chemical Society, Copyright 2019.

separation conditions.[96] Guest-free MIL-53 stays at the sp or narrow-pore (np) state with pore size of 2.6 × 13.6 Å^2 and lp state with pore size of 8.5 × 8.5 Å^2 at low temperature and high temperature, respectively. H_2 adsorption isotherms at 20, 40 and 77 K showed pore-opening pressures of <0.01 mbar, 10 mbar and >1 bar, respectively. The TDS experiment carried out at 40 K and 10 mbar gave a D_2–H_2 selectivity of 10.5 and a D_2 uptake of 2.3 mmol g^{-1}. Under the same pressure at 20 K for the lp state, the D_2–H_2 selectivity and D_2 uptake were determined as 2.83 and 22.1 mmol g^{-1}, respectively. For the np state at 77 K, the D_2–H_2 selectivity and D_2 uptake were determined as 2.05 and 0.002 mmol g^{-1}, respectively (Figure 2.13). The optimal separation performance at the conditions between the np and lp states was attributed to continuous expansion of the pores during gradual increase of adsorption, which kept providing the maximal host–guest confinement effects. In other words, the high adsorption selectivity was based on the kinetic mechanism, while the pore-opening actions increased the uptake without decreasing the selectivity.

2.4.2 Flexible MOFs for Adsorptive Separation of the Main Components of Air

Nitrogen (N_2, 78%), oxygen (O_2, 21%) and argon (Ar, 1%), as the main components of air, are generally separated by cryogenic distillation in industry. Adsorptive separation using zeolite adsorbents can be used for applications

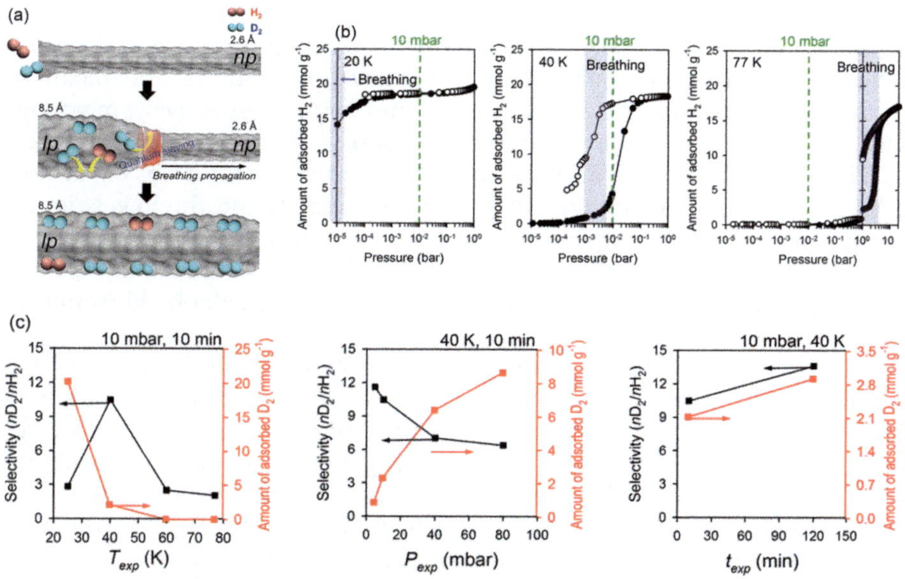

Figure 2.13 H_2–D_2 adsorption–separation properties of MIL-53. (a) Schematic views of H_2–D_2 adsorption in the 1D channel with different sizes. (b) Adsorption isotherms of H_2. (c) TDS measurements for a 1 : 1 H_2–D_2 mixture under different conditions. Reproduced from ref. 96 with permission from American Chemical Society, Copyright 2017.

with lower requirements (purity and amount). According to the molecular sizes of N_2, O_2 and Ar (Table 2.1), the adsorption rates follow O_2 > Ar > N_2, which is the basis of kinetic separation mechanism. Because the size differences of these gases are quite small, conventional adsorbents such as zeolites A and X show low O_2–N_2 selectivities of only *ca.* 10. Thermodynamically, porous materials prefer adsorption of O_2, because N_2 possesses the lowest boiling point and Ar possesses the lowest polarity, but these differences are very small.

Many unique N_2/O_2/Ar adsorption behaviours have been reported for MOFs. For example, the equilibrium adsorption behaviours of these gases can be very different, because OMSs can selectively and chemically bind O_2 or N_2. For physical adsorption, MOFs can also show largely different N_2/O_2/Ar adsorption isotherms based on pore-opening or gating actions. The pore-opening pressures generally follow the order of O_2 < N_2 < Ar,[83,97] which can be attributed to the thermodynamic properties of the gases (Table 2.1). At low temperatures (*e.g.*, 77 K), some MOFs show normal O_2 adsorption isotherm (no special shape) and completely exclude N_2,[94,98] which are similar to those of molecular sieves dominated by the gating apertures. Nevertheless, these phenomena are all observed at low temperatures.

Based on a gating MOF, the O_2–Ar adsorption selectivity can be reversed by varying temperature.[99] [Cu(OPTz)] (H_2OPTz = 5-(5,5-dioxido-10*H*-phenothiazin-10-yl)isophthalic acid) is a 3D network based on the paddle-wheel di-copper units and tripodal ligands, which defines a 3D **bcu** pore system consisting of large cavities (6.5 × 7.2 × 7.8 Å^3) and small apertures/gates (*ca.* 3 Å). Single-component gas adsorption isobars showed that the O_2 and Ar uptakes both increased first and then decreased as the temperature rose, indicating that gating actions have a dominant effect at low temperatures. The maximal O_2 and Ar uptakes were observed at 180 K and 250 K, respectively, meaning that the effective size of O_2 is smaller than that of Ar. At 100–240 K, O_2 uptake was higher than that of Ar, and the highest O_2–Ar uptake ratio was 6.1 at 180 K. At 250–370 K, O_2–Ar uptake ratio was lower than 1, and the lowest is *ca.* 0.54 at 370 K. The batch adsorption experiment performed at 180 K for 1 : 1 and 5 : 95 O_2–Ar mixtures showed high selectivities of 24 and 353.5, respectively (Figure 2.14), which were much higher than those calculated from the single-component uptakes.

For the main components of air, framework flexibility is critical for physical adsorbents to show large adsorption differences. So far, these properties were observed at very low temperatures and/or high pressures. Considering that the smaller O_2 can induce pore-opening easier, co-adsorption of the larger N_2 could be avoided if the opened pores have suitable aperture sizes. To separate O_2 from air, leakage of O_2 is also not a problem. Nevertheless, the pore-opening pressure of O_2 at room temperature should be reduced to improve the adsorption efficiency. However, relatively rigid MOFs with higher gating energy barriers should be targets for effective kinetic separation at room temperature.

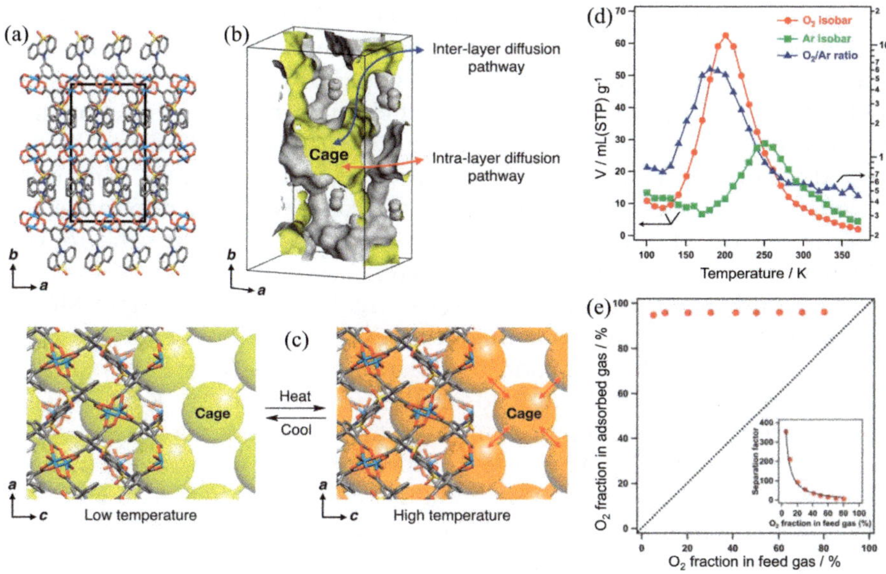

Figure 2.14 O_2–Ar adsorption–separation properties of [Cu(OPTz)]. (a) Framework structure. (b) Pore structure. (c) Schematic diagram of gating pore system. (d) Adsorption isobars at 1 bar and the O_2–Ar uptake ratio. (e) McCabe–Thiele diagram for O_2–Ar separation at 180 K. Reproduced from ref. 99 with permission from The American Association for the Advancement of Science.

2.4.3 Flexible MOFs for CO Capture

As an important chemical raw material, carbon monoxide (CO) is mainly obtained from syngas and the metallurgy industry, coexisting with H_2, CH_4, CO_2, N_2, *etc.* Having almost the same physical properties, CO and N_2 cannot effectively be separated by distillation or physical adsorption. Cu(I)-modified zeolites can efficiently separate CO from other gases by chemical adsorption, but Cu(I) is unstable toward H_2O and O_2. Moreover, the isotherms for chemical adsorption of CO are so steep at the low pressures, which is not beneficial for desorption. In this context, introducing pore-opening flexibility to chemical adsorbents can be a promising strategy.

Chemical adsorption of CO by a flexible MOF was first reported for the Cu(II) active site.[100] [Cu(aip)(H_2O)] (H_2aip = 5-azidoisophthalic acid) is a 2D coordination network consisting of the classic paddle-wheel di-copper units and a bended dicarboxylate ligand. After removal of the coordinated water molecules, the 2D layers interconnect to form a 3D network, accompanying with the reduction of void ratio from 38% to 25%. A single-component CO adsorption isotherm at 120 K showed a pore-expanding action, during which the uptake suddenly jumped from 63 cm^3 (STP) g^{-1} at 5 kPa to 175 cm^3 (STP) g^{-1} at 80 kPa, corresponding to 1.52 and 4.2 CO molecules per Cu_2 unit, respectively. There was an obvious adsorption–desorption hysteresis, and during desorption the closed-loop pressure was 0.1 kPa. Under the same conditions, N_2 showed normal physical adsorption with the uptakes of 50 and 71 cm^3 (STP) g^{-1} at 5 and 80 kPa, respectively (Figure 2.15). The crystal structure of CO-loaded [Cu(aip)] was determined by synchrotron powder diffraction, in which CO coordinated with the Cu(II) centre (Cu–C 2.280 Å; C–O 1.126 Å) and converted the coordination network back to the layered structure. The very low temperature for the pore-expanding action can be attributed to the relatively weak Cu(II)–CO interaction. At room temperature, similar behaviours may be realised by using CO_2 as the adsorbate.

The cooperative spin transition of Fe(II) ions in a seemingly rigid MOF was utilised to realise pore-opening like CO adsorption behaviours.[101] [Fe_2Cl_2(bbta)] (MAF-X26) is a 3D coordination network with large honeycomb-like 1D channels (11 Å).[102] The CO adsorption isotherm at 298 K showed a sharp increase from 0.5 mmol g^{-1} at 0.17 bar to 5.5 mmol g^{-1} at 0.3 bar, corresponding to 0.1 and 0.9 CO per Fe(II), respectively. The adsorption enthalpies at zero coverage and after pore-opening were calculated as 36 and 66 kJ mol^{-1}, corresponding to weak and strong chemical adsorption on high-spin and low-spin Fe(II), respectively. There was a large adsorption–desorption hysteresis, and during desorption the closed-loop pressure was 0.02 bar. Mössbauer spectra showed that there was a high-spin to low-spin transition

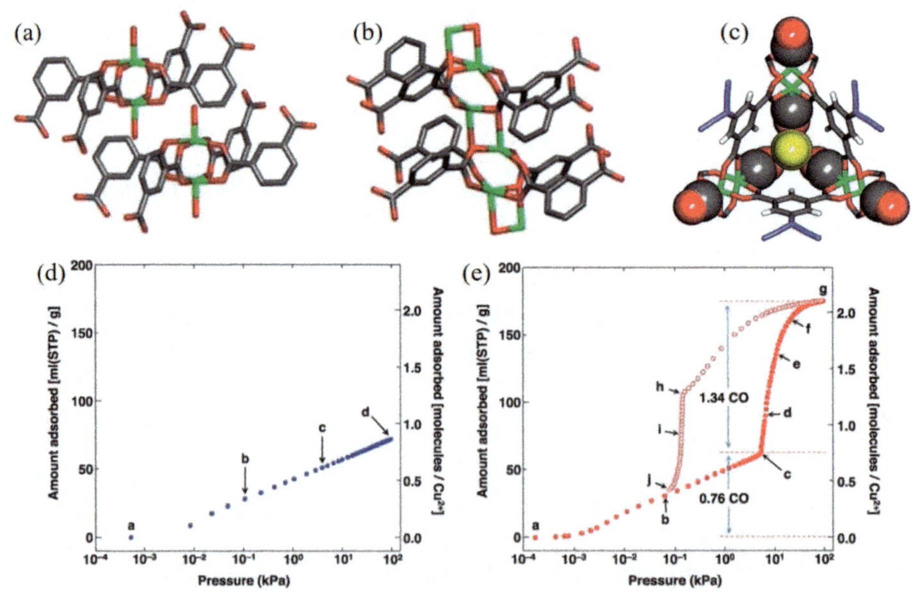

Figure 2.15 CO–N_2 adsorption properties of [Cu(aip)]. (a) Crystal structure of [Cu(aip)(H_2O)]. (b) Crystal structure of [Cu(aip)]. (c) Crystal structure of [Cu(aip)] (stick mode) in CO (space-filling mode). (d) N_2 adsorption isotherm at 120 K. (e) CO adsorption and desorption isotherms at 120 K. Reproduced from ref. 100 with permission from The American Association for the Advancement of Science.

upon CO adsorption. In principle, the spin state of each Fe(II) can be individually controlled by its binding status toward CO. However, according to the adsorption isotherm, the CO-induced spin transitions of the Fe(II) ions should be cooperated in the whole crystallite, which is similar with the typical guest-induced structural transformations of flexible MOFs. In other words, adsorption of a small amount of CO helps all the Fe(II) centres to change their spin state by cooperative structural transformation, which in turn helps the adsorption of additional CO molecules. The CO-bound structure solved by PXRD confirmed the coordination of CO on the Fe(II) centres (Fe–C 1.76 Å; Fe–C–O ~180°). After CO adsorption, the Fe–N and Fe–Cl bond distances decrease from 2.11(2) and 2.385(6) Å to 1.972(5) and 2.259(4) Å, respectively, which are small changes but lead to a large contraction of the unit-cell volume by ~12%. Such significant structural variations are difficult to occur individually in the crystalline lattice. At 298 K and 1 bar, the CO uptake reached 5.9 mmol g^{-1}, while that of N_2 was just ~0.3 mmol g^{-1} (Figure 2.16). Similar CO adsorption behaviours were also observed for the analogue [Fe_2Cl_2(bbta)] consisting of a lengthened ligand.

Using weak chemical adsorption sites, a couple of flexible MOFs have demonstrated pore-opening like CO adsorption behaviours, which may facilitate CO desorption but impede removal/enrichment of low-concentration CO. Quantitative mixture adsorption/separation experiments should be valuable for further optimisation of the CO separation performances.

2.4.4 Flexible MOFs for Adsorptive Separation of Xe–Kr

The noble gases, helium (He, 5.2 ppmv), neon (Ne, 18 ppmv), argon (Ar, 9340 ppmv), krypton (Kr, 1 ppmv) and xenon (Xe, 0.087 ppmv), having widespread applications, are mainly obtained from the air by cryogenic air separation.

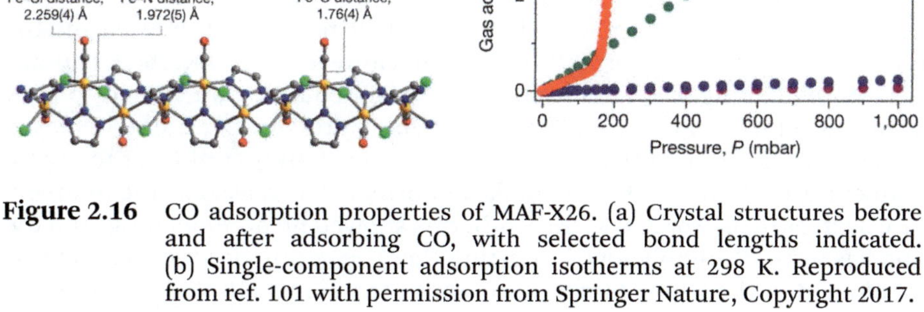

Figure 2.16 CO adsorption properties of MAF-X26. (a) Crystal structures before and after adsorbing CO, with selected bond lengths indicated. (b) Single-component adsorption isotherms at 298 K. Reproduced from ref. 101 with permission from Springer Nature, Copyright 2017.

Adsorptive separation of Xe–Kr is of particular interest, because Xe and Kr are extremely rare and closely related to the nuclear industry. As monoatomic inert molecules, they can only be physically adsorbed. Although OMSs show relatively strong adsorption affinities to Xe, the highest IAST selectivity reported so far is only ~10.[103] In contrast, by decreasing the pore size to fit Xe, the highest IAST selectivity achieved by MOFs can be higher than 100.[104] However, their Xe uptakes are relatively low (<2.7 mmol g^{-1} at 298 K and 0.2 bar) due to the low concentration of OMSs and small pore volume, respectively.[103] With small pores comparable to the guests, MOFs often show framework flexibility. Pore expansion or contraction can be critical for achieving high adsorption Xe affinity/selectivity, but such structural transformations for ultramicroporous materials are usually small and difficult to notice from PXRD.[105]

The interplay between thermodynamic pore-opening/expanding and kinetic gating is noteworthy for Xe–Kr adsorptive separation. While equilibrium adsorption prefers adsorption of the larger Xe molecule, the smaller Kr can be adsorbed in preference by kinetic adsorption. In 2012, the switch between equilibrium and kinetic adsorption of Xe–Kr was observed.[106] The pore aperture size of FMOFCu (3.2×3.5 Å^2) is close to the kinetic diameter of Kr but obviously smaller than that of Xe (Table 2.1). Single-component Xe–Kr isotherms at 203–313 K showed normal shapes without steps or inflection points, ruling out pore-opening behaviour. The Kr uptakes increased normally as the temperature decreased, while those of Xe abnormally decreased, indicating that the gating actions of the apertures were more effective for the larger guests at lower temperatures. The Kr–Xe uptake ratio inversed from 0.45 at 313 K to 36 at 203 K (Figure 2.17), meaning that the adsorption mechanisms were thermodynamic (preferred Xe) and kinetic (preferred Kr), respectively. Note that selective adsorption of Kr over Xe is useful for removal of the low-concentration Kr in the off-gas of used nuclear fuel (40 ppm and 400 ppm for Kr and Xe, respectively). The relatively large amounts of ^{127}Xe with short half-lives (36.3 days) can be released or used, while the small amounts of ^{85}Kr with much longer half-lives (10.8 years) requires long-term storage.

A temperature-dependent switch of MOF flexibility between pore-opening and gating for Xe–Kr adsorption was observed for [Mn(ina)$_2$] (Hina = isonicotinic acid),[107] a 3D coordination network containing 1D channels with size of 3.66 and 4.22 Å at the sp and lp states, respectively. Single-component Xe–Kr adsorption isotherms at 200–298 K showed pore-opening characteristics. As the temperature increased, the Kr uptake normally decreased and the pore-opening pressure normally increased. However, at 200 K, the Kr uptake after pore-opening was abnormally low, which could be attributed to slow diffusion at low temperature. The pore-opening pressure of Xe also increased with the increase of temperature and was lower than that of Kr at the same temperature. However, the Xe uptake abnormally decreased as the temperature decreased. Below 230 K, the isotherms showed no pore-opening, and the uptakes were negligibly low, indicating that the gating or kinetic effect was dominated in the whole temperature range including 298 K. Further, below

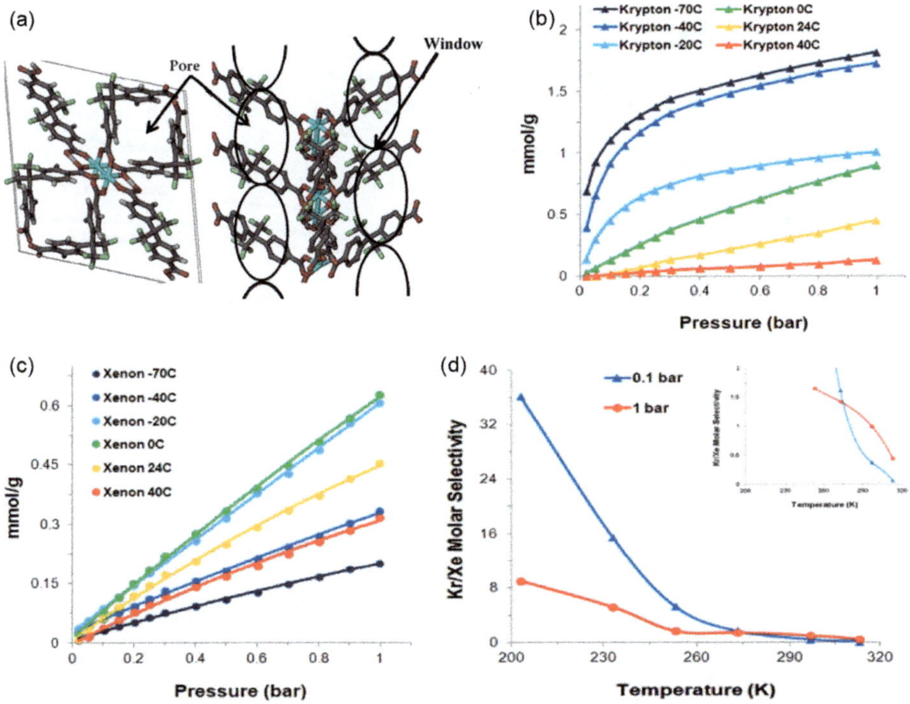

Figure 2.17 Xe–Kr adsorption properties of FMOFCu. (a) Crystal and pore structures. (b) Xe adsorption isotherms. (c) Kr adsorption isotherms. (d) Temperature dependence of Kr–Xe uptake ratio. Reproduced from ref. 106 with permission from American Chemical Society, Copyright 2012.

230 K, there was no pore-opening and no obvious Xe adsorption because the gating aperture became inaccessible. At 230 K and 290 K, the Xe–Kr uptake ratios were ~0.16 and ~10, respectively, meaning that the adsorption process inversed from Kr-selective kinetic adsorption at low temperature to Xe-selective thermodynamic adsorption at high temperature (Figure 2.18). Xe and Kr adsorption enthalpies were calculated as 26.5 and 17.9 kJ mol^{-1}, confirming the thermodynamically dominant separation at high temperature.

Highly effective separation of Xe–Kr might be realised by using the pore-opening action. $[Cu(NbOF_5)(dpa)]$ (ZU-62, dpa = 4,4′-dipyridylacetylene) is an interpenetrated 3D coordination network containing 1D channels with diameter of 3.3 Å, which is obviously smaller than Kr (Table 2.1). Single-component Xe and Kr adsorption isotherms at 273 K showed pore-opening at 0.1 and 0.92 bar, respectively. The Xe and Kr uptakes before pore-opening were 0.25 and 0.4 mmol g^{-1}, respectively. SCXRD showed that after adsorption of Xe, the framework expanded by 4.7%, giving an open channel of 5.0 Å, which was large enough for Xe. Column breakthrough experiment (20 : 80 Xe–Kr at 273 K and 1 bar) showed that, a large amount of high-purity Kr (>99.9%) can be obtained in the adsorption process. This indicates that there

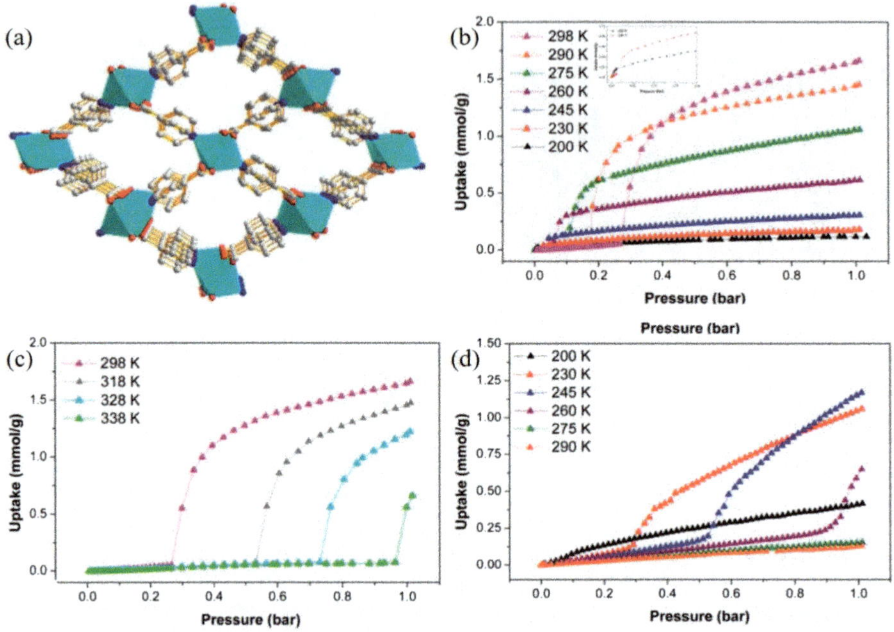

Figure 2.18 Xe–Kr adsorption–separation properties of $[Mn(ina)_2]$. (a) Crystal structure. (b) Xe adsorption isotherms at 200–298 K. (c) Xe adsorption isotherms at 298–338 K. (d) Kr adsorption isotherms at 200–290 K. Reproduced from ref. 107 with permission from American Chemical Society, Copyright 2020.

is low leakage of Xe, meaning that the small amount of Xe adsorption before pore-opening is non-negligible. In addition, large amounts of high-purity Xe (>99.9%) can be obtained during the desorption process carried out by N_2 purge at 298 K, indicating that the co-adsorption of Kr was not significant. In the mixture adsorption/separation process, Kr was considered to be not adsorbed and ascribed to the inability of Kr for pore-opening at this condition.[108] Nevertheless, considering that the opened pore size is larger than Kr, the separation should be dominated by the thermodynamic mechanism.

2.4.5 Flexible MOFs for CO_2 Capture

CO_2 is a major contributor to the greenhouse effect and also an important industrial raw material. Separating CO_2 from syngas/flue gas is of great importance to environment and industry. CO_2 separation processes generally involve the low-boiling-point gases H_2, N_2, CO, O_2 and CH_4. In industry, CO_2 is absorbed chemically by amine solutions, but limited by the relatively low capacity, high energy consumption, and poor stability. Rigid porous materials including MOFs have been widely studied for adsorptive separation of CO_2, and high performances have been achieved by using chemical adsorption of aliphatic amine groups or by using cooperation of multiple weak supramolecular interactions in ultramicropores.[109] In principle, kinetic separation and especially molecular sieving (based on small apertures rather than small pores) can avoid the high energy consumption of the desorption process for chemical adsorption and low adsorption capacity of ultramicropores. Note that thermodynamically CO_2 adsorption selectivity over the relevant low-boiling-point gases is intrinsically high, meaning that assignment of adsorption/separation mechanism, especially for molecular sieving, should be careful.

With low cost and suitable boiling point, CO_2 is one of the most commonly used gases for studying the adsorption and flexibility behaviours of porous materials. Many microporous materials exclude N_2 at 77 K but readily adsorb CO_2 at 195 K, which can be explained by the gating effect dependent on temperature and the host–guest interaction at the gating moment, or simply ascribed to the smaller kinetic diameter of CO_2. Note that the van der Waals sizes of CO_2 are larger than N_2 (Table 2.1). Nevertheless, the gating energy barriers of most MOFs are not high enough to achieve molecular sieving of CO_2 over other gases at ambient temperature.

Post-synthetic modification is a rational strategy for tuning the gating energy barrier, which has been used to achieve CO_2–CH_4 molecular sieving.[110] $[Cu_4(btm)_2]\cdot C_6H_6$ (MAF-42, H_2btm = bis(5-methyl-1*H*-1,2,4-triazol-3-yl)methane) is a 3D coordination framework possessing 1D channels with size of 4.8 × 7.1 to 6.0 × 10.8 Å^2, which shrink to 2.4 × 2.6 to 3.5 × 3.7 Å^2 after removing guests. By virtue of the enzyme-like Cu(I) active sites on the pore surface, the methylene bridges of the btm^{2-} ligands in MAF-42 can be oxidised by air to form the more rigid carbonyl groups. Complete oxidation yielded an isostructural framework $[Cu_4(btk)_2]$ (O100, H_2btk = bis(5-methyl-1,2,4-triazol-3-yl)methanone), which possesses slightly smaller channel size of 2.0 × 4.0 to 3.3 × 4.8 Å^2. At room temperature, MAF-42 showed a CO_2-induced pore-expanding action at 16 bar, and the uptake reached 80 cm^3 (STP) g^{-1} at 30 bar. O100 showed a type-I CO_2 isotherm, with uptake of only 35 cm^3 (STP) g^{-1} at 30 bar, highlighting the reduction of framework flexibility as a result of the ligand oxidation. For CH_4 with lower adsorption enthalpy, MAF-42 showed a type-I isotherm with a high uptake of 24 cm^3 (STP) g^{-1} at 35 bar, while O100 showed a very low uptake of 2.2 cm^3 (STP) g^{-1} at the same condition, indicating that the gating energy barrier after oxidation was too high for CH_4. Mixed-gas adsorption isotherms gave CO_2–CH_4 selectivities of ~20 and ~600 for MAF-42 and O100, which confirmed the molecular sieving effect of the oxidised MOF (Figure 2.19).

The conventional crystal engineering approach can also tune the gating energy barrier, which has been used to achieve molecular sieving separation of CO_2–N_2 and CO_2–CH_4.[33] $[Zn_3(OH)_2(pzdc)(tz)]\cdot$DMA (MAF-91·DMA, H_3pzdc = 3,5-pyrazoledicarboxylic acid, Htz = 1,2,4-triazole, DMA = *N,N*-dimethylacetamide) is a 3D coordination network possessing quasi-discrete/gating pores with cavity size of 4.2 × 7.0 × 8.4 Å^3 and aperture size of 2.1 × 4.0 Å^2. By introducing amino group onto the triazolate ligand, an isostructural MOF

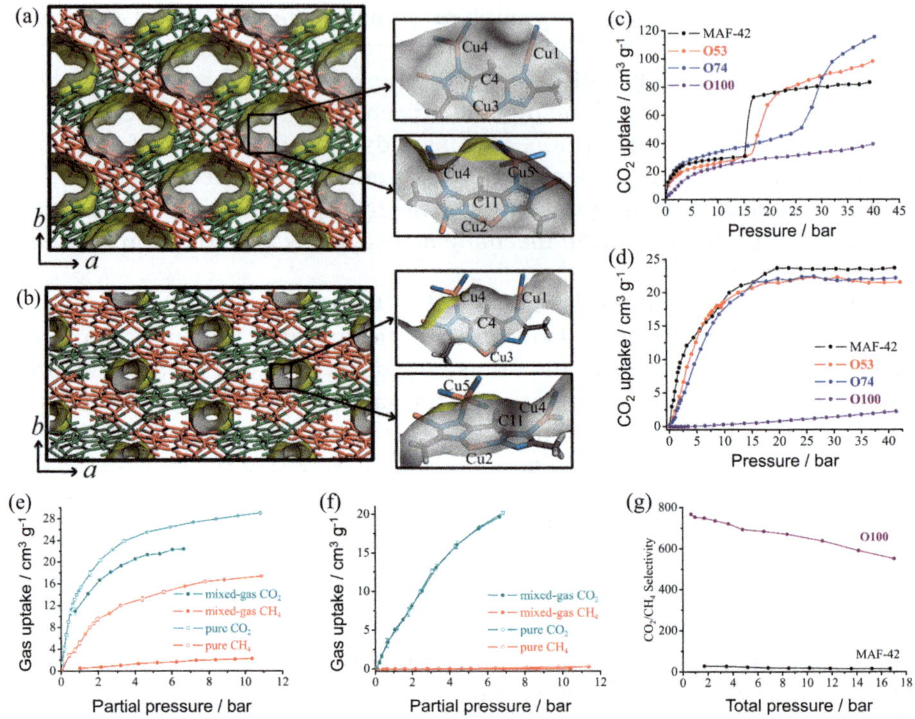

Figure 2.19 CO_2–CH_4 adsorption–separation properties of MAF-42 and its oxidised analogues. (a) and (b) Framework and pore structures of lp (a) and sp (b) phases. (c) and (d) CO_2 (c) and CH_4 (d) adsorption isotherms at 298 K. (e) Mixed-gas adsorption isotherms of MAF-42 for a 40 : 60 CO_2–CH_4 mixture. (f) Mixed-gas adsorption isotherms of O100 for a 40 : 60 CO_2–CH_4 mixture. (g) Experimental CO_2–CH_4 adsorption selectivity. Reproduced from ref. 110 with permission from Springer Nature, Copyright 2015.

$[Zn_3(OH)_2(pzdc)(atz)]$·DMA (MAF-92·DMA, Hatz = 3-amino-1,2,4-triazole) with similar cavity size of 4.4 × 6.8 × 7.8 Å^3 and aperture size of 2.1 × 4.0 Å^2 was also synthesised. Note that their cavities and apertures are larger and smaller than $CO_2/N_2/CH_4$, respectively (Table 2.1). After removing DMA guests, $[Zn_3(OH)_2(pzdc)(tz)]$ (MAF-91) remained almost unchanged but $[Zn_3(OH)_2(pzdc)(atz)]$ (MAF-92) shrunk significantly. There were 1.3% and 18.7% reduction of the unit-cell volumes, and their aperture sizes contracted to 2.0 × 4.4 and 1.0 × 2.5 Å^2, respectively. At low temperatures (77 K and 195 K for N_2 and CO_2, respectively), MAF-91 can readily adsorb N_2–CO_2, while MAF-92 excluded N_2 and adsorbed a small amount of CO_2, indicating negligible and high gating energy barriers, respectively. At room temperature, MAF-91 can strongly adsorb CO_2, N_2 and CH_4, while MAF-92 can normally adsorb CO_2 but showed negligible N_2 and CH_4 uptakes (~0.05 mmol g^{-1} at 1 bar), indicating that the gating energy barrier of MAF-92 was suitable to distinguish CO_2 and N_2–CH_4. At lower temperatures, the N_2 and CH_4 uptakes of MAF-92 did not increase, meaning that their low uptakes were determined by kinetic rather than thermodynamic factors. *In situ* PXRD further confirmed that CO_2 adsorption in the whole pressure range did not involve the pore-opening action. Column breakthrough experiments (10 : 90 CO_2–N_2, 10 : 90 CO_2–CH_4 and 10 : 90 CO_2–He mixtures at room temperature) showed that N_2, CH_4 and He outflowed together, meaning that the N_2–CH_4 adsorption can be ignored just like the non-adsorbing gas He, which confirmed the molecular sieving effect. The N_2 and CH_4 uptakes were quantitatively determined as 0.002(6) and 0.009(10) mmol g^{-1}, respectively (Figure 2.20).[33]

When compared with relevant gases, CO_2 usually shows much lower pore-opening pressures, which is potentially useful for the molecular sieving separation of CO_2, but co-adsorption and leakage may be problems. In 2013, the leakage of CO_2 for pore-opening behaviour was experimentally observed.[111] [Zn(5NO_2-ip)(bpy)] (CID-5, 5NO_2-H_2ip = 5-nitroisophthalic acid) is a 2D coordination network with 1D channels, which transforms into a non-porous structure after removing guests. Single-component CO_2 adsorption isotherms at 273 K showed pore-opening at 1 bar and high uptake of 57 cm^3 (STP) g^{-1} at 10 bar, while CH_4 cannot induce pore-opening, with low

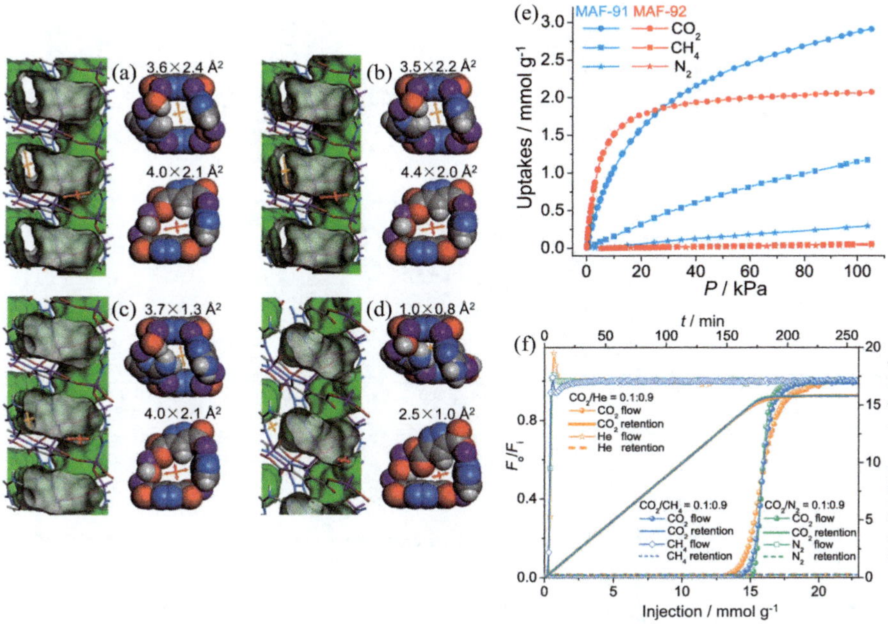

Figure 2.20 CO_2/N_2/CH_4 adsorption–separation properties of MAF-91 and MAF-92. (a)–(d) Framework, pore, and aperture structures of MAF-91·DMA (a), MAF-91 (b), MAF-92·DMA (c) and MAF-92 (d). (e) CO_2, CH_4 and N_2 adsorption isotherms at 298 K. (f) Column breakthrough curves of MAF-92 for 10 : 90 CO_2–N_2, 10 : 90 CO_2–CH_4 and 10 : 90 CO_2–He mixtures at 298 K and 1 bar. Reproduced from ref. 33 with permission from Elsevier, Copyright 2021.

uptake of <5 cm^3 (STP) g^{-1} at 10 bar. Column breakthrough experiment (40 : 60 CO_2–CH_4 mixture at 273 K and 8 bar) showed preferential adsorption of CO_2, but CH_4 and CO_2 flowed out at the same time, and there was a plateau at partial pressure 0.8 bar for the CO_2 breakthrough curve. Later, they improved the CO_2–CH_4 separation performance by rational tuning the framework flexibility.[111] [Zn(5MeO-ip)(bpy)] (CID-6, 5MeO-H_2ip = 5-methoxyisophthalic acid) is isostructural to CID-5, but it can retain the porous structure after removal of guests. CID-6 can readily adsorb CO_2 and CH_4 at 273 K from very low pressures and show high uptakes of 58 and 45 cm^3 (STP) g^{-1} at 10 bar, respectively. A column breakthrough experiment for CID-6 showed that there was no leakage of CO_2, and CO_2 flowed out later than CH_4, but the difference was small, being consistent with the single-component adsorption isotherms. Interestingly, the solid-solution framework [Zn(5NO_2-ip)$_{0.9}$(5MeO-ip)$_{0.1}$(bpy)] (CID-5/6) showed a low CO_2 pore-opening pressure of 0.2 bar. While the CO_2 uptake of CID-5/6 at 10 bar was similar to that of CID-5, its CH_4 uptake (~30 cm^3 (STP) g^{-1}) was in between those of CID-5 and CID-6. The column breakthrough experiment showed significantly improved difference between CO_2 and CH_4 and no leakage of CO_2 (Figure 2.21). Increasing the doping ratio

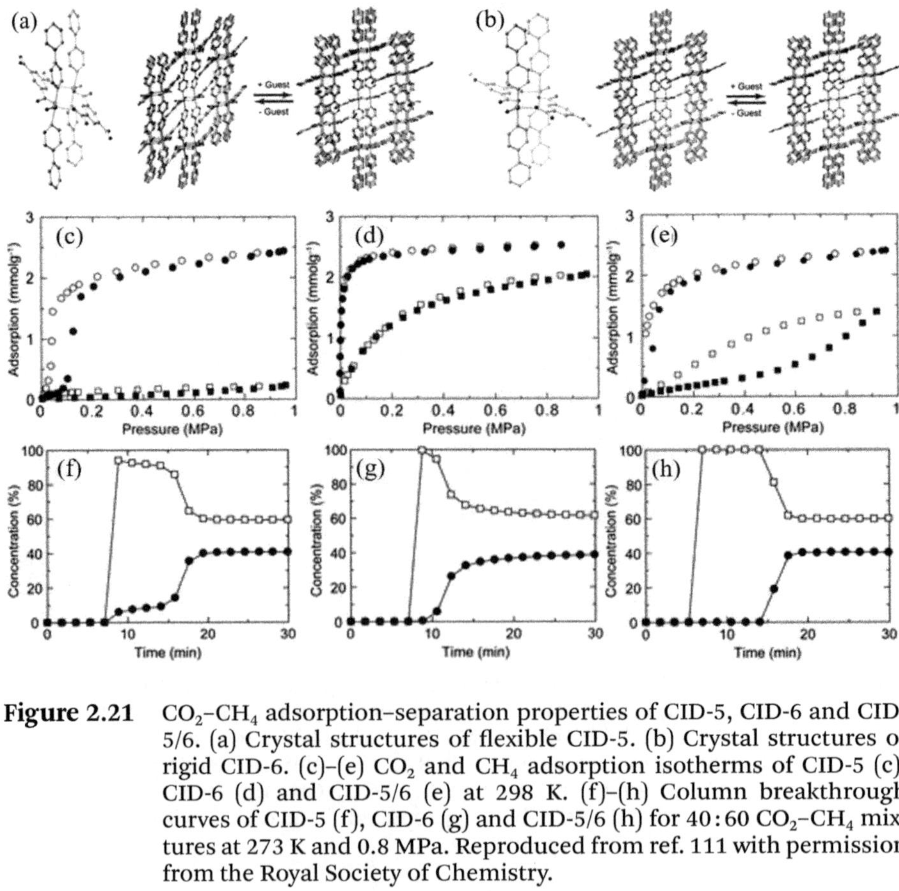

Figure 2.21 CO_2–CH_4 adsorption–separation properties of CID-5, CID-6 and CID-5/6. (a) Crystal structures of flexible CID-5. (b) Crystal structures of rigid CID-6. (c)–(e) CO_2 and CH_4 adsorption isotherms of CID-5 (c), CID-6 (d) and CID-5/6 (e) at 298 K. (f)–(h) Column breakthrough curves of CID-5 (f), CID-6 (g) and CID-5/6 (h) for 40 : 60 CO_2–CH_4 mixtures at 273 K and 0.8 MPa. Reproduced from ref. 111 with permission from the Royal Society of Chemistry.

of 5MeO-ip^{2-} gave lower CO_2–CH_4 separation performances, which can be attributed to the increase of framework rigidity and CH_4 uptake.

The switching between molecular sieving and co-adsorption of CO_2–CH_4 at different gas pressures was observed for [Co(bdp)].[112] A single-component CO_2 adsorption isotherm at 298 K showed multiple pore-opening/pore-expanding at 2, 4.3, 5.9 and 17.2 bar, while CH_4 only showed one-step pore-opening at 18 bar. Equimolar CO_2–CH_4 mixture adsorption experiments showed that, when the total pressure was lower than 25 bar, the CO_2 uptakes were consistent with those of the single-component isotherm, and CH_4 uptakes were negligible, meaning molecular sieving. At a total pressure of 58.6 bar (94 : 6 CH_4–CO_2), the uptakes of CO_2 and CH_4 were 8.5 and 2.1 mmol g^{-1}, respectively, corresponding to a CO_2–CH_4 selectivity of 61, confirming the co-adsorption. *In situ* PXRD patterns in pure CO_2 at 3.6 bar and in 50 : 50 CO_2–CH_4 mixture at 7.2 bar (corresponding to the first CO_2 isotherm step) were almost the same (Figure 2.22). The crystal structure determined by Rietveld refinement exhibited as an intermediate state (3.43 Å) between the non-porous state (~2.6 Å) and the fully opened state (~9 Å), which was larger than CO_2 but smaller than CH_4. Increasing the pressure further resulted in the expansion of the framework to a state similar to or larger than that of the CH_4-saturated structure, which explained the co-adsorption phenomenon.

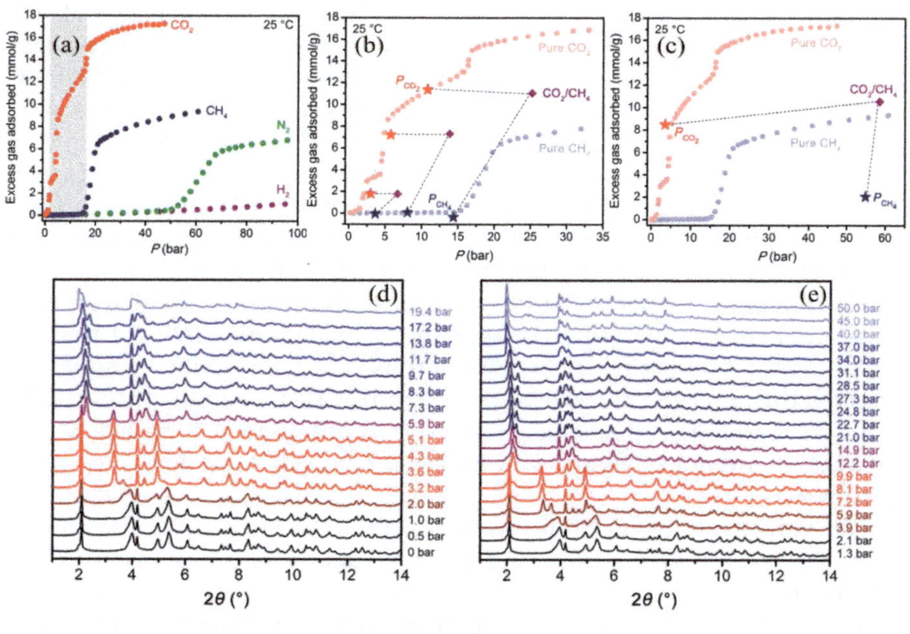

Figure 2.22 CO_2–CH_4 adsorption–separation properties of [Co(bdp)]. (a) Single-component adsorption isotherms. (b) and (c) Multicomponent adsorption experiments for equimolar CO_2–CH_4 mixtures at 6.7, 13.9 and 25.3 bar (b), and a 6 : 94 CO_2–CH_4 atmosphere (c). (d) and (e) PXRD patterns in pure CO_2 (d) and in a 50 : 50 CO_2–CH_4 mixture (e). Reproduced from ref. 112 with permission from American Chemical Society, Copyright 2018.

2.4.6 Flexible MOFs for Adsorptive Separation of C_2 Hydrocarbons

As an important raw material of the petrochemical industry, ethylene/ethene (C_2H_4) is mainly produced by steam cracking and thermal decomposition of naphtha or ethane (C_2H_6), which also yields various by-products. In industry, C_2H_2 and C_2H_6 are separated by an energy-intensive distillation process carried out at low temperature and high pressure.

Hydrophilic porous materials can preferentially adsorb C_2H_4, the most effective mechanism of which is the coordination between OMS and C_2H_4, but the selectivity is not very high because physical adsorption of C_2H_6 is relatively strong.[76] Not only hydrophobic but also hydrophilic adsorbents may show stronger C_2H_6 adsorption than C_2H_4, because C_2H_6 can form more C–H⋯X weak hydrogen bonds with pores with suitable sizes and shapes.[113] Note that selective adsorption of C_2H_6 rather than C_2H_4 is generally more useful, because the C_2H_6 concentration is relatively low in the C_2 stream of cracking gases.

Theoretically, pore-opening can be selectively induced by chemical adsorption of C_2H_4, but so far this phenomenon has not been reported. Based on physical adsorption, examples of C_2H_4/C_2H_6-induced pore-opening examples are also rare.[33,114–118] SOD-[Zn(bim)$_2$] (MAF-3/ZIF-7, Hbim = benzimidazole) is the first MOF with a zeolite topology.[39,119] The first example of alkane/alkene adsorption selectivity was observed by using MAF-3.[118] After guest removal, the effective pore diameter of MAF-3 shrinks from *ca.* 8 Å to 6 Å.[120] Single-component C_2H_4 and C_2H_6 isotherms at 298 K showed pore-opening at 40 and 10 kPa, respectively, indicating a stronger adsorption affinity toward C_2H_6. A column breakthrough experiment (40 : 40 : 20 C_2H_4–C_2H_6–H_2 at 298 K and 1 bar) confirmed C_2H_6–C_2H_4 selectivity but showed poor C_2H_4 purification ability, and co-adsorption of C_2H_4 can be clearly observed in the desorption process. Later, it was observed that the C_2H_6–C_2H_4 separation and C_2H_4 purification performances of MAF-3 were lower than MAF-4, which were both much lower than [Zn(batz)] (MAF-49, Hbatz = bis(5-amino-1*H*-1,2,4-triazole-3-yl)methane) possessing ultramicroporous 1D hydrophilic channels.[113] Although the pore aperture sizes of MAF-3 and MAF-4 are both much smaller than C_2H_4 and C_2H_6, their kinetic adsorption differences in MAF-3, MAF-4 and MAF-49 are not obvious, indicating the relatively high flexibility of the apertures.

Decreasing/avoiding flexibility is useful for achieving molecular sieving effect. [Ca(C_4O_4)(H_2O)] (UTSA-280, $H_2C_4O_4$ = squaric acid) is a 3D coordination network containing two types of 1D channels with aperture sizes of 3.2 × 4.5 and 3.8 × 3.8 Å^2. Considering that the cross-sectional areas of 14.4 Å^2 lie between those of C_2H_4 (13.7 Å^2) and C_2H_6 (15.5 Å^2), UTSA-280 was selected as a potential molecular sieve (Table 2.1).[121] More importantly, the apertures are surrounded by $C_4O_4^{2-}$ linkers, which are already parallel to the channels, meaning that the aperture sizes would not increase due to the rotation of the linkers. Single-component isotherms showed C_2H_4 and C_2H_6 uptakes of 2.5

and 0.098 mmol g^{-1}, respectively, at 298 K and 1 bar. Mixture breakthrough experiments showed that C_2H_6 flowed out quickly similar to the inert gas H_2 and the larger gases C_3H_6 and C_3H_8, while C_2H_4 was retained for a remarkably long time. The adsorbent can be regenerated by applying vacuum at 298 K, during which the purity of C_2H_4 can reach 99% (Figure 2.23). Note that activation of this MOF at higher temperature or for longer times would remove the coordinated water, leading to framework collapse and loss of the separation behaviour.

C_2H_4–C_2H_6 adsorption selectivity in a flexible MOF can be inversed before and after pore-opening.[33] The cavity of MAF-92 ($3.8 \times 4.8 \times 7.5$ Å^3) can accommodate only the small C_2H_4 molecule, while that of MAF-92·DMA ($4.4 \times 6.8 \times 7.8$ Å^3) is also large enough for C_2H_6 (Figure 2.20 and Table 2.1). Single-component C_2H_4 and C_2H_6 uptakes of MAF-92 at 298 K suddenly increased from 0.17 mmol at 26 kPa and 0.017 mmol g^{-1} at 15 kPa, which were identified as pore-expanding and pore-opening by *in situ* PXRD. A column breakthrough experiment (equimolar C_2H_4–C_2H_6 at 298 K and 1 bar) showed that, at the pore-opened/expanded condition, both C_2H_4 and C_2H_6 can be adsorbed, and C_2H_6 was adsorbed stronger with a C_2H_6–C_2H_4 selectivity of 1.9. For the 10:10:80 C_2H_4–C_2H_6–He mixture, which was not able to induce pore-opening/expanding, the equilibrium C_2H_4 and C_2H_6 uptakes were determined as 0.077

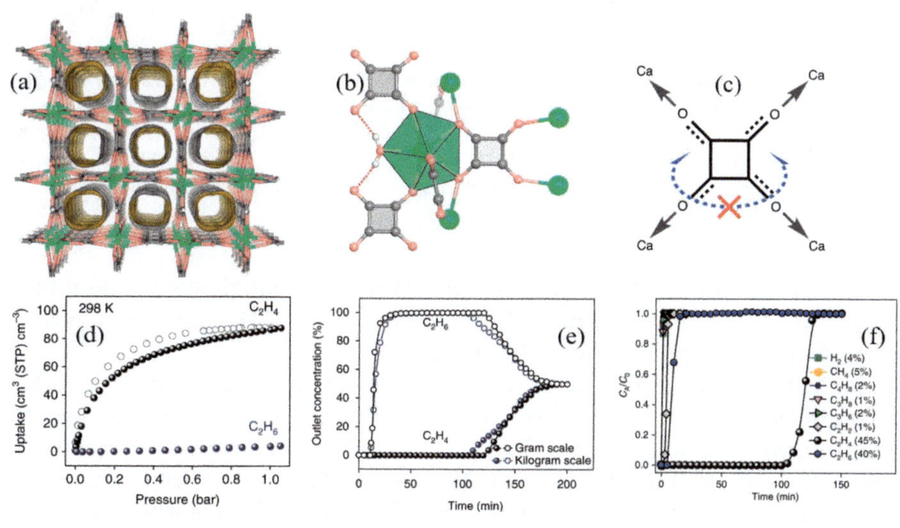

Figure 2.23 C_2H_4–C_2H_6 adsorption–separation properties of UTSA-280. (a) Framework and pore structures of guest-free UTSA-280. (b) The local coordination environments of $C_4O_4^{2-}$ and Ca^{2+}. (c) A schematic diagram showing the coordination constraints applied on the $C_4O_4^{2-}$ linker. (d) Single-component adsorption and desorption isotherms. (e) Column breakthrough curves for a 50:50 C_2H_4–C_2H_6 mixture at 298 K and 1 bar. (f) Multi-component breakthrough curves for a mixture of H_2–CH_4–C_2H_2–C_2H_4–C_2H_6–C_3H_6–C_3H_8–C_4H_8 (4:5:1:45:40:2:1:2) at 298 K and 1 bar. Reproduced from ref. 121 with permission from Springer Nature, Copyright 2018.

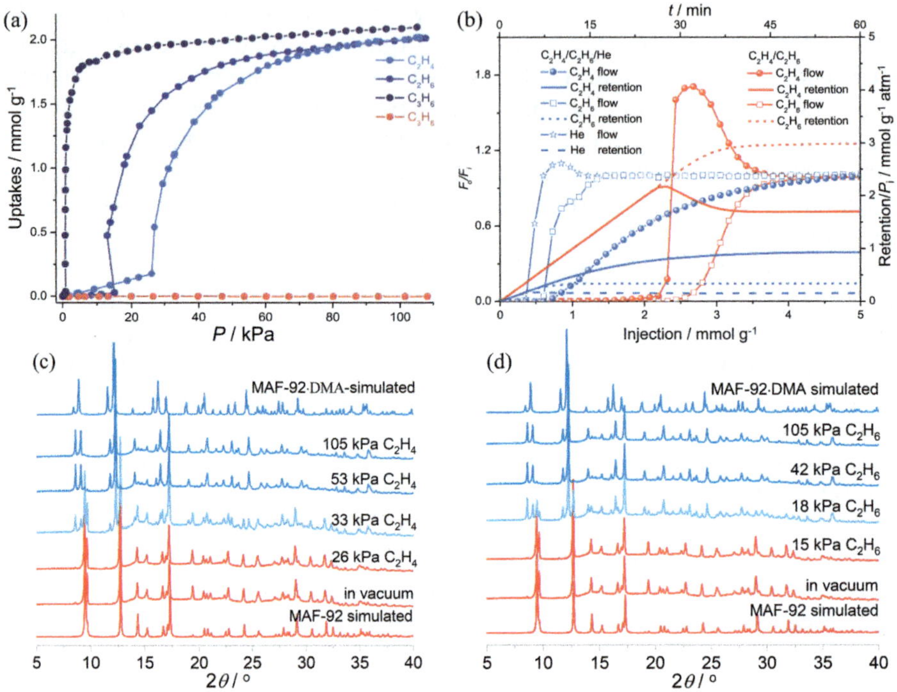

Figure 2.24 C_2H_4–C_2H_6 adsorption–separation properties of MAF-92. (a) Single-component adsorption isotherms at 298 K. (b) Breakthrough curves for 50:50 C_2H_4–C_2H_6 and 10:10:80 C_2H_4–C_2H_6–He mixtures at 298 K and 1 bar. (c) and (d) PXRD patterns in C_2H_4 (c) and in C_2H_6 (d). Reproduced from ref. 33 with permission from Elsevier, Copyright 2021.

and 0.018 mmol g^{-1}, which were consistent with the single-component isotherms (0.073 and 0.017 mmol g^{-1}), indicating C_2H_4–C_2H_6 molecular sieving dominated by the gating effect (Figure 2.24).

The concentration of C_2H_2 impurity is a critical parameter for C_2H_4 polymerisation. In industry, C_2H_2 can be readily removed by catalytic hydrogenation, which also increases the C_2H_4 productivity, but excessive hydrogenation can convert both C_2H_2 and C_2H_4 to C_2H_6. With much higher polarity and hydrogen-bonding ability, C_2H_2 can be selectively adsorbed by rigid microporous materials to achieve high selectivity and capacity. Flexible adsorbents showed similar C_2H_2–C_2H_4 separation performances, but examples are relatively rare.[122,123]

C_2H_2, as a major raw feedstock for various polymers and chemicals, is mainly produced by partial CH_4 combustion and thermal cracking of hydrocarbons and coexists with CO_2. Considering the highly similar physical properties and molecular shapes of C_2H_2 and CO_2 (Table 2.1), adsorptive separation of C_2H_2–CO_2 has attracted a lot of attention. Due to the relatively higher polarity of C_2H_2, rigid porous materials generally show C_2H_2–CO_2

selectivity. CO_2–C_2H_2 adsorption selectivity can be achieved by using CO_2-selective chemisorption sites such as aliphatic amines, or by appropriate electrostatic field of the channel which fits the quadrupole moment of CO_2.[124,125]

For C_2H_2 or CO_2 adsorption, flexible porous materials usually show pore-opening actions at low temperature such as 195 K.[122,126] Depending on the compatibility to the electrostatic field of the opened pores, either CO_2 or C_2H_2 with opposite quadrupole moments may show a lower pore-opening pressure, and sometimes only one of them can induced the structural transformation.[78,127,128]

A cooperative C_2H_2/CO_2-induced pore-opening action was observed for [Cu(L1)$_2$(SiF$_6$)]·guest (NTU-65, L1 = 1,4-di(1*H*-imidazol-1-yl)benzene), a 3D coordination network containing 2D channels with cavity size of 7.4 × 11.5 × 12.2 Å^3.[122] PXRD showed structural transformation after guest removal. At 195 K, *in situ* PXRD/sorption measurements showed that the pore-opening pressure followed $C_2H_2 < CO_2 < C_2H_4$. At 263 K, the pore-opening pressures of C_2H_2, CO_2 and C_2H_4 were 1.2 kPa, 50 kPa and > 100 kPa, respectively. A column breakthrough experiment (90 : 1 : 9 C_2H_4–C_2H_2–CO_2 at 263 K and 1 bar) showed that a large amount of high-purity (>99.99%) C_2H_4 can be obtained in the adsorption process. Considering that the partial pressures of the three gases were all lower than their single-component pore-opening pressures, the efficient adsorption could be attributed to the cooperative pore opening.[92,129]

Efficient separation of C_2H_2 and CO_2–C_2H_4 in a flexible MOF showing pore-opening behaviour was reported for [Zn(dps)$_2$(SiF$_6$)] (UTSA-300, dps = 4,4′-dipyridylsulfide), a complicated 2D coordination network with pores inside and between the layers.[126] SCXRD showed that, after removing guests, there were only discrete pores of 2.6 × 3.9 × 4.3 Å^3 inside the layers, which were significantly smaller than the molecular sizes of C_2H_2/CO_2/C_2H_4. Single-component isotherms at 195 and 273 K showed that both C_2H_2 and CO_2 can induce pore-opening, but C_2H_4 cannot. At 273 K, the C_2H_2 uptake abruptly jumped from 3.6 cm^3 (STP) g^{-1} at ~0.07 bar to 70 cm^3 (STP) g^{-1} at 0.32 bar, and finally reached 76.4 cm^3 (STP) g^{-1} at 1 bar, while the CO_2 and C_2H_4 uptakes were only 4.1 and 0.92 cm^3 (STP) g^{-1} at 1 bar, respectively (Figure 2.25). Neutron powder diffraction showed that the host returned to the as-synthesised state after adsorbing C_2D_2. Column breakthrough experiments (equimolar mixture at 298 K and 1 bar) showed long retention times for C_2H_2, giving large amounts of CO_2 or C_2H_4 with purity >99.9%, which indicated very low C_2H_2 leakage, meaning that the small uptake before formal pore opening cannot be ignored.

Thermodynamically/kinetically different structural responses of a flexible MOF towards adsorption of CO_2 and C_2H_2 were observed for [Mn(bdc)(dpe)]·DMF (dpe = 1,2-di(4-pyridyl)ethylene), a two-fold interpenetrated **pcu** network possessing 1D channels with cavities of 4.2 × 6.5 × 7.1 Å^3 and apertures of 3.8 Å, which shrink to 3.3 × 3.5 × 5.6 Å^3 and 2.8 Å after guest removal.[127] At 195 K, the C_2H_2 isotherm showed a pore-opening behaviour, during which the uptake abruptly jumps from 6.7 cm^3 (STP) g^{-1} at 1.45 kPa to 80.0 cm^3 (STP) g^{-1} at 3 kPa, and reached 86.2 cm^3 (STP) g^{-1} at 100 kPa. At the same temperature, the CO_2 isotherm showed the type-I characteristic, and

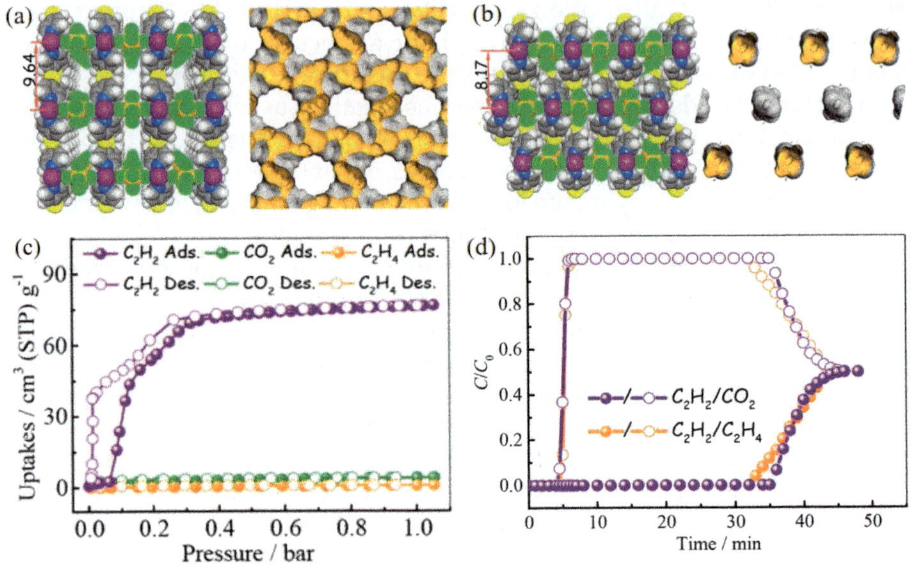

Figure 2.25 $C_2H_2/CO_2/C_2H_4$ adsorption–separation properties of UTSA-300. (a) and (b) Framework and pore structures before (a) and after (b) removing guests. (c) Single-component adsorption and desorption isotherms. (d) Breakthrough curves for equimolar mixtures at 298 K and 1 bar. Reproduced from ref. 126 with permission from American Chemical Society, Copyright 2017.

the CO_2 uptakes at all pressures were higher than those of C_2H_2. *In situ* PXRD showed that after adsorbing CO_2 or C_2H_2, the host experienced small or large expansion, respectively. Obviously, the shrunk cavities are too small for C_2H_2 but large enough for CO_2, while the opened cavities are large enough for both gases (Table 2.1). Therefore, only C_2H_2 need pore-opening for adsorption. Nevertheless, the ultra-small aperture size of the guest-free MOF should show gating actions to allow CO_2 adsorption. At 273 K, the pore-opening pressure of C_2H_2 was higher than 1 bar, so that the C_2H_2 uptakes below 1 bar were all significantly lower than those of CO_2 (Figure 2.26). The batch adsorption experiment (50 : 50 CO_2–C_2H_2 mixture at 273 K and 100 kPa) showed a selectivity of 5.7, which was slightly lower than the isotherm uptake ratio of 6.4.

2.4.7 Flexible MOFs for Adsorptive Separation of C_3 Hydrocarbons

Propylene/propene (C_3H_6) is used and produced in similar way to C_2H_4. High-pressure distillation separation of C_3H_6 and propane (C_3H_8) is more energy-intensive than for C_2H_4 and C_2H_6. OMSs also play similar roles for adsorptive separation of C_3H_6 and C_3H_8, but they are less effective because of the stronger physical adsorption of C_3H_8.

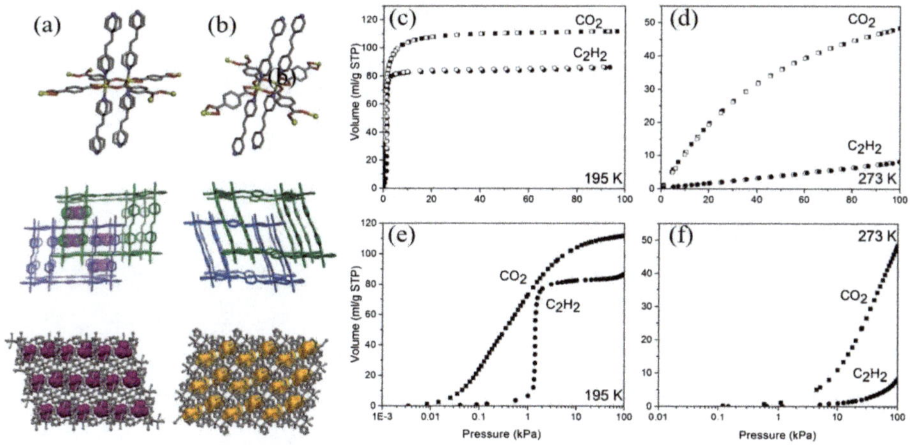

Figure 2.26 CO_2–C_2H_2 adsorption–separation properties of [Mn(bdc)(dpe)]. (a) Framework and pore structures of [Mn(bdc)(dpe)]·DMF. (b) Framework and pore structures of [Mn(bdc)(dpe)]. (c)–(f) Single-component adsorption and desorption isotherms. Reproduced from ref. 127 with permission from American Chemical Society, Copyright 2016.

Contrary to the trend of C_2H_4 and C_2H_6, examples of preferential C_3H_8-induced pore-opening are rare.[118] In most cases, C_3H_6 shows stronger host–guest interactions and lower pore-opening pressures than C_3H_8.[33,130,131] This means the size/shape differences between C_3H_6 and C_3H_8 are smaller than those between C_2H_4 and C_2H_6, so that polarity plays a greater role. Because alkanes are larger than the corresponding alkenes, in principle, co-adsorption of alkanes could be avoided after pore-opening induced by the corresponding alkenes, if the sizes of the opened pores were suitable (Figure 2.10). Nevertheless, the size differences between C_2/C_3 alkanes and C_2/C_3 alkenes are very small, as indicated by the difficulty in achieving molecular sieving of these molecules. According to our knowledge, all pore-opening adsorbents showed leakage and co-adsorption for C_3H_6/C_3H_8 adsorptive separation.[33,115,130,131] For example, single-component C_3H_6–C_3H_8 adsorption isotherms of MAF-92 at 298 K showed sharp difference like molecular sieves (Figure 2.24).[33] The C_3H_6 and C_3H_8 uptakes at 1 bar were 2.0 and 0.001 mmol g^{-1}, respectively. However, the seemingly type-I C_3H_6 isotherm actually showed pore-opening at 1 kPa. Single-component breakthrough experiments showed that C_3H_8 behaved like the non-adsorbing gas He, and the equilibrium C_3H_6 uptake was also consistent with the isotherm. However, 1 : 1 C_3H_6–C_3H_8 breakthrough experiments showed that C_3H_6 and C_3H_8 flowed out simultaneously with poor separation performance. Moreover, the abnormally dropped portion of the C_3H_8 curve indicated that C_3H_8 can be adsorbed after C_3H_6-induced pore-opening. The equilibrium C_3H_6 and C_3H_8 uptakes were determined as 1.35 and 0.42 mmol g^{-1}, respectively, which quantitatively illustrated the co-adsorption behaviour (Figure 2.27).

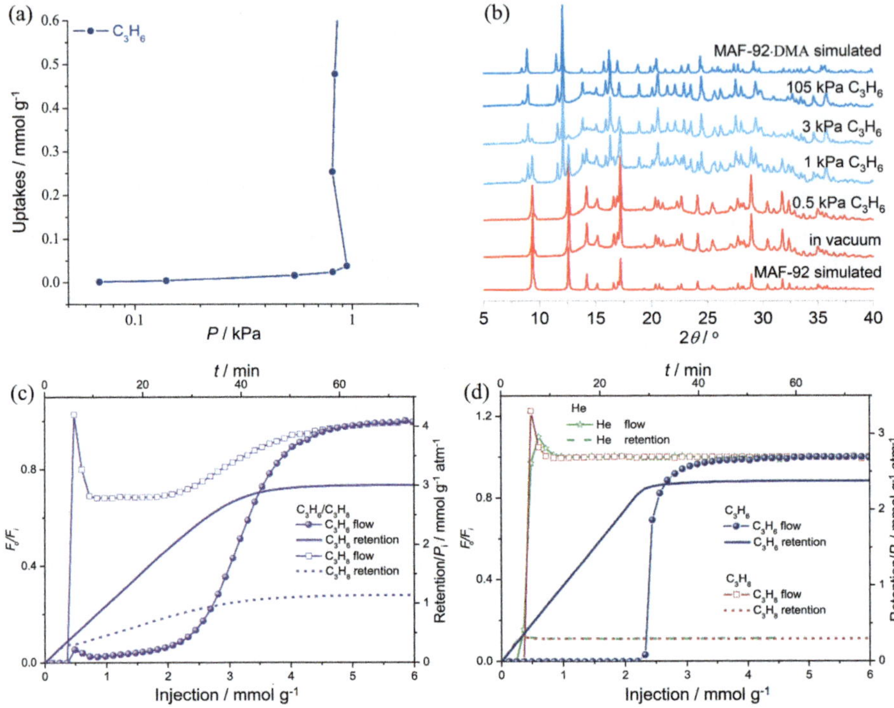

Figure 2.27 C_3H_6–C_3H_8 separation properties of MAF-92. (a) Single-component C_3H_6 adsorption isotherm in the low-pressure region. (b) PXRD patterns in C_3H_6. (c) Breakthrough curves for a 50 : 50 C_3H_6–C_3H_8 mixture at 298 K and 1 bar. (d) Breakthrough curves for pure gases at 298 K and 1 bar. Reproduced from ref. 33 with permission from Elsevier, Copyright 2021.

The pore-opening and co-adsorption behaviours can be explained by comparison of the molecular sizes of C_3H_6 and C_3H_8 (Table 2.1) with the shrunk and expanded cavities (Figure 2.20).

Pore-opening and pore-expanding are sometimes difficult to distinguish. Because the sp state can readily adsorb guests, pore-expanding actions can avoid the leakage problem of pore-opening actions. The different effects of pore-opening and pore-expanding on the adsorptive separation of C_3H_6 and C_3H_8 have been demonstrated by two isostructural ultramicroporous MOFs with small difference of flexibility.[130] [Mn(mpba)$_2$]·guest (MCF-56, Hmpba = 3-(3-methylpyridin-4-yl)benzoic acid) and [Co(mpba)$_2$]·guest (MCF-57) possess very similar 1D channels with effective sizes of 4.8 × 7.5 and 4.6 × 7.4 Å^2, which shrink differently to 3.0 × 3.1 and 3.9 × 5.4 Å^2 after removing guests, respectively. GCMC simulations showed that, among the four pore sizes, only the smallest one cannot accommodate C_3H_6 and C_3H_8. Single-component C_3H_6–C_3H_8 isotherms at 298 K showed pore-opening and pore-expanding characteristics for MCF-56 and MCF-57, respectively. The pore-opening/expanding pressures of C_3H_6 were lower than those of C_3H_8, but the pore-opening adsorbent MCF-56 showed obviously larger isotherm differences. Column breakthrough experiments (1:1:2 C_3H_6–C_3H_8–He mixture at 298 K and 1 bar) showed that, for MCF-56, C_3H_6, C_3H_8 and He flowed out together with small difference on the breakthrough curves. Moreover, before reaching equilibrium (breakthrough curves stabilise at $F_o/F_i = 1$), the C_3H_6 and C_3H_8 breakthrough curves showed long plateaus with heights (flow rates) consistent with the values predicted from their pore-opening pressures. For MCF-57, C_3H_8 flowed out later than He, and C_3H_6 flowed out much later than C_3H_8, indicating that there was no leakage like MCF-56 did. The equilibrium C_3H_6–C_3H_8 adsorption selectivity for MCF-56 and MCF-57 were determined as 1.64 and 1.81, respectively (Figure 2.28).

Efficient separation of C_3H_6 and C_3H_8 was recently reported for a flexible MOF showing pore-expanding behaviour. [Co(mptbdc)] (JNU-3, H_2mptbdc = 5-(3-methyl-5-(pyridin-4-yl)-4*H*-1,2,4-triazol-4-yl)-1,3-benzenedicarboxylic acid) possesses 1D channels (4.5 × 5.3 Å^2) decorating orthogonally arranged pocket-like pores (pocket aperture size: 3.7 Å, pocket size 4.14 × 4.38 × 6.33 Å^3).[132] The molecular sizes of C_3H_6 and C_3H_8 were considered to be smaller than the 1D channel but larger than the pocket-like pores, respectively. Single-component C_3H_6–C_3H_8 sorption isotherms at 273–323 K showed typical pore-expanding-like behaviours, indicating that the guest molecules are first adsorbed in the 1D channel before structural transformation and then in the pocket-like pores after structural transformation. Due to the higher adsorption enthalpy, the pore-expanding pressures of C_3H_6 (29.3 kJ mol^{-1}) are obviously lower than those of C_3H_8 (16.1 kJ mol^{-1}), giving rise to large uptake differences at particular temperature and pressure ranges. C_3H_6/C_3H_8-loaded

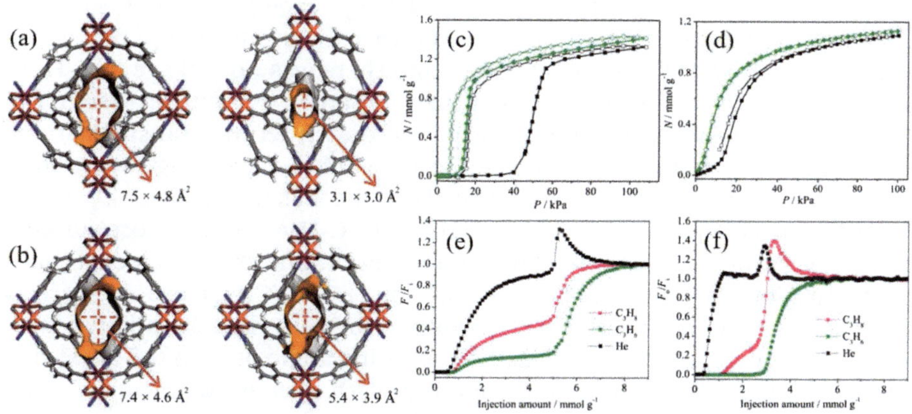

Figure 2.28 C_3H_6–C_3H_8 adsorption–separation properties of MCF-56 and MCF-57. (a) and (b) Framework and pore structures of MCF-56 (a) and MCF-57 (b) before (left) and after (right) removing guests. (c) and (d) Single-component C_3H_6 (green) and C_3H_8 (black) sorption isotherms of MCF-56 (c) and MCF-57 (d) at 298 K. (e) and (f) Breakthrough curves of MCF-56 (e) and MCF-57 (f) for a 1:1:2 C_3H_6–C_3H_8–He mixture at 298 K and 1 bar. Reproduced from ref. 130 with permission from American Chemical Society, Copyright 2020.

crystal structures showed that both C_3H_6 and C_3H_8 can be adsorbed in the expanded pocket-like pores (size: 4.48 × 5.38 × 7.17 Å^3). Column breakthrough experiments (1:1 C_3H_6–C_3H_8 mixture at 303 K and 1 bar) showed that C_3H_8 flowed out much earlier than C_3H_6, giving the C_3H_8 with purity of 99.99% in the adsorption process. The adsorbed gases can be obtained by He sweeping, during which C_3H_8 can be considered to be almost completely eluted in ~41 min, large amount of C_3H_6 with purity of >99.5% can be collected in the next ~39 min, and finally complete desorption/activation can be achieved in ~33 min (C_3H_6 purity >95%). Furthermore, JNU-3 can maintain performance even after 50 adsorption–desorption cycles and work under humid conditions. The equilibrium C_3H_6–C_3H_8 uptakes were calculated as 54.9 and 18.7 cm^3 (STP) g^{-1}, corresponding to a selectivity of 2.9. For comparison, column breakthrough experiments for [Ni(NbOF$_5$)(pz)] (KAUST-7, pz = pyrazine) and ($H_2N(CH_3)_2)_2$[$Y_6(OH)_8$(abtc)$_3(H_2O)_6$] (Y-abtc, H_4abtc = 3,3′,5,5′-azobenzene-tetracarboxylic acid) were carried out under the same conditions, which gave highest C_3H_6 purities of *ca.* 90% and C_3H_6–C_3H_8 selectivities of 2.3 and 2.7, respectively (Figure 2.29).

For pore-expanding adsorbents, the adsorption–separation performance can be sensitive to the operation condition that determines the pore sizes. Temperature-dependent C_3H_6–C_3H_8 adsorption–separation properties were observed for a classic flexible ultramicroporous MOF [Cu_2(pzdc)$_2$(pz)] (CPL-1, H_2pzdc = pyrazine-2,3-dicarboxylic acid).[133] Single-component adsorption isotherms at 273, 278 and 283 K showed abrupt increase of C_3H_6 uptake from ~0.4 mmol g^{-1} at 40, 45 and 65 kPa to 1.8, 1.6 and 1.4 mmol g^{-1} at 100 kPa, respectively, which were assigned as pore-opening transitions. In contrast, C_3H_8 cannot induce the pore-opening transition, and the uptakes were just 0.2–0.3 mmol g^{-1} at 1 bar. Column breakthrough experiments (50:50 C_3H_6–C_3H_8 at 273 and 298 K and 1 bar) showed that the separation performance at 273 K was higher than that at 298 K, and there was no leakage of C_3H_6 at the two conditions. Considering that the C_3H_6 and C_3H_8 uptakes were quite high before the isotherm inflection points, the structural transformations in the adsorption processes were more likely pore-expanding rather than pore-opening, which should be critical about the leakage of C_3H_6.

With molecular sizes larger than C_2H_4/C_2H_6, C_3H_6/C_3H_8 are more suitable to utilise gating apertures for kinetic adsorption and even molecular sieving separation.[134–137] So far, molecular sieving of C_3H_6/C_3H_8 has been scarcely reported.[135,136] As a representative example, KAUST-7 is a 3D pillared-layer structure with 1D square channels.[135] The tilting angle of pz in the guest-free framework is 29.2°, giving a small aperture with effective diameter of 3.047 Å. Considering the rotation of pz to a tilting angle of 0°, the theoretically maximal aperture size was claimed as 4.752 Å, which lies between the kinetical diameters of C_3H_6 and C_3H_8 (Table 2.1). Single-component isotherms at 298 K showed that the C_3H_6 and C_3H_8 uptakes at 1 bar were 58 and 3 mg g^{-1}, respectively, and there was no sign of structural transformation. The total gas uptakes for a 50:50 C_3H_6–C_3H_8 mixture adsorption isotherm almost overlapped with those of the single-component C_3H_6 isotherm, which was

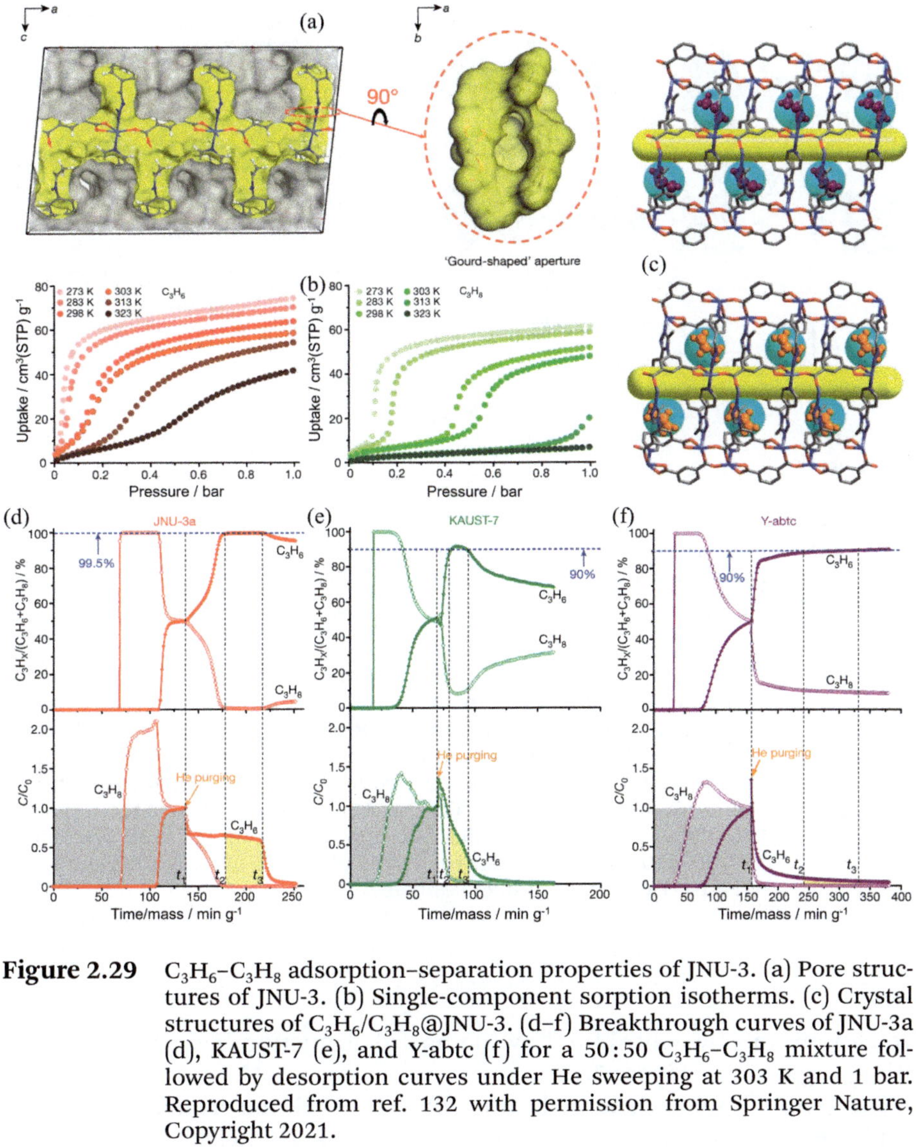

Figure 2.29 C_3H_6–C_3H_8 adsorption–separation properties of JNU-3. (a) Pore structures of JNU-3. (b) Single-component sorption isotherms. (c) Crystal structures of C_3H_6/C_3H_8@JNU-3. (d–f) Breakthrough curves of JNU-3a (d), KAUST-7 (e), and Y-abtc (f) for a 50:50 C_3H_6–C_3H_8 mixture followed by desorption curves under He sweeping at 303 K and 1 bar. Reproduced from ref. 132 with permission from Springer Nature, Copyright 2021.

considered as a sign of no co-adsorption of C_3H_8. Column breakthrough experiment (50:50 C_3H_6–C_3H_8 at 1 bar) showed large difference between the two gases, which further confirmed the effective separation. Later, the structural dynamism of these host-guest systems was studied by DFT simulations.[138] The results showed that changing the pz tilting angle by 9.4° and 15.3° can increase the effective aperture diameter from 2.72 Å to 2.93 Å and 3.07 Å to allow diffusion of C_3H_6 and C_3H_8, respectively. The gating energy barriers for C_3H_6 and C_3H_8 were calculated as 18 and 44 kJ mol^{-1}, respectively,

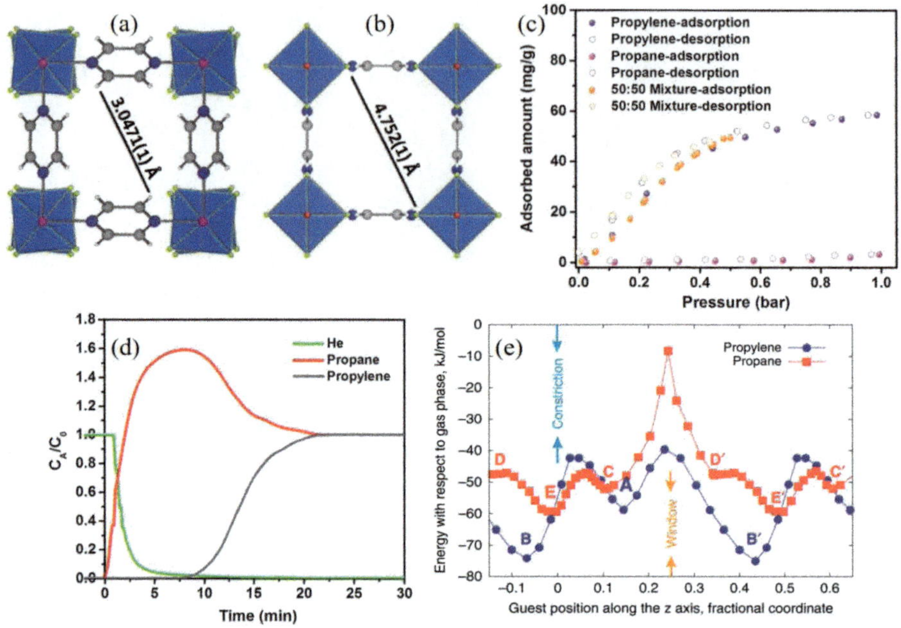

Figure 2.30 C_3H_6–C_3H_8 adsorption–separation properties of KAUST-7. (a) Crystal structure at 100 K. (b) Simulation of the maximum open structure. (c) Single-component and mixture adsorption/desorption isotherms at 298 K. (d) Breakthrough curves for a 50:50 C_3H_6–C_3H_8 mixture at 298 K and 1 bar. Reproduced from ref. 135 with permission from The American Association for the Advancement of Science. (e) The simulated diffusion profiles. Reproduced from ref. 138, 10.1038/s41467-020-19207-9, under the terms of the CC BY 4.0 license http://creativecommons.org/licenses/by/4.0/.

which were mostly contributed by the host deformation (*i.e.*, changing the titling angle of pz) (Figure 2.30).

The gating/diffusion energy barrier for C_3H_6–C_3H_8 adsorptive separation can be reasonably regulated by various ways. In 2009, efficient separation of C_3H_6–C_3H_8 by the kinetic adsorption mechanism was first demonstrated by using MAF-4.[134] The aperture size of MAF-4 is obviously smaller than C_3H_6 and C_3H_8. The single-component C_3H_6–C_3H_8 adsorption isotherms measured at 303–393 K were quite similar and show no sign of structural transformations. MAF-4 also adsorbed C_3H_8 slightly stronger than C_3H_6. However, the C_3H_6 adsorption rates were obviously higher than those of C_3H_8, giving a ratio of C_3H_6–C_3H_8 diffusion rate coefficients of 125 at 303 K. As an analogue of MAF-4 with a slightly larger aperture size by *ca.* 0.1 Å, [Zn(2-cim)$_2$] (2-cimH = 2-chloroimidazole) showed a much lower ratio of C_3H_6–C_3H_8 diffusion rate coefficients of 60 at the same condition.

[Al(OH)(1,3-bdc)] (CAU-10, 1,3-H_2bdc = 1,3-benzenedicarboxylic acid) is a rigid 3D structure possessing 1D channels.[139] Flexible $-OCH_3$ ligand side groups were introduced in CAU-10 to construct 1D gating channels with crystallographic aperture size of 2.5 Å.[140] Single-component isotherms of various gases showed that the effective aperture size can gradually increase from 3.6 Å at 87 K to 5.2 Å at 343 K, meanwhile, no sign of structural transformations was observed. The maximum single-component C_3H_6 and C_3H_8 uptakes appeared at 273 and 323 K, respectively, being consistent with their size difference. The C_3H_6 and C_3H_8 uptakes at 273 and 298 K and 1 bar were 32 and 28 cm^3 (STP) g^{-1} and 6 and 7 cm^3 (STP) g^{-1}, respectively. A column breakthrough experiment (equimolar C_3H_6–C_3H_8 at 298 K and 1 bar) confirmed the preferential adsorption of C_3H_6.

The gating energy barrier can be rationally increased by post-synthetic modification to achieve efficient adsorption separation of C_3H_6–C_3H_8.[137] [Zn_2(btm)$_2$] (MAF-23, H_2btm = bis(5-methyl-1*H*-1,2,4-triazol-3-yl)methane) is a flexible 3D coordination network containing 1D channels with aperture size of 3.6 Å.[141] It can show adaptive structural transformation during CO_2 adsorption, and can adsorb C_2/C_3/C_4 hydrocarbons larger than its aperture very fast,[142] indicating high flexibility of the aperture. Interestingly, heating MAF-23 in oxygen flow yielded [Zn_2(btm)(btk)] (MAF-23-O). The selective oxidation of half of the ligands was confirmed by nuclear magnetic resonance spectroscopy (NMR) and SCXRD. Although the pore shape and size were basically unchanged, the oxidised ligand was more rigid and hydrophilic, which can increase the gating energy barrier and host–guest recognition ability.[110] Single-component C_3H_6–C_3H_8 isotherms of MAF-23-O at ambient temperatures showed slightly larger uptake differences than for MAF-23. More importantly, MAF-23-O showed obviously slower adsorption rates, especially for C_3H_8. Single-component adsorption kinetics measurements confirmed that the diffusion rates of C_3H_6–C_3H_8 were significantly different, and the kinetic selectivity was calculated as 71. Quantitative breakthrough experiments (1 : 1 C_3H_6–C_3H_8 at 298 K and 1 bar) gave C_3H_6–C_3H_8 selectivities of 1.45 and 15 for MAF-23 and MAF-23-O, respectively (Figure 2.31). Under the same conditions, the C_3H_6–C_3H_8 selectivities of [Co_2(dobdc)] (Co-MOF-74, H_4dobdc = 2,5-dioxido-1,4-benzenedicarboxylic acid) and KAUST-7 were determined as 6.48 and 12, respectively. The MAF-23-O column loaded with C_3H_6/C_3H_8 adsorbed can be regenerated by He sweeping and heating to 373 K, during which C_3H_8 completely eluted in ~40 min and high-purity C_3H_6 (>99.8%) was collected in the next ~140 min. When MAF-23 and MAF-23-O were set as flexible, molecular dynamics (MD) simulations gave C_3H_6–C_3H_8 diffusion coefficients of 3.3 and 2.8, and 0.82 and 0.0073 × 10^{-10} m^2 s^{-1}, respectively. Assuming that the frameworks were rigid, no diffusion of C_3H_6–C_3H_8 could be observed. MD-PDFT simulations showed that, at the gating moment for C_3H_6 diffusion, the host distortion of MAF-23 and MAF-23-O were similar, but the energy increase of the latter was much higher, which was consistent with the increased rigidity of the oxidised ligand.

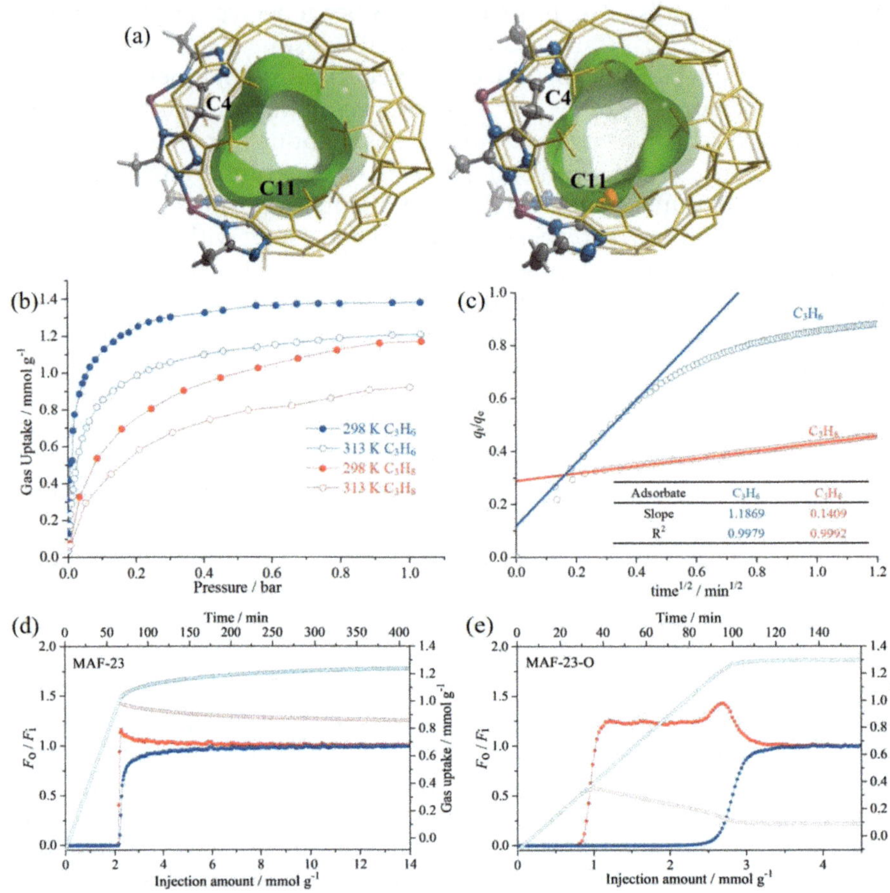

Figure 2.31 C_3H_6–C_3H_8 adsorption–separation properties of MAF-23 and MAF-23-O. (a) Framework and pore structures of MAF-23 (left) and MAF-23-O (right). (b) Single-component sorption isotherms of MAF-23-O. (c) Adsorption kinetics profiles (point) and linear fittings (line) of MAF-23-O. (d) and (e) Breakthrough curves for a 50:50 C_3H_6–C_3H_8 (blue/red) mixture at 298 K and 1 bar. Reproduced from ref. 137 with permission from John Wiley & Sons, Copyright 2019.

2.4.8 Flexible MOFs for Adsorptive Separation of C_4 Hydrocarbons

With the increase in the number of carbon atoms, the structural complexity of hydrocarbons increases exponentially, and it is more and more difficult to separate their homologues. For example, the common C_4 hydrocarbons, 1,3-butadiene, *n*-butene, *iso*-butene, *n*-butane, *iso*-butane and 2-butenes coexist in the C_4 streams of cracking gas. In the petrochemical industry, butadiene, *n*-butene and *iso*-butene, as important raw materials, are separated by complicated distillation and/or chemical conversion processes. The other remaining C_4 hydrocarbons are even more difficult to separate and are usually directly used as fuels.

Having suitable boiling points close to room temperature, butanes are suitable to serve as the guest for studying the adsorption and flexibility of adsorbents. Nevertheless, when compared with C_2 and C_3 hydrocarbons, C_4 hydrocarbons have received less attention for separation probably because of the complexity and difficulty. Among the common C_4 hydrocarbons, butadiene is generally the most easily adsorbed one for its high polarity, small molecular size, and presence of more coordinative groups.[142–152]

In 2016, flexible MOFs were first employed for adsorptive separation of C_4 hydrocarbons.[143] [Zn(5NO$_2$-ip)(dpe)] (SD-65) is a three-fold interpenetrated **dmp** coordination network containing quasi-discrete pores with size of 6.0 × 2.8 × 2.8 Å^3, which is too small for any C_4 hydrocarbons. Single-component adsorption isotherms at 298 K showed that butadiene can induce pore-opening at 50 kPa, and the adsorption amount at 101 kPa can reach 40 cm^3 (STP) g^{-1}. In contrast, other homologues showed negligible adsorption amounts less than 2.5 cm^3 (STP) g^{-1} at 101 kPa. Column breakthrough curves for a mixture of butadiene, butenes, and butanes showed a butadiene selectivity of 6.6 with obvious co-adsorption of butenes/butanes and leaking of butadiene. *In situ* PXRD/adsorption measurements showed that the host completely transformed to an expanded structure in butadiene at 65 kPa, at which the adsorption amount was 27 cm^3 (STP) g^{-1}, meaning that there are some opened yet unloaded pores for adsorption of other guests (Figure 2.32).

For large molecules such as C_4 hydrocarbons, molecular flexibility may play a role in the adsorption–separation process.[34,142,153–155] The guest conformations can be controlled by quasi-discrete pores and the separation can take advantages from the gating apertures.[142] The energy increase from the *trans*-conformation to the *cis*-conformation requires 14.1 kJ mol^{-1} for butadiene, which is much higher than those of *n*-butene (2.4 kJ mol^{-1}) and *n*-butane (3.5 kJ mol^{-1}). The energy penalties of conformation changes of C_4 hydrocarbons may be used to lower the adsorption affinity of butadiene, which is beneficial for purification of butadiene from the C_4 stream. Discrete pores with the target guest sizes/shapes are necessary for restricting the flexible guest molecules in the energy unfavourable *cis*-conformation, but they cannot allow guest adsorption/diffusion. Therefore, quasi-discrete pores connected by gating apertures should be the best choice. Ten MOFs with four typical characteristics, *i.e.*, hydrophilic with large channels (UiO-66, HKUST-1, Co-MOF-74, and Mg-MOF-74), hydrophilic with ultra-small channels (SIFSIX-Cu-3), hydrophobic (MAF-4, MAF-5, and MAF-6), and hydrophilic with quasi-discrete pores (MAF-23/Zn-BTM and MAF-49/Zn-BTA), were used for comparison. The adsorption selectivities were first qualitatively studied by column breakthrough curves of an equimolar butadiene–*n*-butene–*n*-butane–*iso*-butene mixture. All the hydrophilic MOFs with continuous channels (large or ultra-small) showed butadiene selectivity. The hydrophobic MOFs showed *n*-butane selectivity and the weakest butadiene adsorption affinity, but the separation performance was negligibly low. Both possessing

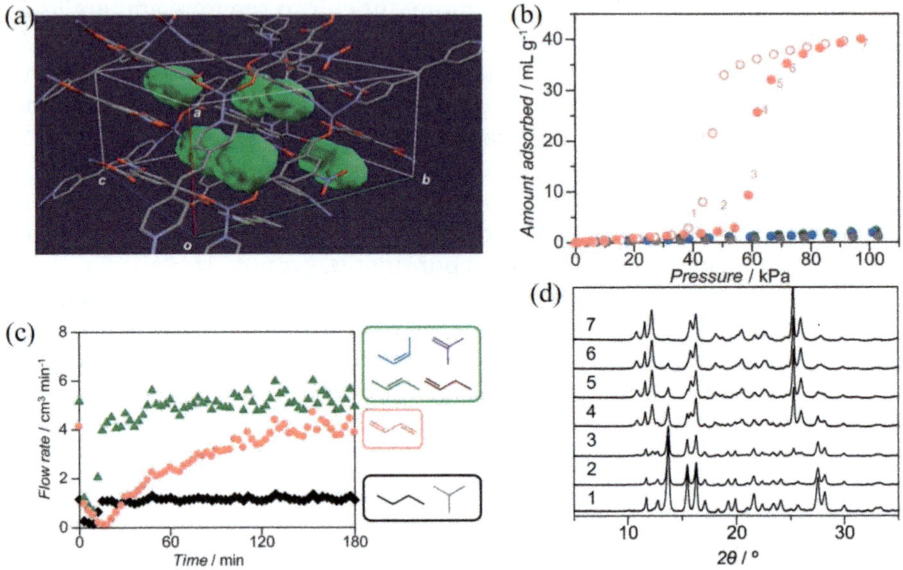

Figure 2.32 C_4-Hydrocarbon adsorption–separation properties of SD-65. (a) Crystal structure of as-synthesised/guest-free SD-65. (b) Single-component adsorption isotherms for 1,3-butadiene (marked with numbers for *in situ* PXRD) and other typical C_4 hydrocarbons at 25 °C. (c) Breakthrough curves of butadiene, butenes, and butanes at 25 °C with a total pressure of 230 kPa and a total flow rate of 10 cm^3 min^{-1}. (d) *In situ* PXRD patterns for 1,3-butadiene adsorption. Reproduced from ref. 143 with permission from John Wiley & Sons, Copyright 2016.

quasi-discrete pores, MAF-23 and MAF-49 showed contrasting breakthrough curves. MAF-23 with relatively large aperture size (3.6 Å) showed distinctive elution sequence of butadiene < *n*-butane < *n*-butene < *iso*-butene with quite different breakthrough times. Adsorption enthalpies calculated from single-component adsorption isotherms followed the same sequence, meaning that the adsorption selectivity is based on the thermodynamic mechanism. MAF-49 with relatively small aperture size (3.0 Å) cannot adsorb any C_4 hydrocarbons, as indicated by the completely overlapped breakthrough curves (Figure 2.33). After increasing the butadiene ratio to mimic the C_4 stream in industry, MAF-23 showed significantly enhanced butadiene purification performance, while hydrophobic MAF-6 still showed very poor separation performance.

GCMC simulations showed that, MAF-23 preferred binding of butadiene and *n*-butene in the high-energy *cis*-conformation but can confine just *n*-butene in this conformation considering the adsorption enthalpy. These simulation results were further confirmed by SCXRD. In MAF-49, the host-guest binding energies, and adsorption enthalpies of all flexible C_4 hydrocarbons favour the high-energy *cis*-conformation, and the adsorption enthalpy of butadiene was obvious smaller than other C_4 hydrocarbons. The simulation

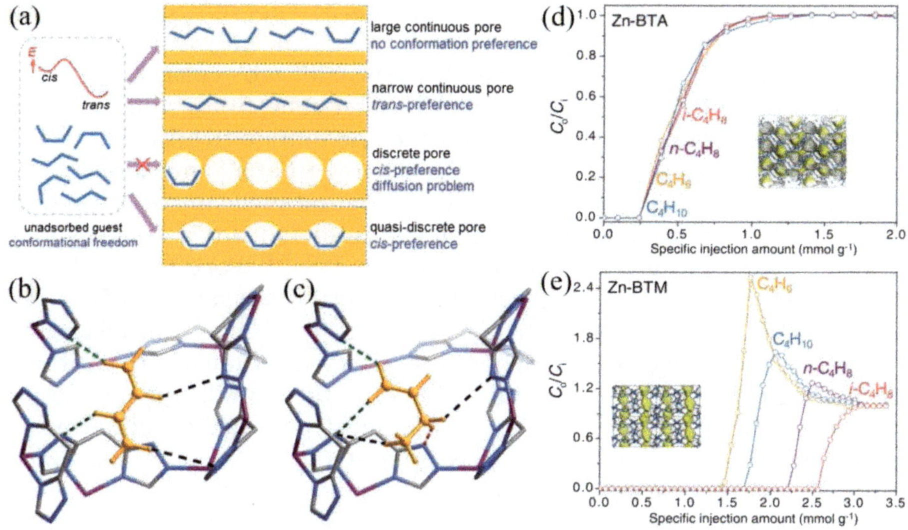

Figure 2.33 C4-hydrocarbon adsorption–separation of gating MOFs. (a) The effects of pore size, shape and dimensionality for controlling the conformations of flexible C4 hydrocarbons. (b), (c) Crystal structure of MAF-23 loaded with butadiene (b) and *n*-butene (c). (d), (e) Column breakthrough curves of an equimolar butadiene–*n*-butene–*iso*-butene–*n*-butane mixture for MAF-49 (d) and MAF-23 (e). Reproduced from ref. 142 with permission from The American Association for the Advancement of Science.

results showed that, in the thermodynamic point-of-view, the pore size/shape of MAF-49 were more suitable than that of MAF-23 to control the guest conformation. MD simulations gave self-diffusion coefficients $<1 \times 10^{-11}$ m^2 s^{-1} or $>1 \times 10^{-9}$ m^2 s^{-1} when MAF-23 was set rigid or flexible, respectively, meaning that the gating aperture was necessary for the adsorption/diffusion. However, the self-diffusion coefficients in flexible MAF-49 were $<1 \times 10^{-11}$ m^2 s^{-1}, meaning that the gating barrier was too high for the adsorption/diffusion. These results demonstrate that to simultaneously maximise host–guest confinement and guest diffusion, the pore shape, size and flexibility need further optimisation.

2.4.9 Flexible MOFs for Adsorptive Separation of Alkane Isomers

Butane and *iso*-butane have widespread applications, but they coexist in cracking gas and butane isomerisation products. Without separation, butane/*iso*-butane mixtures are used as gas fuels. C_5–C_{12}, C_9–C_{16}, and C_{10}–C_{19} alkanes are the main components of gasoline, kerosene, and diesel, respectively, the liquid fuels for internal combustion engines. The research octane number (RON) value of alkanes increases with the branching degree, but alkane isomers have very similar boiling points and are difficult to separate by distillation. When compared with alkane/alkene/alkyne homologues, alkane isomers show relatively large differences of molecular sizes/shapes (Tables 2.1 and 2.2). Zeolites can efficiently separate linear and branched alkanes but cannot separate monobranched and dibranced alkanes.[156,157]

Many MOFs have been studied for the separation of alkane isomers. Few examples are based on thermodynamic separation mechanism using rigid MOFs.[158–161] Most examples separate alkane isomers using a kinetic gating mechanism based on ultra-small pore apertures.[157,162–172] For example, the gating apertures of MAF-4 are useful for kinetic adsorption separation of not only C_3H_6 and C_3H_8 but also many other larger molecules.[134,157,172] The adsorptive separation behaviour of MAF-4 for hexane isomers was studied by quantitative column breakthrough experiments.[157] At 373 K, the equilibrium adsorption amounts of hexane/2-methlypentane/3-methlypentane/2,2-dimethlybutane/2,3-dimethlybutane were determined as 1.127/0.210/0.123/0.020/0.006 mmol g^{-1} and 1.580/0.315/0.196/0.044/0.017 mmol g^{-1} at 0.1 bar and 0.5 bar, respectively. The larger increases of adsorption amount for the dibranched isomers indicate that they could be adsorbed. At 473 K and 0.5 bar, the adsorption amount decreased to 0.357/0.174/0.104/0.032/0.013 mmol g^{-1}, in which the dibranched isomers showed smaller decreases, indicating that the adsorption process was mainly controlled by kinetic factors, *i.e.*, the gating apertures.

Although the thermodynamic difference of alkane isomers for adsorption is smaller than the kinetic difference, the thermodynamically controlled pore-opening can be used to attain suitable pore sizes for highly efficient kinetic separation and even molecular sieving. Because linear alkane isomers have larger surface areas and higher polarisabilities than branched isomers, they show higher boiling points, stronger intermolecular interactions, and lower pore-opening pressures.

Besides the thermodynamic difference of alkane isomers, pore-opening actions can be controlled by other means. For example, the pore sizes and pore-opening behaviours of MIL-53(Fe) were tuned by a crystal engineering approach to realise efficient adsorptive separation of hexane isomers.[173] Guest-free MIL-53(Fe) adopts the sp or narrow pore (np) structure with a pore size of 1.8 Å, which is virtually non-porous. After pore-opening, the pore size increases to 7.4 Å, which is too large for strong confinement of hexane isomers. Introducing methyl groups onto the organic ligand of MIL-53(Fe) slightly changes the pore sizes to 3.0 Å and 7.3 Å, respectively. In contrast, introducing trifluoromethyl groups not only significantly reduces the pore size to 5.6 Å for the lp state, but also restricts framework shrinkage to give a larger pore size of 4.2 Å for the sp state (Figure 2.34). Single-component adsorption isotherms and PXRD showed that hexane, 3-methylpentane, and 2,2-dimethylbutane can all expand the host to the lp state. However, they showed significantly different self-diffusion coefficients of 10.5×10^{-11}, 9×10^{-11} and $<10^{-12}$ m^2 s^{-1}, respectively, indicating that the pore sizes were suitable for kinetic adsorption separation. A column breakthrough experiment

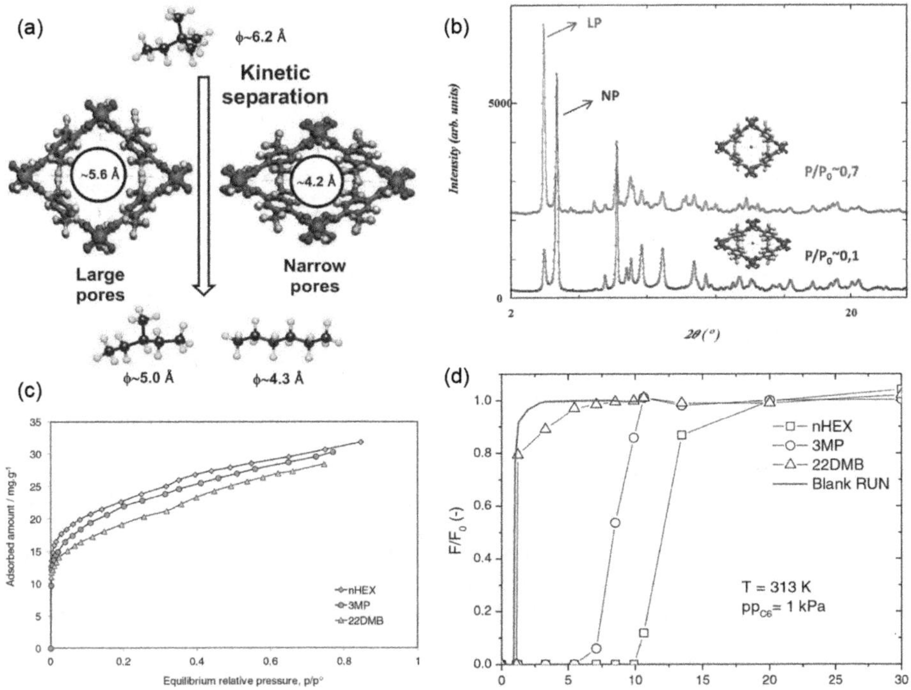

Figure 2.34 Hexane isomer adsorption–separation properties of MIL-53(Fe) functionalised by trifluoromethyl groups. (a) Size comparison of the pores and the guests. (b) PXRD in hexane vapour. (c) Single-component adsorption isotherms at 313 K. (d) Breakthrough curves at 313 K and 1 kPa. Reproduced from ref. 173 with permission from John Wiley & Sons, Copyright 2014.

showed efficient separation of 2,2-dimethylbutane/3-methylpentane/hexane with equilibrium uptakes of $0/6.1/9.6 \times 10^{-2}$ mmol g^{-1}. For the parent and the methyl decorated MIL-53(Fe), poor separation performances with very low equilibrium uptakes for all three isomers were observed at the same conditions, indicating that the host retained the sp structures.

The pore-opening behaviour can also be regulated by external physical stimuli. Tuneable molecular sieving of hexane isomers was achieved by changing the temperature.[174] $[Ca(H_2tcpb)]\cdot xH_2O$ (H_4tcpb = 1,2,4,5-tetrakis(4-carboxyphenyl)-benzene) is a 3D framework containing 1D channels with effective sizes of 3.6×4.8 to 6.2×7.5 and 3.8×4.0 to 5.7×6.2 Å^2. After guest removal, SCXRD showed slight shrinkage of the channels to 3.5×4.3 to 5.9×6.2 and 3.5×3.6 to 5.6×5.7 Å^2. Single-component adsorption isotherms measured from ~20 to ~100 torr showed that hexane can completely or almost completely open the pores at 30–120 °C, while 3-methylpentane and 2,2-dimethylbutane can open the pores below 90 °C and 30 °C, respectively (Figure 2.35). PXRD showed that, after saturated adsorption, the unit-cell volume change

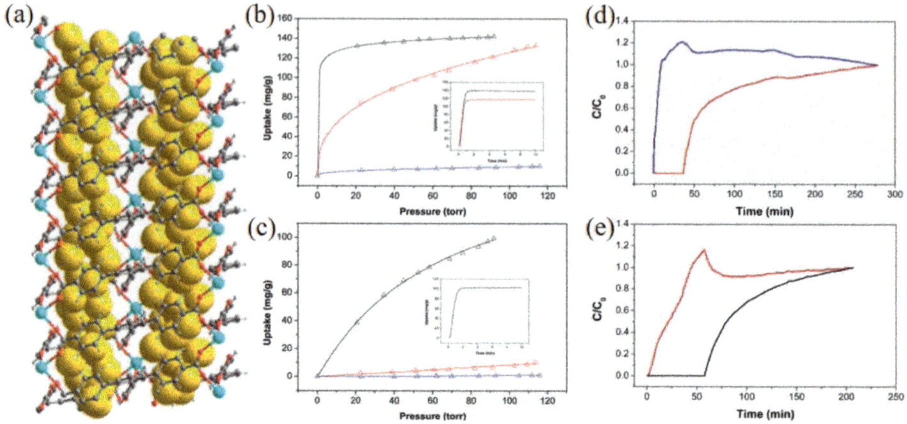

Figure 2.35 Hexane isomer adsorption–separation properties of [Ca(H_2tcpb)]. (a) Crystal structure. (b), (c) Adsorption isotherms (blue: 2,2-dimethylbutane, red: 3-methylpentane, black:hexane) at 60 °C (b) and 120 °C (c). (d), (e) Column breackthrough curves of equimolar binary mixtures (blue: 2,2-dimethylbutane, red: 3-methylpentane, black:hexane) at 60 °C (d) and 120 °C (e). Reproduced from ref. 174 with permission from the Royal Society of Chemistry.

followed 2,2-dimethylbutane > 3-methylpentane > hexane > as-synthesised > guest-free, indicating that the host is adaptive to the guest size. In other words, the pore size after pore-opening may be too small to accommodate the larger isomers, which can avoid co-adsorption (Figure 2.10). Column breakthrough experiments for a 3-methylpentane–hexane mixture at 120 °C and a 3-methylpentane–2,2-dimethylbutane mixture at 60 °C both showed that, the larger isomers were not adsorbed, and the adsorption amounts of the smaller isomers were consistent with the isotherms. Further, passing the mixture through two columns at 120 °C and 60 °C consequently can efficiently separate the three isomers.

In some cases, the pores opened by adsorption of linear isomers are large enough for the branched isomers. For example, molecular-sieving-like butane–*iso*-butane adsorption–separation behaviours with slight co-adsorption were observed for a pore-opening adsorbent.[170] [Ag_2(*o*-Hmpba)$_2$(*o*-H_2mpba)$_2$]·DMF (MAF-44, *o*-H_2mpba = 2-(3,5-dimethyl-1*H*-pyrazol-4-yl) benzoic acid) is a 2D network containing quasi-discrete pores (4.9 × 7.2 × 7.6 Å^3) connected by ultra-small apertures (1.8 × 4.4 Å^2), which slightly shrinks (pore 4.2 × 6.0 × 7.5 Å^3, aperture 1.1 × 4.5 Å^2) after guest removal (Figure 2.36). Single-component adsorption isotherms at 298 K showed that, the butane uptake suddenly jumped from 0.04 mmol g^{-1} at 10 kPa to 0.78 mmol g^{-1} at 12 kPa, and reached 0.9 mmol g^{-1} at 1 atm, while *iso*-butane uptake was only 0.012 mmol g^{-1} at 1 atm. The butane adsorption and desorption isotherms almost completely overlap, which are relatively rare fore pore-opening adsorbents and indicate small energy barrier of the pore-opening action.[175]

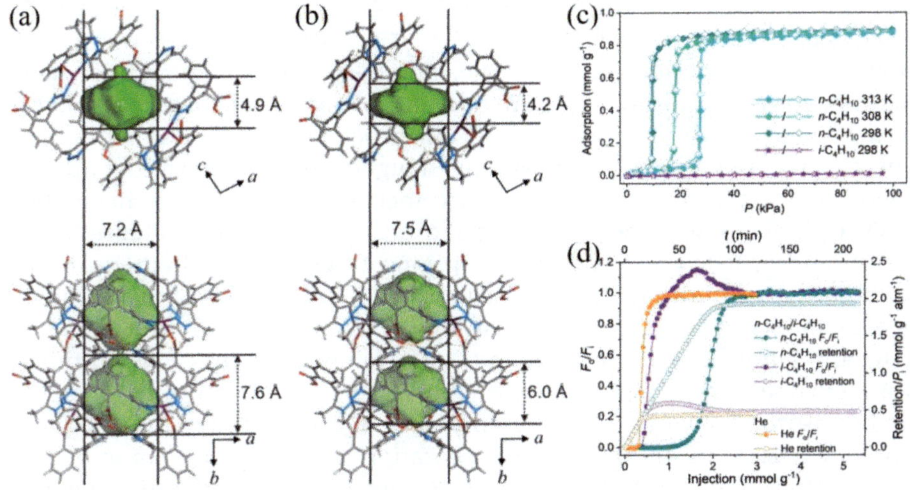

Figure 2.36 Butane isomer adsorption–separation properties of MAF-44. (a), (b) Crystallographic pore sizes and shapes of the as-synthesised (a) and the guest-free (b) structures. (c) Single-component adsorption isotherms. (d) Column breakthrough curves and retention curves. Reproduced from ref. 170 with permission from John Wiley & Sons, Copyright 2020.

A quantitative column breakthrough experiment (1 : 1 butane–*iso*-butane mixture at ambient condition) showed that relatively large amounts of purified *iso*-butane flowed out during the adsorption process, meaning that the leakage of butane was not serious. The *iso*-butane breakthrough curve did not directly reach equilibrium like the non-adsorbing gas He, indicating that it was adsorbed. The equilibrium uptakes of butane and *iso*-butane were calculated as 0.75(1) and 0.033(18) mmol g^{-1}, being lower and higher than the single-component isotherm uptakes of 0.88 and 0.0007 mmol g^{-1}, respectively. Although the experimental selectivity of 23 was high, the results confirmed co-adsorption of *iso*-butane. Configurational bias Monte Carlo (CBMC) simulations showed that both isomers were able to enter the pores at the as-synthesised state rather than the guest-free state, and *in situ* PXRD/adsorption measurements confirmed that the as-synthesised structure was restored after butane adsorption. DFT simulations for butane–*iso*-butane adsorption in the as-synthesised structure gave similar adsorption enthalpies of 60.8 and 57.0 kJ mol^{-1}, indicating that the high adsorption selectivity was mainly contributed by kinetic effects. MD simulations showed largely different butane–*iso*-butane diffusion constants of 64.5 and 5.27×10^{-6} cm^2 s^{-1} when the host was set flexible, and no diffusion of both gases can be observed when the host was set rigid, highlighting the critical role of gating apertures.

There are also some examples showing thermodynamic separation after pore-opening. $[Zn_2(PdTCPP)(bpa)]\cdot DMF$ (PdTCPP = [5,10,15,20-tetrakis-(4-methoxycarbonylphenyl)porphyrinato]-Pd^{II}; bpa = 9,10-bis(4-pyridyl)-anthracene) is a two-fold interpenetrated pillared-column MOF containing a 3D pore system with a pore size of 5.7 × 9.8 $Å^2$.[176] After solvent exchange, SCXRD revealed a shrinkage of pore size to 3.7 × 8.1 $Å^2$. After guest removal, PXRD showed further framework shrinkage. Single-component adsorption isotherms showed relatively low pore-opening pressure for hexane, and the maximum uptake did not decrease obviously below 50 °C. For 3-methlypentane, the pore-opening pressure was much higher, and the maximum uptake obviously decreased as the temperature increased and became negligible at 60 °C, indicating no pore-opening above this temperature (Figure 2.37). SCXRD and PXRD revealed that adsorption of hexane or 3-methlypentane can both restore the as-synthesised structure. Column mixture breakthrough measurements at 65 °C showed that 3-methylpentane outflowed immediately but had obvious adsorption and the difference of breakthrough times of 3-methylpentane and hexane was small. At 50 °C, 3-methylpentane still can be adsorbed obviously, but the difference of breakthrough times significantly increased, meaning a higher separation performance than at 65 °C. These results indicate that, at both 50 °C and 65 °C, 3-methylpentane can be adsorbed after pore-opening induced by adsorption of hexane. At 50 °C, although the difference (ratio) of single-component uptakes is smaller than that at 65 °C, the hexane adsorption amount is much higher, meaning that the adsorption–separation is dominated by a thermodynamic mechanism.

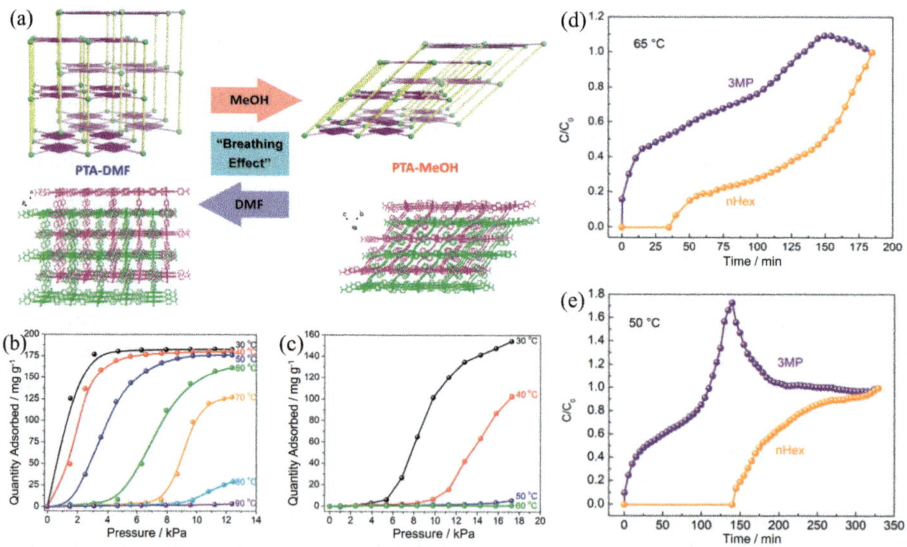

Figure 2.37 Hexane isomer adsorption–separation properties of [Zn_2(PdTCPP)(bpa)]. (a) Guest-induced structural transformations. (b), (c) Single-component adsorption isotherms of hexane (b) and 3-methylpentane (c). (d), (e) Column breakthrough curves at 65 °C (d) and 50 °C (e). Reproduced from ref. 176 with permission from John Wiley & Sons, Copyright 2021.

2.4.10 Flexible MOFs for Adsorptive Separation of C_6 Cyclic Hydrocarbons

In industry, cyclohexane and cyclohexene are mainly produced by Bz hydrogenation. Because the boiling points of Bz, cyclohexene and cyclohexane are very close (Table 2.2), they are separated by extractive distillation and azeotropic distillation in industry. Thermodynamically, preferential adsorption of Bz (π ligand, larger polarity) and cyclohexane can be achieved by using hydrophilic (OMS or polar groups) and hydrophobic pores, respectively,[177] but the adsorption difference should not be very large, because they can both form strong dispersion interactions. Large adsorption difference could be realised by utilising the flexibility of MOFs.[32,178]

In 2007, the role of pore-opening for Bz–cyclohexane separation was studied.[37] [Zn(μ_4-TCNQ-TCNQ)(bpy)]·1.5benzene (TCNQ = 7,7,8,8-tetracyano-*p*-quinodimethane) is a 3D framework consisting of a unique ligand, the electron-rich radical dimer TCNQ-TCNQ^{2-}. As guest or template, Bz molecules showed C–H···π interaction with the TCNQ-TCNQ^{2-} ligand. PXRD showed reversible structural transformations by removal and adsorption of Bz. Single-component Bz adsorption isotherm at 298 K showed a unique shape. An abrupt increase of uptake was accompanied with an obvious decrease of the pressure from ~0 cm^3 (STP) g^{-1} at P/P_0 = 0.25 to ~15 cm^3 (STP) g^{-1} at P/P_0 = 0.20, indicating that the pore-opening transformation possessed a relatively high energy barrier (Figure 2.38). After that, the adsorption amount of Bz rapidly increased to ~60 cm^3 (STP) g^{-1} at P/P_0 = 0.3 and reached 80.6 cm^3 (STP) g^{-1} at P/P_0 ~ 1.0. There was a large adsorption–desorption hysteresis, and obvious desorption started below P/P_0 ~ 0.05. The cyclohexane uptake at P/P_0 = 0.8 was only ~4.5 cm^3 (STP) g^{-1} but increased to ~20 cm^3 (STP) g^{-1} at P/P_0 = 0.98. Separation experiments were carried out by using 1:1, 1:3 and 1:10 Bz–cyclohexane vapour mixtures. NMR spectroscopy of the digested MOF samples showed that the adsorption ratios (96:4 to 95:5) were similar. The concentration-independent adsorption ratios imply that the selectivity was mainly govern by kinetic effect.

So far, most studies have focused on the single-component adsorption behaviours of Bz and cyclohexane.[32,178,179] In 2016, a huge Bz–cyclohexane adsorption difference was observed for [M_2(dobdc)] (M-MOF-74, M = Mg, Mn, Fe, Co, Ni, Cu, Zn), which have a 3D honeycomb-like network containing hexagonal 1D channels with diameter of 11 Å (Figure 2.39).[52] Taking Mn-MOF-74 as an example, single-component adsorption isotherms showed contrasting Bz (P/P_0 = 0.95) and cyclohexane (P/P_0 = 0.92) uptakes of 9.38 mmol g^{-1} and 0.25 mmol g^{-1} at 12 kPa, respectively. Other M-MOF-74 analogues showed similar results, except that the Bz uptake decreased along with the decrease of ionic potential, which was attributed to the π-complexation between Bz and the OMSs. The exceptionally low cyclohexane adsorption was explained by the lack of π-complexation ability. Note that the Bz adsorption isotherms of M-MOF-74 showed negligible adsorption at P/P_0 < 0.001~0.01, which is a sign of pore-opening action. The negligibly low adsorption of cyclohexane

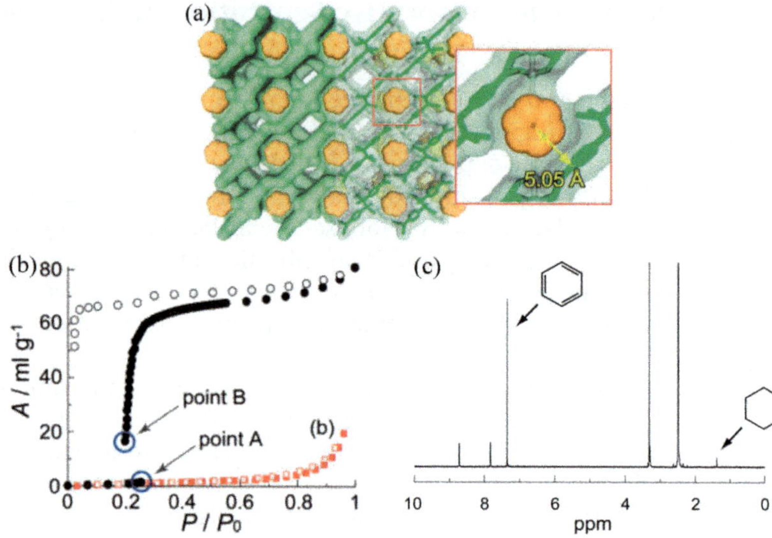

Figure 2.38 Bz–cyclohexane adsorption–separation properties of [Zn(μ_4-TCNQ-TCNQ)(bpy)] (a) Crystal structure (benzene loaded). (b) Single-component adsorption isotherms (black: benzene, red: cyclohexane) at 298 K. (c) NMR spectra of the digested MOF sample after exposing to 1:1 Bz–cyclohexane vapour. Reproduced from ref. 37 with permission from American Chemical Society, Copyright 2007.

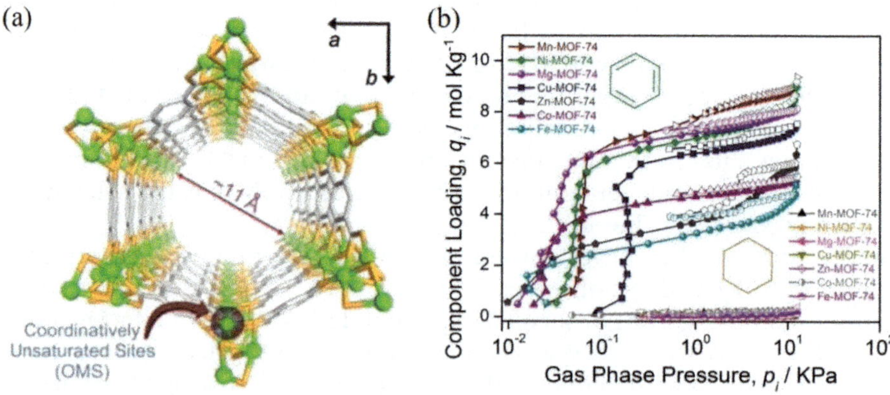

Figure 2.39 Benzene and cyclohexane adsorption properties of M-MOF-74. (a) Crystal structure. (b) Single-component adsorption isotherms. Reproduced from ref. 52 with permission from the Royal Society of Chemistry.

also indicated a nonporous adsorbent structure rather than weak adsorption. As indicated by the high boiling point of cyclohexane, dispersive intermolecular interactions are able to achieve saturated adsorption far below the saturation vapour pressure. However, the pore sizes of M-MOF-74 are significantly larger than both Bz and cyclohexane, and no structural flexibility was ever reported for M-MOF-74, meaning that the pore-opening phenomena should not arise from the common crystal-to-crystal structural transformation. Instead, defects on the crystal surfaces may serve as the block for guest diffusion.[51]

2.4.11 Flexible MOFs for Adsorptive Separation of C_8 Aromatics

Xylenes, (*p*-xylene (*p*X), *m*-xylene (*m*X) and *o*-xylene (*o*X)) are important raw materials in chemical industry, among which *p*-xylene has the highest demand. Xylenes and their isomer ethylbenzene (EB) are mainly produced by catalytic reforming and pyrolysis procedures. Zeolites have been used for the separation of xylene isomers, but the adsorption selectivity, adsorption capacity, and adsorption rate are not high. For example, MFI-type and LTA-type zeolites show *p*X selectivity of 20 and 7 and *p*X capacity of 1.3 and 1.0 mmol g^{-1}, respectively, and they need over 100 min to reach *p*X adsorption equilibrium at ambient temperatures.[180–183]

MOFs have demonstrated various adsorptive separation mechanisms for xylene isomers. Large-pore structures generally show thermodynamic separation by different interactions with the OMSs, ligands, and/or guest molecules, which can achieve high adsorption capacity, but low selectivity. For example, [V(OH)(bdc)] (MIL-47) exhibited a high *p*X uptake of 4.55 mmol g^{-1} but low *p*X selectivity of 1.1.[184] For MOFs with small pore apertures, *p*X (although with the lowest boiling point) is generally the most easily adsorbed one for its smallest cross-section size (Table 2.2). For example, MAF-4 has been widely studied for adsorptive separation of xylene isomers, which showed *p*X capacity of 3 mmol g^{-1} at room temperature,[185] and experimental *p*X/*m*X and *p*X/*o*X selectivities of 1.2–13 and 1.5–19, respectively.[186–189] In fact, adsorbents showing *p*X selectivity over 10 are mostly based on size/shape matched ultramicropores, which are usually quasi-discrete or 1D pores with low capacities of less than 2 mmol g^{-1}.[190–193]

The MOF flexibility can be rationally tuned to compare the rigid and pore-opening behaviours for adsorption–separation of xylene isomers.[194] Three isostructural Hoffmann-type MOFs [M(pz)Ni(CN)$_4$] (FePzNi, CoPzNi, NiPzNi for M = Fe, Co, Ni) were synthesised. Single component *p*X/*m*X/*o*X adsorption isotherms at 298 K showed rigid pores of FePzNi, with adsorption amount following *p*X ≫ *m*X > *o*X (Figure 2.40). CoPzNi showed selective pore-opening behaviour for *p*X at P/P_0 = 0.18–0.22, and the maximum *p*X/*m*X /*o*X uptakes were similar to those of FePzNi. NiPzNi was basically nonporous, as indicated by the very low maximum *p*X/*m*X/*o*X uptakes. In *p*X vapour, PXRD patterns of FePzNi and NiPzNi showed negligible changes, being consistent with their rigid behaviours revealed by the adsorption isotherms. For CoPzNi, diffraction peaks related to the interlayer separation shifted to lower angles and fitted to those of FePzNi, meaning that the pore-opening resulted from the increase of the interlayer distance. DFT and Monte Carlo (MC) simulations showed that *p*X could be adsorbed after pz rotation. FePzNi had the largest interlayer distance and lowest rotational barrier, so that *p*X could be

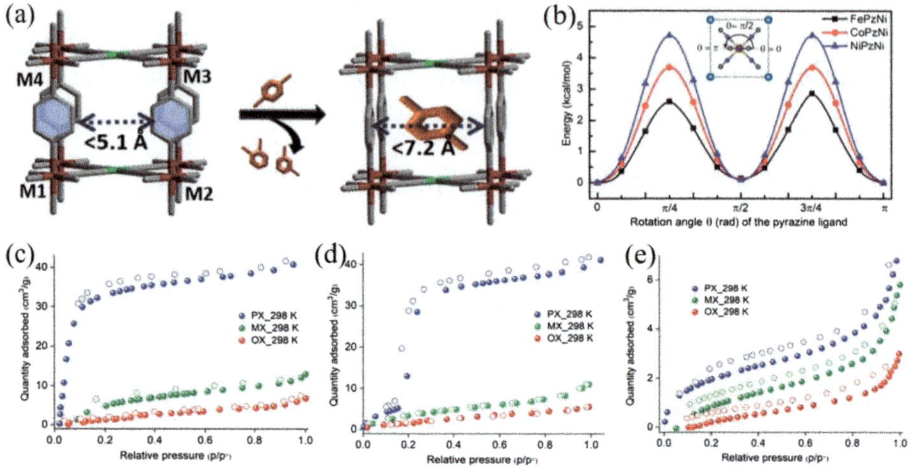

Figure 2.40 Xylene isomer adsorption–separation properties of MPzNi. (a) Structure transformation induced by adsorption of *p*X. (b) Rotation barriers of the pz ligands in MPzNi. (c)–(e) Single-component xylene isomer adsorption isotherms of FePzNi (c), CoPzNi (d) and NiPzNi (e) at 298 K. Reproduced from ref. 194 with permission from the Royal Society of Chemistry.

directly adsorbed without pore-opening. NiPzNi had the shortest interlayer distance and highest rotational barrier, corresponding to the low adsorption amounts. CoPzNi showed intermediate interlayer distance and rotational barrier, for which adsorption can occur after pore-opening based on increase of interlayer distance and rotation of the pz pillars. Separation experiments were carried out in binary vapour mixtures and analysed by NMR. The rigid framework FePzNi adsorbed only *p*X in *p*X/*m*X but showed a *p*X/*o*X selectivity of only 2. The pore-opening framework CoPzNi showed *p*X/*m*X and *p*X/*o*X selectivities of 1.2 and 1.15, respectively, indicating obvious co-adsorption.

High *p*X adsorption selectivity has been reported for a pore-opening MOF $[Zn_4O(L)_3(DMF)_2]\cdot$Guest (DynaMOF-100, H_2L = 4,4′-[[4-(1,1-dimethylethyl)-1,2-phenylene]bis(oxy)]bis-benzoic acid),[195] a 2D coordination network with 1D channels showing obvious framework shrinkage after guest removal (Figure 2.41). Single-component vapour adsorption isotherms showed high *p*X uptake of 64 cm^3 (STP) g^{-1} with pore-opening characteristics, while *m*X, *o*X and EB uptakes were just 5, 1.5 and 9.4 cm^3 (STP) g^{-1}. PXRD showed that, after exposed in single-component and different mixtures of xylene isomers, the as-synthesised structure can be restored just in the presence of *p*X. The selective pore-opening behaviour was attributed to the smallest cross section of *p*X (kinetic reason). NMR measurements of the digested MOF samples suggested that, after exposed in various combinations of equimolar binary vapour mixtures, only *p*X was adsorbed. Gas chromatography (GC) analyses of the supernatant solutions showed that, in various combinations of equimolar ternary mixtures, only the concentration of *p*X decreased. Although

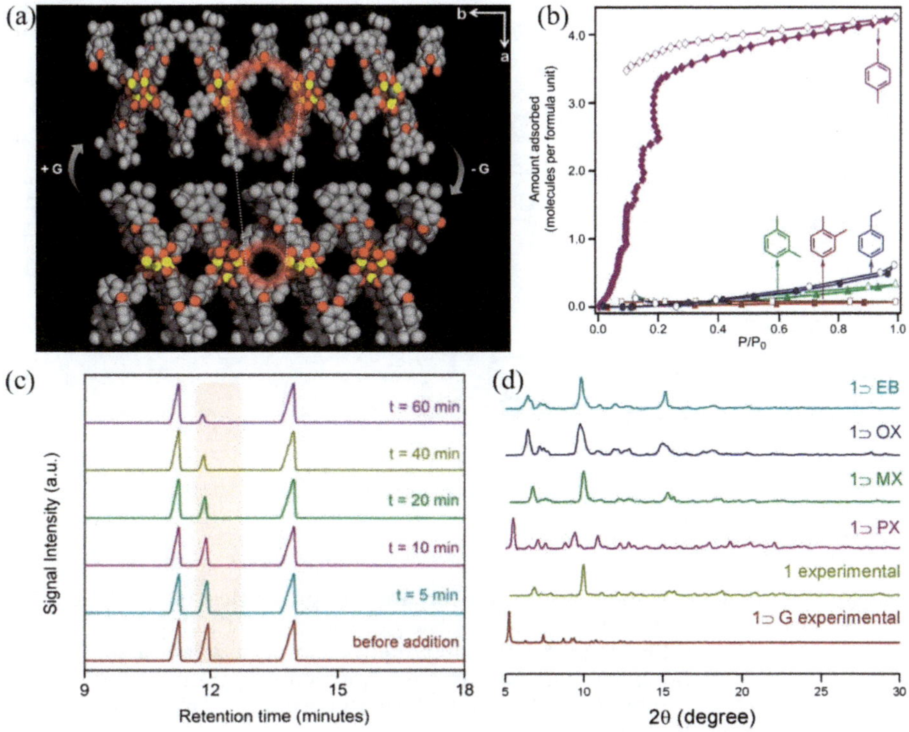

Figure 2.41 Xylene isomer adsorption–separation properties of DynaMOF-100. (a) Guest-induced structural transformation. (b) Single-component adsorption isotherms at 298 K. (c) GC chromatogram of the supernatant solutions of various multicomponent recorded at the specified time intervals. (d) PXRD after immersed in xylene and ethlybenzene. Reproduced from ref. 195 with permission from Springer Nature, Copyright 2014.

the experimental mixture adsorption amounts and detection limits were not reported, these results still demonstrated very high *p*X selectivities.

High *o*X adsorption selectivity was observed for a flexible MOF showing pore-opening behaviour.[196] $[Co(bpy)_2(NCS)_2]\cdot 2TFT$ (**sql-1-Co-NCS**, TFT = α,α,α-trifluorotoluene) is a 2D coordination network with **sql** topology. In the as-synthesised state, the **sql** nets are nearly square with a relatively long interlayer separation, which define 2D channels within and between the layers. After guest removal, the **sql** nets shrink to a rhombic state with a relatively short interlayer separation, giving a non-porous structure. Single-component *o*X/*p*X/*m*X/EB vapour adsorption isotherms at 298 K showed pore-opening at $P/P_0 = 0.25/0.3/0.5/0.7$ and uptake of 8/8/6.1/3.3 mmol g^{-1} at $P/P_0 = 0.95$, indicating high *o*X adsorption affinity (Figure 2.42). SCXRD showed that the **sql** nets were restored to square after adsorption of the guests. Moreover, *o*X/*p*X/*m*X not only fully occupied the pores between the **sql** nets but also those inside the nets, while EB only occupied the interlayer spaces. Besides the

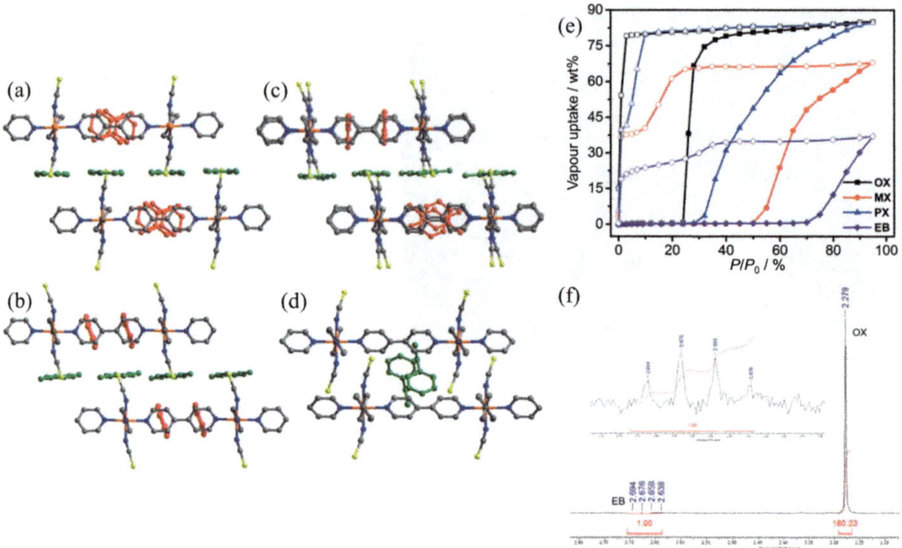

Figure 2.42 Xylene isomer adsorption–separation properties of **sql-1-Co-NCS**. (a)–(d) Crystal structures of the MOF loaded with *o*X (a), *m*X (b), *p*X (c) and EB (d). (e) Single-component adsorption isotherms. (f) NMR of $CDCl_3$ extractions for mixture adsorption–separation in equimolar *o*X–EB. Reproduced from ref. 196 with permission from John Wiley & Sons, Copyright 2019.

order of guest boiling points, the high *o*X adsorption affinity can be attributed to the presence of more host-guest C–H···π interactions as compared with those for *p*X/*m*X/EB. Thermogravimetry curves showed two-step guest removal for *o*X/*p*X/*m*X and only one step for EB, being consistent with the SCXRD results. The relatively low *m*X and EB isotherm uptakes compared with those observed in SCXRD indicated slow guest diffusion. Adsorption kinetics showed that, *m*X could reach an uptake of 7.5 mmol g^{-1} in 1 h but needed another 2 h to reach 7.8 mmol g^{-1}, while *o*X could reach saturation in 2 h. NMR measurements of the digested MOF samples after immersed in various combinations of the isomers gave *o*X–*p*X, *o*X–*m*X and *o*X–EB selectivities of 9.6, 7.5 and 60.1, respectively.

A flexible MOF may show two pore-opening processes with very different adsorption–separation behaviours.[190] $[Cu_2(pypz)_2]\cdot 0.5$*p*X (MAF-36, Hpypz = 4-(1*H*-pyrazol-4-yl)pyridine) is a 2D coordination network containing 1D channels with aperture sizes of 3.4 × 6.4 $Å^2$. The cavity sizes (4.1 × 5.7 × 7.3 $Å^3$) fit the sizes of the template *p*X molecule well. After guest removal, MAF-36 first turned to a metastable state with smaller cavity sizes of 3.3 × 5.4 × 7.0 $Å^3$ and aperture sizes of 1.3 × 6.4 $Å^2$, which can transform back to the as-synthesised structure by adsorption of *p*X. A more stable, non-porous, guest-free structure could be obtained by further increasing the temperature, which could also transform back to the as-synthesised structure by adsorption of

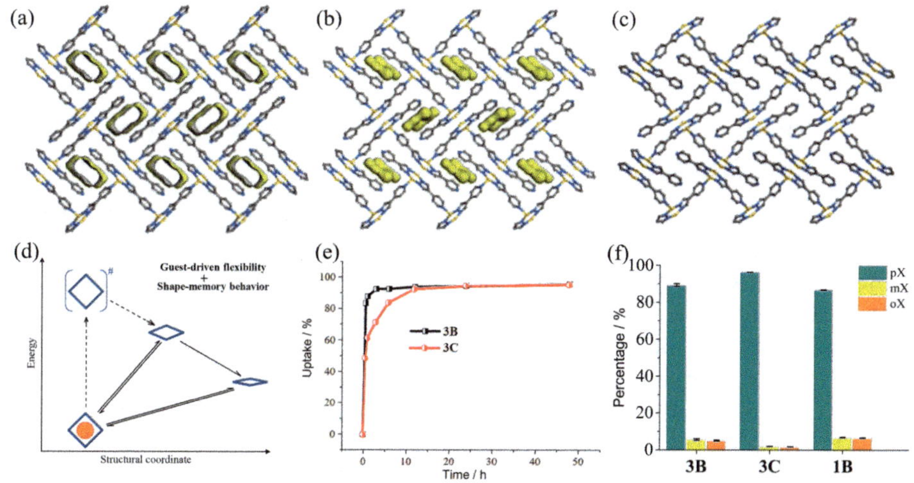

Figure 2.43 Xylene isomer adsorption–separation properties of MAF-36. (a)–(c) Crystal structures of the as-synthesised (a), metastable guest-free (b) and stable guest-free (c) structures. (d) Coexisting of guest-driven flexibility and shape-memory behaviour. (e) Ternary mixture adsorption kinetics. (f) Uptake ratios for ternary mixture. Reproduced from ref. 190, https://doi.org/10.34133/2019/9463719, under the terms of the CC BY 4.0 license https://creativecommons.org/licenses/by/4.0/.

*p*X. In other words, the thermodynamic controlled guest-driven structural transformation and kinetically controlled shape-memory behaviour coexist in MAF-36 (Figure 2.43). The transformation temperature from the metastable to the stable guest-free phase decreased from 240 °C to 90 °C, when the crystal size increased from $0.5 \times 0.5 \times 1\ \mu m^3$ to $100 \times 100 \times 500\ \mu m^3$. By virtue of the large crystal size, the details of structural transformations were visualised by SCXRD. The results showed coordination bond reconstitution and chirality inversion mechanisms for the phase transition, which provided a sufficiently high energy barrier to stabilise the metastable phase without the need of downsizing to the nanoscale. The large crystals of metastable and stable guest-free phases, denoted as **3B** and **3C**, respectively, showed similar *p*X/*m*X/*o*X adsorption amounts of 1.3/0.7/0.65 mmol g^{-1}. PXRD also showed that they both cannot transform back to the as-synthesised structure by adsorption of *m*X and *o*X, which may be attributed to the mismatching sizes/shapes. In this context, MAF-36 could be regarded as a crystalline and flexible adsorbent material molecular-imprinted by *p*X. Equimolar *p*X/*m*X/*o*X mixture batch adsorption experiments gave *p*X selectivity of 16 for **3B** and 51 for **3C**, but the large crystals needed 3 h and 12 h to reach adsorption equilibrium, respectively. This means a more difficult pore-opening action may increase adsorption selectivity but decrease adsorption rate. Nevertheless, but using small crystals of the metastable phase (denoted as **1B**), the adsorption could rapidly reach equilibrium in 2 min, and the *p*X selectivity remained 15. Note that the pore apertures of all the structures of MAF-36 are significantly smaller than xylene isomers, meaning that the adsorption–separation processes are mainly dominated by the gating apertures.

2.4.12 Flexible MOFs for Adsorptive Separation of Styrene and Ethylbenzene

As the most important aromatic monomer in industry, styrene (ST) is produced by the dehydrogenation of EB. Limited by unfavourable thermodynamics, the conversion rate of this reaction is limited, and 20–40% reactant EB is mixed with the product ST. Because of their similar boiling points (Table 2.2) and the high reactivity of ST (easy to polymerise), in industry they are separated by vacuum/extractive distillation. Rigid adsorbents tend to bind the less saturated molecule ST stronger, but the ST–EB selectivity is relatively low because EB is large and also unsaturated.[197–200] Flexible MOFs have been rarely studied for ST–EB separation.[154,200–202]

Besides adsorption–separation of xylene isomers, the pore-opening behaviour of DynaMOF-100 was also used for ST–EB separation.[202] The ethyl group and phenyl ring were considered non-coplanar, meaning that EB is much thicker than ST (Table 2.2). Single-component vapour adsorption isotherms at 298 K showed ST-induced pore-opening at $P/P_0 \sim 0.1$, and the uptake reached 86 cm^3 (STP) g^{-1} at $P/P_0 \sim 0.95$ (Figure 2.44). In contrast, the EB uptake gradually increased to only 9.4 cm^3 (STP) g^{-1} at $P/P_0 \sim 0.95$. PXRD confirmed that adsorption of ST can restore the as-synthesised structure. GC analyses of the supernatant showed that, only the concentration of ST in the

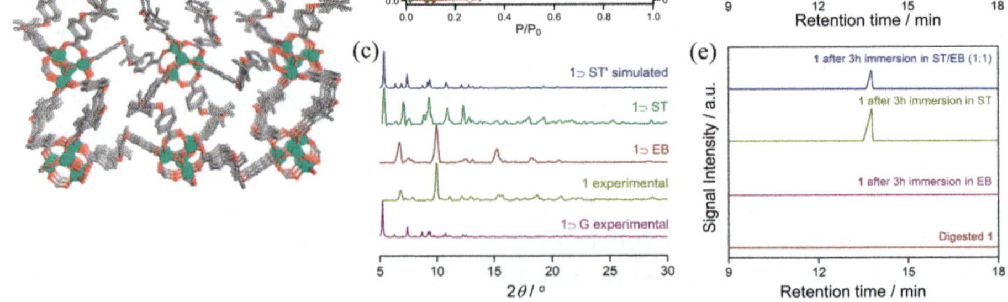

Figure 2.44 ST–EB adsorption–separation properties of DynaMOF-100. (a) Crystal structure (guest-free state). (b) Single-component adsorption isotherms. (c) PXRD after immersed in ST and EB. (d), (e) GC chromatogram of supernatant solutions (d) and digested samples (e) after immersed in ST–EB. Reproduced from ref. 202 with permission from American Chemical Society, Copyright 2015.

ST–EB mixture decreased, indicating that only ST was adsorbed. GC analyses of the digested MOF sample showed only the presence of ST. Although the experimental mixture adsorption amounts and detection limits were not reported, these results indicated high ST–EB adsorption selectivity.

Toluene (Tol) and Bz are also by-products of ST production. Although their boiling points are somewhat lower than those of C_8 aromatics, the additional distillation process still increase the complexity and cost for ST separation. Combining the pore-opening and gating flexibility, molecular sieving of ST over not only the larger molecule EB but also the smaller molecules Tol and Bz was achieved.[154] $[Cu_2(fbdim)]\cdot pX$ (MAF-41, H_2fbdim = 2,6-ditrifluoromethyl-benzodiimidazole) is a hinged-fence-like coordination network with **cds** topology. In the as-synthesised state, large cavities ($4.2 \times 9.8 \times 10.1$ Å^3) are connected by small apertures (3.9×6.7 Å^2) to form 1D sugar-coated-berry type channels. After removing the *p*X template, the host significantly shrunk, in which the cavities ($1.0 \times 6.0 \times 6.1$ Å^3) and apertures (0.7×6.0 Å^2) are too small for any guest. Considering that the aromatic ligand is quite rigid, the aperture size defined by the hinged-fence action may vary from 4.6×6.1 Å^2 to 1.0×6.0 Å^2 (Figure 2.45). DFT calculations gave a very low rotation barrier of 3 kJ for the ethyl group in EB, meaning that the molecule can temporally adopt the coplanar configuration (Table 2.2). However, the smallest cross-section of flexible EB is still smaller than the

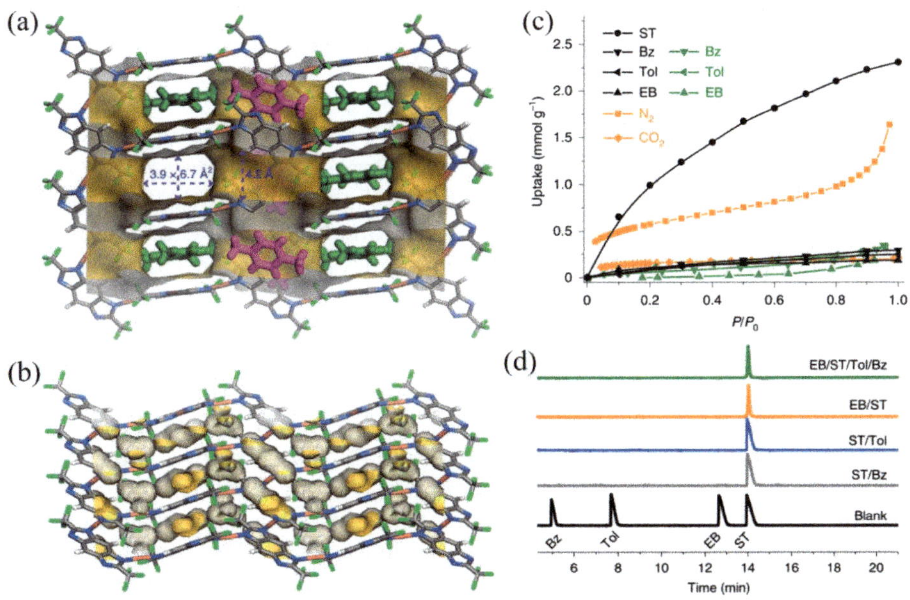

Figure 2.45 ST adsorption–separation properties of MAF-41. (a), (b) Crystal structures at the as-synthesised state (a) and the guest-free state (b). (c) Single-component gas–vapour adsorption isotherms. (d) GC analyses of MeOH extractions for mixture adsorption–separation. Reproduced from ref. 154 with permission from Springer Nature, Copyright 2019.

rhombus aperture at any state. Therefore, in kinetics, ST instead of EB can pass through the aperture. Single-component vapour–liquid adsorption isotherms showed high ST uptake of 2.31 mmol g^{-1} and low EB/Bz/Tol uptakes of 0.2–0.3 mmol g^{-1} at $P/P_0 \sim 1$. PXRD showed that the MOF sample restored the as-synthesised structure after immersed in ST and retained the guest-free structure in EB/Bz/Tol. Interestingly, the ST adsorption isotherm did not show pore-opening characteristics. The adsorption of ST and exclusion of EB can be easily understood by the common kinetic mechanism (although the flexibilities of the host and guest were considered). However, exclusions of the smaller molecules Tol and Bz were explained by the thermodynamic mechanism. DFT simulations showed that the ST–EB and Tol–Bz adsorption enthalpies were higher and lower than the energy needed for the pore-opening structural transformation. Batch separation experiments showed selectivity of 1250 in ST–EB, 5000 in ST–Bz and ST–Tol, and 3300 in ST–Bz–Tol–EB, meaning that ST could be produced with a purity of 99.9%+ in a single adsorption–desorption cycle. The ultrahigh mixture adsorption selectivities indicated negligible co-adsorption after the ST-induced pore-opening, which can be easily understood for the larger molecule EB, but difficult for the smaller ones Tol and Bz. In principle, this can be achieved by non-periodic or non-cooperative pore-opening actions, which is consistent with the ST isotherm shape. MD simulation also showed that when the host and guest were set flexible, the cavities can be independently opened and occupied only by ST.

2.4.13 Flexible MOFs for Adsorptive Separation of Other Gases

Adsorptive separation is also useful for recovery of resources or removal of pollutants from the air. Generally, the main component of air, *i.e.*, N_2/O_2, is much more difficult to be adsorbed, meaning that the adsorptive separation performance mainly depends on the adsorption affinity and capacity of the target molecules. Conventional adsorbents such as activated carbons have been widely used for capturing volatile organic compounds (VOCs).

Separation/enrichment of CH_4 from the coalbed gases (CH_4 20–45%, N_2 40–60%) is very important for energy utilisation and reducing greenhouse gas emissions. Nevertheless, adsorptive enrichment of CH_4 is not easy at ambient conditions. Activated carbons and zeolites show relatively low CH_4–N_2 selectivities (up to 4) and CH_4 adsorption amounts (up to 1 mmol g^{-1}). With much stronger CH_4 adsorption abilities, MOFs have attracted more and more attention in recent years. Most examples are based on rigid materials, and the highest CH_4–N_2 selectivity and CH_4 uptake at ambient conditions were reported to be 13.1 and 2.9 mmol g^{-1}, respectively.[203,204] Flexible materials can show pore-opening actions for CH_4 adsorption. However, the pore-opening pressures are usually very high at room temperature, which is not beneficial for CH_4 recovery. Further, considering that N_2 is smaller than CH_4 (Table 2.1), it is difficult to avoid the co-adsorption of N_2 after

CH_4-induced pore-opening. Gating channels show faster adsorption of the smaller N_2 molecules, meaning that the use of adsorbents with ultra-small pore size should be avoided.

For ultra-low-concentration resources/pollutants, such as noble gases and VOCs, adsorption affinity becomes predominantly important for the separation/enrichment performance. In this case, the tailorable pore characteristics of MOFs show advantages compared with conventional adsorbents. Moreover, the adaptive structural changes and/or gating pores of flexible MOFs can be particularly useful to maximise the host–guest confinement effect.[105,205,206]

2.5 Summary and Outlook

The flexibility and dynamic behaviours of MOFs depend on not only their structural characteristics but also the physical and chemical environment/stimuli. As the core of the chemical environment, the type and pressure/concentration of guest play a decisive role in controlling the structures of most flexible MOFs. Flexible MOFs show many special equilibrium and non-equilibrium adsorption behaviours, which are not only critical for storage and separation performances but also because they can be controlled by external chemical/physical stimuli. This chapter has discussed and analysed the advantages and disadvantages of the thermodynamically controlled pore-opening (commonly reported but controversial) and the kinetically controlled gating (common for ultramicropores but usually neglected) behaviours for adsorptive storage and separation. Induced-fit or adaptive pores could be also common for ultramicroporous MOFs and useful for enhancing separation/enrichment performances, but such effects are difficult to study and verify.

The flexibility of MOFs can also be used for other adsorption-related functions. For chemical sensing, the optical/electrical/magnetic responses toward the analytes can be amplified by the guest-selective structural transformations.[207–210] For heterogeneous catalysis, flexible MOFs may provide enzyme-like active sites by induced-fit, adaptive, and/or cooperative structural transformations.[211,212] Flexible properties responsible to physical stimuli are also useful for many other functions. When compared with conventional materials, MOFs can show much larger and/or special structural responses toward the change of temperature, mechanical pressure, *etc.*[213–216] The local dynamism of ionic components are critical for proton conductivity.[217,218]

Regardless of the type of function and/or flexibility, there are still very few in-depth studies on the structure–property relationships for flexible MOFs and similar molecule-based materials. In addition to more comprehensive *in situ* structure, energy, and spectroscopy characterisations, this research field can benefit from the developments/applications of universal computational simulation methods for the thermodynamic and kinetic mechanisms and quantitative measurements.

Acknowledgements

The support from the National Natural Science Foundation of China (22231012) and the XPLORER PRIZE are acknowledged.

References

1. S. R. Batten and R. Robson, *Angew. Chem., Int. Ed.*, 1998, **37**, 1460.
2. M. Eddaoudi, D. B. Moler, H. Li, B. Chen, T. M. Reineke, M. O'Keeffe and O. M. Yaghi, *Acc. Chem. Res.*, 2001, **34**, 319.
3. S. Kitagawa, R. Kitaura and S.-i. Noro, *Angew. Chem., Int. Ed.*, 2004, **43**, 2334.
4. J.-P. Zhang, Y.-B. Zhang, J.-B. Lin and X.-M. Chen, *Chem. Rev.*, 2012, **112**, 1001.
5. J.-R. Li, J. Sculley and H.-C. Zhou, *Chem. Rev.*, 2012, **112**, 869.
6. L. E. Kreno, K. Leong, O. K. Farha, M. Allendorf, R. P. Van Duyne and J. T. Hupp, *Chem. Rev.*, 2012, **112**, 1105.
7. K. Sumida, D. L. Rogow, J. A. Mason, T. M. McDonald, E. D. Bloch, Z. R. Herm, T.-H. Bae and J. R. Long, *Chem. Rev.*, 2012, **112**, 724.
8. N. Stock and S. Biswas, *Chem. Rev.*, 2012, **112**, 933.
9. H. Wu, Q. Gong, D. H. Olson and J. Li, *Chem. Rev.*, 2012, **112**, 836.
10. P. Horcajada, R. Gref, T. Baati, P. K. Allan, G. Maurin, P. Couvreur, G. Férey, R. E. Morris and C. Serre, *Chem. Rev.*, 2012, **112**, 1232.
11. C. Wang, T. Zhang and W. Lin, *Chem. Rev.*, 2012, **112**, 1084.
12. H. Furukawa, E. Cordova Kyle, M. O'Keeffe and M. Yaghi Omar, *Science*, 2013, **341**, 1230444.
13. W. Lu, Z. Wei, Z.-Y. Gu, T.-F. Liu, J. Park, J. Park, J. Tian, M. Zhang, Q. Zhang, T. Gentle III, M. Bosch and H.-C. Zhou, *Chem. Soc. Rev.*, 2014, **43**, 5561.
14. Y. He, B. Li, M. O'Keeffe and B. Chen, *Chem. Soc. Rev.*, 2014, **43**, 5618.
15. A. J. Howarth, Y. Liu, P. Li, Z. Li, T. C. Wang, J. T. Hupp and O. K. Farha, *Nat. Rev. Mater.*, 2016, **1**, 15018.
16. J. Li, X. Wang, G. Zhao, C. Chen, Z. Chai, A. Alsaedi, T. Hayat and X. Wang, *Chem. Soc. Rev.*, 2018, **47**, 2322.
17. M. Alhamami, H. Doan and C.-H. Cheng, *Materials*, 2014, **7**, 3198.
18. A. Schneemann, V. Bon, I. Schwedler, I. Senkovska, S. Kaskel and R. A. Fischer, *Chem. Soc. Rev.*, 2014, **43**, 6062.
19. J.-P. Zhang, P.-Q. Liao, H.-L. Zhou, R.-B. Lin and X.-M. Chen, *Chem. Soc. Rev.*, 2014, **43**, 5789.
20. S. K. Elsaidi, M. H. Mohamed, D. Banerjee and P. K. Thallapally, *Coord. Chem. Rev.*, 2018, **358**, 125.
21. A. J. Fletcher, K. M. Thomas and M. J. Rosseinsky, *J. Solid State Chem.*, 2005, **178**, 2491.
22. J. D. Evans, V. Bon, I. Senkovska, H.-C. Lee and S. Kaskel, *Nat. Commun.*, 2020, **11**, 2690.
23. J.-P. Zhang, H.-L. Zhou, D.-D. Zhou, P.-Q. Liao and X.-M. Chen, *Natl. Sci. Rev.*, 2018, **5**, 907.

24. F.-X. Coudert, M. Jeffroy, A. H. Fuchs, A. Boutin and C. Mellot-Draznieks, *J. Am. Chem. Soc.*, 2008, **130**, 14294.
25. J. Fonseca, T. Gong, L. Jiao and H.-L. Jiang, *J. Mater. Chem. A*, 2021, **9**, 10562.
26. T. D. Bennett and A. K. Cheetham, *Acc. Chem. Res.*, 2014, **47**, 1555.
27. N. Ma and S. Horike, *Chem. Rev.*, 2022, **122**, 4163.
28. Y. Li, X.-Y. Cao, K. Zheng, X.-W. Zhang, D.-D. Zhou, W.-X. Zhang, X.-M. Chen and J.-P. Zhang, *CCS Chem.*, 2021, **3**, 1532.
29. R. E. Morris and L. Brammer, *Chem. Soc. Rev.*, 2017, **46**, 5444.
30. J.-P. Zhang and S. Kitagawa, *J. Am. Chem. Soc.*, 2008, **130**, 907.
31. R.-B. Lin, Y. He, P. Li, H. Wang, W. Zhou and B. Chen, *Chem. Soc. Rev.*, 2019, **48**, 1362.
32. J.-P. Zhang and X.-M. Chen, *J. Am. Chem. Soc.*, 2008, **130**, 6010.
33. X.-W. Zhang, D.-D. Zhou and J.-P. Zhang, *Chem*, 2021, **7**, 1006.
34. S. Krause, N. Hosono and S. Kitagawa, *Angew. Chem., Int. Ed.*, 2020, **59**, 15325.
35. S. Krause, V. Bon, I. Senkovska, U. Stoeck, D. Wallacher, D. M. Többens, S. Zander, R. S. Pillai, G. Maurin, F.-X. Coudert and S. Kaskel, *Nature*, 2016, **532**, 348.
36. Y. Sakata, S. Furukawa, M. Kondo, K. Hirai, N. Horike, Y. Takashima, H. Uehara, N. Louvain, M. Meilikhov, T. Tsuruoka, S. Isoda, W. Kosaka, O. Sakata and S. Kitagawa, *Science*, 2013, **339**, 193.
37. S. Shimomura, S. Horike, R. Matsuda and S. Kitagawa, *J. Am. Chem. Soc.*, 2007, **129**, 10990.
38. X.-C. Huang, Y.-Y. Lin, J.-P. Zhang and X.-M. Chen, *Angew. Chem., Int. Ed.*, 2006, **45**, 1557.
39. K. S. Park, Z. Ni, A. P. Côté, J. Y. Choi, R. Huang, F. J. Uribe-Romo, H. K. Chae, M. O'Keeffe and O. M. Yaghi, *Proc. Natl. Acad. Sci. U. S. A.*, 2006, **103**, 10186.
40. Z. Bao, G. Chang, H. Xing, R. Krishna, Q. Ren and B. Chen, *Energy Environ. Sci.*, 2016, **9**, 3612.
41. J.-P. Zhang, A.-X. Zhu, R.-B. Lin, X.-L. Qi and X.-M. Chen, *Adv. Mater.*, 2011, **23**, 1268.
42. J. A. Rackers and J. W. Ponder, *J. Chem. Phys.*, 2019, **150**, 084104.
43. J.-P. Zhang, A.-X. Zhu and X.-M. Chen, *Chem. Commun.*, 2012, **48**, 11395.
44. J.-R. Li, R. J. Kuppler and H.-C. Zhou, *Chem. Soc. Rev.*, 2009, **38**, 1477.
45. P. Canepa, N. Nijem, Y. J. Chabal and T. Thonhauser, *Phys. Rev. Lett.*, 2013, **110**, 026102.
46. Q. Gao, J. Xu, D. Cao, Z. Chang and X.-H. Bu, *Angew. Chem., Int. Ed.*, 2016, **55**, 15027.
47. G. Li, J. Shang, Q. Gu, R. V. Awati, N. Jensen, A. Grant, X. Zhang, D. S. Sholl, J. Z. Liu, P. A. Webley and E. F. May, *Nat. Commun.*, 2017, **8**, 15777.
48. J. Shang, G. Li, R. Singh, Q. Gu, K. M. Nairn, T. J. Bastow, N. Medhekar, C. M. Doherty, A. J. Hill, J. Z. Liu and P. A. Webley, *J. Am. Chem. Soc.*, 2012, **134**, 19246.

49. S. Aguado, G. Bergeret, C. Daniel and D. Farrusseng, *J. Am. Chem. Soc.*, 2012, **134**, 14635.
50. S. Ma, D. Sun, X.-S. Wang and H.-C. Zhou, *Angew. Chem., Int. Ed.*, 2007, **46**, 2458.
51. J. I. Feldblyum, M. Liu, D. W. Gidley and A. J. Matzger, *J. Am. Chem. Soc.*, 2011, **133**, 18257.
52. S. Mukherjee, B. Manna, A. V. Desai, Y. Yin, R. Krishna, R. Babarao and S. K. Ghosh, *Chem. Commun.*, 2016, **52**, 8215.
53. L. Peng, S. Yang, S. Jawahery, S. M. Moosavi, A. J. Huckaba, M. Asgari, E. Oveisi, M. K. Nazeeruddin, B. Smit and W. L. Queen, *J. Am. Chem. Soc.*, 2019, **141**, 12397.
54. K. J. Korman, G. E. Decker, M. R. Dworzak, M. M. Deegan, A. M. Antonio, G. A. Taggart and E. D. Bloch, *ACS Appl. Mater. Interfaces*, 2020, **12**, 40318.
55. US Department of Energy, ARPA, https://arpa-e-foa.energy.gov/.
56. S. Krause, V. Bon, H. Du, R. E. Dunin-Borkowski, U. Stoeck, I. Senkovska and S. Kaskel, *Beilstein J. Nanotechnol.*, 2019, **10**, 1737.
57. J. A. Mason, J. Oktawiec, M. K. Taylor, M. R. Hudson, J. Rodriguez, J. E. Bachman, M. I. Gonzalez, A. Cervellino, A. Guagliardi, C. M. Brown, P. L. Llewellyn, N. Masciocchi and J. R. Long, *Nature*, 2015, **527**, 357.
58. C. Zhang, J. A. Gee, D. S. Sholl and R. P. Lively, *J. Phys. Chem. C*, 2014, **118**, 20727.
59. S. A. Moggach, T. D. Bennett and A. K. Cheetham, *Angew. Chem., Int. Ed.*, 2009, **48**, 7087.
60. A. Ghoufi, K. Benhamed, L. Boukli-Hacene and G. Maurin, *ACS Cent. Sci.*, 2017, **3**, 394.
61. J. Park, D. Yuan, K. T. Pham, J.-R. Li, A. Yakovenko and H.-C. Zhou, *J. Am. Chem. Soc.*, 2012, **134**, 99.
62. R. Lyndon, K. Konstas, B. P. Ladewig, P. D. Southon, P. C. J. Kepert and M. R. Hill, *Angew. Chem., Int. Ed.*, 2013, **52**, 3695.
63. F. Luo, C. B. Fan, M. B. Luo, X. L. Wu, Y. Zhu, S. Z. Pu, W.-Y. Xu and G.-C. Guo, *Angew. Chem., Int. Ed.*, 2014, **53**, 9298.
64. N. Yanai, T. Uemura, M. Inoue, R. Matsuda, T. Fukushima, M. Tsujimoto, S. Isoda and S. Kitagawa, *J. Am. Chem. Soc.*, 2012, **134**, 4501.
65. S.-J. Lee and Y.-S. Bae, *J. Phys. Chem. C*, 2014, **118**, 19833.
66. S. K. Bhatia and A. L. Myers, *Langmuir*, 2006, **22**, 1688.
67. C. M. Simon, J. Kim, D. A. Gomez-Gualdron, J. S. Camp, Y. G. Chung, R. L. Martin, R. Mercado, M. W. Deem, D. Gunter, M. Haranczyk, D. S. Sholl, R. Q. Snurr and B. Smit, *Energy Environ. Sci.*, 2015, **8**, 1190.
68. C. M. Simon, J. Kim, L.-C. Lin, R. L. Martin, M. Haranczyk and B. Smit, *Phys. Chem. Chem. Phys.*, 2014, **16**, 5499.
69. J.-M. Lin, C.-T. He, Y. Liu, P.-Q. Liao, D.-D. Zhou, J.-P. Zhang and X.-M. Chen, *Angew. Chem., Int. Ed.*, 2016, **55**, 4674.
70. Z.-W. Mo, H.-L. Zhou, D.-D. Zhou, R.-B. Lin, P.-Q. Liao, C.-T. He, W.-X. Zhang, X.-M. Chen and J.-P. Zhang, *Adv. Mater.*, 2018, **30**, 1704350.

71. R.-B. Lin, T.-Y. Li, H.-L. Zhou, C.-T. He, J.-P. Zhang and X.-M. Chen, *Chem. Sci.*, 2015, **6**, 2516.
72. T. A. Makal, J.-R. Li, W. Lu and H.-C. Zhou, *Chem. Soc. Rev.*, 2012, **41**, 7761.
73. Y. He, W. Zhou, G. Qian and B. Chen, *Chem. Soc. Rev.*, 2014, **43**, 5657.
74. M. P. Suh, H. J. Park, T. K. Prasad and D.-W. Lim, *Chem. Rev.*, 2012, **112**, 782.
75. M. T. Kapelewski, T. Runčevski, J. D. Tarver, H. Z. H. Jiang, K. E. Hurst, P. A. Parilla, A. Ayala, T. Gennett, S. A. FitzGerald, C. M. Brown and J. R. Long, *Chem. Mater.*, 2018, **30**, 8179.
76. H. Li, L. Li, R.-B. Lin, W. Zhou, Z. Zhang, S. Xiang and B. Chen, *EnergyChem*, 2019, **1**, 100006.
77. Y. Kubota, M. Takata, R. Matsuda, R. Kitaura, S. Kitagawa, K. Kato, M. Sakata and T. C. Kobayashi, *Angew. Chem., Int. Ed.*, 2005, **44**, 920.
78. C.-T. He, Z.-M. Ye, Y.-T. Xu, D.-D. Zhou, H.-L. Zhou, D. Chen, J.-P. Zhang and X.-M. Chen, *Chem. Sci.*, 2017, **8**, 7560.
79. P. F. Liu, J. K. Chu, S. J. Hou, P. Xu and J. Y. Zheng, *Renewable Sustainable Energy Rev.*, 2012, **16**, 1817.
80. H. Wu, W. Zhou and T. Yildirim, *J. Am. Chem. Soc.*, 2009, **131**, 4995.
81. Z.-H. Zhang, H. Fang, D.-X. Xue and J. Bai, *ACS Appl. Mater. Interfaces*, 2021, **13**, 44956.
82. Q.-Y. Yang, P. Lama, S. Sen, M. Lusi, K.-J. Chen, W.-Y. Gao, M. Shivanna, T. Pham, N. Hosono, S. Kusaka, J. J. Perry Iv, S. Ma, B. Space, L. J. Barbour, S. Kitagawa and M. J. Zaworotko, *Angew. Chem., Int. Ed.*, 2018, **57**, 5684.
83. R. Kitaura, K. Seki, G. Akiyama and S. Kitagawa, *Angew. Chem., Int. Ed.*, 2003, **42**, 428.
84. T. Kundu, B. B. Shah, L. Bolinois and D. Zhao, *Chem. Mater.*, 2019, **31**, 2842.
85. T. Kundu, M. Wahiduzzaman, B. B. Shah, G. Maurin and D. Zhao, *Angew. Chem., Int. Ed.*, 2019, **58**, 8073.
86. T. Waitz, K. Tsuchiya, T. Antretter and F. D. Fischer, *MRS Bull.*, 2009, **34**, 814.
87. J.-P. Zhang and X.-M. Chen, *J. Am. Chem. Soc.*, 2009, **131**, 5516.
88. R. Matsuda, R. Kitaura, S. Kitagawa, Y. Kubota, R. V. Belosludov, T. C. Kobayashi, H. Sakamoto, T. Chiba, M. Takata, Y. Kawazoe and Y. Mita, *Nature*, 2005, **436**, 238.
89. Z. Zhang, S. Xiang and B. Chen, *CrystEngComm*, 2011, **13**, 5983.
90. M. Bonneau, C. Lavenn, J.-J. Zheng, A. Legrand, T. Ogawa, K. Sugimoto, F.-X. Coudert, R. Reau, S. Sakaki, K.-i. Otake and S. Kitagawa, *Nat. Chem.*, 2022, **14**, 816.
91. B. Chen, C. Liang, J. Yang, D. S. Contreras, Y. L. Clancy, E. B. Lobkovsky, O. M. Yaghi and S. Dai, *Angew. Chem., Int. Ed.*, 2006, **45**, 1390.
92. F.-X. Coudert, C. Mellot-Draznieks, A. H. Fuchs and A. Boutin, *J. Am. Chem. Soc.*, 2009, **131**, 11329.

93. D.-W. Lim, J. Ha, Y. Oruganti and H. R. Moon, *Mater. Chem. Front.*, 2021, **5**, 4022.
94. J. Teufel, H. Oh, M. Hirscher, M. Wahiduzzaman, L. Zhechkov, A. Kuc, T. Heine, D. Denysenko and D. Volkmer, *Adv. Mater.*, 2013, **25**, 635.
95. L. Zhang, S. Jee, J. Park, M. Jung, D. Wallacher, A. Franz, W. Lee, M. Yoon, K. Choi, M. Hirscher and H. Oh, *J. Am. Chem. Soc.*, 2019, **141**, 19850.
96. J. Y. Kim, L. Zhang, R. Balderas-Xicohténcatl, J. Park, M. Hirscher, H. R. Moon and H. Oh, *J. Am. Chem. Soc.*, 2017, **139**, 17743.
97. H. Kajiro, A. Kondo, K. Kaneko and H. Kanoh, *Int. J. Mol. Sci.*, 2010, **11**, 3803.
98. D.-D. Zhou, Y.-T. Xu, R.-B. Lin, Z.-W. Mo, W.-X. Zhang and J.-P. Zhang, *Chem. Commun.*, 2016, **52**, 4991.
99. C. Gu, N. Hosono, J.-J. Zheng, Y. Sato, S. Kusaka, S. Sakaki and S. Kitagawa, *Science*, 2019, **363**, 387.
100. H. Sato, W. Kosaka, R. Matsuda, A. Hori, Y. Hijikata, V. Belosludov Rodion, S. Sakaki, M. Takata and S. Kitagawa, *Science*, 2014, **343**, 167.
101. D. A. Reed, B. K. Keitz, J. Oktawiec, J. A. Mason, T. Runčevski, D. J. Xiao, L. E. Darago, V. Crocellà, S. Bordiga and J. R. Long, *Nature*, 2017, **550**, 96.
102. P.-Q. Liao, X.-Y. Li, J. Bai, C.-T. He, D.-D. Zhou, W.-X. Zhang, J.-P. Zhang and X.-M. Chen, *Chem. – Eur. J.*, 2014, **20**, 11303.
103. D. Banerjee, C. M. Simon, S. K. Elsaidi, M. Haranczyk and P. K. Thallapally, *Chem*, 2018, **4**, 466.
104. J. Pei, X.-W. Gu, C.-C. Liang, B. Chen, B. Li and G. Qian, *J. Am. Chem. Soc.*, 2022, **144**, 3200.
105. Z. Niu, Z. Fan, T. Pham, G. Verma, K. A. Forrest, B. Space, P. K. Thallapally, A. M. Al-Enizi and S. Ma, *Angew. Chem., Int. Ed.*, 2022, **61**, e202117807.
106. C. A. Fernandez, J. Liu, P. K. Thallapally and D. M. Strachan, *J. Am. Chem. Soc.*, 2012, **134**, 9046.
107. H. Wang, M. Warren, J. Jagiello, S. Jensen, S. K. Ghose, K. Tan, L. Yu, T. J. Emge, T. Thonhauser and J. Li, *J. Am. Chem. Soc.*, 2020, **142**, 20088.
108. Q. Wang, T. Ke, L. Yang, Z. Zhang, X. Cui, Z. Bao, Q. Ren, Q. Yang and H. Xing, *Angew. Chem., Int. Ed.*, 2020, **59**, 3423.
109. D.-D. Zhou, X.-W. Zhang, Z.-W. Mo, Y.-Z. Xu, X.-Y. Tian, Y. Li, X.-M. Chen and J.-P. Zhang, *EnergyChem*, 2019, **1**, 100016.
110. P.-Q. Liao, A.-X. Zhu, W.-X. Zhang, J.-P. Zhang and X.-M. Chen, *Nat. Commun.*, 2015, **6**, 6350.
111. S. Horike, Y. Inubushi, T. Hori, T. Fukushima and S. Kitagawa, *Chem. Sci.*, 2012, **3**, 116.
112. M. K. Taylor, T. Runčevski, J. Oktawiec, J. E. Bachman, R. L. Siegelman, H. Jiang, J. A. Mason, J. D. Tarver and J. R. Long, *J. Am. Chem. Soc.*, 2018, **140**, 10324.
113. P.-Q. Liao, W.-X. Zhang, J.-P. Zhang and X.-M. Chen, *Nat. Commun.*, 2015, **6**, 8697.
114. Y. Hijikata, S. Horike, M. Sugimoto, M. Inukai, T. Fukushima and S. Kitagawa, *Inorg. Chem.*, 2013, **52**, 3634.

115. K. Kishida, Y. Watanabe, S. Horike, Y. Watanabe, Y. Okumura, Y. Hijikata, S. Sakaki and S. Kitagawa, *Eur. J. Inorg. Chem.*, 2014, **2014**, 2747.
116. N. Nijem, H. Wu, P. Canepa, A. Marti, K. J. Balkus, T. Thonhauser, J. Li and Y. J. Chabal, *J. Am. Chem. Soc.*, 2012, **134**, 15201.
117. Y. Chen, K. B. Idrees, F. A. Son, X. Wang, Z. Chen, Q. Xia, Z. Li, X. Zhang and O. K. Farha, *ACS Appl. Mater. Interfaces*, 2021, **13**, 16820.
118. C. Gücüyener, J. van den Bergh, J. Gascon and F. Kapteijn, *J. Am. Chem. Soc.*, 2010, **132**, 17704.
119. X.-C. Huang, J.-P. Zhang and X.-M. Chen, *Chin. Sci. Bull.*, 2003, **48**, 1531.
120. P. Zhao, G. I. Lampronti, G. O. Lloyd, M. T. Wharmby, S. Facq, A. K. Cheetham and S. A. T. Redfern, *Chem. Mater.*, 2014, **26**, 1767.
121. R.-B. Lin, L. Li, H.-L. Zhou, H. Wu, C. He, S. Li, R. Krishna, J. Li, W. Zhou and B. Chen, *Nat. Mater.*, 2018, **17**, 1128.
122. Q. Dong, X. Zhang, S. Liu, R.-B. Lin, Y. Guo, Y. Ma, A. Yonezu, R. Krishna, G. Liu, J. Duan, R. Matsuda, W. Jin and B. Chen, *Angew. Chem., Int. Ed.*, 2020, **59**, 22756.
123. L. Li, R. Krishna, Y. Wang, X. Wang, J. Yang and J. Li, *Eur. J. Inorg. Chem.*, 2016, **2016**, 4457.
124. Y. Xie, H. Cui, H. Wu, R.-B. Lin, W. Zhou and B. Chen, *Angew. Chem., Int. Ed.*, 2021, **60**, 9604.
125. Z. Zhang, S. B. Peh, R. Krishna, C. Kang, K. Chai, Y. Wang, D. Shi and D. Zhao, *Angew. Chem., Int. Ed.*, 2021, **60**, 17198.
126. R.-B. Lin, L. Li, H. Wu, H. Arman, B. Li, R.-G. Lin, W. Zhou and B. Chen, *J. Am. Chem. Soc.*, 2017, **139**, 8022.
127. M. L. Foo, R. Matsuda, Y. Hijikata, R. Krishna, H. Sato, S. Horike, A. Hori, J. Duan, Y. Sato, Y. Kubota, M. Takata and S. Kitagawa, *J. Am. Chem. Soc.*, 2016, **138**, 3022.
128. J.-P. Zhang, S. K. Ghosh, J.-B. Lin and S. Kitagawa, *Inorg. Chem.*, 2009, **48**, 7970.
129. F.-X. Coudert, *Phys. Chem. Chem. Phys.*, 2010, **12**, 10904.
130. X.-Y. Tian, H.-L. Zhou, X.-W. Zhang, C. Wang, Z.-H. Qiu, D.-D. Zhou and J.-P. Zhang, *Inorg. Chem.*, 2020, **59**, 6047.
131. X. Wang, R. Krishna, L. Li, B. Wang, T. He, Y.-Z. Zhang, J.-R. Li and J. Li, *Chem. Eng. J.*, 2018, **346**, 489.
132. H. Zeng, M. Xie, T. Wang, R.-J. Wei, X.-J. Xie, Y. Zhao, W. Lu and D. Li, *Nature*, 2021, **595**, 542.
133. Y. Chen, Z. Qiao, D. Lv, C. Duan, X. Sun, H. Wu, R. Shi, Q. Xia and Z. Li, *Chem. Eng. J.*, 2017, **328**, 360.
134. K. Li, D. H. Olson, J. Seidel, T. J. Emge, H. Gong, H. Zeng and J. Li, *J. Am. Chem. Soc.*, 2009, **131**, 10368.
135. A. Cadiau, K. Adil, P. M. Bhatt, Y. Belmabkhout and M. Eddaoudi, *Science*, 2016, **353**, 137.
136. H. Wang, X. Dong, V. Colombo, Q. Wang, Y. Liu, W. Liu, X.-L. Wang, X.-Y. Huang, D. M. Proserpio, A. Sironi, Y. Han and J. Li, *Adv. Mater.*, 2018, **30**, 1805088.

137. Y. Wang, N.-Y. Huang, X.-W. Zhang, H. He, R.-K. Huang, Z.-M. Ye, Y. Li, D.-D. Zhou, P.-Q. Liao, X.-M. Chen and J.-P. Zhang, *Angew. Chem., Int. Ed.*, 2019, **58**, 7692.
138. D. Antypov, A. Shkurenko, P. M. Bhatt, Y. Belmabkhout, K. Adil, A. Cadiau, M. Suyetin, M. Eddaoudi, M. J. Rosseinsky and M. S. Dyer, *Nat. Commun.*, 2020, **11**, 6099.
139. H. Reinsch, M. A. van der Veen, B. Gil, B. Marszalek, T. Verbiest, D. de Vos and N. Stock, *Chem. Mater.*, 2013, **25**, 17.
140. Q. Tan, H. Huang, Y. Peng, Y. Chang, Z. Zhang, D. Liu and C. Zhong, *J. Mater. Chem. A*, 2019, **7**, 26574.
141. P.-Q. Liao, D.-D. Zhou, A.-X. Zhu, L. Jiang, R.-B. Lin, J.-P. Zhang and X.-M. Chen, *J. Am. Chem. Soc.*, 2012, **134**, 17380.
142. P.-Q. Liao, N.-Y. Huang, W.-X. Zhang, J.-P. Zhang and X.-M. Chen, *Science*, 2017, **356**, 1193.
143. K. Kishida, Y. Okumura, Y. Watanabe, M. Mukoyoshi, S. Bracco, A. Comotti, P. Sozzani, S. Horike and S. Kitagawa, *Angew. Chem., Int. Ed.*, 2016, **55**, 13784.
144. H. Liu, Y. He, J. Jiao, D. Bai, D.-l. Chen, R. Krishna and B. Chen, *Chem. – Eur. J.*, 2016, **22**, 14988.
145. Z. Chen, L. Feng, L. Liu, P. M. Bhatt, K. Adil, A.-H. Emwas, A. H. Assen, Y. Belmabkhout, Y. Han and M. Eddaoudi, *Langmuir*, 2018, **34**, 14546.
146. J. Chen, J. Wang, L. Guo, L. Li, Q. Yang, Z. Zhang, Y. Yang, Z. Bao and Q. Ren, *ACS Appl. Mater. Interfaces*, 2020, **12**, 9609.
147. Z. Zhang, Q. Yang, X. Cui, L. Yang, Z. Bao, Q. Ren and H. Xing, *Angew. Chem., Int. Ed.*, 2017, **56**, 16282.
148. M. Lange, M. Kobalz, J. Bergmann, D. Lässig, J. Lincke, J. Möllmer, A. Möller, J. Hofmann, H. Krautscheid, R. Staudt and R. Gläser, *J. Mater. Chem. A*, 2014, **2**, 8075.
149. Z.-M. Ye, C.-T. He, Y.-T. Xu, R. Krishna, Y. Xie, D.-D. Zhou, H.-L. Zhou, J.-P. Zhang and X.-M. Chen, *Cryst. Growth Des.*, 2017, **17**, 2166.
150. A. H. Assen, T. Virdis, W. De Moor, A. Moussa, M. Eddaoudi, G. Baron, J. F. M. Denayer and Y. Belmabkhout, *Chem. Eng. J.*, 2021, **413**, 127388.
151. Y. Tian, B. Mu, B. Li, X. Li, W. Xu and Y. Lin, *ChemistrySelect*, 2019, **4**, 3841.
152. W. Lu, H. Huang, Z. Hejin, C. Yanjiao, G. Xiangyu, Y. Fan and C. Zhong, *AlChE J.*, 2022, **68**, e17568.
153. N. Oe, N. Hosono and T. Uemura, *Chem. Sci.*, 2021, **12**, 12576.
154. D.-D. Zhou, P. Chen, C. Wang, S.-S. Wang, Y. Du, H. Yan, Z.-M. Ye, C.-T. He, R.-K. Huang, Z.-W. Mo, N.-Y. Huang and J.-P. Zhang, *Nat. Mater.*, 2019, **18**, 994.
155. X. Lian, Y. Fang, E. Joseph, Q. Wang, J. Li, S. Banerjee, C. Lollar, X. Wang and H.-C. Zhou, *Chem. Soc. Rev.*, 2017, **46**, 3386.
156. D. Peralta, G. Chaplais, A. Simon-Masseron, K. Barthelet and G. D. Pirngruber, *Ind. Eng. Chem. Res.*, 2012, **51**, 4692.
157. A. Henrique, A. E. Rodrigues and J. A. C. Silva, *Ind. Eng. Chem. Res.*, 2019, **58**, 378.

158. H. Wang and J. Li, *Acc. Chem. Res.*, 2019, **52**, 1968.
159. R. Herm Zoey, M. Wiers Brian, A. Mason Jarad, M. van Baten Jasper, R. Hudson Matthew, P. Zajdel, M. Brown Craig, N. Masciocchi, R. Krishna and R. Long Jeffrey, *Science*, 2013, **340**, 960.
160. P. A. P. Mendes, F. Ragon, A. E. Rodrigues, P. Horcajada, C. Serre and J. A. C. Silva, *Microporous Mesoporous Mater.*, 2013, **170**, 251.
161. P. A. P. Mendes, A. E. Rodrigues, P. Horcajada, J. Eubank, T. Devic, C. Serre and J. A. C. Silva, *Adsorpt. Sci. Technol.*, 2014, **32**, 475.
162. S. Mukherjee, A. V. Desai and S. K. Ghosh, *Coord. Chem. Rev.*, 2018, **367**, 82.
163. B. L. Suh and J. Kim, *RSC Adv.*, 2020, **10**, 22601.
164. A. Henrique, A. E. Rodrigues and J. A. C. Silva, *Sep. Purif. Technol.*, 2020, **238**, 116419.
165. E. Velasco, S. Xian, H. Wang, S. J. Teat, D. H. Olson, K. Tan, S. Ullah, T. M. Osborn Popp, A. D. Bernstein, K. A. Oyekan, A. J. Nieuwkoop, T. Thonhauser and J. Li, *ACS Appl. Mater. Interfaces*, 2021, **13**, 51997.
166. D. Wang, X. Dong, Y. Han and Y. Liu, *Nano Res.*, 2021, **14**, 526.
167. H. Wang, X. Dong, J. Ding, K. Wang, L. Yu, S. Zhang, Y. Han, Q. Gong, A. Ma and J. Li, *Chem. – Eur. J.*, 2021, **27**, 11795.
168. M. I. H. Mohideen, Y. Belmabkhout, P. M. Bhatt, A. Shkurenko, Z. Chen, K. Adil and M. Eddaoudi, *Chem. Commun.*, 2018, **54**, 9414.
169. L. Chen, S. Yuan, J.-F. Qian, W. Fan, M.-Y. He, Q. Chen and Z.-H. Zhang, *Ind. Eng. Chem. Res.*, 2016, **55**, 10751.
170. Z.-M. Ye, X.-W. Zhang, P.-Q. Liao, Y. Xie, Y.-T. Xu, X.-F. Zhang, C. Wang, D.-X. Liu, N.-Y. Huang, Z.-H. Qiu, D.-D. Zhou, C.-T. He and J.-P. Zhang, *Angew. Chem., Int. Ed.*, 2020, **59**, 23322.
171. H. Wang, X. Dong, J. Lin, S. J. Teat, S. Jensen, J. Cure, E. V. Alexandrov, Q. Xia, K. Tan, Q. Wang, D. H. Olson, D. M. Proserpio, Y. J. Chabal, T. Thonhauser, J. Sun, Y. Han and J. Li, *Nat. Commun.*, 2018, **9**, 1745.
172. A. F. P. Ferreira, M. C. Mittelmeijer-Hazeleger, M. A. Granato, V. F. D. Martins, A. E. Rodrigues and G. Rothenberg, *Phys. Chem. Chem. Phys.*, 2013, **15**, 8795.
173. P. A. P. Mendes, P. Horcajada, S. Rives, H. Ren, A. E. Rodrigues, T. Devic, E. Magnier, P. Trens, H. Jobic, J. Ollivier, G. Maurin, C. Serre and J. A. C. Silva, *Adv. Funct. Mater.*, 2014, **24**, 7666.
174. H. Wang, X. Dong, E. Velasco, D. H. Olson, Y. Han and J. Li, *Energy Environ. Sci.*, 2018, **11**, 1226.
175. R. Numaguchi, H. Tanaka, S. Watanabe and M. T. Miyahara, *J. Chem. Phys.*, 2014, **140**, 044707.
176. Q. Chen, S. Xian, X. Dong, Y. Liu, H. Wang, D. H. Olson, L. J. Williams, Y. Han, X.-H. Bu and J. Li, *Angew. Chem., Int. Ed.*, 2021, **60**, 10593.
177. A. A. Lysova, D. G. Samsonenko, P. V. Dorovatovskii, V. A. Lazarenko, V. N. Khrustalev, K. A. Kovalenko, D. N. Dybtsev and V. P. Fedin, *J. Am. Chem. Soc.*, 2019, **141**, 17260.
178. B. Joarder, S. Mukherjee, A. K. Chaudhari, A. V. Desai, B. Manna and S. K. Ghosh, *Chem. – Eur. J.*, 2014, **20**, 15303.

179. S. Mukherjee, D. Sensharma, O. T. Qazvini, S. Dutta, L. K. Macreadie, S. K. Ghosh and R. Babarao, *Coord. Chem. Rev.*, 2021, **437**, 213852.
180. M. Rasouli, N. Yaghobi, S. Chitsazan and M. H. Sayyar, *Microporous Mesoporous Mater.*, 2012, **150**, 47.
181. Y. Yang, P. Bai and X. Guo, *Ind. Eng. Chem. Res.*, 2017, **56**, 14725.
182. V. R. Choudhary, V. S. Nayak and T. V. Choudhary, *Ind. Eng. Chem. Res.*, 1997, **36**, 1812.
183. M. Minceva and A. E. Rodrigues, *Chem. Eng. Res. Des.*, 2004, **82**, 667.
184. J. A. Gee, K. Zhang, S. Bhattacharyya, J. Bentley, M. Rungta, J. S. Abichandani, D. S. Sholl and S. Nair, *J. Phys. Chem. C*, 2016, **120**, 12075.
185. M. Tu, S. Wannapaiboon, K. Khaletskaya and R. A. Fischer, *Adv. Funct. Mater.*, 2015, **25**, 4470.
186. D. Peralta, G. Chaplais, A. Simon-Masseron, K. Barthelet, C. Chizallet, A.-A. Quoineaud and G. D. Pirngruber, *J. Am. Chem. Soc.*, 2012, **134**, 8115.
187. D. Peralta, G. Chaplais, J.-L. Paillaud, A. Simon-Masseron, K. Barthelet and G. D. Pirngruber, *Microporous Mesoporous Mater.*, 2013, **173**, 1.
188. D. M. Polyukhov, A. S. Poryvaev, S. A. Gromilov and M. V. Fedin, *Nano Lett.*, 2019, **19**, 6506.
189. D. Peralta, K. Barthelet, G. Pirngruber, G. Chaplais, A. Simon and J. Patarin, *Int. Pat. Appl.*, WO2013011210A2013011211, 2013.
190. X. Yang, H.-L. Zhou, C.-T. He, Z.-W. Mo, J.-W. Ye, X.-M. Chen and J.-P. Zhang, *Research*, 2019, **2019**, 9463719.
191. M. du Plessis, V. I. Nikolayenko and L. J. Barbour, *J. Am. Chem. Soc.*, 2020, **142**, 4529.
192. L. Yang, H. Liu, D. Yuan, J. Xing, Y. Xu and Z. Liu, *ACS Appl. Mater. Interfaces*, 2021, **13**, 41600.
193. J. Lannoeye, B. Van de Voorde, B. Bozbiyik, H. Reinsch, J. Denayer and D. De Vos, *Microporous Mesoporous Mater.*, 2016, **226**, 292.
194. M. Shivanna, K.-i. Otake, J.-J. Zheng, S. Sakaki and S. Kitagawa, *Chem. Commun.*, 2020, **56**, 9632.
195. S. Mukherjee, B. Joarder, B. Manna, A. V. Desai, A. K. Chaudhari and S. K. Ghosh, *Sci. Rep.*, 2014, **4**, 5761.
196. S.-Q. Wang, S. Mukherjee, E. Patyk-Kaźmierczak, S. Darwish, A. Bajpai, Q.-Y. Yang and M. J. Zaworotko, *Angew. Chem., Int. Ed.*, 2019, **58**, 6630.
197. K. J. Hartlieb, J. M. Holcroft, P. Z. Moghadam, N. A. Vermeulen, M. M. Algaradah, M. S. Nassar, Y. Y. Botros, R. Q. Snurr and J. F. Stoddart, *J. Am. Chem. Soc.*, 2016, **138**, 2292.
198. M. Maes, F. Vermoortele, L. Alaerts, J. F. M. Denayer and D. E. De Vos, *J. Phys. Chem. C*, 2011, **115**, 1051.
199. S.-I. Kim, A.-R. Kim, S.-Y. Kim, J.-Y. Lee and Y.-S. Bae, *Sep. Purif. Technol.*, 2020, **242**, 116758.
200. M. Maes, F. Vermoortele, L. Alaerts, S. Couck, C. E. A. Kirschhock, J. F. M. Denayer and D. E. De Vos, *J. Am. Chem. Soc.*, 2010, **132**, 15277.

201. T. Remy, L. Ma, M. Maes, D. E. De Vos, G. V. Baron and J. F. M. Denayer, *Ind. Eng. Chem. Res.*, 2012, **51**, 14824.
202. S. Mukherjee, B. Joarder, A. V. Desai, B. Manna, R. Krishna and S. K. Ghosh, *Inorg. Chem.*, 2015, **54**, 4403.
203. M. Chang, Y. Zhao, D. Liu, J. Yang, J. Li and C. Zhong, *Sustainable Energy Fuels*, 2020, **4**, 138.
204. Z. Niu, X. Cui, T. Pham, P. C. Lan, H. Xing, K. A. Forrest, L. Wojtas, B. Space and S. Ma, *Angew. Chem., Int. Ed.*, 2019, **58**, 10138.
205. M.-H. Yu, B. Space, D. Franz, W. Zhou, C. He, L. Li, R. Krishna, Z. Chang, W. Li, T.-L. Hu and X.-H. Bu, *J. Am. Chem. Soc.*, 2019, **141**, 17703.
206. L.-H. Xie, X.-M. Liu, T. He and J.-R. Li, *Chem*, 2018, **4**, 1911.
207. N. Yanai, K. Kitayama, Y. Hijikata, H. Sato, R. Matsuda, Y. Kubota, M. Takata, M. Mizuno, T. Uemura and S. Kitagawa, *Nat. Mater.*, 2011, **10**, 787.
208. X.-L. Qi, R.-B. Lin, Q. Chen, J.-B. Lin, J.-P. Zhang and X.-M. Chen, *Chem. Sci.*, 2011, **2**, 2214.
209. S. Chong, S. M. J. Rogge and J. Kim, *Chem. Mater.*, 2022, **34**, 254.
210. S. M. Neville, B. Moubaraki, K. S. Murray and C. J. Kepert, *Angew. Chem., Int. Ed.*, 2007, **46**, 2059.
211. M. Zhong, H. Li, J. Chen, L. Tao, C. Li and Q. Yang, *Chem. – Eur. J.*, 2017, **23**, 11504.
212. J. Li, H. Huang, W. Xue, K. Sun, X. Song, C. Wu, L. Nie, Y. Li, C. Liu, Y. Pan, H.-L. Jiang, D. Mei and C. Zhong, *Nat. Catal.*, 2021, **4**, 719.
213. H.-L. Zhou, R.-B. Lin, C.-T. He, Y.-B. Zhang, N. Feng, Q. Wang, F. Deng, J.-P. Zhang and X.-M. Chen, *Nat. Commun.*, 2013, **4**, 2534.
214. H.-L. Zhou, Y.-B. Zhang, J.-P. Zhang and X.-M. Chen, *Nat. Commun.*, 2015, **6**, 6917.
215. X. Yu, J. Zhou, H. Liang, Z. Jiang and L. Wu, *Prog. Mater. Sci.*, 2018, **94**, 114.
216. Z.-M. Ye, K.-P. Xie, C. Wang, X.-W. Zhang, H.-L. Zhou and J.-P. Zhang, *Inorg. Chem.*, 2021, **60**, 11893.
217. S. Bureekaew, S. Horike, M. Higuchi, M. Mizuno, T. Kawamura, D. Tanaka, N. Yanai and S. Kitagawa, *Nat. Mater.*, 2009, **8**, 831.
218. S. Horike, D. Umeyama and S. Kitagawa, *Acc. Chem. Res.*, 2013, **46**, 2376.
219. F. P. Billingsley and M. Krauss, *J. Chem. Phys.*, 1974, **60**, 2767.
220. A. Halkier and S. Coriani, *Chem. Phys. Lett.*, 1999, **303**, 408.
221. V. W. Couling and S. S. Ntombela, *Chem. Phys. Lett.*, 2014, **614**, 41.
222. C. J. Werkhoven, P. A. Geldof, M. F. M. Post, J. Langelaar, R. P. H. Rettschnick and J. D. W. van Voorst, *Chem. Phys. Lett.*, 1971, **9**, 6.
223. M. Mantina, A. C. Chamberlin, R. Valero, C. J. Cramer and D. G. Truhlar, *J. Phys. Chem. A*, 2009, **113**, 5806.
224. Z. Zhang, S. B. Peh, C. Kang, K. Chai and D. Zhao, *EnergyChem*, 2021, **3**, 100057.
225. G. Zhenbin, Z. Yuanyuan and F. Xiao, *Acta Chim. Sinica*, 2020, **78**, 397.
226. P. Dagaut and M. Cathonnet, *Combust. Flame*, 1999, **118**, 191.

227. P. Ghosh, K. J. Hickey and S. B. Jaffe, *Ind. Eng. Chem. Res.*, 2006, **45**, 337.
228. L. Yang, S. Qian, X. Wang, X. Cui, B. Chen and H. Xing, *Chem. Soc. Rev.*, 2020, **49**, 5359.
229. M. Gehre, Z. Guo, G. Rothenberg and S. Tanase, *ChemSusChem*, 2017, **10**, 3947.
230. Y. Matsumura, K. Hashimoto, H. Kobayashi and S. Yoshida, *J. Chem. Soc., Faraday Trans.*, 1990, **86**, 561.

CHAPTER 3

The Dynamic View: Multiscale Characterisation Techniques for Flexible Frameworks

V. BON[a], K. ROZTOCKI[b,c], P. IACOMI[d], C. L. HOBDAY[e], I. SENKOVSKA[a], A. PÖPPL[f], E. BRUNNER[g] AND S. KASKEL[a*]

[a]Chair of Inorganic Chemistry I, Faculty of Chemistry and Food Chemistry, TU Dresden, 01062 Dresden, Germany; [b]Faculty of Chemistry, Adam Mickiewicz University, Uniwersytetu Poznańskiego 8, 61-614 Poznań, Poland; [c]Center for Advanced Technologies, Adam Mickiewicz University, Uniwersytetu Poznańskiego 10, 61-614 Poznań, Poland; [d]ICGM University of Montpellier, CNRS ENSCM, F-34095 Montpellier, France; [e]School of Chemistry and Centre for Science at Extreme Conditions, The University of Edinburgh, King's Buildings, David Brewster Road, Edinburgh, EH9 3FJ, Scotland, UK; [f]Felix-Bloch-Institut für Festkörperphysik, Fakultät für Physik und Geowissenschaften, Universität Leipzig, Linnéstraße 5, 04103 Leipzig, Germany; [g]Chair of Bioanalytical Chemistry, Faculty of Chemistry and Food Chemistry TU Dresden, 01062 Dresden, Germany
*E-mail: stefan.kaskel@tu-dresden.de

3.1 Introduction

For dynamic framework materials the term 'structure' expands into multiple dimensions as it becomes dependent on thermodynamic variables such as gas pressure, external pressure, temperature *etc.* in a much wider sense as compared to traditional rigid porous solids. Colossal structural

Inorganic Materials Series No. 13
Flexible Metal–Organic Frameworks: Structural Design, Synthesis and Properties
Edited by Susumu Kitagawa

Published by the Royal Society of Chemistry, www.rsc.org

transformations even involving topological transformations may occur along an adsorption isotherm or during the application of novel flexible metal–organic frameworks (MOFs) in gas storage, catalysis, separation *etc.* Such structural changes often remain undetected due to the limited availability of parallelised (*in situ*) characterisation techniques.[1–3] Along the transformation trajectory initiated by guest inclusion a whole realm of structures may open up, but only advanced characterisation tools in parallel to controlled adsorption instrumentation can unravel them. Both, global and local *in situ* techniques are important. In particular, diffraction techniques give insights into crystal structure changes in periodic frameworks (X-ray, electron and neutron), total scattering provides information on cluster configurations, but also spectroscopic techniques are important (NMR, UV-vis, EPR) to detect local geometric and electronic changes of linker and cluster configurations.[4–7] More importantly, they offer distinctive insights into host–guest interactions. Deeper insight into lattice vibrations is obtained from Raman spectroscopy or inelastic neutron scattering.[8,9] In recent years instrumentation to couple adsorption equipment with the above-mentioned techniques has been significantly advanced for powders and single crystals by leading experts worldwide. In the following, we highlight the most important advanced methods as they are relevant for dynamic MOFs. In the following, the term '*in situ*' characterises the parallelisation of adsorption measurements with other characterisation methods while '*ex situ*' is the classical static observation of a single state before or after a transformation. Some '*in situ*' techniques only control the gas pressure but do not measure the gas uptake. Advanced methods allow to measure a full adsorption isotherm while the sample is in the beam by coupling an automated adsorption instrument to the diffractometer or spectrometer. The term '*operando*' refers to a porous material being investigated under real operation conditions. Especially, in catalysis '*operando*' and '*in situ*' techniques differ significantly. In adsorption processes, in academia adsorbents are frequently tested under idealised laboratory conditions, hence in these cases the term '*in situ*' is more appropriate, despite both terms being often used synonymously.

An important aspect of framework flexibility is also time-dependence.[10] In many cases, metastable states and hystereses in the isotherms are observed and the transformations have to be classified as 'irreversible' in a thermodynamic sense. The latter implies that sample properties become strongly history dependant, which is often a challenge. Frustrations may arise in such cases as the observations appear not reproducible, however, in fact, they are a result of a poor understanding of the factors affecting sample history. Capturing the complete spatiotemporal evolution of dynamic frameworks requires new developments involving time-resolved methods on time scales ranging from nanoseconds to hours or even years. This field is still in its infancy and this chapter can only describe these early beginnings. We hope the comprehensive chapter will be a useful guide for students and experts and may inspire them to advance the field further. The highly interdisciplinary field profits from understandings in chemistry, physics, and computational methods. The 'dynamic view' on materials is a rapidly evolving field with bright prospects.

3.2 Indicators of Framework Flexibility *Via 'Ex Situ'* Characterisation

3.2.1 Desolvation of the Framework

Usually, the guest molecules present in the pores of MOFs after synthesis are removed in order to access porosity. The removal, however, can cause changes to the framework and at this step, first evidence about the potential flexibility may be observed. The framework changes upon desolvation are diverse: (i) irreversible amorphisation (first generation of MOFs); (ii) amorphisation, which can be reversed by resolvation/adsorption, (iii) structural transition (reversible or irreversible in nature) under formation of new crystalline phase (third generation of MOFs).[11,12,12–14]

The removal of guest molecules from the pores can be realised using different methods, mostly differing in the stress acting on the framework upon the desorption and kinetics of desorption process, which also influences the state of the framework afterwards. The stress acting on the framework during desolvation depends on the surface tension and capillary forces imposed on the structure by the liquid-to-gas phase transformation of included solvent molecules, which are high, especially when the solvent has a high boiling point and/or high surface tension.[15]

Traditional desolvation includes heating the MOF material under dynamic vacuum (so-called thermal treatment), implying strong capillary forces causing high strain during solvent desorption. Fragile frameworks with low bulk moduli tend to collapse upon classical desolvation. Flexible frameworks, however, often respond to their desolvation stress by structural rearrangement.

One alternative procedure,[16] with the advantage of avoiding high temperature, is to exchange an initial high-boiling point solvent, remaining in the pores after synthesis, by a solvent with a lower boiling point/lower surface tension such as dichloromethane, chloroform, acetone, *n*-heptane, methanol or ethanol followed by evacuation under relatively low temperature. Reduced stress leads, in some cases, to the preservation of the initial structure (metastability), even if it is less energetically favourable.

Very gentle desolvation techniques include supercritical drying (SCD).[17] SCD avoids the liquid-to-gas phase transition and, thus, any capillary forces. The first step in SCD usually (but not necessary) involves solvent exchange, where the guest molecules used in the synthesis (usually DMF) are exchanged with a solvent of good miscibility with liquid CO_2. Ethanol, acetone, or amyl acetate are often used for this purpose. In the second step, the selected solvent is exchanged again by liquid CO_2. After the framework is completely soaked with the CO_2, the system can be heated and pressurised above the supercritical conditions for CO_2 and the supercritical fluid can be released.

After supercritical drying, when the temperature approaches room temperature, the pores are filled with gaseous carbon dioxide, which can be easily removed prior to the adsorption experiments. In some cases, the interaction between CO_2 and the framework is significant even at ambient conditions, resulting in a framework transition when the vacuum is applied to remove CO_2. In these cases, it is advisable to exchange CO_2 with nitrogen or argon prior to evacuation.

To some extent, depending on the guest molecules present in the pore and the desolvation technique applied, the forces acting on the framework can be varied, and indirect information about the robustness of the framework can be gleaned. For the flexible frameworks, some evidence about the flexibility, as well as an energetic barrier between the phases can be gained. Moreover, by gentle solvent removal, the metastable phases can be captured (see Figure 3.1).

In DUT-8, a representative of pillared layer switchable MOFs, containing Ni, Co, Zn or Cu paddle wheels, the phase present after desolvation is strongly dependent on the metal forming the paddle wheel, as well as on the desolvation conditions. DUT-8(M) (M = Ni, Co, Zn) MOF ($[M_2(2,6\text{-ndc})_2(\text{dabco})]_n$, 2,6-ndc = 2,6-naphthalenedicarboxylate, dabco = 1,4-diazabicyclo-[2.2.2]-octane, DUT = Dresden University of Technology), belongs to the 'gate pressure' MOFs, switching from the less energetically stable open pore (op) form to the more favourable closed pore (cp) form in the course of classical desolvation.[18–21] If all compounds were subjected to SCD, the DUT-8(Ni) sample transforms completely to the cp form, whereas DUT-8(Zn) and DUT-8(Co) show a mixture of the cp and op phases, where the op phase is dominating for Zn based sample and cp phase prevails for Co. DUT-8(Cu) is

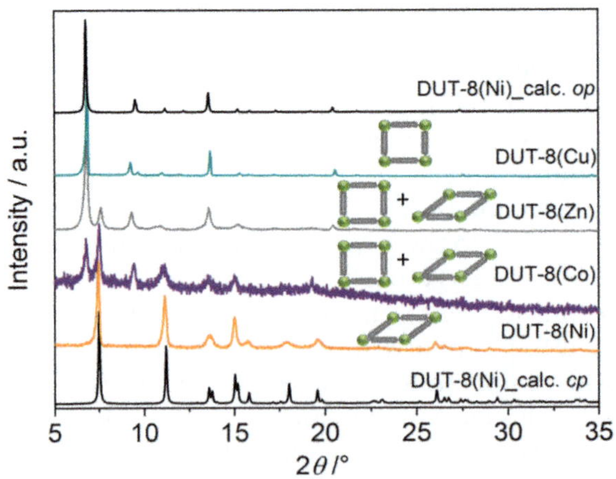

Figure 3.1 PXRD patterns of DUT-8(M) (M = Ni, Co, Zn, Cu) after SCD desolvation in comparison to the theoretical patterns, calculated from the crystal structures of op and cp phases.

preserved in the op state. Since the compounds are isomorphous, the distribution of the dried phases is attributed to the rigidity/softness of the corresponding paddle wheels.

Similar observations were made for the mesoporous breathing DUT-49(M) framework, based on a paddle wheel containing metal–organic polyhedra.[22,23] Mn, Fe, Co, Zn, Cd containing DUT-49 frameworks show no crystallinity and very low remaining porosity after SCD. DUT-49(Ni) stays crystalline and shows moderate porosity and flexibility, whereas DUT-49(Cu) shows the highest porosity and flexibility with NGA.[24]

Brammer *et al.* showed that desolvation of $\{Me_2NH_2[In(abdc)_2]\}_n$ (SHF-61, abdc = 2-aminobenzene-1,4-dicarboxyliate) from two different solvents (chloroform or $DMF–H_2O$) leads to two polymorphic-activated forms with very different pore openings and markedly different gas-adsorption properties.[25] Removal of $CHCl_3$ leads to little change in the host framework, which exhibits a high porosity. Desolvation from $DMF–H_2O$ leads to significant framework contraction and a low porosity phase. Interestingly, the framework contraction during removal of $DMF–H_2O$ is continuous, and partial desolvation produces partially contracted phases.

It should be mentioned that in non-flexible MOFs the diffusion pathway for solvent removal is persistent for the whole activation period, ensuring the continuous movement of guests from the crystal inside to the surface. In flexible compounds, however, where pores change their geometry during solvent removal, the residence time of the solvent and the diffusion kinetics through the channels might play an additional crucial role in the activation process and should not be overlooked. The concerted synchronous flexing of the network and fast contraction of the pores may lead to grain boundary formation or even cracking of the crystals, but also the obstruction of the diffusion pathways and, as a result, presence of regions with trapped solvent molecules are conceivable.

The trapped molecules can lead to evaporation-induced nanoconfined cavitation effects and consequently to high evaporation rates.[26,27] Cavitation is the phenomenon of vapour bubbles formation beyond a certain limit of reduced pressure. Cavitation events caused spots of enormous pressure and temperature increase. If the stress level is high enough and the rate of material deformation is fast enough, the crystals would undergo damage, cracking, fragmentation, and formation of defects. Since the particle size, shape and texture are critical for flexibility, the meso/macroscopic texture after solvent removal may also determine the flexible properties of resulting materials.[28–32]

Moreover, the simultaneous switching and solvent removal may lead to surface phenomena since the unit cells nearby the surface may transform faster than the unit cells situated in the bulk, leading to phase co-existence in the single particle.

Krautscheid *et al.*, showed the impact of solvent removal and cracking of the crystals on the flexibility using the framework with composition $[Cu(Me\text{-}4py\text{-}trz\text{-}ia)]_n$ (Me-4py-trz-ia = 3-methyl-5-pyridine-4-yl substituted triazolyl isophthalate).[33] Water molecules at the Cu^{2+} centres in the as-synthesised material and in material obtained after incomplete solvent exchange with methanol cause the formation of the intermediate narrow pore phase during the activation under vacuum before the fully activated state is reached. Consequently, not only longer activation times or higher temperatures are needed to fully desolvate the MOF but also fragmentation of the crystals occurs. If a complete exchange of all solvent molecules by methanol was performed before desolvation, the transformation to the intermediate phase is not observed and the MOF can be activated at significantly lower temperatures without fragmentation.

Abylgazina *et al.*, have shown for DUT-8(Zn) that the solvent removal process and desolvation stress contracting the framework, significantly controls the closed pore–open pore phase ratio after desolvation and, subsequently, the adsorption induced switchability characteristics of the desolvated samples.[21]

Interesting phenomena regarded as 'shape memory effect' were reported for pillared layered interpenetrated MOFs by Kitagawa *et al.*, for $[Cu_2(bdc)_2(bpy)]_n$[28] and Zaworotko *et al.*, for $[Zn_2$(4,4′-biphenyldicarboxylate)$_2$(1,4-bis(4-pyridyl)benzene)$]_n$.[34] The op structures of regular large MOF crystals could be transformed to the cp phases by solvent removal accompanied by thermal treatment. In smaller crystals a kinetically stabilised open pore phase is obtained. Conversion of the metastable empty op phase to the cp phase requires additional energy input to overcome the activation barrier, which can be realised by heating.

The profound impact of guest species in the pores on structure was also shown by Rosseinsky *et al.* for flexible $[Zn_7O_2(csa)_{10}L_2]$ containing the asymmetric *N*-(4-carboxyphenyl)succinamate (csa) linkers and zinc oxo-clusters involving two coordinated terminal ligands (L).[35] The full DFT optimisation predicts the larger structural change upon removal of only uncoordinated guests for L = DMSO, when compared to L = DMF. The overall mechanism leads to a large reduction in the accessible void space (Figure 3.2).

Ehrling *et al.* reported on the adaptive response of DUT-8(Ni) to the solvent exchange in the pores through reversible linker based disorder–disorder transitions.[36] Solvent exchange stimulates the formation of distinct disordered frameworks containing monoclinic and tetragonal domains (Figure 3.3), as demonstrated by high-resolution transmission electron microscopy and diffraction techniques. The guest exchange was shown to stimulate repeatable switching transitions between distinct disorder states.

The theoretical calculations show that the open pore phase to the closed pore phase transition is energetically favourable for the monoclinic domains only, therefore upon desolvation, all domains should be transformed to monoclinic first, before the framework can transform to the cp state.[37] Indeed, the *in situ* XRD experiments conducted upon desolvation prove this hypothesis[36] (Figure 3.3(d)).

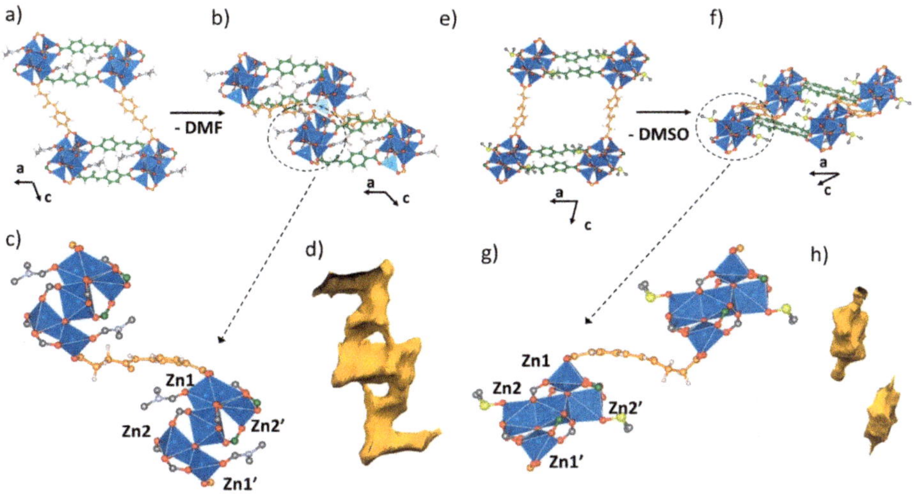

Figure 3.2 (a)–(d) View on $[Zn_7O_2(csa)_{10}DMF_2]n$. (a), (b) View along the *b* axis of predicted DMF structure before (a) and after (b) guest removal. (c) Two SBUs of desolvated structure connected by a single linker. (d) Connolly surface of desolvated structure showing the 1D channel remaining after structure collapse. (e)–(h) View on $[Zn_7O_2(csa)_{10}DMSO_2]n$. (e), (f) View along the *b* axis before (e) and after (f) DMSO removal. (g) Two SBUs connected by a single linker. (h) Connolly surface of showing the removal of all channels. Reproduced from ref. 35, http://doi.org/10.1021/acs.cgd.9b00558, under the terms of the CC BY 4.0 license https://creativecommons.org/licenses/by/4.0/.

Thus, the solvent removal process is one of the more crucial steps in flexible MOFs handling. For DUT-8(Ni), the analysis of 50 isotherms shows that the main discrepancies in the collected isotherms originate from the solvent exchange and removal process. Only an accurate control of the sample handling provides the ability to reproduce the physisorption data.[38] Therefore, it is of primary importance to exactly record and report the history of the samples and all steps undertaken upon solvent removal as well as desolvation conditions.[39]

3.2.2 Characterisation by PXRD

Flexibility in MOFs can be recognised at the early stage of the framework characterisation using conventional PXRD phase analysis.[40] After the synthesis, the pore system of MOFs in most of the cases is filled with guest molecules. Therefore, measurement of PXRD patterns on the MOF samples before and after desolvation may be considered as first indication of flexibility. In case of the formation of the contracted or closed pore phase, completely different reflection patterns are observed and smaller interplanar distances in cp structure result in a high-angle shift of reflections (Figure 3.4(d)). If the particle size distribution in the sample is too broad or desolvation is

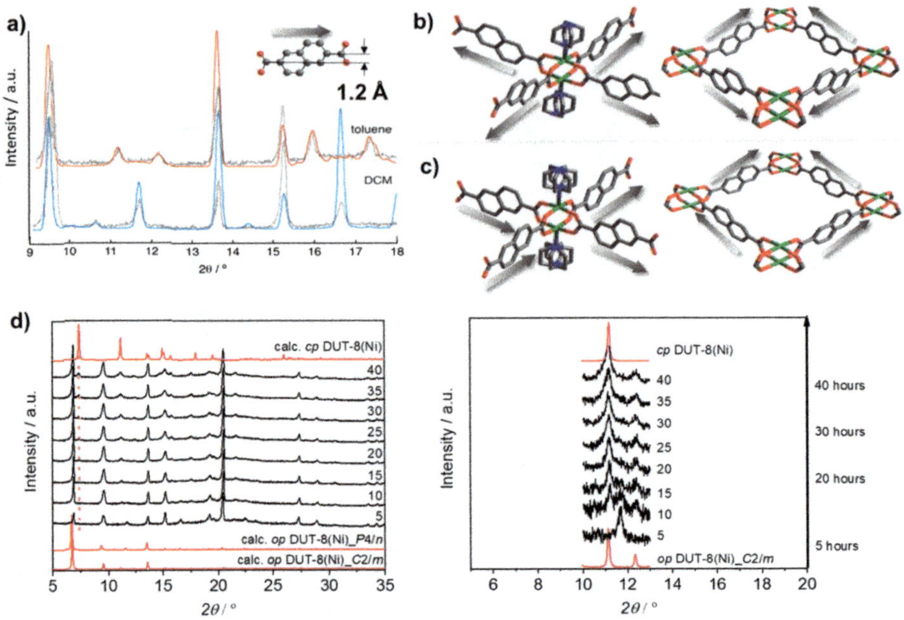

Figure 3.3 (a) Experimental (in grey) and calculated (in colour) X-ray diffraction patterns of DUT-8(Ni) containing toluene (blue) or dichloromethane (orange) in the pores. The differences result from different disordered states of the frameworks. Inset shows the step-function of the 2,6-ndc linker causing the framework disorder. (b), (c) Paddle wheel cluster showing two possible linker configurations: (b) the all linkers showing in top direction (regarding the step-function of the linker) as well as resulting configuration of the pore walls and tetragonal domains, (c) two linkers showing to the top and two linker pointing down and corresponding pore walls configuration of monoclinic domains; (d) desolvation of DUT-8(Ni) containing DCM in the pores monitored by PXRD. DUT-8(Ni) loaded with DCM shows the signature of the tetragonal domains shown in b. After *ca.* 12 h three reflections in the range between 10 and 13° are detected indicating structural transformation to monoclinic domains shown in *c*. After 25 h only the monoclinic phase is visible and a small peak at 7.52°, which can be assigned to the cp phase of DUT-8(Ni). In the further course the network closes further on, as evident from the intensity increase of the peak at 7.52°.

incomplete for some reason, the mixture of op and cp phases is observed in PXRD patterns (Figure 3.4(c)). The measurement of the desolvated phase should be conducted in inert conditions to avoid the adsorption of water vapours or other impurities from the air, which may cause either decomposition or reopening of the frameworks. Scotch tape often used for PXRD sample preparation contains water in the adhesive and is not suitable for high-quality data recording.

In some cases, cp phase is thermodynamically more stable than the guest-filled op phase, the flexibility is most likely irreversible, and the structure

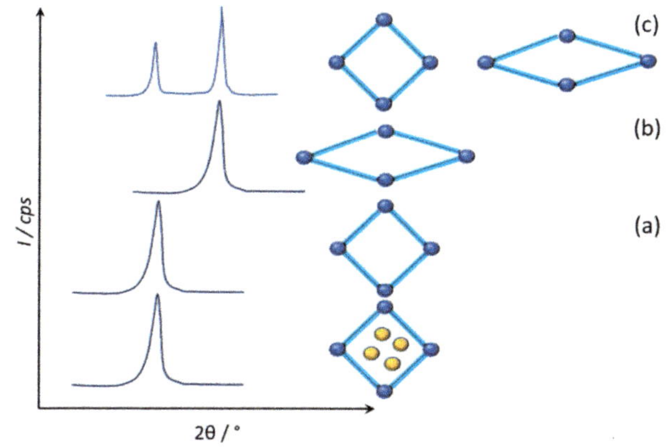

Figure 3.4 Schematic representation of the peak shift in PXRD patterns upon desolvation of rigid (a) and flexible MOFs (b), (c).

cannot be returned in the initial guest-filled phase because of high activation barrier. A most effective way to prove the reversibility of guest-induced transitions is to soak the activated framework in the solvent used in the synthesis. However, only flexibility modes with large 'gating' amplitude can be unravelled using this technique. In case of minor changes in the unit cell, ligand flipping or breathing, in which the guest-filled and desolvated structures are similar, no signs of flexibility can be observed in *ex situ* measured PXRD patterns and further characterisation techniques such as gas physisorption or *in situ* PXRD are necessary to analyse the flexible behaviour.

3.2.3 Adsorption of Gases and Vapours

Adsorption is the enrichment of one or more components in an interfacial layer.[41] In classical understanding it is based on solid–fluid interactions (which are dependent on the pore surface functionality and the pore diameter in a rigid adsorbent) and on the fluid–fluid interactions.

The kinetics and thermodynamics of adsorption in flexible MOFs, however, is coupled to the structural transformation of the adsorbent (mostly associated with the changes in the pore and pore window size), leading to complex and hitherto unexpected adsorption isotherms.[1,14,42] Depending on the type of the transformation (Figure 3.5) there are different isotherm profiles expected, but in principle, one can summarise some general attributes typical for the isotherms of flexible materials.

Most of the structural transitions in the MOFs reported so far, belong to the first-order phase transitions, exhibiting a discontinuity in the first derivative of the free energy with respect to some thermodynamic variables. In case of adsorption induced transitions, the structural change is associated with specific gas pressure (chemical potential).

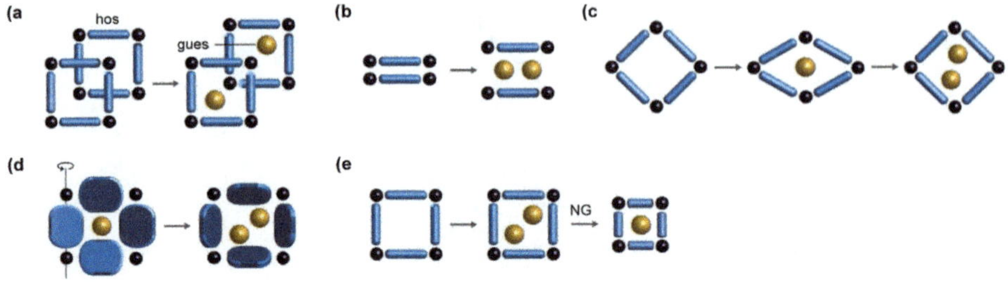

Figure 3.5 Schematic representations of the modes of adsorption-induced deformation in flexible MOFs: (a) subnetwork displacement, (b) gating, (c) breathing, (d) linker rotation, and (e) contraction showing the NGA. Reproduced from ref. 43 with permission from Elsevier, Copyright 2019.

Thus, the adsorption isotherm of a flexible MOF material is expected to be a combination of the isotherms for the individual crystalline phases possessing different pore volumes,[43,44] where the switch from one isotherm to another (discontinuous change in adsorbed amount, further referred to as step) takes place at a characteristic gas pressure of phase transition.

The first attribute of the flexible material isotherms, often observed, is the presence of the steps, which do not originate from the condensation of the adsorbate. The structural changes in the crystalline solid lead to the various adsorption isotherm profiles, depending on the type of the pore volume change upon transitions.

Type 1: If $\frac{\Delta V}{V} > 0$ (where $\Delta V = V - V_0$, is the difference between the pore volume after transition (V) and initial pore volume (V_0)). Such a transformation pathway is most frequently referred to as gate opening (one-step opening, closed/contracted pore (cp) phase to open pore phase (op) transition). In this case, cp is energetically below op ($F_{cp} < F_{op}$, F = Helmholtz free-energy of the empty host).[10] Gate opening occurs if the guest stabilises the op form. The isotherm is characterised by a stepwise increase in adsorbed amount (Figure 3.6(b)). Such a feature in the isotherm is characteristic for subnetwork displacement,[45,46] gating,[47–50] linker rotation[51–53] and is included as the second step in the breathing transition (Figure 3.5).[54–57] Successive gate opening events would lead to the multi-step isotherms (Figure 3.6(c)).[58,59]

Type 2: If $\frac{\Delta V}{V} < 0$ upon gas pressure increase, such transformation pathway can be considered as contraction (gate closing). In this case the empty open pore framework is energetically below the empty contracted pore phase (breathing), or the op phase is metastable. The contraction occurs if the guest stabilises the cp form with reduced but non-zero porosity (point marked by a circle in Figure 3.6(d)). This stabilisation is driven by the enhanced adsorption enthalpy ($-\Delta_{ads}H$) of the guest for the cp host ($|\Delta_{ads}H_{np}| > |\Delta_{ads}H_{op}|$). The typical adsorption isotherm exhibits a plateau with uptake lower than

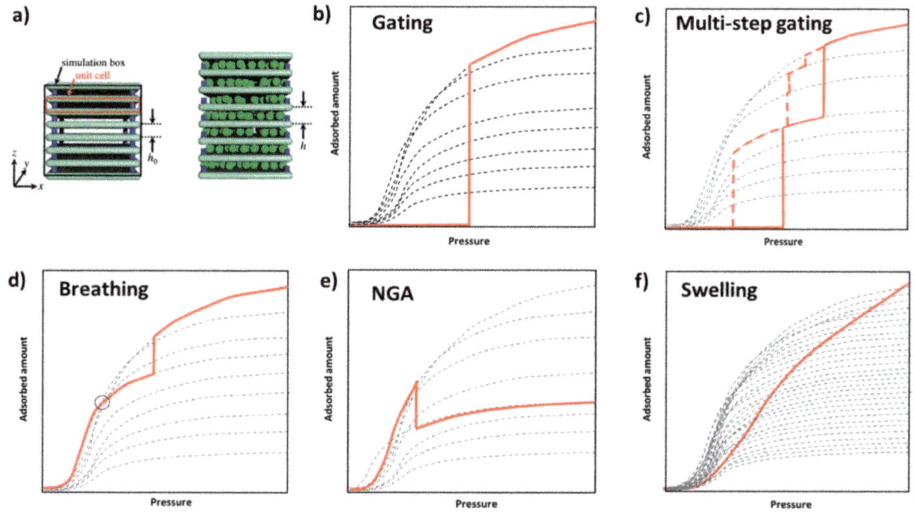

Figure 3.6 (a) Simplified model (closed pore phase) with interlayer distance h_0 (left); and of the open pore phase with guest molecules with interlayer distance h_0 (right) ($h_0 < h$). Adapted from ref. 265 with permission from American Institute of Physics, Copyright 2013. (b)–(f) Schematic representations of adsorption isotherms, which are typical for a simple model mimicking the stacked-layer flexible MOF shown in (a), are presented in grey. Red solid lines show hypothetical adsorption isotherms for: gate opening behaviour (a), multistep gate opening (b), breathing: op to cp followed by cp to op (d), NGA (e), and swelling (f). The dashed red line in (c) shows the desorption branch.

expected for the op form, therefore it is difficult to recognise from the adsorption isotherm profile only. The shape is very similar to the type I isotherms, characteristic for non-flexible MOFs, therefore such transition can be easily overlooked. Indeed, there are few reports in the literature addressing flexible MOFs exhibiting contraction without subsequent reopening. Contraction is a typical first structural transition in breathing MOFs, such as MIL-53,[54–57] where it is followed by re-opening.

Type 3: Successive gate closing and gate opening transitions upon gas pressure increase are referred to as breathing (Figure 3.6(d)).[54–57]

Type 4: The special case of contraction (gate closing) transitions is the 'negative gas adsorption' (NGA) profile.[23,60–63] In this case the contraction (op–cp) has a high activation barrier leading to a material that pushes out previously adsorbed guest molecules, resulting in the negative step in uptake on the adsorption isotherm.

The ideal isotherm of the perfect flexible single crystallite (single grain) of MOF undergoing first order phase transition would contain steps of rectangular shape,[64] where the only change in porosity is due to structure transition events (Figure 3.7(a)). But the steps in adsorption isotherms usually have

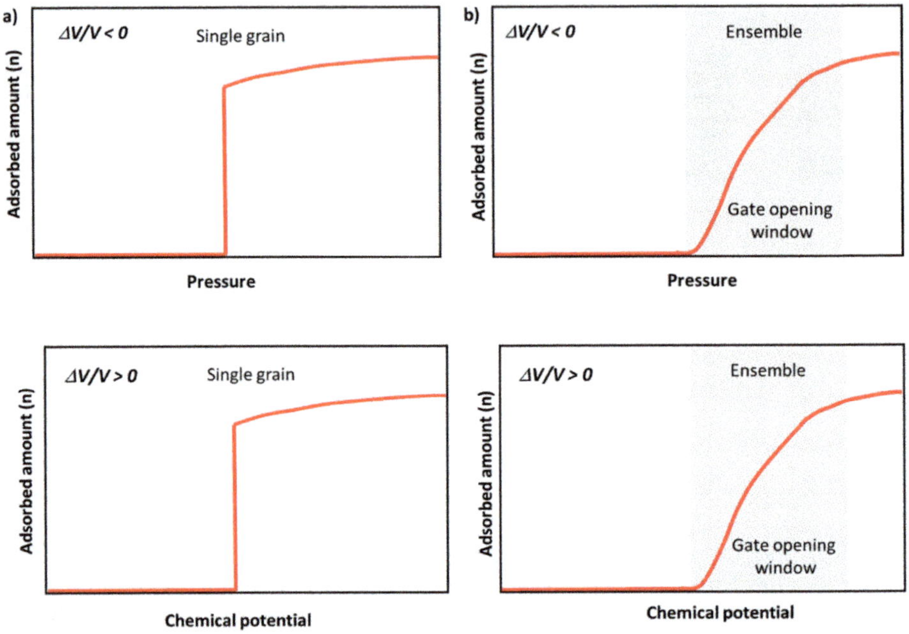

Figure 3.7 (a) Ideal gating profile. (b) In the case of adsorption induced phase-transition, the 'slope' of gate opening shares a trade-off relationship with the width of the gate opening window, roughly described as $d(\mathrm{n})/d(\mathrm{p})$.

a more complex shape. In a real crystal, there are always lattice defects and as a consequence, the transition pressure is non-uniform. In addition to that, the grain boundaries carry interfacial energies, so that the local effects due to the changes of the lattice curvature should not be neglected. Moreover, the adsorption measurements are usually performed on powder (ensemble of single crystallite and their agglomerates), representing quasi 'non-interacting grains'. Thus, the distribution of nucleation barriers of the crystallites controls the transition pressure spread over a wider pressure range (gate opening window). This is typically reflected by the less-steep adsorption steps in the phase transition regions (Figure 3.7(b)).[31,32]

The second reason for broadening of the phase transition window can be the change from first-order to second-order-like transition (*Type 5*). These materials are characterised by a gradual increase in the unit-cell volume and gas uptake with increasing gas pressure, and the phenomenon is referred to as swelling (Figure 3.6(f)).[25,65]

The next aspect, which should be addressed with respect to the isotherm shape obtained upon adsorption of fluids is the restriction due to the construction of the adsorption instruments. The stepwise shape of hypothetical isotherms shown in Figure 3.6 would be valid if the volume of the gas phase approaches infinity. Such isotherms can be recorded by gravimetric instruments working in the gas flow conditions, for example. The majority of the isotherms reported, however, were recorded by means of volumetric (manometric) devices with a finite volume of the gas phase reservoir (manifold). The phase transition and change in the porosity of the adsorbents significantly influence the thermodynamics of the gas phase, leading to a jump in the overall pressure in the measuring system, as illustrated in Figure 3.8. The so-called 'negative pressure' adsorption profile is sometimes observed for gating materials[46,66] and is associated with the stepwise, sudden gas uptake in a limited system dosing volume, and consequent pressure drop due to the adsorption event. The opposite effect (pressure amplification) is observed for NGA, which is caused by massive desorption of the gas upon structural contraction.[67] Both phenomena rely on metastable states, the equilibrium chemical potential at which the transition should occur is significantly surpassed and the step is highly irreversible. In this sense, the pressure amplification reflects the activation barriers involved.

The second attribute of the isotherms of flexible MOFs, and a consequence of metastable states, is the presence of hysteresis between adsorption and desorption branches, because of the kinetic barriers controlling the nucleation of the new phase (Figure 3.6(c)).[10] The hysteresis width depends on the barrier.[68] As in type IV isotherms, the exponential influence of barriers causes lifetimes of metastable states to be orders of magnitudes higher compared

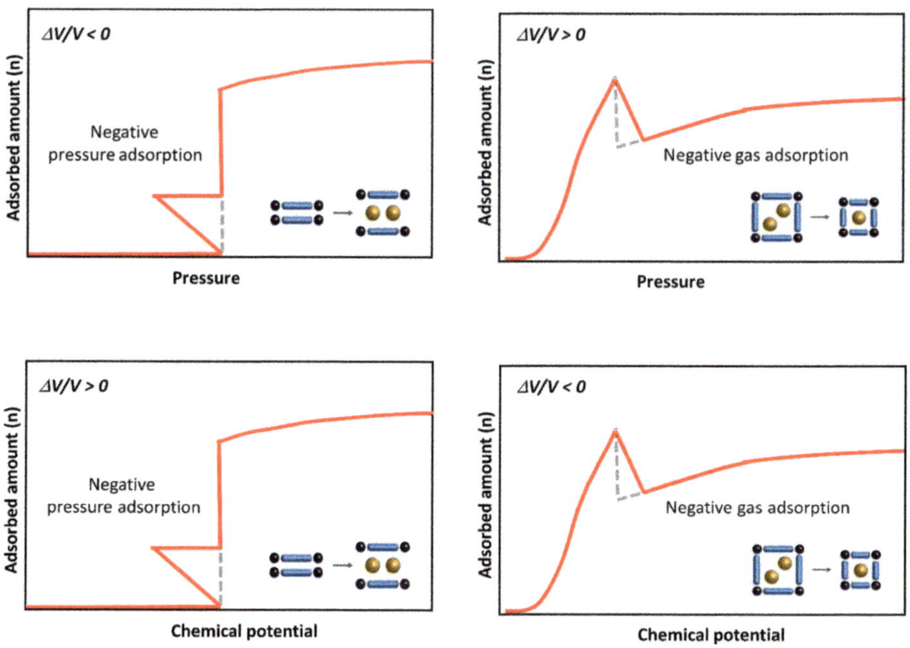

Figure 3.8 Typical deviation (red line) from the expected isotherm profiles (grey dotted line) in case of negative pressure adsorption (left) and negative gas adsorption (right).

to the experiment time. Hence, the rate at which the thermodynamic conditions change within a given experiment (apparent equilibration time) has only a minor influence.

Thus, as it was discussed in Section 3.2.1, the first pieces of evidence of the flexibility can be collected already during desolvation of the frameworks (solvent desorption) by analytical methods revealing the structural changes (XRD, IR, NMR, Raman *etc.*). But the phase transition upon desolvation is not the ultimate requirement for flexibility, since the breathing, framework displacement or linker rotation often evolve upon adsorption at specific conditions. The flexibility in such type of materials can be unambiguously identified only by the parallelised XRD/adsorption or other *in situ* experiments.

It is quite expectable that flexibility for many MOF materials remained so far unrecognised because it appears only for the specific adsorbate at specific thermodynamic conditions. The same is true for 'shape memory' materials exhibiting non-striking isotherm profiles, but this does not prove or even suggest a shape-memory effect, which can only be identified by measuring multiple cycles of adsorption–desorption or collecting PXRD data after adsorption.[34]

3.3 Advanced *In Situ* Characterisation Techniques

3.3.1 *In Situ* Diffraction Techniques in Parallel to Gas and Vapour Physisorption

The crystallinity of most of the flexible MOFs before and after guest-induced phase transitions motivated researchers to use existing and develop new diffraction techniques for the monitoring of the changes in the framework structure at the atomic level. The mechanisms and aspects to be studied define the choice of experimental technique and also type of radiation, used in experiments. The use of X-rays in combination with gas or vapour physisorption in most cases provide information on the evolution of host framework upon adsorption and desorption of the fluid, whereas neutron diffraction is mostly used for determination of guest molecules' positions, often containing deuterium atoms. In both cases, the development of these techniques occurs at the large-scale facilities such as synchrotrons and research reactors, mainly because of the complex customised instrumentations and *in situ* cells, requiring temperature and pressure stability in a broad range. Herein we review the different types of existing instrumentation for the study of flexibility in MOFs by *in situ* X-ray and neutron diffraction and most representative results and examples of applying these techniques for unravelling the flexibility mechanisms in MOFs. We also refer the readers to the earlier reported dedicated review articles and book chapters, which contain extended information on the topic.[1–3]

3.3.1.1 *Instrumentation for* In Situ *Synchrotron X-ray and Neutron Diffraction in Controlled Pressure and Temperature Conditions*

The development of *in situ* PXRD techniques for monitoring flexibility in MOFs started in the early 2000s at synchrotron beamlines because such facilities offer the high photon fluxes and wide X-ray energy ranges required for avoiding X-ray absorption, fluorescence, and speed-up the data collection, but mostly because of flexible and adjustable sample environment or even *in situ* and *operando* cells, earlier used for catalytic studies. The set of requirements from both X-ray diffraction and gas adsorption standpoints should be met in order to conduct *in situ* PXRD experiments at variable pressures and temperatures. The cell should be transparent for X-rays, vacuum and gas pressure tight, and the temperature of the sample holder should be stable and homogeneous. The possibility of oscillation or spinning of the cell in the beam is desired. Considering all these technical requirements two different cell types have been developed: (1) capillary based *in situ* cells; and (2) adsorption chambers.

In the typical capillary cell, a commercially available glass or quartz capillary is filled with MOF powder and connected to the customised gas handling system using a compatible gas line connection. The capillary with the sample is further centred in the synchrotron beam and the temperature is regulated by a nitrogen cryo stream, available at most beamlines. This system is compact, relatively simple in construction and it has many advantages such as mobility, low attenuation for X-rays, the possibility of spinning or oscillation in beam and therefore provide high-quality PXRD data. However, as a rule, a low amount of the sample in the capillary does not allow for measurement of the isotherm and, therefore, the temperature control of the sample in the beam is crucial in order to attribute the measured pressure points to the *ex situ* measured isotherm. The second setup is based on a dome or adsorption chamber positioned in the beam. Typically, the adsorption chamber is mounted on the cryostat for precise temperature control. The available space in the adsorption chamber is usually used for the allocation of larger amounts of sample, sufficient for a volumetric physisorption measurement. The connection of such an *in situ* cell to the volumetric adsorption apparatus allows us to measure PXRD patterns at selected points of the isotherm, making this kind of setup more precise in terms of adsorption thermodynamics. However, the setup is usually bulkier and cannot be oscillated in the beam or adapted to other beamlines easily. Another drawback of the setup is the beam attenuation and parasitic scattering or diffraction, originating from the adsorption chamber, which is usually made of beryllium or polyimide films. Capillary-based setups are currently available at numerous synchrotron beamlines, such as SNBL[69] and BL-25[70] at ESRF, MSPD station[71] at ALBA, KMC-2[3] at BESSY-II, BL02B2[72] at SPRING-8, I19[25] at Diamond light source, P02.1[73] at PETRA III and many others. Dome-based *in situ* chambers are more complicated and expensive in the construction and therefore are not as common. Such setups are available at KMC-2[3] at BESSY-II and BL9[74] at DELTA *etc.*

In the case of *in situ* neutron powder diffraction studies, larger amounts of the sample are usually required and therefore, predominantly the setups of the second type with aluminium or vanadium cans are used. In case of adsorption of fluids with low saturated pressures, also known as vapours, in particular at temperatures higher than ambient temperature, more complicated adsorption instrumentation with heatable gas lines and special cell should be used in order to avoid the condensation of the fluid in the capillary. Additional stimuli such as electric field or UV-vis irradiation can be introduced. Caro *et al.* studied the permeability of ZIF-8 membranes using a specially designed *in situ* cell, allowing an electric field to be applied in addition to the gas mixtures.[75] Kaskel *et al.* developed *in situ* PXRD cells, based on adsorption chambers, in which the sample can be irradiated by UV-vis light in addition to the gas adsorption (Figure 3.9).[3]

3.3.1.2 In Situ *PXRD in Parallel to Fluid Sorption on Flexible MOFs*

PXRD patterns, measured on a flexible MOF at variable gas pressure and temperature conditions contain exact information about the phase composition, evolution of the unit cell metric and crystal structure of each phase. Typically, either temperature or gas pressure can be varied unravelling the behaviour of MOF under isobaric or isothermal conditions. Kitagawa *et al.* pioneered *in situ* PXRD studies with adsorption of oxygen on CPL-1 (CPL = coordination polymer) with a pillared-layer structure. The oxygen isobar (p = 80 kPa) was measured in the temperature range between 300 and 90 K and revealed the phase transition at 130 K. Analysis of PXRD patterns using the Rietveld method determined the positions of oxygen guest molecules in the channels of CPL-1. A similar experiment was conducted by the same group using acetylene guest molecules.[76] Since acetylene is explosive at high pressures, the *in situ* PXRD isobar was measured at 10 kPa in the temperature range of 393–170 K. The authors found that during the cooling process, a new diffraction peak at $2\theta = 4.78°$ appears at $T = 320$ K (Figure 3.10), indicating the formation of the op phase. In this case, not only positions but also the orientation of the acetylene molecules in the pores could be extracted from the Rietveld analysis of synchrotron PXRD patterns. Close packing of the acetylene in the pores of CPL-1 opens the possibility for safe storage of acetylene at a density 200 times larger compared to the safe compression limit of free acetylene at ambient temperature.

At a later stage, the development of dedicated *in situ* instruments at the synchrotron beamlines allows *in situ* PXRD studies on flexible MOFs in the isothermal regime. These experiments provide a direct structural evidence and providing an explanation for all steps in adsorption isotherms. Llewellyn *et al.* conducted a series of *in situ* PXRD studies on MIL-53(Cr) and MIL-53(Fe) in parallel to the physisorption of different gases and their mixtures aiming

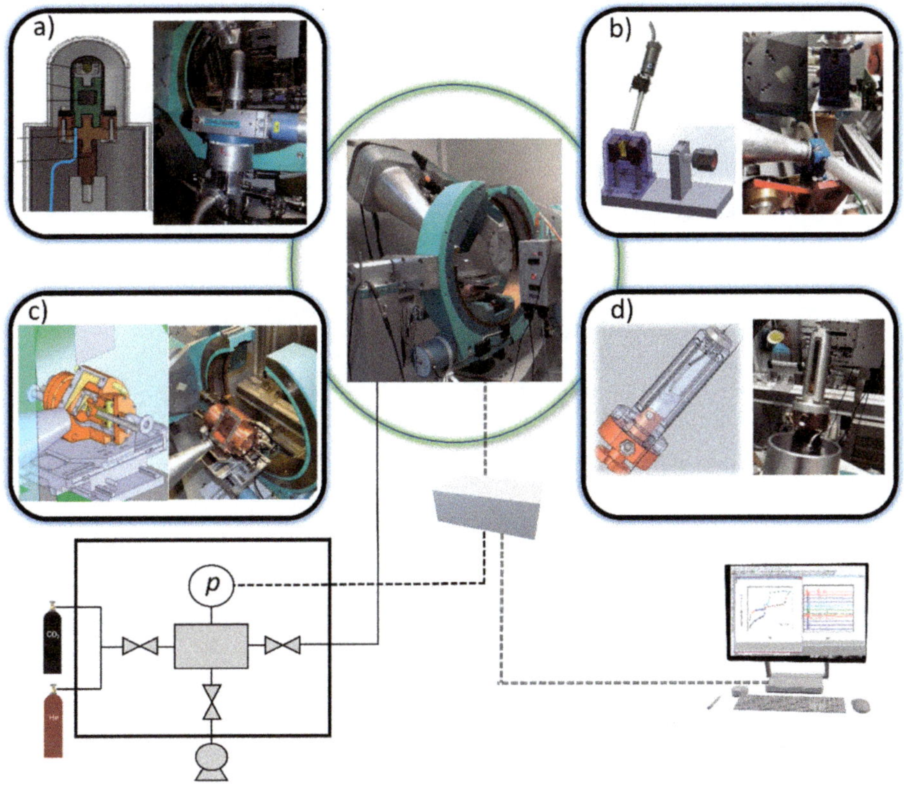

Figure 3.9 *In situ* instrumentation for parallelised PXRD and gas–vapour adsorption measurements at KMC-2 beamline of BESSY II synchrotron: (a) *in situ* cell for the low-pressure gas adsorption (p = 0–1 bar, T = 10–450 K); (b) *in situ* cell for high-pressure gas physisorption (p = 1–100 bar, T = 200–300 K); (c) *in situ* cell for vapour physisorption (p = 1–10 bar, T = 280–320 K); (d) *in situ* cell for the low-pressure gas adsorption (p = 0–1 bar, T = 10–450 K) and light irradiation (λ = 365–850 nm).

to explain the breathing behaviour in these solids.[56] One of the most representative *in situ* PXRD studies address the influence of the ligand functionalisation on the adsorption of CO_2 on MIL-53(Fe)-X (X = H, Cl, Br, CH_3, NH_2, $(CO_2H)_2$) solids at 230 K[77] under variable pressures of CO_2 (Figure 3.11). As a result, frameworks bearing $-CO_2H$ and $-NH_2$ functional groups remain in the cp phase in the entire pressure range from 0 to 9 bar. In contrast, the MIL-53(Fe)-X (X = Cl, Br, CH_3) solids adsorb a significant amount of CO_2 followed by cp–np and np–lp structural transformations.

Kaskel *et al.* studied the adsorption-induced phase transition in DUT-8(Ni), a 'gate pressure' MOF showing ~250% change of unit cell volume upon physisorption of gases. Adsorption of nitrogen (77 K), CO_2 (195 K), *n*-butane (273 K), ethane (185 K) and ethene (169 K) was followed by synchrotron *in situ*

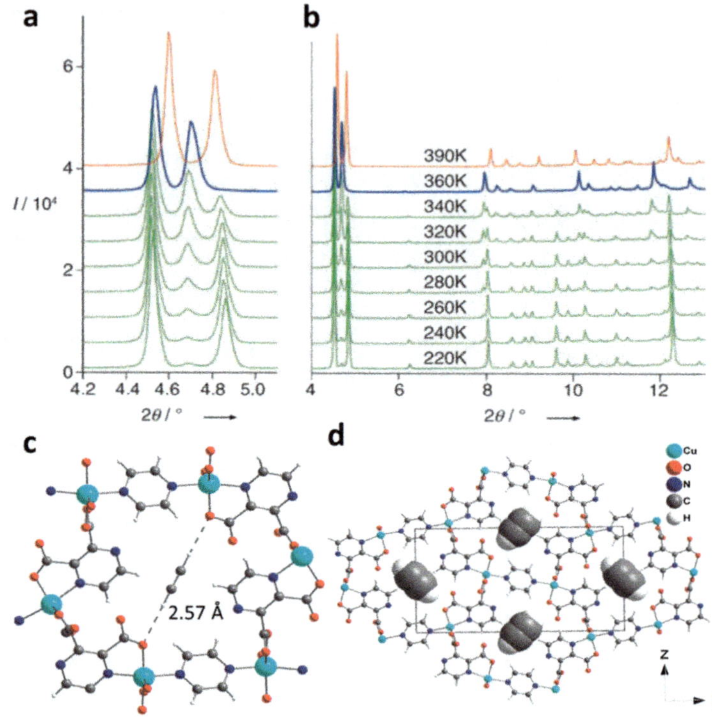

Figure 3.10 *In situ* synchrotron powder diffraction patterns of CPL-1 with adsorption of acetylene at 150 kPa. The 2θ ranges were (a) from 4.2° to 5.18° and (b) from 4.0° to 13.08°; Adapted from ref. 266 with permission from John Wiley & Sons, Copyright 2006. Crystal structure of C_2H_2@CPL-1 complex at 170 K obtained from MEM/Rietveld analysis: (c) illustration of the pore aperture with adsorbed C_2H_2; (d) representation of the crystal structure along [100] direction.

PXRD indicating the formation of intermediate phases upon adsorption and desorption of ethane and ethene (Figure 3.12).[78]

Tanaka *et al.* applied *in situ* PXRD technique in parallel to adsorption of CO_2 aiming the explanation of gating mechanism in the 2D square-grid framework $[Cu(BF_4)_2(bpy)_2]_n$ (bpy = 4,4′-bipyridine), also known as ELM-11.[72,79] The guest-filled structure could be precisely modelled and visualised by a new structure refinement method, which combines Rietveld analysis of *in situ* synchrotron X-ray powder diffraction data with molecular simulations. The crystallographic data of the open framework structure of ELM-11 enable grand canonical Monte Carlo (GCMC) simulations of CO_2 adsorption, required for calculations of Helmholtz free energy change of the host framework during the structural transition.

In situ PXRD studies were of paramount importance in the discovery of the NGA phenomenon in DUT-49 and related mesoporous frameworks.[23] *In situ* PXRD upon physisorption of methane at 111 K enormously helped to

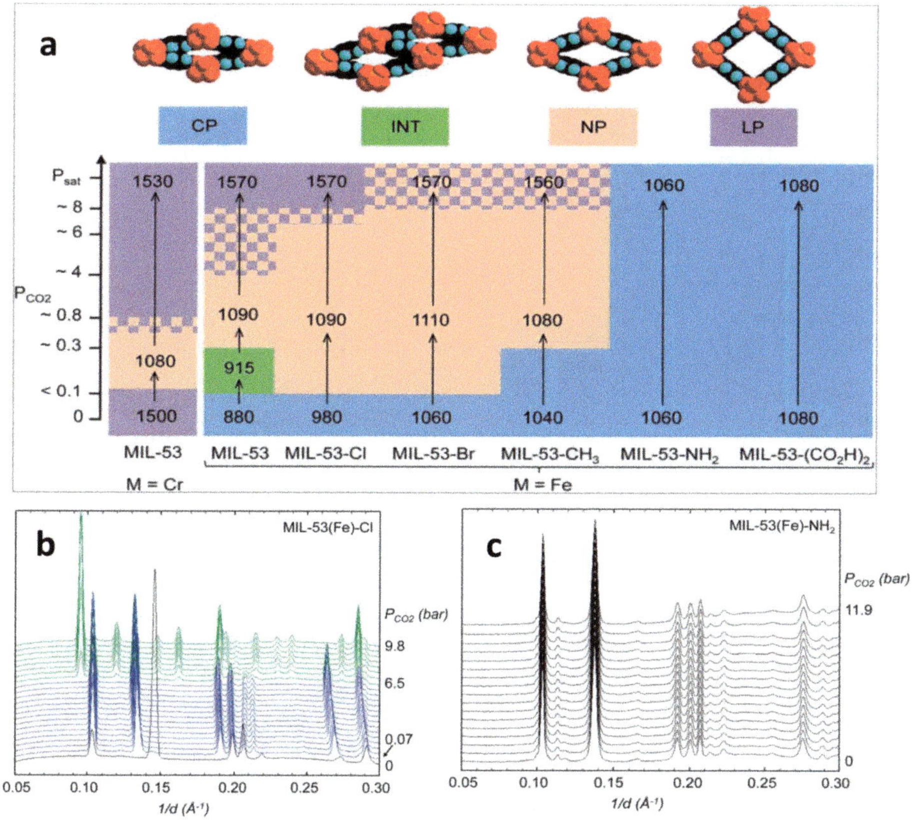

Figure 3.11 (a) Evolution of the pore opening and unit cell volume with the CO_2 pressure at 230 K for the MIL-53(Cr) and MIL-53(Fe)–X solids (CP: closed pore phase, INT: intermediate triclinic phase; NP: narrow pore phase, LP: large pore phase). Adsorption of CO_2 on MIL-53(Fe)–Cl (b) and MIL-53(Fe)–NH_2 (c) at 230 K followed by PXRD. The colour corresponds to the major phase observed at the given pressure. Black: anhydrous CP form; blue: NP form, green: LP form. Adapted from ref. 77 with permission from the Royal Society of Chemistry.

unwrap the mechanism of the contraction and reopening (Figure 3.13). The contracted phase and intermediate phases were directly solved and refined from *in situ* measured PXRD data. Moreover, *in situ* PXRD patterns measured on DUT-49 samples with different crystallite sizes upon nitrogen physisorption at 77 K were used for the determination of the phase composition, which enormously helped in the explanation of crystallite size dependency and hysteresis width in the isotherms.

In situ PXRD patterns, measured in parallel to methane, nitrogen, *n*-butane and carbon dioxide adsorption were crucial for uncovering the adsorption-induced breathing and even pressure amplification phenomena in several mesoporous frameworks.[61,80,81]

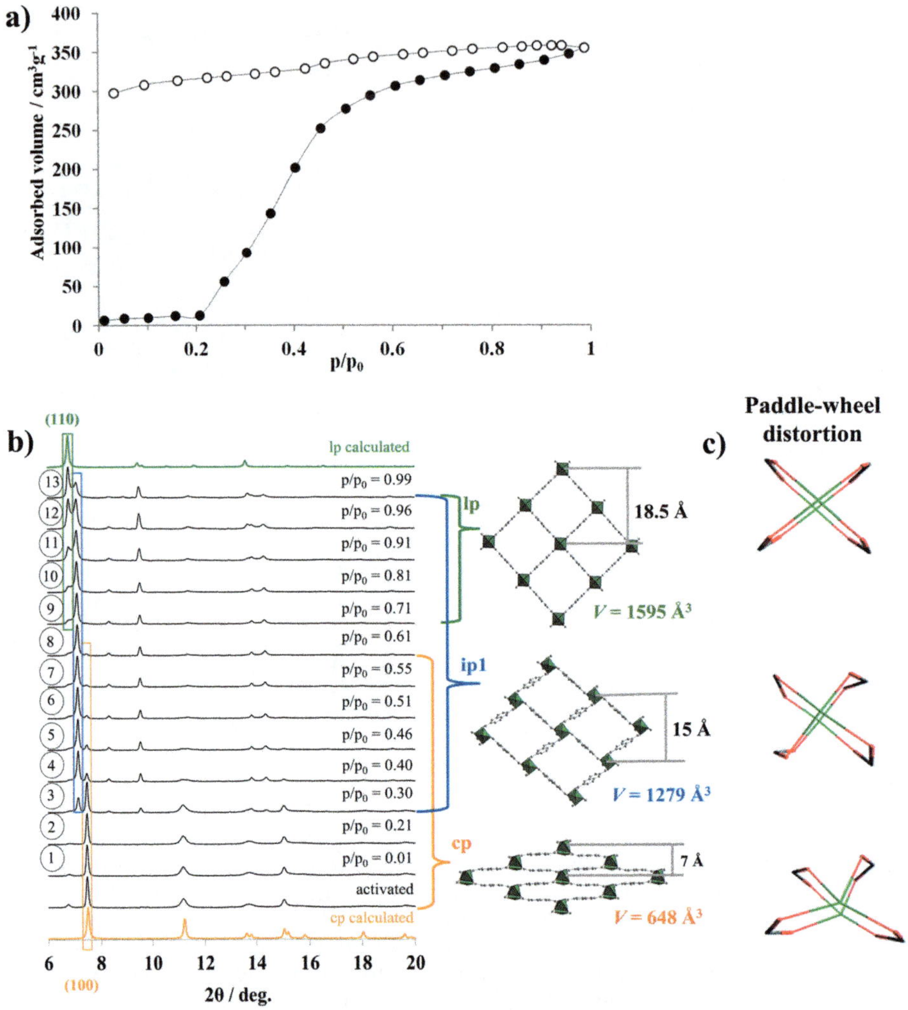

Figure 3.12 Adsorption of C_2H_6 (184.6 K) on DUT-8(Ni) (a) monitored *in situ* by powder XRD (b); (c) evolution of the crystal structure and paddle-wheel distortion in DUT-8 cp, IP1 and lp. Adapted from ref. 78 with permission from the Royal Society of Chemistry.

In summary, PXRD patterns, measured in parallel to gas physisorption, play a paramount role in the understanding of guest-induced flexibility in MOFs on the atomic scale. This information includes the mechanism of the framework deformation, changes in porosity, and in some particular cases, the guest loading mechanism. The obtained crystal structures are often used as an input for further *in silico* studies, such as GCMC calculation of adsorption isotherms for selected phases and guests, but also molecular dynamics and thermodynamics calculations, which shine the light on the driving force of the guest-induced transitions. The knowledge, obtained from *in situ* PXRD

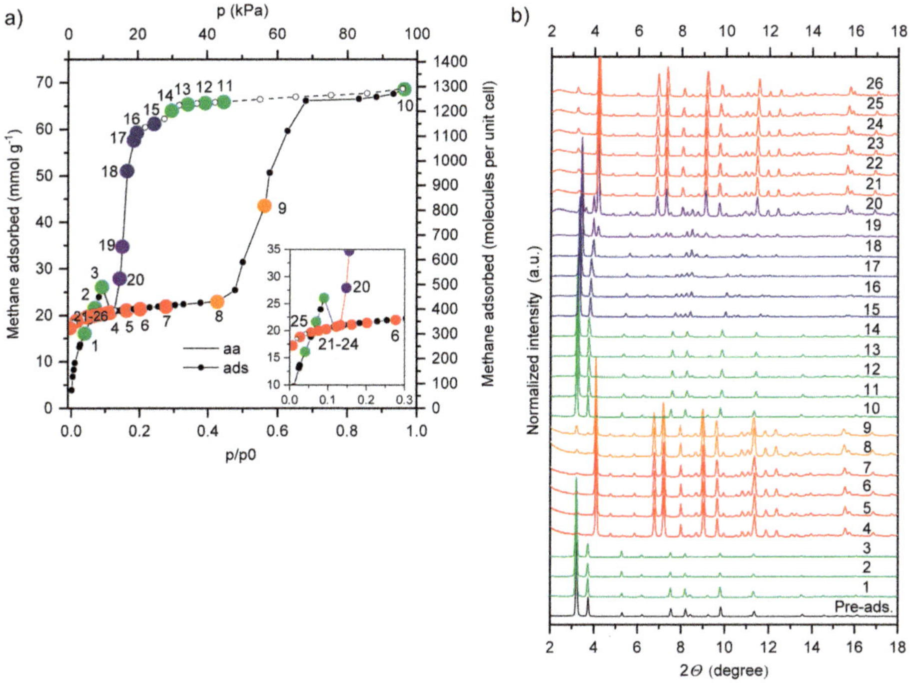

Figure 3.13 *In situ* PXRD in parallel to adsorption. (a) Adsorption isotherm of methane at 111 K. (b) PXRD pattern obtained at the corresponding adsorption points. Colour code: DUT-49op before measurement (black), DUT-49op (green), DUT-49cp (red), DUT-49ip (blue), phase mixture (orange, purple). Adapted from ref. 23 with permission from Springer Nature, Copyright 2016.

experiments, is often useful for judicious choice of the best possible application of the MOF. So far, the technique is mainly explored at the synchrotrons, however, several laboratory implementations of *in situ* adsorption-PXRD systems have been realised *e.g.*, by Rigaku (Kyoto University), Panalytical (TU Dresden) and STOE Stadi P (University of Leipzig).

3.3.1.3 In Situ *NPD in Parallel to Fluid Sorption on Flexible MOFs*

In contrast to X-rays, which are scattered by the electron shell of the atom, neutrons interact with the atomic nuclei, therefore the scattering length is strongly dependent on the nucleus composition and is isotope-dependent. In some cases, light atoms, embedded in structure or adsorbed in the pores are more 'visible' in the presence of heavy elements. Thus, the scattering length of deuterium isotope is higher than those for ^{12}C and even some 3d-metals. This makes neutron diffraction techniques attractive for the studies of the pore-filling mechanism in crystalline porous solids using deuterium containing molecules or even combining these studies with adsorbed-induced phase transitions. However, the number of reports on the elucidation of flexibility in MOFs is still rather small.

Chen *et al.* used *in situ* neutron powder diffraction (NPD) for explaining the high separation efficiency of [Cu(bpy)$_2$(OTf)$_2$] (ELM-12) in removing trace C_2H_2 from important raw C_2H_2–C_2H_4 mixtures under ambient conditions.[82] Rietveld analysis of high-resolution NPD uncovered were two preferable adsorption places of C_2D_2 in the op structure of ELM-12, both showing weak C–D···O and C–D···F weak H-bonding interactions (Figure 3.14).

The same group reported a similar study on propyne–propylene separation using ELM-12.[83] NPD measurements were conducted *in situ* for explaining the mechanism of the high selectivity of the MOF towards propyne. Similar to the acetylene, two preferable adsorption sites for propyne have been found.

Recently, Oh *et al.* discovered the isotope-selective gate opening of MIL-53(Al).[84] The authors found that both hydrogen and deuterium induce the

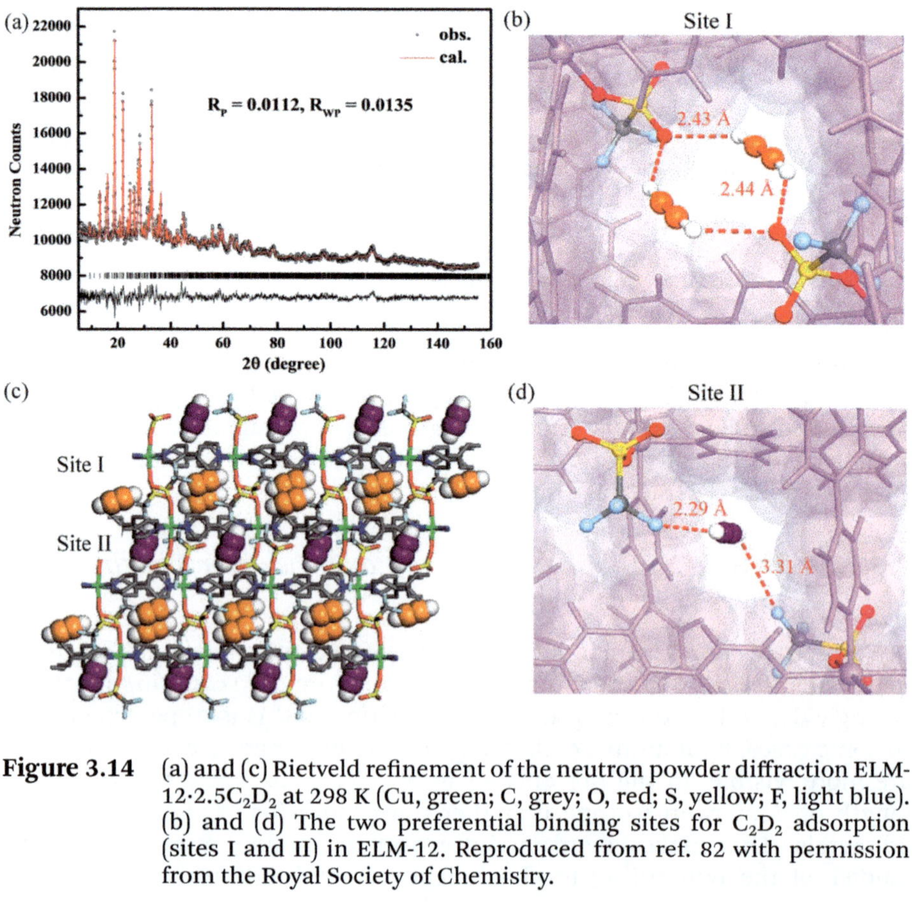

Figure 3.14 (a) and (c) Rietveld refinement of the neutron powder diffraction ELM-12·2.5C_2D_2 at 298 K (Cu, green; C, grey; O, red; S, yellow; F, light blue). (b) and (d) The two preferential binding sites for C_2D_2 adsorption (sites I and II) in ELM-12. Reproduced from ref. 82 with permission from the Royal Society of Chemistry.

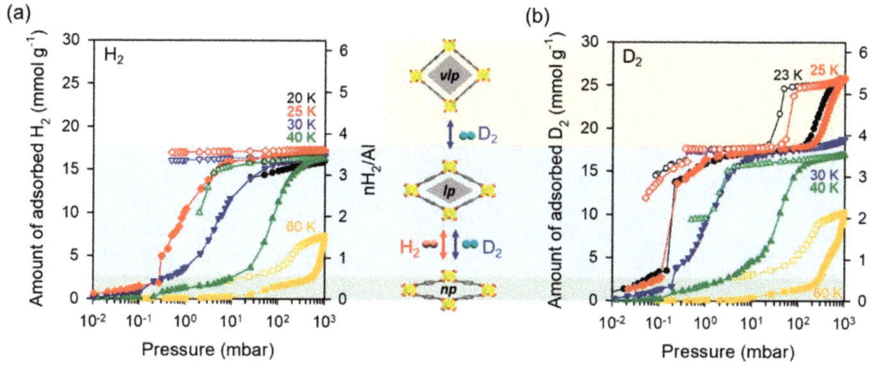

Figure 3.15 Isotherms for (a) H_2 and (b) D_2 sorption on MIL-53(Al) measured at 20 K (H_2)/23 K (D_2) (black), 25 K (red), 30 K (blue), 40 K (green), and 60 K (yellow), with pressure shown in the logarithmic scale. Filled and open symbols correspond to adsorption and desorption, respectively. Each background colour of the sorption graphs represents the dominant pore structure of MIL-53 in that region; light green for narrow pores (np), light blue for large pores (lp), and light orange for very large pores (vlp). Reproduced from ref. 84 with permission from American Chemical Society, Copyright 2020.

structural transition in the narrow pore phase, however, only deuterium reopens the structure at higher pressures, which can be used for sustainable and efficient isotope separation. The mechanism was proven by *in situ* NPD studies at different deuterium loadings (Figure 3.15).

To derive the impact of different pore sizes on the NGA phenomenon in mesoporous flexible DUT-49, the NPD patterns were measured at varied loadings before the NGA event (Figure 3.16).[61] Rietveld analysis and GCMC simulations revealed the smallest cuboctahedral voids to be filled first with methane reaching saturation at around 1 kPa. In the range of 0.7–2 kPa tetrahedral and octahedral voids are starting to be filled with methane molecules adsorbing on the linker backbone framing these pores. At higher pressures, the decrease in adsorption enthalpy can be correlated with reduced solid–fluid interactions due to multilayer adsorption on the surface of the tetrahedral and octahedral voids.

3.3.2 *In Situ* High Pressure Crystallography on Flexible MOFs

3.3.2.1 Introduction

Pressure is a powerful thermodynamic parameter when compared to temperature, with a range of four more orders of magnitude. Nonetheless, it is a much less explored avenue to understand structure–property relationships. A search of the Cambridge Structural Database showed that while over 107 000 crystal structures of MOFs have been determined since 2005,

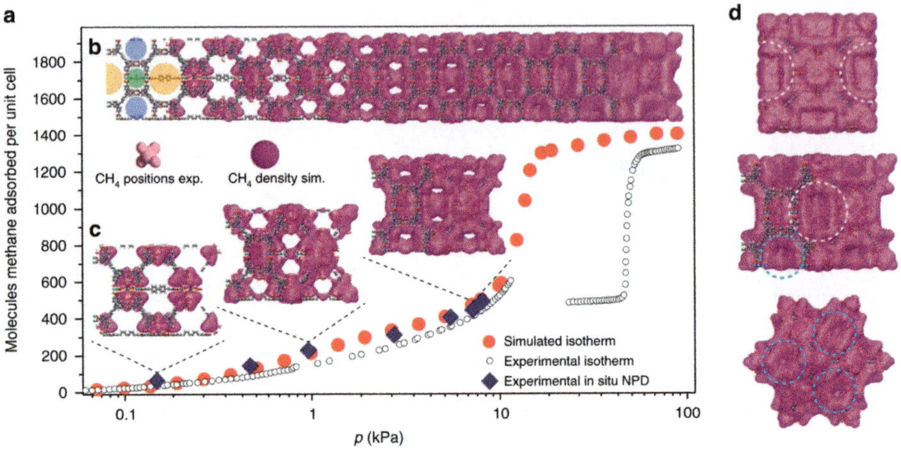

Figure 3.16 Methane distribution in DUT-49. (a) Simulated (red circles) and experimental (open symbols) methane adsorption isotherm of DUT-49 at 111 K including points at which NPD patterns were recorded (blue diamonds). (b) Structure of DUT-49 with the increasing population of pores with methane upon pressure increase, including on the left trimodal pore system: octahedral (oct, orange), tetrahedral (tet, blue), cuboctahedral (cub, green). Experimentally refined methane positions in light pink, simulated methane density in pink. (c) Visualisation of DUT-49 unit cell at selected loadings which correspond to experimentally investigated pressure. (d) Illustration of methane-filled pores of DUT-49 unit cell in the region of NGA viewed along (100) direction (top), (110) direction (middle), and (111) direction (bottom). Reproduced from ref. 61, https://doi.org/s41467-019-11565-3, under the terms of the CC BY 4.0 license https://creativecommons.org/licenses/by/4.0/.

829 high-pressure crystal structures have been determined, and only 246 of those are unique MOFs.[85] There has been a rise in the number of high-pressure X-ray diffraction studies on MOFs, not only to examine their mechanical properties but also as a means to understand the intra and intermolecular interactions at play within the framework itself and with guest molecules. Pressure provides a powerful tool to explore the free energy landscape of MOFs. These studies become even more important as the research community realises that MOFs are much more flexible than initially perceived. Pressure studies have shown that MOFs can have interesting high-pressure responses such as negative linear and area compressibility, phase transitions and amorphisation. These pressure responses are exciting for the field, as it opens the door to applications for MOFs beyond the traditional gas separation or sequestration application and into the area of sensors or actuators, artificial muscles, and switchable materials. This section will first explore the instrumentation and experimental methods used to generate high-pressure for X-ray diffraction studies, as well as provide some guidance over best practices for high-pressure diffraction studies, before moving on to discuss some

exemplary case studies of MOFs which demonstrate peculiar pressure-responsive behaviour.

3.3.2.2 High-pressure Crystallography Instrumentation

3.3.2.2.1 Diamond Anvil Cell. High pressure diffraction was first attempted soon after the discovery of X-rays to determine crystal structures, with Cohn *et al.* carrying out high pressure powder diffraction in 1933.[86] The apparatus designed for creating pressure on the sample consisted of a steel pressure bomb, which allowed for the compression of a gas (usually helium) and the sample. The cell was equipped with two beryllium windows to allow the incident and diffracted beams to pass through. These first experiments yielded poor data due to the primitive equipment. The data was poor as the diffraction angle was limited by the design of the windows and the reflections had to be strong enough to register against the noisy background arising from the pressure cell itself. The pressures the apparatus could obtain were also quite modest (≈0.1 GPa or 1000 bar). Since these first investigations, great improvements to the apparatus have been made, with the most notable being the development of diamond anvil-based pressure cells. The diamond anvil cell (DAC) was first designed in 1958 at the National Institute of Standards and Technology (NIST), as an apparatus for high-pressure Raman spectroscopy, however the cell could also be used effectively in X-ray diffraction studies.[87,88] The original design of the cell has been modified throughout the years to increase the quality of data produced.[89,90] The schematic diagram in Figure 3.17 shows the components of the modified Merril-Bassett DAC. The cell is composed of two steel plates which have a conical hole at the centre, with a half opening angle of 40°, to permit optical viewing of the sample and diffraction of X-rays. The two plates house tungsten carbide backing seats that hold the opposing diamond anvils in place. A metal gasket, made from materials such as tungsten or Inconel (which has a cylindrical hole drilled through it of 200–300 μm in diameter) is placed between flat-faced 'culet' diamond anvils in order to create a cylindrical sample chamber. The pressure inside the sample chamber is increased by forcing the diamond faces closer together, which is achieved by tightening three Allen screws which force the steel plates together, alternatively, the pressure can be controlled remotely *via* a metal membrane. The sample chamber usually contains a single crystal (or powder) surrounded by a small amount of pressure transmitting medium (PTM), which is used to apply pressure hydrostatically (that is, isotropically) to the sample when the two diamonds are brought together. The cell loading depends on the PTM of choice, which can either be a liquid, gas or even a soft solid. Liquid media are simply pipetted into the chamber, whilst gases can be loaded using more complex methodologies such as gas loading or cryoloading.[91,92] Along with the sample, a ruby chip is also placed inside the chamber (Figure 3.17) which serves as a pressure calibrant through the well-established ruby fluorescence method, as the fluorescence transition of ruby is dependent on the pressure applied.[93,94]

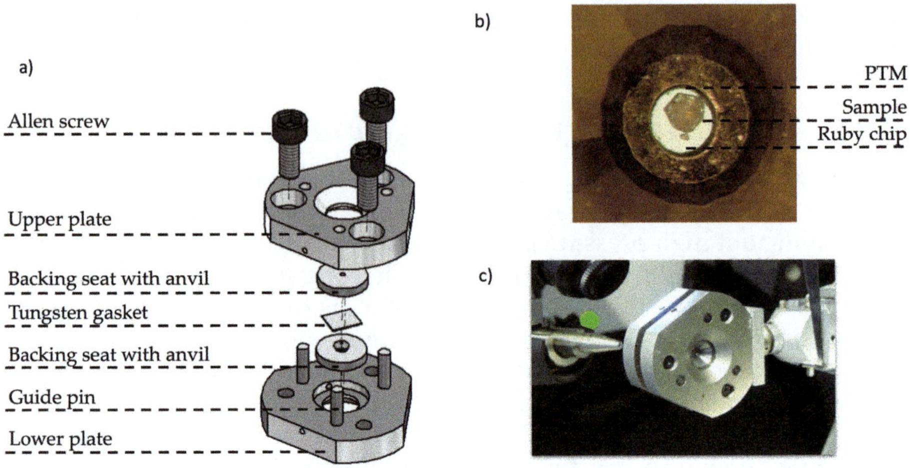

Figure 3.17 (a) Exploded view of a Merril-Basset DAC used for single-crystal X-ray diffraction. Reproduced from ref. 90 with permission from the International Union of Crystallography, copyright, 2008. (b) View of sample chamber containing single crystal sample, ruby chip and PTM. (c) Picture of DAC mounted on diffractometer. Reproduced from ref. 95 with permission from the International Union of Crystallography, copyright 2004.

Due to the cumbersome nature of the DAC, X-ray data collection strategising the methodology involved in X-ray data collection is invaluable in minimising shading, improve the completeness of low symmetry samples and optimise data collection times. On fixed χ, three circle diffractometers, data collections are usually ω scans performed at different values of φ and 2θ to ensure the angle between the incident beam and the cell axis is less than 40° and less than 80% of the detector is shaded.[95] With the move towards four-circle diffractometers, especially at central facilities, adding the detectors did not lead to improved completeness of samples, but rather improved data quality.[96] In addition to non-traditional data collection, the processing of high-pressure single crystal data is non-standard compared to routine single-crystal diffraction experiments. The images also contain powder rings from the tungsten carbide backing seats, intense reflections from the diamond anvil and also the ruby chip, which all must be subtracted from the data pertaining to the sample. Once indexed, the sample data must be integrated with dynamic masks, which allow for areas of the images that are shaded by the body of the cell to be excluded from the integration of the reflections. Once all reflections have been integrated, an absorption correction is applied for the DAC and shadowing from the gasket, using programs such as ABSORB[97] or SHADE.[98] These programs omit reflections which are poorly measured due to partial shading of the cell or overlap with diamond reflections must be omitted.

Whilst the DAC has been the work horse of high-pressure crystallography within the MOF community since the 2000s,[51,99] these high-pressure structural studies have shown that MOFs are very soft materials. As a result, in some cases the pressure regime of DAC is too high, with a closing pressure of >0.1 GPa, therefore lower pressure devices must be used. A few of these devices are described below.

3.3.2.2.2 Sapphire Capillary Cell. The sapphire capillary cell (SCC) has an operating range between 0 and 0.15 GPa, and therefore is the ideal cell to bridge the gap between ambient pressure and the closing pressure of the DAC.[100,101] The SCC has a simple design and can be described by three main components: (1) the pressure cell where the synthetic, pristine single crystal of sapphire capillary and sample is housed; (2) experiments can be done with other gas pumps beside just hand pumps; and (3) a phase separator which links components (1) and (2) together as outlined in Figure 3.18.

The simple design of the has many advantages over the DAC in terms of X-ray data collection, the first is due to using a capillary instead of a stainless-steel body within a DAC, there is little or no restriction to available reciprocal space which allows for the study of low symmetry systems. Additionally, processing the diffraction images is relatively straightforward compared to DAC

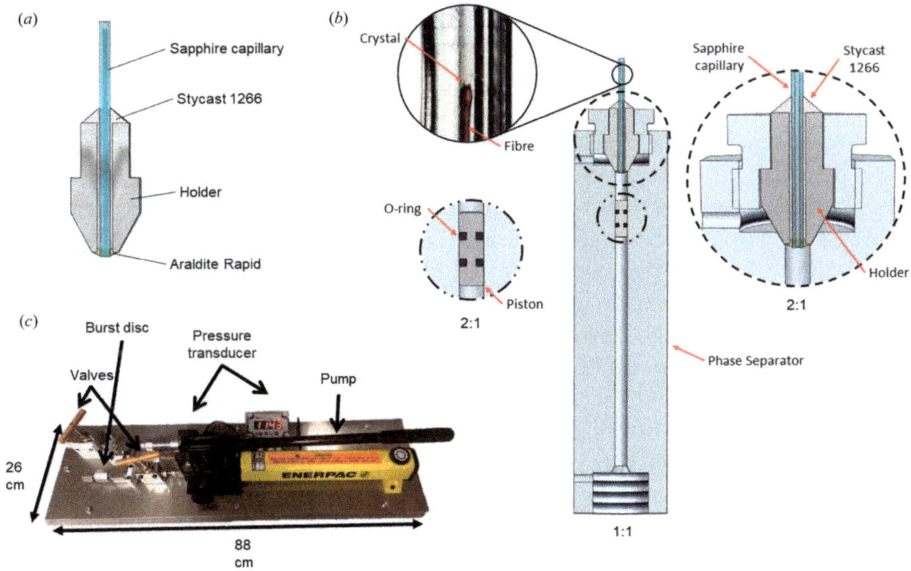

Figure 3.18 (a) Sapphire capillary cell. (b) CAD-model (1:1) of the sapphire capillary and phase separator, with the crystal fibre, the cone-type metal-to-metal seal between the cell body and phase separator, and the movable piston enlarged for clarity. (c) Application of pressure to the SCC *via* a hand-pump. Reproduced from ref. 100 with permission from the International Union of Crystallography, copyright 2020.

experiments, as the SCC does not shade the detector, nor are there any gasket powder rings. While the SCC is composed of a single crystal of sapphire which contribute broad and intense reflections from the crystalline capillary itself, these domains can be easily removed. The final advantage of the SCC compared to the DAC, and possibly where it has the biggest opportunity for the exploration of high-pressure MOF behaviour, is the variety of PTMs which can be used. The phase separator allows for the pressure in the pump to be generated *via* hydraulic fluid and for another fluid to be used within the sample environment, directly applying pressure to the sample. This allows the use of volatile or harsh solvents (which can decompose the glue which holds the diamonds in place in DACs) as PTMs. While the SCC has been little used within the MOF community, the set-up has huge potential for the exploration of host–guest interactions in MOFs, as well as accurate tracking of lattice parameters within the low-pressure regime to follow the structural changes that occur within phase changes of MOFs.[102]

3.3.2.2.3 Pressure Jump Cell. Another cell that provides access to the low-pressure regime (up to 0.4 GPa) is the pressure jump cell (PJC). Originally designed for SAXS and WAXS measurements at the Diamond Light Source beamline I22, where the jumps in pressure were very finely controlled (accuracy of 0.001 GPa) and quickly executed (5 ms), allowing for time-resolved experiments.[103] As shown in Figure 3.19, the PJC is composed of a large stainless-steel block, with two perpendicular bores, one runs along the X-ray path and is capped with diamond windows with opening half-angles of 21°, the other is a loading port for the sample. The sample is held within a Kapton capillary as a finely ground powder additionally oil is added as the PTM. Pressure is applied to the sample through a water pump.

Recently, the PJC has been used on DLS beamline I15, for high-pressure powder studies and has been very efficient.[104] The positives of the PJC are the fine pressure control that can be controlled remotely. This allows for accurate tracking of the lattice parameters and accurate bulk modulus determinations. The experiments are also much quicker than the SCC, as acquisition of powder diffraction images is much quicker than that of single crystal diffraction. There are a few downsides to using the PJC to collect high-pressure powder diffraction data. The small opening half-angle of the diamond windows is only 21°, this limits the resolution of X-ray data which can be collected, compounded with the high background, from both the water and liquid PTM, this can limit any structural refinements beyond indexing the diffraction pattern. Finally, one quality which the PJC possesses which can be seen as both a positive and negative is the fact that the cell is completely stationary during data collection. This means that the cell does not need to be mounted on a goniometer which means it is easily portable. The downside is that the sample cannot be rocked during data collection. The rocking helps with the averaging of powder rings, especially if samples cannot be ground to fine powders. Grinding of MOF crystals, especially those which are soft, has been shown to alter their mechanical behaviour.[105] The PJC is a relatively

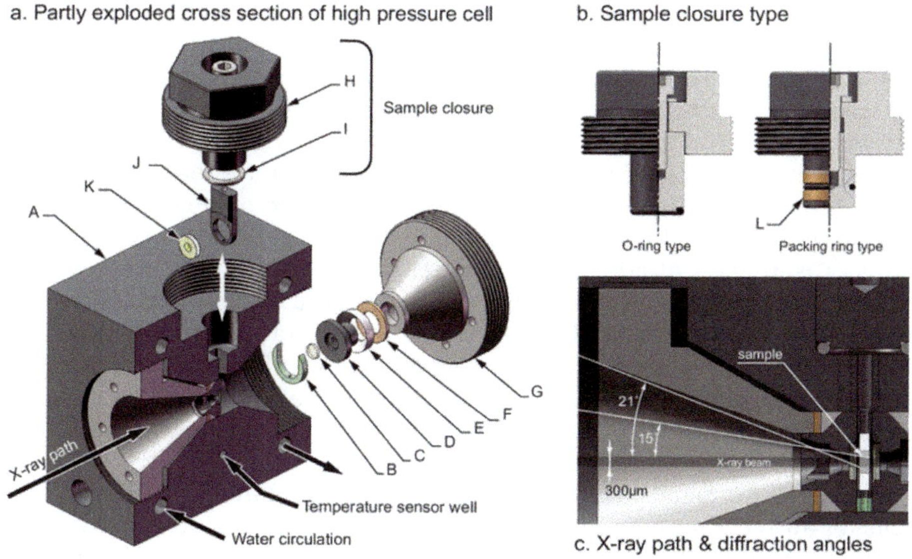

Figure 3.19 (a) Cross section of pressure jump cell. A: cell body; B: window spacer and sample carrier guide; C: diamond window; D: window support; E: window seal; F: anti-extrusion washer; G: window plug; H: sample plug; I: sealing o-ring; J: sample loader; K: sample Teflon spacer carrier. (b) Option of two sample closure type. (c) X-ray path and accessible diffraction angles for a 300 μm diameter beam. Reproduced from ref. 103 with permission from AIP Publishing, Copyright 2010.

recent development for the instrumentation of high pressure studies, and has been used in powder diffraction studies of MOFs, and has been successfully employed to uncover the low-pressure behaviour of defects in MOFs[104] and provide clarity on phase changes in the sub-DAC pressure regime.[106] The PJC has great potential for low pressure studies of flexible MOFs, and as a novel technique it provides opportunity for further development in terms of improving the data quality, or for the data to be combined with computational methods to improve structural models.[107]

3.3.2.2.4 Gas Cell Capillary. The final cell discussed here, is the gas cell capillary. This cell is composed of a capillary, either of glass, fused quartz, or sapphire inside the capillary. The single crystal is mounted on a fibre or powder packed and there is an inlet for the gas. These cells have variable pressure limits depending on the strength of the capillary, usually reaching an upper limit of 300 bar. For example, Chupas *et al.* developed (at the Advanced Photon Source (APS), USA) an open-ended capillary supported in a frame that allowed gasses to flow over crystalline powdered samples at pressures up to 140 bar.[108] Heating elements were also placed close to the capillary to allow heating up to 973 K. Jensen *et al.* (at MAX-II, MAXLAB, Sweden) extended the pressure and temperature range of this cell to 300 bar and 1273 K.[69] A lower pressure version rated to 100 bar using a quartz sealed capillary has also been used extensively. Andrieux *et al.* (at the European Synchrotron Radiation Facility (ESRF), France), also developed a similar designed cell rated to 200 bar and 1273 K, but with improved gas flow. This cell also allowed gas mixtures to be studied.[109] Combining gas cells with a gas control rig allows for the remote control of a selection of gases and flow rates that are delivered to the cell *via* mass flow controllers. Within the context of flexible MOFs, the gas cell has provided useful experimental settings to provide crystal structures of isotherms of adsorption. Gas cells, in particular the pressure gas cells, offer a great amount of control of the pressure but they are commonly employed for *in situ* gas adsorption experiments rather than hydrostatic compression studies that would be comparable to typical DAC experiments. The highest useable pressure for a gas cell is normally limited by the liquefaction of the gas under pressure but the highest rated cells are in the region of 300 bar at most.

This is by no means an exhaustive list of all the kinds of high-pressure cells for diffraction but is a substantial overview of those which are used the most or have great potential within the pressure range of studying the properties of flexible MOFs. For a comprehensive overview of the field, readers are directed to reviews on high-pressure cells by Katrusiak[96] and an up-to-date review of MOFs under hydrostatic pressure by Vervoots *et al.*[110]

3.3.2.3 Pressure-responsive Behaviour in Flexible MOFs

Through the use of the high-pressure cells outlined in the previous section, flexible MOFs have been studied in great detail. This next section will explore some examples of the kinds of flexibility that high-pressure X-ray diffraction studies have uncovered in flexible MOFs including, variable pressure response as a function of PTM, breathing behaviour, swing effect and pressure driven phase changes.

3.3.2.3.1 PTM-dependent Behaviour. The first few studies of MOFs at pressure demonstrated the importance of the PTM for high-pressure crystallographic studies. The choice of PTM in a high-pressure crystallographic study is important. Within the high-pressure community, typically gases (for example, He, Ar, N_2) are used as they provide the highest hydrostatic limits (the point where the pressure across the experimental volume becomes inhomogeneous and differential). More often in high-pressure MOF studies, liquid PTMs are employed as they are simple to load into the DAC, and generally have much more modest hydrostatic limits, as outlined in Table 3.1.

High-pressure PXRD experiments were carried out by Chapman *et al.*, on HKUST-1 and demonstrated for the first time by using PTMs of increasing size (such as methanol–ethanol–water (MEW), isopropyl alcohol (IPA) and fluorinert FC-70), that the compressibility of the MOF was strongly dependent on the PTM.[99] In both MEW and IPA, there is a two-step compression with an initial 'hard' regime, where the bulk modulus, K, measures ~118 GPa

Table 3.1 Pressure transmitting media and their hydrostatic limit[267]

Pressure transmitting medium		Hydrostatic limit/GPa
Iso-*n*-pentane mixtures	1:1 Iso-*n*-pentane	7.4
	5:1 Iso-*n*-pentane	7.4
Methanol–ethanol mixtures	4:1 Methanol–ethanol	~10.5
	16:3:1 Methanol–ethanol–water	~10.5
Fluorinert	1:1 FC84-FC87 Fluorinert	2.3
Oils	Daphne 7474	3.7
	Silicone oil	3
Gases	Nitrogen	~13
	Neon	~15
	Argon	~10
	Helium	~5

and a secondary 'soft' regime where K measures ~30 GPa, occurring at 2.2 and 0.8 GPa for MEW and IPA, respectively. This transition was attributed to a change from hyperfilling to a pore emptying mechanism. In comparison, in the non-penetrating FC-70 media, direct compression occurred, and the sample became amorphous at significantly lower pressures than with the penetrating media. The pressures at which HKUST-1 becomes amorphous is also dependent on the size of the PTM, where MEW allows the framework to retain its crystallinity until 7.5 GPa, IPA until 4.2 GPa and FC-70 until 2 GPa. Since the high-pressure PXRD study did not report any structural data, Graham *et al.* carried out a comparable single-crystal high-pressure X-ray diffraction experiment with MEW to clarify the exact nature of the structural transitions between the hard and soft regions.[111] In agreement with Chapman *et al.*, it was confirmed that the pores of HKUST-1 filled up until 3.9 GPa.[112] After 3.9 GPa, the electron density in the pores decreased which coupled with a change in the Cu–O bond lengths. Cu–O1 equatorial bonds decreased in length, while the axial Cu–O2 bonds elongated. This shows that the pore emptying mechanism was associated with compression of the equatorial Cu–O1 bonds, resulting in the reduction in volume during the pore emptying transition. These initial studies demonstrate that the PTM plays an important role even in relatively in-flexible MOFs.

Recently, a rigorous study was carried out to understand the effect of multiple PTMs (soft solids and 'non-penetrating' liquids) on the mechanical properties of mesoporous MOF MIL-101. The liquid PTMs were able to enter the pores and stabilise the framework against compression and amorphisation up to 13 GPa, meanwhile MIL-101 became amorphous at 0.4 GPa when soft solid NaCl was used. This demonstrated how (even historically) non-penetrating liquid PTMs can alter the mechanical properties of porous MOFs to a certain extent.

A final example showing the effect of PTM on flexible MOF structures is the case of CuPyr-I (bis[1-(4-pyridyl)butane-1,3-dione]copper(II)).[113,113] CuPyr-I is highly unusual as it exhibits unique properties in the high pressure regime within the same framework, which can be tuned as a function of both PTM and pressure range. These include a single-crystal to single-crystal phase transition, a switching of the Jahn–Teller (JT) axis that depends on the PTM used to compress the crystal, piezochromism and negative linear compressibility (NLC) behaviour. CuPyr-I was compressed in a penetrating (MeOH and *n*-pentane) and non-penetrating (FC-70) PTM to 4.90 and 1.56 GPa, respectively. In both pressure experiments, the JT axis, which extends along the Cu–N pyridyl bond, steadily compresses and then switches to lie along one of the Cu–O dione bonds. Curiously, the JT switching can be delayed *via* compression in MeOH due to the PTM entering the framework pores, increasing the JT switching pressure to 1.47 GPa, compared with 0.64 GPa during compression in FC-70. The penetrated MeOH PTM also interacts with the framework, prompting a twisting of the ligand backbone, which is not observed in the absence of adsorbed guest. Additionally, the suppression of the JT axis is accompanied by a piezochromic colour change, which is followed by an isosymmetric phase transition at 2.15 GPa. Contrastingly, this phase transition occurs at much lower pressure (0.77 GPa) in *n*-pentane, pensively associated to the favourable hydrophobic interactions between the *n*-pentane and hydrophobic channels, which has been seen in other hydrophobic pores in MOFs.[114]

3.3.2.3.2 Penetrating PTM Interactions With Flexible MOFs. As outlined in the previous example, PTMs that are penetrating can exhibit guest–host interactions with the framework, causing rearrangements of the framework structure. One key example of this in the MOF field is the family of SOD topology zeolitic imidazolate frameworks (ZIFs). One of the first ZIFs studied under pressure was ZIF-8 ($[Zn(mIm)_2]_n$, where mIm = 2-methyl imidazole). At ambient pressure and temperature, ZIF-8 contains one central nano-sized pore per unit cell, with a volume of ~2500 Å^3 and pore diameter of 11.6 Å. Connecting these large nanopores are eight six-membered ring (6MR) windows *ca.* 3.0 Å in diameter and six smaller four-membered ring (4MR) windows of *ca.* 0.8 Å. Originally, the structure seemed promising for molecular sieving due to the small window diameters, as a 3.0 Å window should be able to selectively adsorb H_2 over CH_4, with kinetic diameters of 2.85 Å and 3.80 Å, respectively. However, in practice Bux *et al.* showed that many molecules with kinetic diameters larger than the 6MR window could permeate through ZIF-8, including O_2, N_2 and CH_4, the authors could speculate on why this observed behaviour occurred.[115] In 2009, Moggach *et al.* demonstrated in a high-pressure diffraction experiment using a DAC and a 4:1 methanol–ethanol PTM that ZIF-8 undergoes a phase transition. Upon increasing the pressure to 0.18 GPa, the hydrostatic media entered the pores causing the unit cell volume to initially expand, and then the framework compresses to 0.90 GPa. Upon further increasing the pressure (to 1.47 GPa), a single-crystal to single-crystal phase transition took place (from ZIF-8-AP to ZIF-8-HP, with AP and HP denoting the ambient and high-pressure phases, respectively). This transition was characterised by a rotation of the mIm linker, quantified by defining an angle of rotation (θ, defined as the angle between the (100)

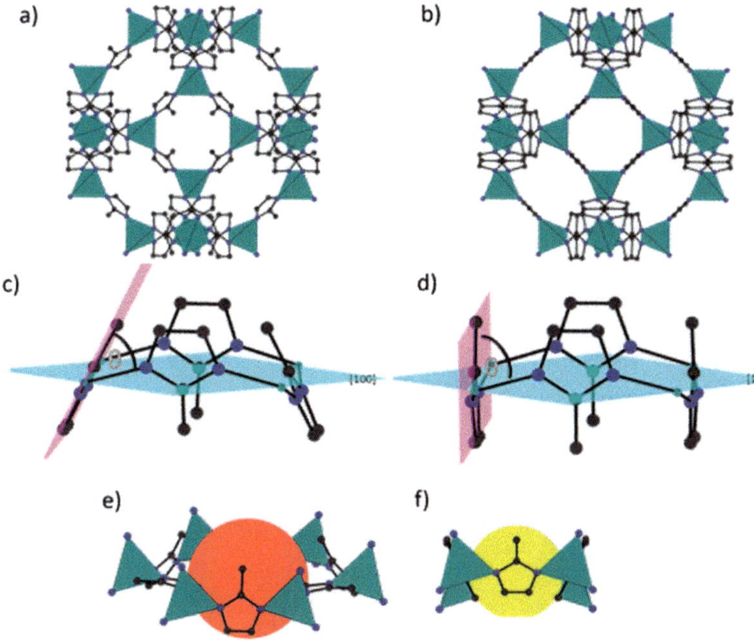

Figure 3.20 (a) ZIF-8-AP with 4-membered ring windows in the faces and 6-membered ring windows in the body diagonal. (b) ZIF-8-HP with the mIm linkers rotated ~30°. (c) 4MR window with opening angle of 68° for ZIF-8-AP. (d) 4MR window showing an opening angle of 89° for ZIF-8-HP. (e) 6MR window with pore diameter of 3.0 Å and (f) 4MR window with pore diameter of 0.8 Å.

plane and the mean plane passing through the mIm ring), which increases from 64.3° at ambient pressure to 89.7° at 1.47 GPa, see Figure 3.20.

This linker rotation increased the size of the windows that connect the nanopores in the centre of the unit cell, as the 4MR and 6MR window diameters increased from 0.8 Å to 2.2 Å, and 3.0 Å to 3.6 Å, respectively, which allowed more MeOH and EtOH molecules to enter the pores. As a result, the phase transition to the HP structure has also been used to explain the uptake of other gases including CO_2 and C1–C4 alkanes.[116]

Replacement of the $-CH_3$ substituent on the mIm linker with either –CHO or $-NO_2$ groups leads to the isostructural and isosymmetric ZIF-90 ($[Zn(ICA)_2]_n$, ICA = imidazolate-2-carboxyaldehyde) and ZIF-65 ($[Zn(nIm)_2]_n$, nIm = 2-nitroimidazolate), respectively. The chemical functionalisation results in different linker orientations and therefore different values for θ and the 4MR and 6MR diameters. When ZIF-90 and ZIF-65 were compressed in MeOH–EtOH (4 : 1), different pressure responses were found, see Figure 3.21.[117]

ZIF-90 undergoes a phase transition to a 'gate open' HP phase; however, the degree of rotation of the Im linker is less than that of ZIF-8. In addition, it was demonstrated that ZIF-65 undergoes a transition to a more 'gate closed'

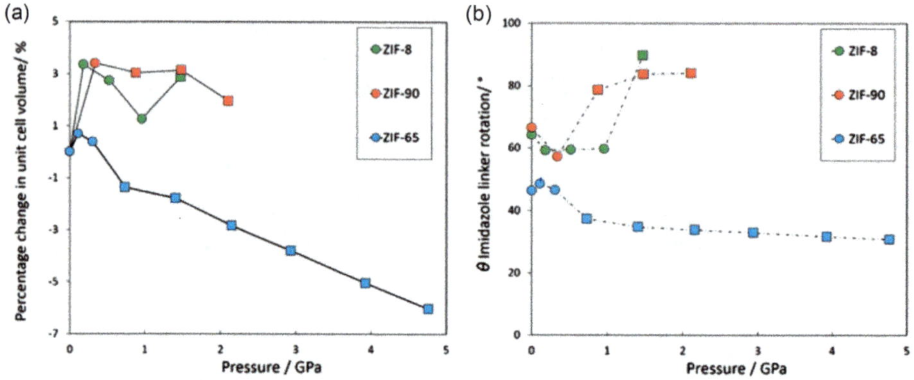

Figure 3.21 Evolution of cell volumes (a) and θ (b) with pressure for ZIF-8 (green), ZIF-90 (red), and ZIF-65 (blue) in MeOH–EtOH (4 : 1), where AP and HP phases are represented by circles and squares, respectively. ZIF-8 data from original study by Moggach *et al.*[51] Reproduced from ref. 117 with permission from American Chemical Society, Copyright 2018.

structure upon application of high pressure. The rationale behind the differences was attributed to the change in polarity of the functional group on the ZIF and the interaction with the penetrating PTM. This example demonstrates how we can use pressure to control flexible MOFs.

3.3.2.3.3 Non-penetrating PTM Interactions With Flexible MOFs. The PTM in a high-pressure diffraction experiment does not need to ingress into the framework structure to cause structural transitions. Non-penetrating PTMs can be used to apply hydrostatic pressure to a flexible MOF, causing a phase transition within the material *via* mechanical pressure instead of chemical interaction. The majority of pressure induced flexible MOFs studied tend to adopt the 'wine-rack' structure, such as MIL-53 [55] or DUT-8.[48] Originally, the diamond shaped pores in MIL-53 were found to contract from op to cp due to hinge-like movement in the bdc linkers, upon the uptake of certain guest molecules. However, through the application of hydrostatic pressure, the cp phases were also found *via* mechanical compression,[118,119] and both the effect of metal centre and functionalisation could be investigated *via* high-pressure crystallographic studies.[120]

As the community discovers softer flexible MOFs, it is important that their phase behaviour is accurately characterised at the low-pressure regime. One such example is ZIF-4(Zn). Under ambient conditions, ZIF-4(Zn) crystallises in an orthorhombic crystal system with an open pore structure, ZIF-4(Zn)_op, that displays a distorted cag topology. A high-pressure phase ZIF-4(Zn)_cp has been described in the monoclinic space-group $P2_1/c$; however, the pressure of the phase transition has been reported between a broad pressure range of 0.075 to 0.530 GPa. The discrepancies came from using different experimental methods to apply pressure. The lower pressure recorded came from Hg intrusion experiments, and found the phase transition irreversible, whereas the higher recorded phase transition pressure was from single crystal DAC experiments and found the transition reversible.[121] Recently, high-pressure diffraction experiments in the region of 0–0.4 GPa were carried out on ZIF-4(Zn) using the PJC. The phase transition was recorded at 0.1 GPa, overall, in agreement with the Hg intrusion experiment, however the transition was found to be reversible.[106] Reversibility and the nature of hysteretic behaviour is important for the use of MOFs in pressure-induced energy storage processes.

3.3.2.4 *Conclusions and the Future of High-pressure Crystallography on Flexible MOFs*

As stated in the introduction, the field is in its infancy with high-pressure crystal structures representing 0.8% of all MOF crystal structures within the CSD.[122] Nonetheless, this section is intended to demonstrate that high-pressure diffraction is an effective tool in which to study the behaviour of flexible MOFs. These fundamental studies have serendipitously expanded the potential applications of MOFs, from their initial origins as gas storage or separation materials, to wider applications such as shock absorbers,[123] energy storage springs,[124] barocaloric materials,[125] or ferroelectric materials[126] through their intriguing pressure response. One can only hope our understanding of flexible MOFs will continue to increase as the methodology for carrying out high-pressure crystallographic experiments improve.

3.3.3 *In Situ* NMR Techniques

3.3.3.1 *Principles of NMR Spectroscopy*

Nuclear magnetic resonance (NMR) spectroscopy[127,128] relies on the detection of nuclei with non-zero nuclear spin, *i.e.*, with a nuclear spin quantum number $I = \frac{1}{2}, 1, \frac{3}{2}, 2, \frac{5}{2}$, *etc.* Important examples with respect to MOFs are ^{1}H and ^{13}C ($I = \frac{1}{2}$), ^{2}H ($I = 1$), ^{17}O ($I = \frac{5}{2}$), $^{63/65}Cu$ ($I = \frac{3}{2}$) or ^{27}Al ($I = \frac{5}{2}$). Nuclear spins serve as element-selective, local probes and are mainly influenced by their close spatial environment *via* so-called internal magnetic interactions within distances lower than 1 nm. Such interactions are for example the spatially anisotropic homonuclear and heteronuclear magnetic dipole–dipole interactions between neighbouring spins. Another example is magnetic shielding. It gives rise to the isotropic chemical shift, δ_{iso}, and the chemical shift anisotropy (CSA). Spatially anisotropic interactions like CSA or dipole interactions average for rapidly reorienting molecules in isotropic solutions or gases. The resulting highly resolved 'liquid-state' NMR spectra reflect these isotropic averages, *i.e.*, interactions like the isotropic chemical shift and indirect spin–spin coupling (*J*-coupling). Figure 3.22 schematically demonstrates the

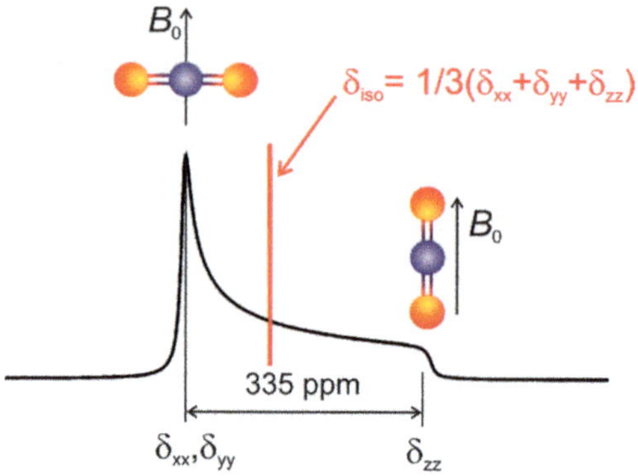

Figure 3.22 Calculated ^{13}C NMR spectrum for fully immobilised and randomly oriented $^{13}CO_2$ (powder spectrum). Note the characteristic line shape dominated by the axially symmetric CSA tensor of the CO_2 molecule with the principal values $\delta_{xx} = \delta_{yy} = \delta_{\perp}$ (perpendicular to the symmetry axis of CO_2) and $\delta_{zz} = \delta_{||}$ (parallel to the symmetry axis). The chemical shift anisotropy is then given by $\Delta\delta = \delta_{\perp} - \delta_{||}$ and amounts to 335 ppm for CO_2. Rapid and isotropic molecular reorientation in the free gas averages the CSA and a narrow signal would be observed at the isotropic chemical shift δ_{iso} indicated by the vertical red line.

CSA-induced line broadening for the ^{13}C NMR spectrum[129,130] of rigid $^{13}CO_2$ and the expected narrow signal at the isotropic chemical shift.

In solids,[131] spatially anisotropic interactions like CSA or dipole interactions result in severe line broadening. They can be suppressed or at least reduced by the magic angle spinning (MAS) technique.[132] It relies on the fast sample rotation around an axis tilted by the magic angle $q_m = \arccos\frac{1}{\sqrt{3}} = 54.7^\circ$ with respect to the external magnetic field of the spectrometer. The combination of the MAS technique with the cross-polarisation (CP) experiment[133,134] is standard for proton-bearing molecules because it enhances the sensitivity and reduces the measurement time significantly. As an example, Figure 3.23 shows the ^{13}C CP MAS NMR spectrum of a linker molecule. Note, however, that polarisation transfer experiments like CP complicate or prevent quantitative analyses.[135]

In summary, NMR experiments provide the following types of information. (1) For directly excited spectra, NMR is quantitative because the signal intensity is then directly proportional to the number of spins, *i.e.*, NMR 'counts' spins. In contrast to optical methods, intensities of different signals are thus directly comparable with each other. (2) NMR spectra provide structure-relevant parameters. Examples are chemical shifts, *J*-couplings and distance-dependent or spatially anisotropic parameters like dipolar couplings, chemical shift anisotropies and electric field gradient tensors. (3) NMR spectroscopy allows studying dynamic processes like thermal motions of molecules including diffusion, exchange processes and others.

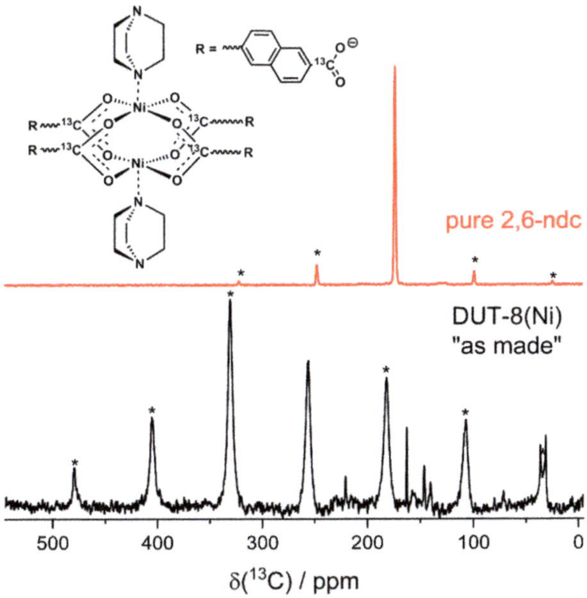

Figure 3.23 ^{13}C CP MAS NMR spectra of 2,6-ndc (2,6-naphtalene dicarboxylic acid) with selectively ^{13}C-enriched carboxylic acid carbons and of the as-synthesised DUT-8(Ni) with ^{13}C-labelled 2,6-ndc linkers. Asterisks (*) denote spinning sidebands that occur in addition to the signal at the isotropic chemical shift at distances corresponding to integer multiples of the sample spinning rate. The structure of 2,6-ndc and of the Ni_2 paddle wheel unit is shown in the insert. Reproduced from ref. 135 with permission from the Royal Society of Chemistry.

These advantageous properties are the reason why solid-state NMR spectroscopic techniques are widely used to study MOFs.[5,136–139]

3.3.3.2 *General Aspects of* In Situ *NMR Spectroscopy*

The term '*in situ*' originates from medicine meaning surgery without moving the operated organ from its place in the human body. In spectroscopy, it denotes experiments monitoring processes like synthesis reactions, adsorption, and diffusion phenomena *etc.* The term '*operando*' is in part synonymous and denotes spectroscopic experiments coupled with the ongoing process of interest like catalytic reactions or to study operating devices like batteries, supercapacitors, and others. *In situ* NMR spectroscopic approaches were meanwhile developed for various purposes[3,140–144] as summarised in Figure 3.24.

In situ techniques can help to visualise and study structural changes of flexible MOFs under the influence of external stimuli like gas pressure or temperature changes. *In situ* NMR studies of gas adsorption processes are thus of special interest and will be considered in detail in the next section.

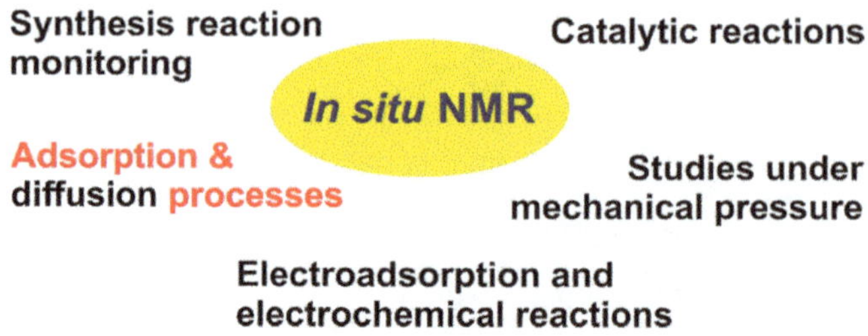

Figure 3.24 Scheme summarising the applications of *in situ* NMR spectroscopy.

It should, however, be emphasised that *in situ* NMR can also be applied to study synthesis reactions of various materials including MOFs.[140,141] Furthermore, *in situ* NMR techniques were also developed for the investigation of catalytic reactions,[142,143] for monitoring electro-adsorption processes and electrochemical reactions in electrochemical devices[144–146] as well as for materials studies under high mechanical pressure.[147] Such experimental techniques may become important for studying flexible MOFs in the future as well.

3.3.3.3 In Situ *NMR Spectroscopy of Flexible MOFs*

Framework flexibility in MOFs can occur at different structural levels. On one hand, organic linker molecules in MOFs can exhibit local thermal motions and/or subtle structural changes under the influence of external stimuli. Examples are linker reorientations, minor deformations, or adsorption-induced mobility changes, *e.g.*, of phenyl ring flips. These local phenomena often do not cause global structural changes and are thus hardly detectable by diffraction techniques. In contrast, NMR spectroscopy is sensitive for local changes and can thus be used to visualise such subtle effects.[148,149] A powerful technique to study the linker motions as well as adsorbed molecules is ^{2}H NMR spectroscopy.[150–153] Selective deuteration schemes make this method site-selective and thus particularly powerful.

On the other hand, some special MOFs exhibit very pronounced structure changes under the influence of external stimuli.[1,13,154,155] Such transitions are for example induced by gas adsorption depending on the type of adsorbed gas, pressure and temperature. The flexible DUT-8(Ni)[48] for example exhibits a very pronounced structural transition resulting in a change of the unit cell volume by more than a factor of two. In the following, MOFs capable of undergoing such major structural transitions are denoted as 'flexible'. Flexibility-related phenomena are *in situ* accessible for special NMR techniques for example by studying gas-loaded samples. Two fundamentally different experimental approaches exist for *in situ* gas adsorption NMR studies as discussed in the following sections.

3.3.3.3.1 Sealed Samples With Constant Loading. In such experiments, samples are filled in sealable containers like glass ampoules.[156] After activation, they are loaded with defined adsorbate amounts and sealed. The absolute number of molecules in the sample cell is then constant during the experiment and does not depend on the measurement temperature. This technique provides several advantages. The samples are long-term stable and well isolated from the surrounding environment, *i.e.*, adsorbed species cannot escape. Sealed samples can even be prepared in a manner allowing the application of the line-narrowing MAS technique. However, this method also exhibits disadvantages. Studies of loading-dependences require the preparation of multiple samples at the different loadings of interest and the pressure cannot be adjusted during temperature variation, which limits the comparability of NMR data with volumetric adsorption–desorption isotherms or isobars.

As an example for such studies, we will consider the ^{129}Xe NMR experiments performed by Giovine *et al.*[157] on the well-known flexible compound MIL-53(Al). Depending on the external conditions, this MOF switches between two different structures, the lp and np form. This type of transition is denoted as 'breathing'.[55] The samples were activated, equilibrated with xenon at room temperature and 5×10^4 Pa, and finally sealed. The static ^{129}Xe NMR spectrum measured without MAS at 298 K is shown in Figure 3.25 (top, left).

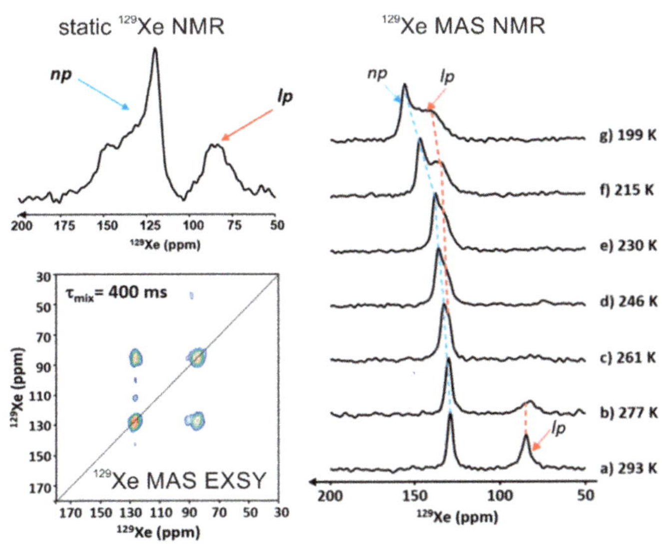

Figure 3.25 ^{129}Xe NMR experiments of xenon-loaded MIL-53(Al). Top, left: static ^{129}Xe NMR spectrum of MIL-53(Al) equilibrated at room temperature with 5×10^4 Pa xenon, sealed and measured at 298 K. Right: temperature-dependent ^{129}Xe MAS NMR spectra. Bottom left: 2D ^{129}Xe EXSY spectrum measured under MAS with 400 ms mixing time. Adapted from ref. 157 with permission from the American Chemical Society, Copyright 2017.

Under these conditions, MIL-53(Al) crystallites in the sample powder coexist in both, the lp and np forms. The sample thus exhibits two different ^{129}Xe NMR signals at isotropic chemical shifts of 90 (lp) and 128 ppm (np). For adsorbed xenon atoms diffusing through the pore system of the material, only one single signal occurs at a chemical shift averaged along its trajectory given by

$$\delta_{\text{iso}} = \delta_{\text{s}} + \delta_{\text{Xe-Xe}}\left(T, \rho_{\text{Xe}}\right) \approx \delta_{\text{s}} + \Delta_{\text{Xe-Xe}}\left(T\right) \times \rho_{\text{Xe}}. \quad (3.1)$$

Here, δ_S summarises the influence of various xenon–wall interactions.[158–163] Commonly, δ_S increases at decreasing average pore size provided the surface chemistry of the compared systems is similar. The term $\delta_{Xe\text{-}Xe}$ describes the influence of xenon–xenon interactions[164,165] and depends on the temperature *T*. At low pressure, it is approximately linear in the xenon density as shown in eqn (3.1) where the parameter $\Delta_{Xe\text{-}Xe}$ denotes the slope and ρ_{Xe} the xenon density in the adsorbed state. The isotropic chemical shift is larger for xenon adsorbed in smaller pores because the influence of xenon–wall interactions represented by δ_S is then bigger. This explains the chemical shift difference observed for xenon in MIL-53(Al) crystallites of np state than in lp state at identical pressures and temperatures (Figure 3.25).

Note the pronounced broadening and characteristic line shape of the np state signal resembling the line shape expected for CSA (*cf.* Figure 3.22). This is for the first glance surprising because ^{129}Xe is an atom and does not exhibit a chemical shift anisotropy. However, the isotropic chemical shift of xenon confined in narrow pores of non-cubic symmetry depends on the orientation of crystallites with respect to the direction of the external magnetic field. For crystallite powders, this results in a similar line shape as observed for CSA, an effect which was first discovered for xenon in zeolites[166] and later also for hyperpolarised ^{129}Xe in channels of peptide crystals[47] and in the narrow pores of MIL-53.[167]

This line broadening is removed by MAS as demonstrated by Figure 3.25 (right). The well-resolved and narrow signals obtained under MAS allowed the authors to perform two-dimensional exchange spectroscopy (2D EXSY) experiments (*cf.* Figure 3.25 (bottom, left)) as well as one-dimensional selective inversion recovery experiments. The latter were used to measure the exchange rate of 43 s^{-1} for xenon exchange processes between the lp and np state.

3.3.3.3.2 *In Situ* NMR With Externally Controlled Gas Pressure. Connecting pressure-stable NMR tubes with an external gas reservoir during an *in situ* NMR experiment allows adjustment and control of the pressure at the given temperature. A corresponding setup enabling high-pressure variable-temperature experiments was described by Hoffmann *et al.*[168] Using this setup, it was for example possible to visualise the counterintuitive negative gas adsorption (NGA) transition which was observed for the flexible compounds DUT-49 and DUT-50.[23,62,63] These MOFs adopt an op state after activation. At higher temperatures, the compounds are non-flexible and stay open

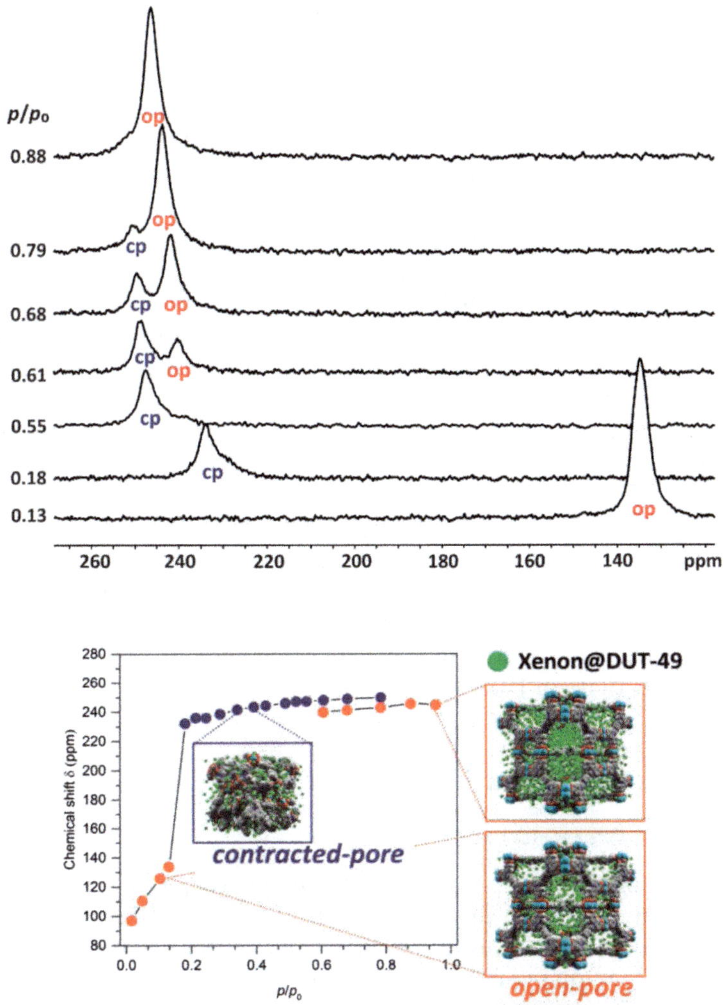

Figure 3.26 High-pressure *in situ* ^{129}Xe NMR experiments on DUT-49. Top: ^{129}Xe NMR spectra of DUT-49 measured at 200 K. Bottom: ^{129}Xe chemical shifts and structure models for the different states. Reproduced from ref. 63, https://doi.org/10.1021/acs.chemmater.9b02003, under the terms of the CC BY-NC-ND 4.0 license https://creativecommons.org/licenses/by-nc-nd/4.0/.

during gas adsorption, *e.g.*, of xenon. DUT-49 becomes flexible at lower temperatures, *i.e.*, 200 K in the experiment shown in Figure 3.26, a structural change occurs after surpassing a specific transition pressure. The MOF then transforms into the cp state.

This transformation results in linker bending causing mesopore contraction and partial release of previously adsorbed gas (NGA). This transition gives rise to a tremendous jump of 100 ppm in the ^{129}Xe chemical shift at relative pressures between 0.13 and 0.18. This sudden change results from the decreasing pore size and increasing xenon density inside the contracted pores (*cf.* eqn (3.1)). At further increasing the xenon pressure, *i.e.*, beyond a relative pressure of 0.6, the structure re-opens and adopts the op state as can be seen from the increasingly intense signal at a slightly lower chemical shift than the cp signal. The op signal has then a considerably higher ^{129}Xe chemical shift than at low relative pressures before the NGA transition. This is due to the larger xenon density at higher relative pressures.

Carbon dioxide is frequently used for *in situ* NMR adsorption studies of porous materials including MOFs.[139,169–173] This linear molecule exhibits a pronounced axially symmetric chemical shift anisotropy. In the adsorbed state, characteristic line shapes occur due to the partial immobilisation of the molecules inside the pore system. The molecules then move under the spatial restrictions of the pores depending on the geometry (symmetry) of the pore system. In the case of crystallite powders, this gives rise to line shapes[63,174] characteristic for CSA. However, the lines are narrower than expected for fully immobilised CO_2 (*cf.* Figure 3.22). This line shape also occurs for flexible DUT-8(Ni) as demonstrated in Figure 3.27. *In situ* NMR spectroscopy can also be used to study gas mixture adsorption, *i.e.*, to evaluate the selectivity. For example, a mixture of ^{13}C-enriched carbon dioxide and methane was used in order to study the selectivity of flexible MOFs.[175,176] An excellent selectivity of DUT-8(Ni) towards CO_2 occurs after pore opening.

3.3.4 *In Situ* EPR Techniques

3.3.4.1 EPR Spectroscopy

Electron paramagnetic resonance (EPR) spectroscopy is the method of choice for the investigation of paramagnetic species such as paramagnetic transition metal ions (TMI) and rare-earth ions (REI) as well as free radicals and photo-excited triplet states.[177,178] Among these the paramagnetic 3d TMI ions Ti^{3+}, V^{4+}, Cu^{2+}, Mn^{2+}, Fe^{3+}, Co^{2+} and Cr^{3+} are of particular importance for EPR studies of flexible MOF materials as they are often found as major constituents of frameworks.[179] Therefore these ions can be employed as local magnetic probes to monitor structural changes of MOFs with responsive frameworks on a microscopic scale.[3] In general, the ground state energies and consequently the continuous wave (cw) EPR spectrum of such a paramagnetic TMI species with electron spin S and central metal ion nuclear spin I in a magnetically diluted system can be described by the spin Hamiltonian[180]

$$\mathcal{H} = \frac{\beta_e \tilde{\boldsymbol{B}}_0 \boldsymbol{g} \boldsymbol{S}}{h} + \tilde{\boldsymbol{S}}\boldsymbol{D}\boldsymbol{S} + \tilde{\boldsymbol{S}}\boldsymbol{A}\boldsymbol{I} \tag{3.2}$$

given in frequency units with Planck's quantum of action h and the Bohr magneton β_e. The electron Zeeman interaction in the first term with the external magnetic field $\boldsymbol{B}_0$ and the electron $\boldsymbol{g}$ tensor comprises the spin–orbit

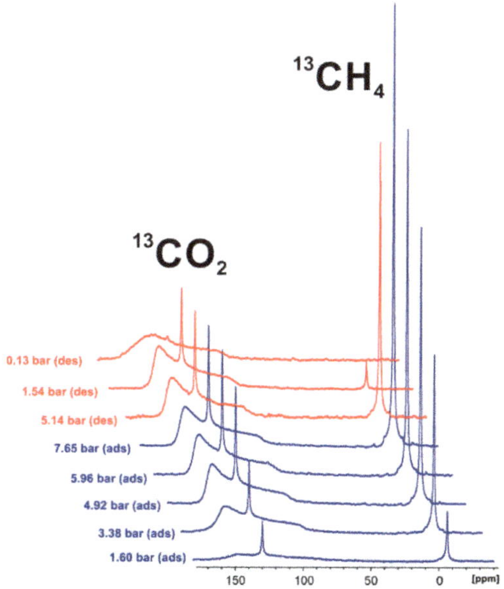

Figure 3.27 Variable-pressure *in situ* ^{13}C NMR experiments of $^{13}CO_2$–$^{13}CH_4$ gas mixture adsorption on DUT-8(Ni). Note the presence of two narrow signals at about 127 ppm and −10 ppm due to gas phase $^{13}CO_2$ and $^{13}CH_4$, respectively. The broad CSA-like signal of *ca.* 50–60 ppm width (slightly depending on pressure) indicates the presence of adsorbed carbon dioxide. Remarkably, the characteristic signal of adsorbed methane which is expected at about −4 to −6 ppm could not be detected indicating an excellent selectivity of DUT-8(Ni) towards carbon dioxide. Adapted from ref. 175 with permission from the American Chemical Society, Copyright 2019.

coupling effects and contains information about the coordination symmetry, the ground state of the paramagnetic centre, and the energy splitting between the latter and the excited sts.[177,178] The second term applies to spin systems with $S > \frac{1}{2}$ and describes the zero-field splitting (ZFS), which originates from electron–electron dipole–dipole interactions and spin–orbit coupling. The traceless ZFS tensor $\boldsymbol{D}$ contains again important information about the coordination symmetry and in addition about bond angles and bond distances.[181] The third term with the hyperfine (HF) interaction tensor $\boldsymbol{A}$ involves the HF interaction between the electron spin and the nuclear spin of the central metal ion nucleus of the paramagnetic centre.[178,181] The tensor $\boldsymbol{A}$ of the nucleus of the TMI contains information about its coordination symmetry and ground state. It is worth emphasising that a thorough and reliable determination of the geometrical and electronic structure of a paramagnetic centre from its spin Hamiltonian parameters, measured by EPR, requires a quantum chemical analysis of the interaction tensors $\boldsymbol{g}$, $\boldsymbol{D}$, $\boldsymbol{A}$ in eqn (3.2) by using for instance density functional theory (DFT) calculations.[182]

3.3.4.2 Application Potential and Experimental Technique of In Situ EPR for Flexible MOFs

So far the majority of *in situ* cw EPR investigations have been devoted to the study of the valence state of TMIs in catalytic reactions over zeolites and various metal oxide surfaces.[183–186] In these experiments a continuous gas flow with the reactants at temperatures up to 813 K[185] is directed through the EPR sample in the *in situ* EPR cell.

However, in current research on MOFs the exploration of adsorption phenomena in MOF materials with flexible frameworks that feature gate pressure and breathing effects[13,187–189] are at the centre of interest and can be explored by EPR spectroscopy on a local microscopic scale. The modification of the coordination geometry of a paramagnetic TMI or REI during an adsorption controlled gate pressure or breathing framework transformation will lead to distinct changes in the EPR interaction tensors g, A, and D or EPR linewidth parameters.[190–193] Changes of these spin Hamiltonian parameters, in turn, lead to characteristic modifications of the EPR spectra, which can be followed in an *in situ* experiment. In that way, the adsorption at specific framework sites and structural changes of the MOF framework at the paramagnetic metal ions can be directly addressed on a molecular level by *in situ* EPR experiments as a local probe technique.

So far, the vast majority of *in situ* EPR cells were constructed for X-band frequencies (9–10 GHz). Today, state of the art X-band EPR spectrometers provide a realistic sensitivity limit of about 10^{13} spins per cm^3; even in case of TMI species in polycrystalline samples,[194] which seems to be fully sufficient for most EPR applications in MOF research. For the specific application case of gas adsorption studies over MOF materials, the *in situ* cell must allow for measurements with controllable static gas pressures between 0.001 mbar and 1330 mbar at low leakage rates and variable temperatures between 4 and 450 K. The size of the X-band EPR cavities enable easy housing of a hose to maintain a static gas pressure. As an example, a typical experimental setup for such *in situ* EPR experiments at X-band frequencies is illustrated in Figure 3.28. It relies on a commercial cylindrical TE_{011} EPR cavity and He flow cryostat.[192] Alternatively, the TE_{011} cavity can be substituted by a rectangular cavity such as TE_{102}. In addition, the He flow cryostat can be substituted by a N_2 flow cryostat for measurements above room temperature. The quartz glass sample tube fixed to the cryostat and cavity is attached *via* an aluminium-made home-built adapter to a stainless-steel hose, which in turn connects directly to a 190 cm^3 gas reservoir. The latter is filled from a gas storage volume by a high precession dosage valve at the desired gas pressure. Capacitive pressure gauges covering a pressure range from $p = 0.001–1330$ mbar allow precise pressure measurements independent of the type of gas. The whole gas loading station except the home-built sample tube adapter is made of DN 10 ISO-KF vacuum components and has a low leakage rate of 0.05 mbar h^{-1} at $p = 10$ mbar.[192] Such low leakage rates are essential because long equilibrium times up to 20 min are typically expected in *in situ* EPR

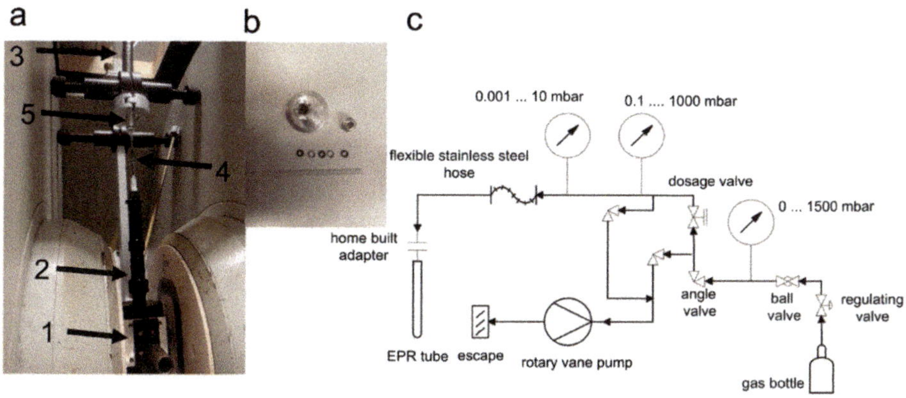

Figure 3.28 Experimental setup for *in situ* X-band EPR experiments: (a) commercial cylindrical cw X-band cavity (1), He flow cryostat (2), stainless-steel hose (3), EPR sample tube (4), sample tube adapter (5), and optical access for light excitation (6). (b) Disassembled adapter with aluminium fittings and o-rings, and (c) process flow diagram of the gas loading apparatus. Adapted from ref. 192 with permission from American Chemical Society, Copyright 2019.

adsorption studies of gate pressure and breathing MOF materials[192] such as those of the DUT-8[20,195] family, pillared layered MOFs[189] and MIL-53(Al).[187]

3.3.4.3 *Selected Examples of* In Situ *EPR Studies of Flexible MOFs*

If the flexible MOF host framework is diamagnetic, application of EPR spectroscopy requires the substitution of a small part (0.1–10%) of the diamagnetic framework ions by suitable paramagnetic probe ions without changing the properties of the MOF significantly. A typical example for this approach is the mixed-metal pillared-layered MOF $[Zn_{1.9}Cu_{0.1}(BME\text{-}bdc)_2(dabco)]_n$, which belongs to the family of so-called soft porous crystals (SPCs) that are flexible and exhibit reversible framework phase transitions from a closed or np phase to an expanded open or lp phase as known for the parent $[Zn_2(BME\text{-}bdc)_2(dabco)]_n$ material.[196] The related swelling and shrinking behaviour of the framework is known as breathing behaviour.[187] In $[Zn_{1.9}Cu_{0.1}(BME\text{-}bdc)_2(dabco)]_n$ a part of the diamagnetic Zn^{2+} ions have been substituted by paramagnetic Cu^{2+} ions with $S = ½$. *In situ* EPR spectroscopy of Cu^{2+} ions is a very suitable probe to monitor the np ↔ lp phase transition on a local molecular level for this MOF during CO_2 adsorption (Figure 3.29) as the coordination geometry of the paramagnetic cupric ions changes during the np ↔ lp transition and the Cu^{2+} g and A tensors are very sensitive with respect to these small deformations of Cu–Zn paddle wheels (PW).[192] Indeed the g_{zz} principal value of the Cu^{2+} g tensor decreases from g_{zz} = 2.362 to 2.355 whereas the A_{zz} principal value of the ^{63}Cu HF interaction tensor A increases from 398 MHz to 418 MHz upon the CO_2 adsorption induced np → lp transition indicating

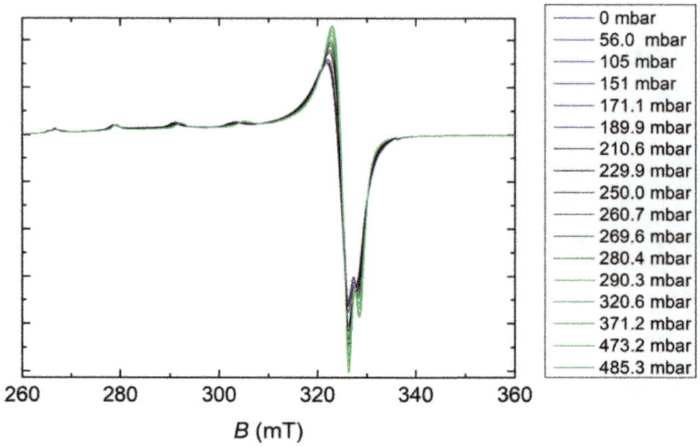

Figure 3.29 Cu^{2+} X-band EPR spectra of the activated $Zn_{1.9}Cu_{0.1}$(BME-bdc)$_2$(dabco) measured at T = 195 K *in situ* during CO_2 adsorption at different gas pressures listed in the caption. Reproduced from ref. 192 with permission from American Chemical Society, Copyright 2019.

an enhanced equatorial ligand field at the cupric ion in the lp phase as illustrated in Figure 3.30. Obviously, the Cu^{2+} species in the np phase transforms into a species belonging to the lp phase with increasing CO_2 pressure where a coexistence of both species is observed for an intermediate pressure range (225–280 mbar).

MIL-53(Al) is another flexible MOF that exhibits breathing behaviour with a reversible np ↔ lp phase transition induced either by temperature variation or adsorption of various gases.[187] Nevjestić *et al.*[191,197] applied *in situ* EPR spectroscopy to explore the adsorption of air, N_2 and O_2 over vanadium doped MIL-53(Al). Here, V^{4+} ions incorporate as paramagnetic VO^{2+} species with S = ½ at diamagnetic Al^{3+} framework sites and can serve as local paramagnetic probes to monitor the reversible np ↔ lp phase transition.[198,199] Figure 3.31 illustrates the effects of increasing air and nitrogen pressures on the room temperature *in situ* measured V^{4+} X-band EPR spectra for the lp phase of vanadium doped MIL-53(Al).[197] Only the spectra measured during air adsorption are affected where a constant increase in line width and decrease in signal amplitude with rising air and consequently oxygen pressure is observed due to an exchange interaction between the VO^{2+} units and the paramagnetic O_2 molecules. Otherwise, the adsorption of diamagnetic N_2 does not influence the V^{4+} EPR spectra. Consequently, line broadening effects of the V^{4+} EPR spectra can be taken to explore the np ↔ lp phase transition of MIL-53(Al) in presence of oxygen.

In case of the switchable porous material DUT-8(M), substitution of Ni^{2+} by Co^{2+} offers the potential to tune gradually the gate opening pressure for this MOF.[200] DUT-8(Ni) shows high responsivity and gate opening from an initially cp into an op phase in response to nitrogen adsorption whereas DUT-8(Co) is not responsive upon N_2 adsorption and remains in its cp phase. In the mixed

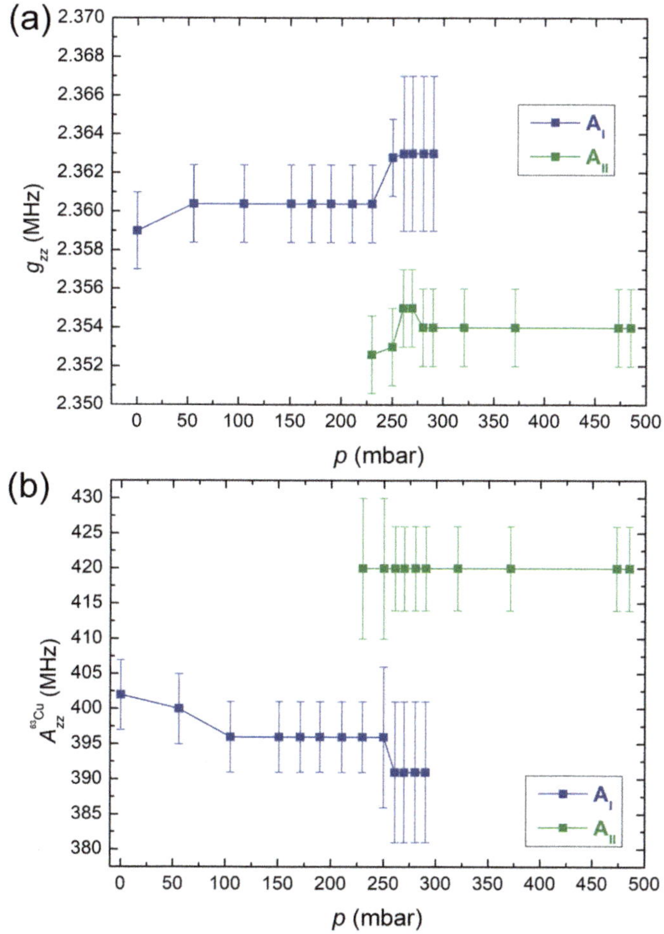

Figure 3.30 EPR parameters g_{zz} (a) and A_{zz} (b) of Cu^{2+} ions in the np (A_I, blue) and lp (A_II, green) phase of activated $Zn_{1.9}Cu_{0.1}(BME\text{-}bdc)_2(dabco)$ at different CO_2 pressures as determined by *in situ* X-band EPR at $T = 195$ K during adsorption. Reproduced from ref. 192 with permission from American Chemical Society, Copyright 2019.

metal system DUT-8($Ni_{1-x}Co_x$) the gating pressure can be controlled by the composition parameter *x*. The parent DUT-8(Ni) and DUT-8(Co) frameworks are EPR silent at X band. Instead for DUT-8($Ni_{1-x}Co_x$) the formed Ni^{2+}–Co^{2+} mixed PW units display an antiferromagnetic coupling between the $S = 1$ Ni^{2+} and $S = 3/2$ Co^{2+} spins resulting in EPR active $S = 1/2$ ground state that provides a characteristic EPR spectrum at X band frequencies (Figure 3.32(a)) in the op phase.[200] The Ni^{2+}–Co^{2+} PW units are EPR silent in the cp phase. Figure 3.32(b) illustrates the intensity of the EPR signal of the Ni^{2+}–Co^{2+} PW units in the op phase in dependence on the nitrogen adsorption and desorption pressure (relative pressures p/p_0) at $T = 71$ K for different composition

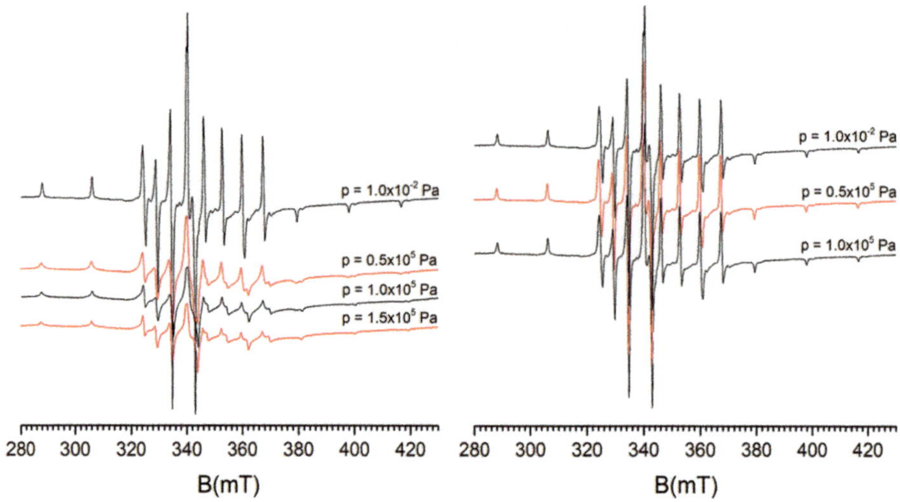

Figure 3.31 Room temperature V^{4+} X-band EPR spectra of activated V-doped MIL-53(Al) in the lp phase illustrating line broadening effects due to exchange interaction between V^{4+} ions and O_2. Left: as a function of air pressure. Right: as a function of N_2 pressure. Reproduced from ref. 197 with permission from American Chemical Society, Copyright 2016.

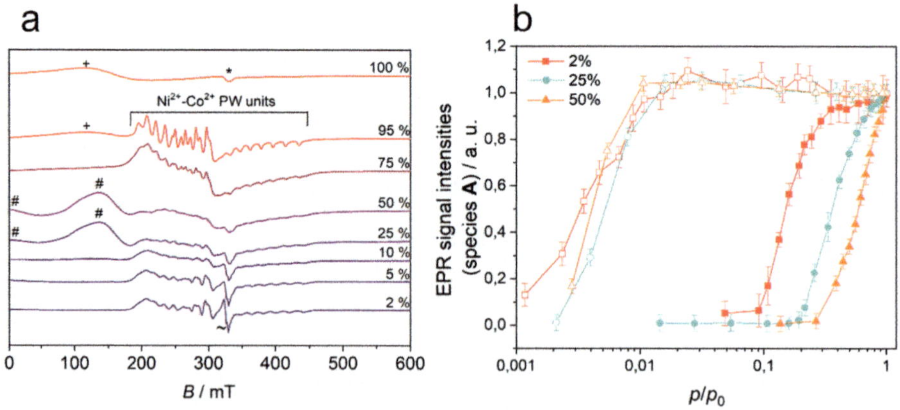

Figure 3.32 (a) X-band EPR spectra of DUT-8(Ni$_{1-}$*x*Co*x*) in the op phase at T = 12 K showing the S = ½ ground state signal of the antiferromagnetically coupled Ni^{2+}–Co^{2+} PW units for various composition parameters x. Additional symbols label impurity signals from Co^{2+} (#) and Cu^{2+} (*) monomers, and from the cryostat (+). (b) EPR signal intensities of Ni^{2+}–Co^{2+} PW units for x = 0.02, 0.25 and 0.50 as a function of relative nitrogen pressures at T = 71 K measured in an EPR *in situ* experiment during N_2 adsorption and desorption. Adapted from ref. 200 with permission from American Chemical Society, Copyright 2020.

parameters x as measured by an *in situ* EPR experiment. These signal intensities of the Ni^{2+}–Co^{2+} PW units can be considered as a measure for the relative volume fraction of the op phase in the material. The cp to op transition is shifted from $p/p_0 = 0.10–0.35$ for 2% Co ($x = 2$) to $p/p_0 = 0.27–1.00$ for 50% Co ($x = 50$). These results agree well with volumetric adsorption data[200] and offer a microscopic view on the gating phenomena for this family of flexible MOFs.

DUT-49(Cu), another material with PW units as a final example and famous for its negative gas adsorption,[23] is one of the rare examples where the original composition of the MOF framework allows straightforward application of EPR spectroscopy without any additional doping by paramagnetic ions. Here, the two cupric ions in the PW units couple antiferromagnetically and provide an EPR active excited state with $S = 1$ that can be observed at $T > 80$ K.[201] Due to the extended linker system in DUT-49 magnetic exchange interactions among the excited $S = 1$ states of the Cu^{2+} PW units are negligible and its ZFS is nicely resolved in the X band EPR spectrum (Figure 3.33).[193]

The principal values of the ZFS tensor D are extremely sensitive to even tiny structural modifications of the structure of the Cu^{2+} PW units leading together with the accompanying variation of the g tensor parameters to characteristic spectral changes in their EPR spectra. As the excited $S = 1$ states of the Cu^{2+} PW units is observable from $T = 70$ K up to room temperature

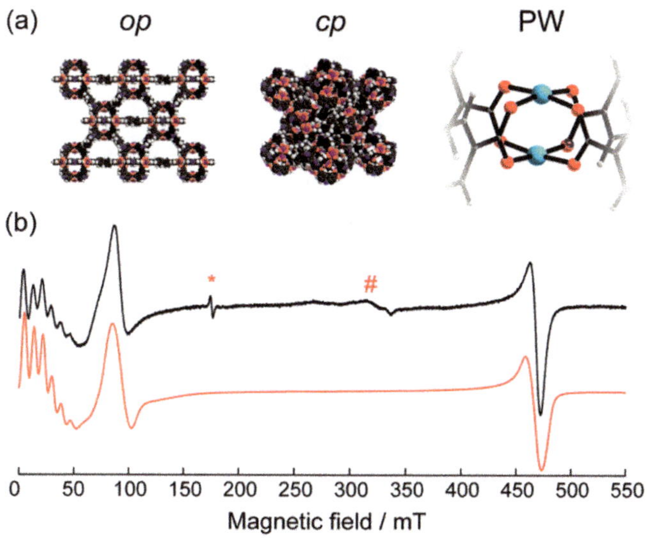

Figure 3.33 (a) Structure of DUT-49(Cu) in the op and cp phase indicating the change of the pore volume during the op ↔ cp transition together with the structure of the Cu^{2+} PW units (Cu, bigger blue balls, O, smaller red balls). (b) Room temperature X-band EPR spectrum of DUT-49(Cu). Symbols * and # indicate impurity signals from cupric ions monomers and from the EPR cavity. Reproduced from ref. 193, https://doi.org/10.1021/acs.jpclett.0c01705, under the terms of the CC BY-NC-ND 4.0 license https://creativecommons.org/licenses/by-nc-nd/4.0/.

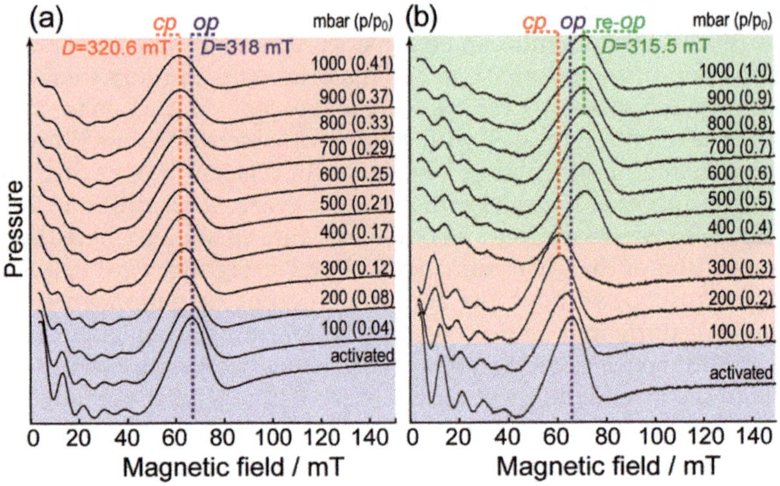

Figure 3.34 X-band *in situ* EPR spectra of DUT-49(Cu) during *n*-butane adsorption at increasing gas pressures: (a) $T = 298$ K with the op → cp phase transition and (b) $T = 278$ K with the op → cp → re-cp phase transitions. The axial ZFS parameter D is of the Cu^{2+} PW units is given for each phase. Differently coloured areas indicate the presence of the three phases. Coloured dotted lines are a guide to the eye for spectral changes in the *in situ* spectra. Reproduced from ref. 193, https://doi.org/10.1021/acs.jpclett.0c01705, under the terms of the CC BY-NC-ND 4.0 license https://creativecommons.org/licenses/by-nc-nd/4.0/.

and above they constitute ideal probes for *in situ* EPR experiments of DUT-49(Cu). Figure 3.34 illustrates X band EPR spectra measured *in situ* during *n*-butane adsorption over DUT-49(Cu) at T = 298 K and 273 K.[193] These *in situ* EPR experiments reveal the gating transitions of the material induced by the adsorption of the *n*-butane in convincing agreement with volumetric adsorption experiments.[193] At $T = 298$ K DUT-49(Cu) exhibits only a transition from the initial op to the cp phase at $p/p_0 = 0.1–0.15$ (Figure 3.34(a)). However, a further transition into a re-opened (re-op) phase could be detected at $p/p_0 = 0.4–0.5$ at a lower temperature $T = 273$ K. Each of the three phases of DUT-49(Cu), op, np, and re-op, observed during the *in situ* EPR *n*-butane adsorption experiments are related to slightly different structures of the Cu^{2+} PW units, even if their structural differences seem to be minor. But they are sufficient to alter their ZFS parameters significantly leading to distinct EPR spectra of the PW units in the three phases of this material and in that way allowing microscopic *in situ* studies of the gating behaviour in this case.

3.3.5 *In Situ* Vibrational Spectroscopy (Infrared and Raman)

Infrared (IR) spectroscopy measures the light that induces vibration in molecules that changes dipole moments, while Raman spectroscopy can be used due to the inelastic scattering of photons by matter, a process driven by the change of a molecule's polarisability. The appearance of signals in both techniques may be predicted from selection rules elaborated from the molecular symmetry of investigated species. The assignment of a signal in vibrational spectroscopy is, essentially, based on the correlation between observed vibrational bands and particular modes of certain functional groups. In many cases, the observed modes may have very similar frequencies, and for resolving complex cases the isotopic substitution was found very useful.[202]

Vibrational spectroscopy (IR and Raman) includes well-known techniques utilised to investigate MOFs, especially their interaction with guest molecules.[9] It enables a distinction to be drawn between the different chemical environments of adsorbates inside the framework,[176] while facilitating monitoring of the activation process,[203] and investigation of the mechanochemical reactivity of the flexible framework.[204] Furthermore, vibrational spectroscopy allows an investigation of very subtle structural changes that are not detectable by PXRD.[205–207]

For one of the first *in situ* vibrational (IR) investigations of flexible MOFs, a reactor cell (Figure 3.35) dedicated for *operando* studies, was used.[56] This reactor cell consists of a cylinder of stainless steel carrying a toroidal sample holder in its centre, where the catalyst is normally placed in the form of a self-supported wafer. However, in the case of flexible MOFs, which are known for their mechanical instability, the sample is dispersed in a solvent and spread on a silicon wafer.

The aforementioned setup was utilised for the detailed investigation of CO_2 adsorption[56] for the flexible MOF of chromium(III) terephthalate[208]

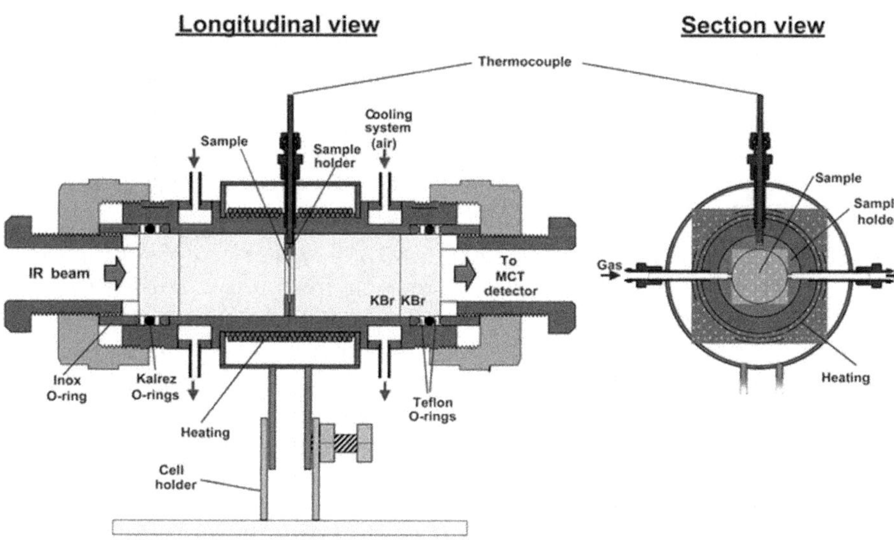

Figure 3.35 Technical view of the operando IR reactor cell used for the investigation of flexible MOFs. Adapted from ref. 268 with permission from the Royal Society of Chemistry.

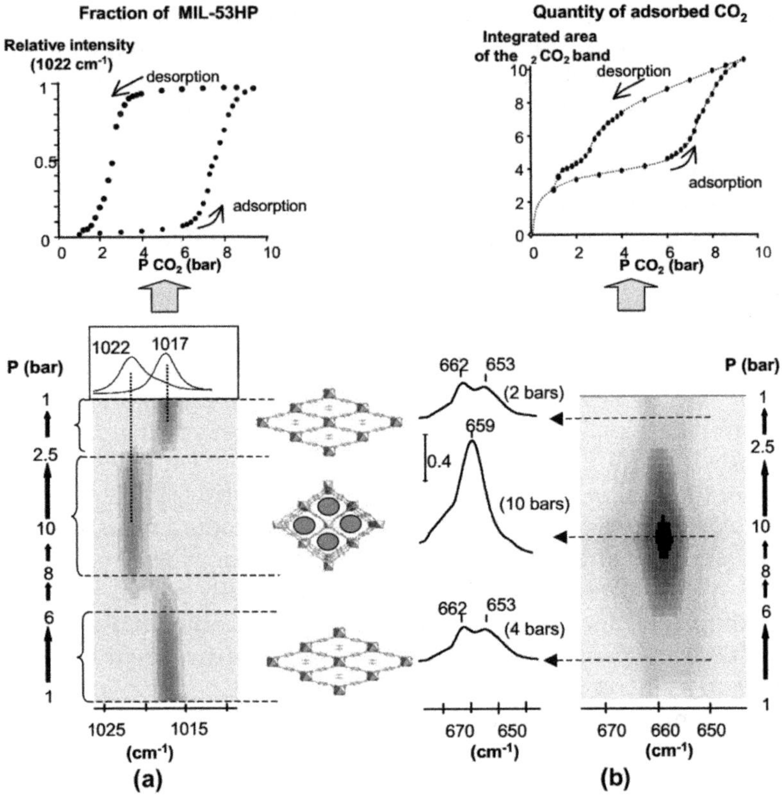

Figure 3.36 Bottom (2D IR Map): (a) variation in the intensities of MIL-53lp (1017 cm^{-1}) and MIL-53HP (1022 cm^{-1}) bands and (b) in the ν_2 CO_2 bands (653, 662 cm^{-1} MIL-53lp; 659 cm^{-1} MIL-53HP) *vs.* CO_2 pressure. Top: hysteresis phenomenon observed during an adsorption–desorption cycle. Curves were obtained from the quantitative analysis of the corresponding IR spectra. HP-high pressure, lp-low pressure. Adapted from ref. 56 with permission from John Wiley & Sons, Copyright 2007.

MIL-53(Cr) [$Cr^{III}(OH)(OOC-C_6H_4-COO)]_n$; MIL = Matériaux de l'Institut Lavoisier]. The authors introduced CO_2 stepwise from one to 10 bar to the activated sample, then progressively removed it (Figure 3.36). By observing the integrating signal intensity of $\nu_2(CO_2)$ bands between 645 and 665 cm^{-1}, they were able to estimate the amount of adsorbed gas. The calculated values were comparable to those obtained from volumetric measurements. Detailed analysis of the spectra during the adsorption–desorption cycle indicated that the *ν*18a ring mode of terephthalate ligand shifted from 1022 cm^{-1} to 1017 cm^{-1} under one bar of CO_2, and steadily returned to its initial position in response to increased pressure. The splitting of $\nu_2(CO_2)$ into two components at 653 and 662 cm^{-1} under one bar CO_2 indicated the creation of a donor–acceptor (EDA) complex between the carbon atom of CO_2 and the electron donor centre of the MIL-53(Cr). Analysis of the perturbations of $\nu(OH)$ and

δ(OH) bands ($\Delta\nu$(OH) = 22 cm^{-1} and $\Delta\delta$(OH) = 30 cm^{-1}) allowed the electron donor centre to be identified as the O atom of the MIL-53(Cr) hydroxyl group.

Further detailed analysis of the interaction of CO_2, CO and CD_3CN molecules with MIL-53(Cr) was provided in a subsequent report.[209] The spectrum of MIL-53(Cr) under p(CO) = 133 Pa displays the band at 2136 cm^{-1} with a shoulder at approximately 2145 cm^{-1}. These low-frequency signals indicate liquid-like CO in 1D channels. Moreover, the absence of any band above 2150 cm^{-1} proves that MIL-53(Cr) does not possess any unsaturated metal sites. However, the lack of significant changes of bands ascribed to the hydroxyl group upon CO adsorption and the downward shift of ν(OH) band (180 cm^{-1}) in response to adsorption of CD_3CN indicates that the acidity of the OH group is weak.

In situ Raman studies of the co-adsorption of the economically important mixture of CO_2–CH_4 on MIL-53(Cr) allows quantification of the fraction of the np and lp form of MIL-53(Cr) at different partial pressures of CO_2/CH_4 mixture.[210] For this evaluation, researchers compared relative intensities of the characteristic bands at ~1432 and 1444 cm^{-1}. The first one is characteristic for the np form, while the second is for lp form. The obtained ratios are comparable to those calculated from *in situ* synchrotron PXRD. Additionally, integration of the Raman signals ascribed to CH_4 and CO_2 allowed the authors to extract the adsorption isotherms. Despite a small overestimation compared to gravimetrical values, the obtained results show that *in situ* Raman spectroscopy may be useful for estimating the amount of the adsorbed phase.

The aforementioned understanding of the spectrochemistry of MIL-53(Cr) allows further development of *in situ* vibrational spectroscopy. In a report from 2011, the authors investigated the adsorption of corrosive gas H_2S in this flexible MOF.[211] The adsorption isotherms were compared with *in situ* IR spectroscopy. The authors again analysed distinct infrared bands of terephthalic entities, v18a ring mode (1022 cm^{-1} and 1017 cm^{-1}, respectively, for the lp and np structure), for the elaboration of the transition mechanism. In response to the first portion of H_2S, the material transforms at 220 K from the lp to np configuration. The initially adsorbed H_2S molecules are hydrogen-bonded with μ_2-OH groups. This provokes, similarly to other adsorbates, the closure of the pores at low vapour or gas loading.[212] Upon further uptake of H_2S (at p range from 2–3 kPa), the reverse phenomenon occurs, with the signal from the np form (1017 cm^{-1}) decreasing in favour of the signal from the lp form (1022 cm^{-1}).

Similar investigations were made for other MIL-53 derivatives,[77,213,214] including iron(III) and aluminium(III). The findings for the MIL-53(Al) materials indicate that during the adsorption of carbon dioxide, in low loading, the CO_2 molecules create OCO···HO adducts, while high coverage provokes the creation of bimolecular $[CO_2]_2$ species. To get inside this process, the authors used *in situ* IR experiments during adsorption of $^{13}CO_2$ and $^{12}CO_2$ mixtures, and their findings were further corroborated with periodic density functional calculations.[214]

Additionally, *in situ* IR and Raman spectroscopy were utilised for investigating the influence of the oxidation state[215] of the metal centre on the flexibility of a MIL-47(V), formulated $[V(O_2C–C_6H_4–CO_2)(OH)]_n$. Treatment of the activated flexible MIL-47(V^{III}) under O_2 leads to the formation of $[V^{IV}(O)(O_2C–C_6H_4–CO_2)]_n$. First, the IR bands (3642, 4511 and 512 cm^{-1}) related to μ_2OH progressively vanish. On the spectrum of the partially oxidised form, the authors recognised a new OH group at 3633 cm^{-1} in addition to an initial band of the OH group (3642 cm^{-1}). Furthermore, severe treatment of the sample with O_2 caused the disappearance of these bands. The authors found that even low amounts of V^{IV} have a dramatic influence on the flexibility of the V^{III} form.

ZIF-8 (zeolitic imidazolate framework), a material constructed from zinc cations and 2-methylimidazole ligands (mIm), possesses a 3D pore system comprising ~11.7 Å cavities connected *via* ~3.4 Å windows. Considering geometrical constraints, molecules with a larger kinetic diameter than ~3.4 Å should not be adsorbed on ZIF-8. However, the mIm linker has rotational freedom allowing changes in the geometry of the windows. This phenomenon, described as gate opening, increases the windows' size, and allows adsorption of geometrically restricted molecules. Moggach *et al.*[51] used high-pressure crystallography to detect two different phases of ZIF-8. Further investigations into the mechanism of gate opening were carried out using temperature-dependent *in situ* Raman spectroscopy[216] and high-resolution neutron and synchrotron vibrational spectroscopy,[217] in conjunction with *ab initio* quantum mechanical modelling (Figure 3.37).

The first report shows that, below 150 K, swinging of the methyl-imidazole group causes opening of the channels. This enables nitrogen and methane gas uptake at low temperatures. In the second report, Ryder *et al.* discovered a soft mode at 0.57 THz (18.9 cm^{-1}) and concluded that this low-frequency vibration could explain the phase transition observed when ZIF-8 is compressed in a diamond-anvil cell. Furthermore, the authors identified a gate-opening mode at 1 THz (33.4 cm^{-1}) and explained in detail how such a lattice motion can allow the gas adsorption on ZIF-8 (Figure 3.37).

Another example of using *in situ* Raman spectroscopy is monitoring the switching transition of single particles of cp and op phases of DUT-8(Ni) under hydrostatic pressure.[218] The authors provide detailed insight into the op → cp phase transformation showing in some pressure ranges characteristic lines that correspond to both cp (60 cm^{-1}) and op (20 cm^{-1}) phases, indicating the coexistence of cp and op domains in a single particle. The authors also demonstrated, that the opening of the cp phase is possible under pressure in methanol, which is not able to provoke structural opening at room temperature and atmospheric pressure.[218] In contrast to methanol, DUT-8(Ni)_cp failed to transform into op in isopropanol in the pressure range up to 8 GPa.[219]

Moreover, *in situ* IR was used to prove flexible modules in "rigid" frameworks, these being subtle changes that cannot be detected with PXRD. Kanno *et al.*[207] systematically designed halomethyl $–CH_2X$ (X = Cl, Br, I)

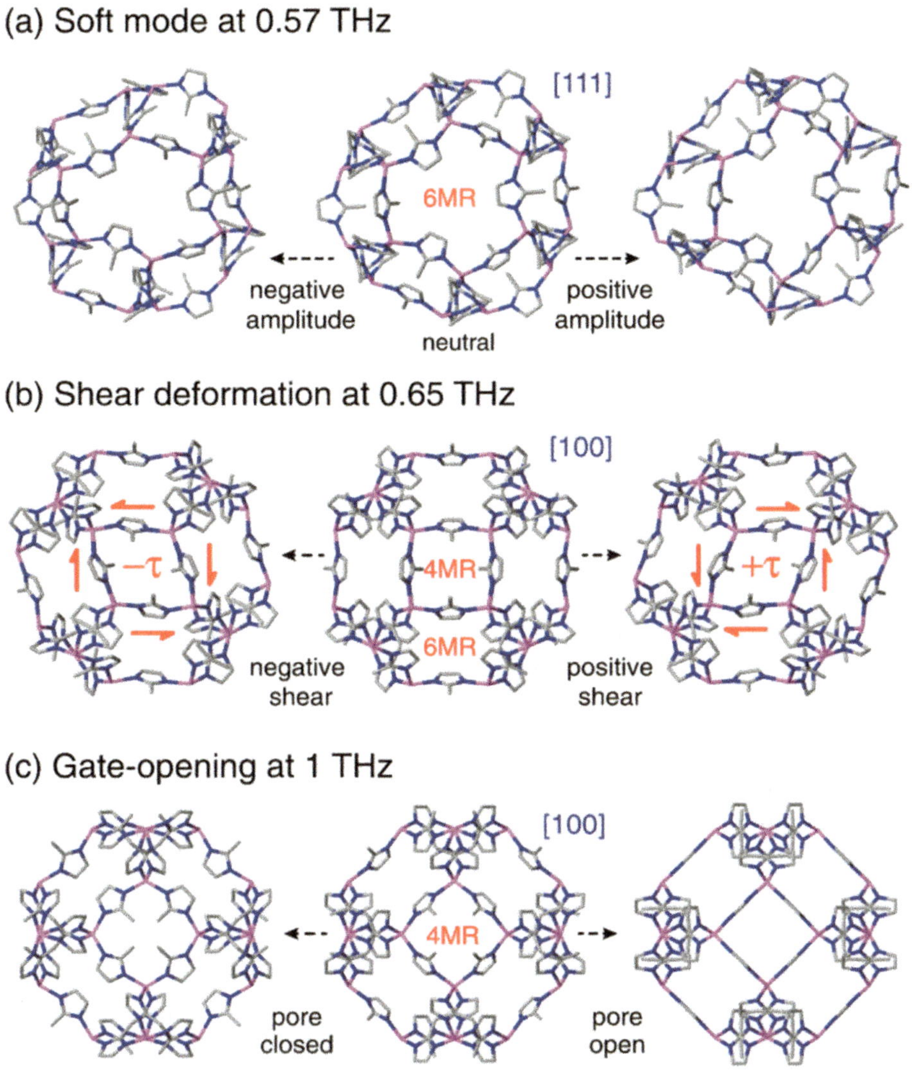

Figure 3.37 Low-frequency lattice dynamics in a sodalite cage of ZIF-8. (a) Soft mode mechanism causing geometrical distortion of 6MR. (b) Shear deformation of 4MR. (c) Gate-opening mechanism of pore apertures. Animated versions are available in the Supplemental Material of ref. 217. Reproduced with permission of the American Physical Society, Copyright 2014.

functionalised MOFs. The CO_2 adsorption at higher pressure is facilitated by two factors: (1) flexible motion of the halomethyl group; and (2) interaction of halogen atoms with carbon dioxide molecules. *In situ* PXRD during CO_2 adsorption on Br indicates no structural changes. Parallel to PXRD, *in situ* IR experiments show a significant evolution of the band at ~550 cm^{-1} corresponding to a Br–C stretch vibration.

During activation, flexible interpenetrated MOFs change from porous to less porous or non-porous phases. This transition is mostly realised by the displacement of sub frameworks connected with the rearrangement of secondary building blocks,[220] but in some rare examples the collective motion of all subnetworks is observed.[25,221]

Seo *et al.* synthesised and investigated a two-fold interpenetrated MOF with the composition $\{[Zn_2(tp)_2L]\cdot 2.5DMF\cdot 0.5H_2O\}_n$ (H_2tp = terephthalic acid, L = 2,3-difluoro-1,4-bis(4-pyridyl)benzene).[220] Upon complete desolvation of this MOF, a break between zinc and oxygen atoms followed by a change of zinc geometry is observed. The Raman spectrum shows double bands at around 1500 cm^{-1} and 1430 cm^{-1}, corresponding to a $\nu_{sym}(COO^-)$ vibration of the typical paddle-wheel. Upon desolvation, the described bands transform to a distinct single band at 1432 cm^{-1}. This observation was used to understand the mechanism of transition triggered by the adsorption of CO_2. Uptake of carbon dioxide causes the reverse phenomenon, in which the distinct single band at 1432 cm^{-1} transforms to double bands. *In situ* Raman spectroscopy confirms the reversible metal–ligand bond formation of the secondary building blocks in response to adsorption of CO_2 molecules.

The eight sub-frameworks of water stable[221] MOF JUK-8 ($[Zn(oba)(pip)]_n$, oba = 4,4′-oxybis(benzenedicarboxylate), pip = 4-pyridyl functionalised benzene-1,3-dicarbohydrazide) breathe upon desolvation and adsorption of CO_2 molecules at different temperature.[176] The flexible diamondoid (dia) sub-frameworks of JUK-8 adhere to each other *via* hydrogen bonds, which freeze their mutual position in response to external stimuli. *Ex situ* SCXRD of open solvated (JUK-8op) and closed desolvated (JUK-8cp) phases explicitly shows that transition from the open to closed phase mostly relies on rearrangements around the Zn^{2+} cations. *In situ* IR during CO_2 adsorption proves that the transition of JUK-8cp to CO_2@JUK-8op involves alteration of the OCO angle and Zn–O bond lengths, which is observable by considerable changes in the OCO asymmetric stretching region of the oba^{2-} carboxylate ligands (Figure 3.38)

Furthermore, comparison of the bands from the acylhydrazone groups (–C(O)=N–NH, *i.e.*, ν(N–H)$_{pip}$ and ν(C=O)$_{pip}$) shows that those groups are involved in the binding of carbon dioxide.

In summary, vibrational spectroscopy is a powerful tool for investigating flexible MOFs. It allows detection of subtle structural transformations and facilitates monitoring of the activation process and their reactivity. Furthermore, *in situ* IR and Raman spectroscopies enable evaluation of the amount of adsorbed gas and the drawing of distinctions between different environments of adsorbed species.

3.3.6 *In Situ* Calorimetric Techniques

One of the fundamental energy transfer mechanisms between thermodynamic systems is heat. Monitoring and quantifying its flow becomes a powerful method for characterising various physicochemical processes taking

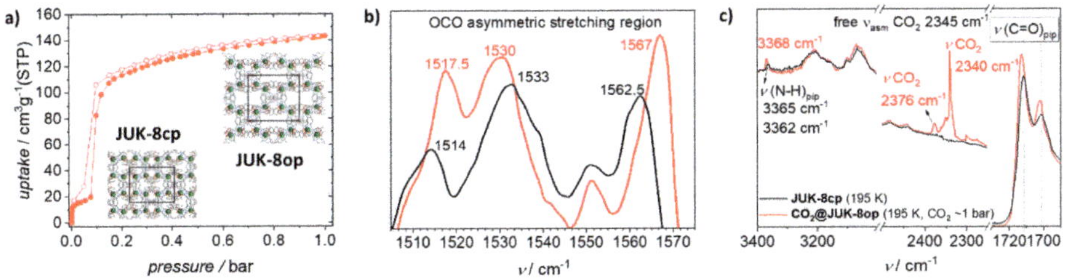

Figure 3.38 (a) CO_2 adsorption measurements for JUK-8 (195 K). *In situ* IR spectra recorded during carbon dioxide adsorption: (b) OCO asymmetric stretching region for the oba^{2-} carboxylates, and (c) characteristic $(N–H)_{pip}$, $(C=O)_{pip}$ and CO_2 regions. Black curve = JUK-8cp (195 K), red curve = CO_2 loaded JUK-8op (195 K; p/p_0 = 0.99). Reproduced from ref. 176, https://doi.org/10.1021/acsami.1c07268, under the terms of the CC BY 4.0 license https://creativecommons.org/licenses/by/4.0/.

place within dynamic materials. This is the goal of calorimetry, a characterisation technique in use since its development in the 1800s by Lavoisier and Laplace.[222]

The principal *in situ* calorimetric technique applicable to flexible coordination polymers is isothermal sorption calorimetry, effectively monitoring heat evolved upon changes in fluid chemical potential at the solid–fluid interface. In this category we include gas and liquid phase sorption calorimetry, but also immersion calorimetry. A second class of *in situ* experiments incorporates temperature as the stimulus, as temperature scanning calorimetry. Finally, an emerging class of experiments detect heat effects during pressure scanning, in effect combining a high-pressure cell (see Section 3.3.2) with a calorimeter.

The heat flow from a calorimetry experiment is necessarily a summation of contributions from all processes occurring in the sample. However, with careful isolation of state variables and thermodynamic analysis, it is possible to deconvolute these signals. This section will first summarise the important features of calorimeters used for *in situ* characterisation, then discuss processes involving structural dynamics in hybrid frameworks that can be followed through calorimetry during application of the above stimuli.

3.3.6.1 Instrumentation for Measuring Heat Flow

Most calorimeters designed for *in situ* monitoring use a diathermal mode of operation.[223–225] This implies that the sample is, ideally, in perfect thermal equilibrium with a surrounding thermostatic block, with a heat flux sensor monitoring heat flow between the two (Figure 3.39). Challenges reside in obtaining an instrumental signal with minimal lag time, and one that is representative of total energy exchanged as heat. Generally, heat transmitted through conduction is targeted, rather than convection or radiation, as it provides a directed and easily measurable heat transfer mechanism.

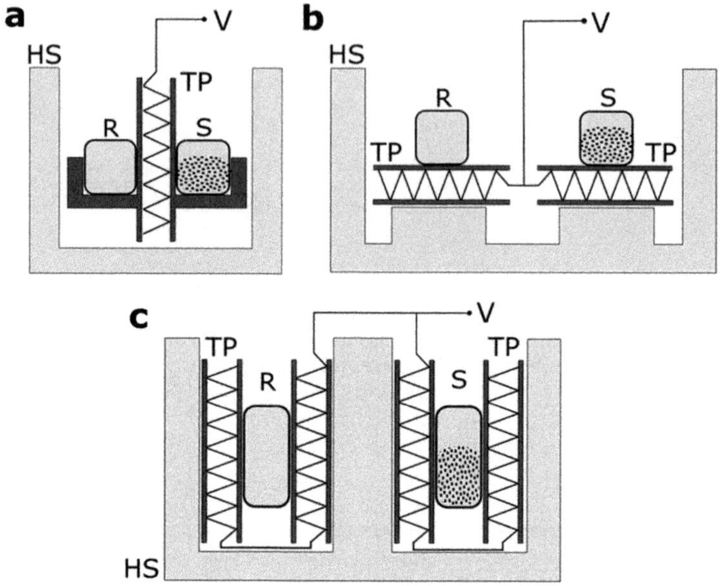

Figure 3.39 Diagrams of several types of diathermal, differentially mounted calorimeter setups using thermopiles. (a) Signal recorded as heat flow between sample and reference (b) Signal recorded as the difference between sample-sink and reference-sink heat flows (c) Tian–Calvet arrangement, where thermopiles completely surround sample and reference cells. Legend: S = sample side, R = reference side, TP = thermopile, HS = heat sink, V = calorimeter signal as voltage.

Therefore, the sample to be studied is placed in an enclosure which combines (i) good sample-wall contact; (ii) good wall thermal conductivity; (iii) radiation-absorbing inner surface and non-emissive outer surface; and (iv) minimal convection. In almost all modern setups, the sample cell is mirrored with an identical reference cell exposed to the same conditions, either empty or with an inert sample inside. The recorded signal is then expressed as a difference between the two sides, to account for effects such as specific heat capacity of the cells, external heat contributions, *etc.*

The heat flux sensors are generally thermopile based with pairs of thermocouples connected in series across a thin area. The voltage output of this arrangement will be proportional to the heat flux through the sensor area. Thermopiles can be placed on one axis of the sample cell, capturing a large percentage (>60%) of the heat evolved from the sample. Alternatively, the thermopiles can be arranged to fully surround a cell in a 3D configuration. A differential version of this setup, commonly called a Tian–Calvet calorimeter, allows over 95% of evolved heat to be accounted for and is likely the most advanced commercially available system.[226]

Alternative configurations also exist, such as power-compensated calorimetry, which instead of heat flow measures the difference in power required to maintain the sample at a target temperature *versus* the reference side. We also mention systems under development where radiative heat is monitored, by recording the infrared thermal emission of the sample with an optical sensor.[227]

3.3.6.2 Fluid Sorption as a Stimulus

Exploring the energetics of sorption is invaluable for flexible porous frameworks, as guest-induced dynamic effects such as breathing or gate opening (see Section 3.2.3) can be investigated.[223,228] In effect, the goal of *in situ* calorimetry during sorption is the determination of an enthalpy of adsorption. It should be noted that enthalpies of adsorption can also be obtained indirectly through the application of the Clausius–Clapeyron method, which presupposes a reversible and non-temperature dependent adsorption mechanism, often an erroneous assumption in the case of dynamic framework phenomena.

Differential adsorption enthalpy ($\Delta_{ads}\dot{h}$), is the enthalpy change associated to an infinitesimal change in adsorbed amount, is usually plotted as a function of loading (or coverage). This graph allows a clearer differentiation between different populations of energetic sites. Integration of this curve yields the total enthalpy change from an empty surface to the specified coverage as the integral enthalpy of adsorption ($\Delta_{ads}H$). Interpreting differential enthalpy curves is a challenge, as due to the global nature of the measurement, enthalpic contributions from multiple effects are combined: (i) guest–host contributions in the form of homogenous or heterogeneous adsorption sites; (ii) guest–guest contributions as lateral interactions of adsorbed molecules; and (iii) contributions from the host framework itself if it undergoes structural or chemical changes. Unambiguous rationalisation of $\Delta_{ads}\dot{h}$ often requires the coupling of additional experimental techniques or computational methods. Figure 3.40 depicts examples of isotherms and enthalpy curves corresponding to typical materials, both rigid and flexible which will be discussed in the following sections.

3.3.6.2.1 Sorption Calorimetry in the Gas Phase. Tian–Calvet calorimeters are highly suitable for coupling adsorption and calorimetry, due to their high sensitivity and stability. They are usually coupled with adsorption volumetry (rather than manometry) owing to the good thermal contact of the sample with the cell walls. As with a standard volumetry experiment, a known small amount of guest is introduced in the sample cell, followed by an equilibration time that considers both pressure stabilisation and recovery of the calorimeter baseline. For each step, the signal peak is integrated to determine evolved heat, from which the work contribution of gas expansion is subtracted to obtain a pseudodifferential enthalpy of adsorption ($\Delta_{ads}\dot{h}$). Full details of the thermodynamic background and exact experimental procedure are available elsewhere.[223,224] It is also possible to introduce the adsorbate in a continuous fashion, working at quasi-equilibrium (provided the flowrate is

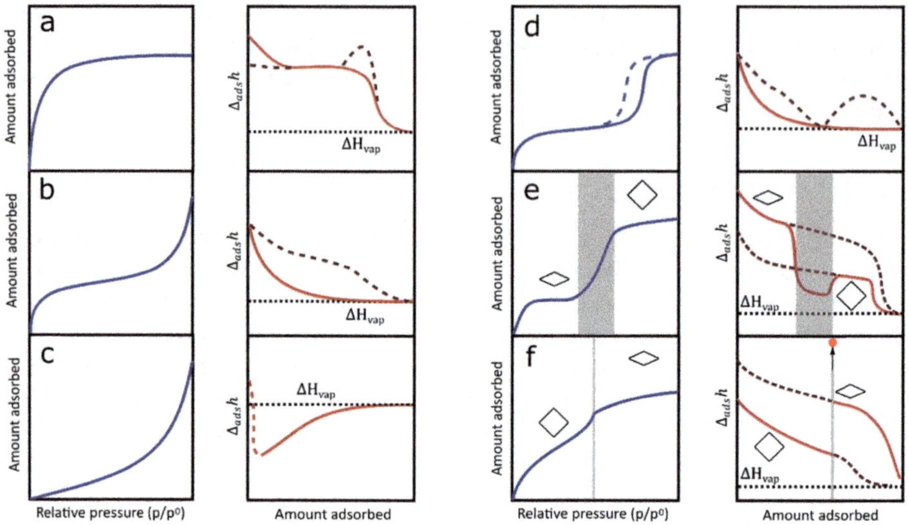

Figure 3.40 A non-exhaustive list of typical isotherms and enthalpy curves. (a) A microporous material exhibiting a type I isotherm. The enthalpy curve may highlight heterogeneity (initial slope) and guest–guest interactions close to full coverage. At complete coverage, differential adsorption enthalpies necessarily converge to the vaporisation enthalpy of the guest, ΔH_{vap}. (b) Type II isotherm of multilayer adsorption. The enthalpy curve shape can be related to the BET C constant. (c) Isotherm for a weakly interacting guest (type III), such as water sorption in a hydrophobic system. Here, clustering drives adsorption at higher coverage, with the differential enthalpy of adsorption characteristically smaller than the fluid vaporisation enthalpy. (d) Type IV isotherm, characteristic of adsorption in mesopores. Enthalpy curve may additionally show a condensation-related peak. (e) Adsorption in a flexible MOF that gradually opens (transition in the shaded region). The measured enthalpy curve switches between theoretical curves for the rigid phases. The energetic contribution of structural transformation lowers the apparent enthalpy of adsorption during the transition. (f) Adsorption in a flexible MOF that suddenly closes. The enthalpy curve similarly reflects the change in guest–host interactions, albeit with transition energetics cumulated to a single point.

small enough), providing a higher resolution in $\Delta_{ads}\dot{h}$ and a better approximation of a reversible system.[223,229]

Before considering the effects of structural dynamics, it is important to detail characteristic enthalpies of adsorption of various structural features in rigid MOFs. The accessibility of open metal sites (OMS) and the nature of the guest–OMS interaction can be evidenced through the magnitude of $\Delta_{ads}\dot{h}$. MOFs incorporating copper paddlewheel metal nodes, such as HKUST-1 (also known as MOF-199), have been observed by calorimetry to coordinate strongly to π-bond containing hydrocarbons like propylene,[230] with an average $\Delta_{ads}\dot{h}$ of 60 kJ mol^{-1}, *vs.* 40 kJ mol^{-1} for propane. Conversely, the enthalpy profile of CO_2 adsorption on the same MOF appears quite homogenous.[231] Closer computational investigation at the DFT level highlighted the existence of several sites of nearly equal interaction strength, cautioning that a flat enthalpy curve does not necessarily entail a single type of adsorption site. The exact nature of the OMS can also be probed through calorimetry. A high initial $\Delta_{ads}\dot{h}$ of carbon monoxide in the aforementioned HKUST-1 (over 60 kJ mol^{-1})[232] highlights the existence of Cu(I) defects in addition to Cu(II) centres. Similarly, in metal timer-based MOFs like MIL-101(M) and MIL-100(M), (M = Fe, Cr), partial reduction of metal nodes from (III) to (II) can be evidenced in the nearly twofold increase of initial $\Delta_{ads}\dot{h}$ of CO and propylene when MIL-100(Fe) is activated at 250 °C.[233] Not only OMS but also adsorption sites created through linker functionalisation or defects can impact $\Delta_{ads}\dot{h}$. For example, the functionalisation of the prototypical UiO-66(Zr) MOF with amino[234] and carboxyl[235] moieties was shown to induce an average increase in the enthalpy of adsorption of ~7 kJ mol^{-1}.

Besides contributions from the above effects, dynamic structural phenomena upon guest adsorption, such as gate opening or breathing can also contribute to the measured enthalpy.[236,237] Depending on the free energy landscape of the {MOF + guest} system, these transitions are often accompanied by thermal events, exothermic or endothermic. We caution that great care should be taken when assigning features in isotherms or enthalpy curves, as they can also correspond to phase changes in the guest itself.[228]

In the case of a continuous, second-order transition, like the opening (or reopening) exhibited by MOFs such as ZIF-7,[238] Co(bdp),[239] MIL-53(M) (M = Cr, Al),[54,55] DUT-49,[23] *etc.*, the enthalpy contribution of framework flexibility will affect $\Delta_{ads}\dot{h}$ throughout the loading (or pressure) range associated with the transition (see Figures 3.40(e) and 3.41(a)). Indeed, this effect is of particular interest from the point-of-view of thermal management of adsorption processes, as it lowers heating effects in the adsorbent bed and reduces the thermal load for activation.[240] However, any enthalpic contributions of flexible MOFs with a first-order transition will only be present in the single dosing event corresponding to the isotherm inflexion point (see Figure 3.40(f)). If differential enthalpies of adsorption can be obtained for a purely rigid version of the framework, for example through GCMC simulations, the difference in integral enthalpies of adsorption ($\Delta_{ads}H$) can be calculated, which represents the enthalpy of transition of the solid.[241–244] Of particular interest are the energetics of metastable systems, such as in DUT-49, where overloading of an unstable state leads to the phenomenon of negative gas adsorption, depicted in Figure 3.41(b).

3.3.6.2.2 Sorption Calorimetry in the Liquid Phase. Analogous to gas phase calorimetry, the energetics of the liquid–solid interface can be followed by *in situ* calorimetry.[223,224] Although similar in principle, the complexity of the liquid phase introduces additional challenges.

If using a pure liquid, the technique is called immersion calorimetry, and measures the total enthalpy change associated with the wetting of an

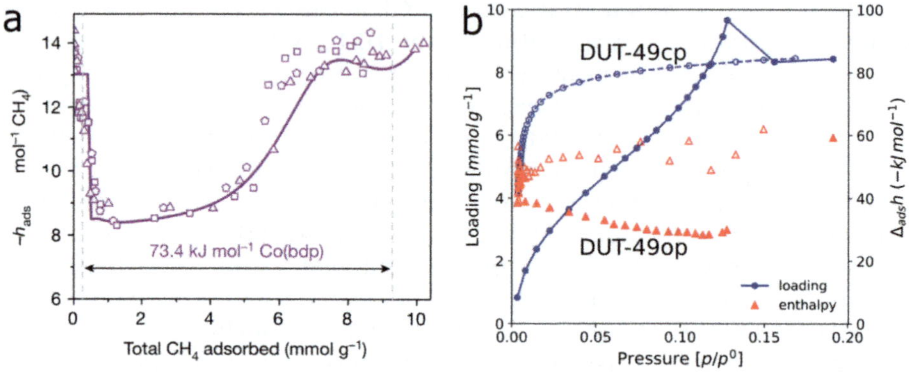

Figure 3.41 (a) Differential enthalpy curve of CH_4 adsorption on Co(bdp) as obtained from microcalorimetry experiments (points) and variable temperature adsorption isotherms (line). During the opening of the MOF structure (0.5–7 mmol g^{-1}), adsorption enthalpy is significantly lower, highlighting an enthalpic contribution from structure dynamics. Reproduced from ref. 240 with permission from Springer Nature, Copyright, 2015. (b) Butane adsorption (closed symbols) and desorption (open symbols) isotherm recorded on DUT-49, and corresponding enthalpy curve. After the massive system contraction and expulsion of gas, the MOF remains in its cp form throughout desorption. Characteristic differential enthalpies of adsorption are observed for both forms and can be integrated and subtracted to obtain an enthalpy of transition. Adapted from ref. 61 with permission from Springer Nature, Copyright, 2019.

adsorbate.[223,224] This parameter is related (but not identical) to the integral enthalpy of adsorption of vapour at full coverage ($\Delta_{ads}H$), with an additional contribution from the bulk liquid–adsorbed film interface. The setup for immersion calorimetry consists in an activated sample in a sealed glass bulb immersed in a bath of fluid in a Tian–Calvet thermopile.[223,224] A brittle tail at the end of the bulb is gently shattered, allowing the fluid to enter and cover the sample. The measured heat signal must be corrected with blank experiments, to avoid contributions from glass breaking or fluid evaporation in the bulb vacuum. Immersion calorimetry can provide a straightforward indication of accessible porosity. By successive immersions in probes of increasing size (and ideally of non-specific nature, like hydrocarbons), the cut-off associated to complete molecular sieving can be identified as a sharp decrease in enthalpy. It is especially helpful in the case of flexible or gate-opening MOFs, obtaining an 'effective pore window', that may significantly differ from the crystallographic structure, as is the case for ZIF-8 [245] and ZIF-7.[246]

Adsorption of a liquid analyte from a solvent solution can also be followed through calorimetry. In this case, the guest displaces previous solvent molecules; we are instead measuring an enthalpy of displacement. Additional contributions should be carefully evaluated, such as enthalpy of dilution, or the mechanical work energy added by an oft-necessary stirrer. In part due to its difficulty, there is a distinct lack of literature regarding solution adsorption calorimetry of flexible frameworks.

3.3.6.3 Temperature as a Stimulus

By changing the temperature of a system, the relative contributions of vibrational and dispersion interactions in a porous material are modified, allowing a dynamic material to explore various configurations in its free energy landscape. Here, *in situ* calorimetry has also seen extensive application, to obtain insight into dynamic phenomena.

Temperature scanning calorimetry, often known just as differential scanning calorimetry (DSC) due to the use of a differential-type calorimeter setup, is the technique most commonly used be used to investigate the temperature-responsivity of flexible materials. A sample is placed in the calorimeter pan, either in a sealed or an open cell. The calorimetric block is then heated or cooled at a known, often constant, rate and the resulting heat flow is plotted as a function of sample temperature.

The base contribution to total heat flow during temperature scanning is equivalent to the heat capacity of the studied sample. Additional contributions or discontinuities in the heat flow curve suggest other transformations in the material, related to chemical reactions, desolvation, phase changes, glass transitions *etc.* By integration of peaks corresponding to a well-defined contribution, the enthalpy change associated with the underlying process can be determined.

Due to the common use of thermogravimetric analysis (TGA) to monitor the desolvation process of an as-synthesised framework (see Section 3.2.1), the coupling of TGA+DSC can yield significant insight into the mechanisms underpinning of observed weight loss processes upon heating. This combination of methods was used to investigate the dehydration of MIL-53-type systems, allowing a direct evaluation of the open-closed structure transition enthalpy.[247] Moreover, DSC can highlight other possible effects which would otherwise be overlooked. Indeed, the discovery of melting in zeolitic imidazolate frameworks was partly attributed to peaks in its TGA+DSC curve (see Figure 3.42(a)).[248]

The use of DSC can also be extended in the low temperature regime, allowing possible flexibility to be hinted upon cooling. Through DSC, ZIF-4 was found to undergo reversible contraction to a dense phase at 139 K (see Figure 3.42(b)), also allowing the enthalpy of transition to be determined at around 8–9 kJ mol^{-1}, in good correspondence to *ab initio* simulations.[249]

3.3.6.4 Pressure as a Stimulus

The use of external stress (isostatic or anisotropic) is a further stimulus which can be used to provoke responses in soft materials, either through the application of stress on the external surface of the material or, in the case of porous solids compressed with a hydrostatic liquid, through the forced intrusion of

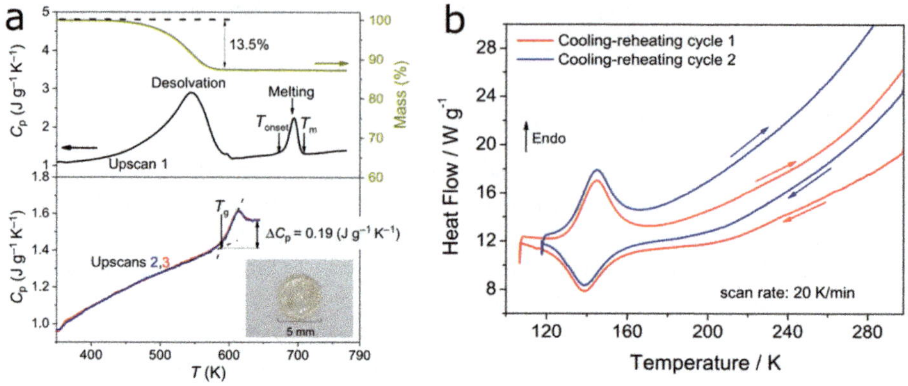

Figure 3.42 (a) Combined TGA–DSC experiment on the glass-forming ZIF-62, highlighting different contributions to heat flow on temperature scanning. (b) Low temperature DSC scans of ZIF-4. Peaks highlight a reversible phase change corresponding to concerted densification of the framework. Adapted from ref. 249 with permission from John Wiley & Sons, Copyright 2015.

the medium in the porous network (see the discussion in Section 3.2.3). The quantification of heat effects during such processes is therefore highly desirable, as it can evaluate the thermal effect of fluid intrusion, phase changes and other dynamic phenomena, with possible applications to mechanical energy storage or barocalorics.

While commercial instrumentation capable of pressure scanning is available (commonly named pressure-DSC, or P-DSC,[250] and often combining the capability to scan temperature), studies investigating thermal effects of porous coordination polymers upon pressure change are relatively lacking.[251] The approach has been first demonstrated by exploring the flexible MIL-53(Al) system on compression with non-intruding silicone oil inside a Tian–Calvet calorimeter. Both mechanical work (from the integration of *PdV*) and the enthalpy of phase transition (from calorimetry) to the narrow pore version were determined, allowing calculation of the internal energy change upon compression.[252] Similarly, an adapted Tian–Calvet calorimeter capable of advanced pressure, volume, and temperature scanning calorimetry[253] was used to evaluate the efficiency of energy conversions in MOFs undergoing combined water intrusion and flexibility effects.

3.4 Time-resolved Techniques

Despite *in situ* techniques applied under adsorption quasi-equilibrium conditions with extended equilibration time providing information on phase composition, crystal structure, chemical bonding, host–guest interactions and thermal effects in the defined points of the isotherm, information about the kinetics of the phase transition as well as cooperativity is limited.[10,254] To gain this information, time-resolved techniques analysing the phase transformation rate are required. Since structural switching in MOFs is a relatively rapid process and proceeds at the sub-second scale, the transition can be followed by the techniques providing required time resolution, such as synchrotron PXRD or TEM. However, since the experimental implementation is not straightforward in both cases, only a few exemplary studies are present in the literature so far.

3.4.1 *In Situ* Synchrotron PXRD

A limited number of reports on time-resolved monitoring of adsorption-induced structural switching in MOFs is available. Only a limited number of beamlines are equipped with the fast detectors and dedicated instrumentations are yet to be developed. Tanaka *et al.* monitored the adsorption induced gating in ELM-11 by PXRD at the synchrotron beamline.[255] Authors show that the structural transition of ELM-11 started immediately after the introduction of CO_2 at 40.8 kPa and 273 K, and was accomplished in ~10 s (Figure 3.43). Moreover, the dependence of the switching rate was studied by conducting the experiments at different pressures and experimentally proven that the structural transition was completed in 5 s when the CO_2 pressure was decreased at the rate of 2.4 kPa s^{-1}. Experiments in the temperature range 227–273 K show clear temperature dependency. The fit of experimental data in Kolmogorov–Johnson–Mehl–Avrami (KJMA) equation suggests that the rate constant only depends on the CO_2 gas pressure and the temperature, which controls the gate-opening pressure, and that the gate-opening rate is amplified by the CO_2 pressure exerted on ELM-11.

Considering the strong dependence of flexibility from the crystal size, Bon *et al.* prepared MIL-53(Al) framework in three different size regimes and conducted time-resolved synchrotron PXRD experiments upon adsorption of *n*-butane at 298 K.[256] As a result, samples with average crystal size of 1.2 μm and 28 μm reveal stark differences in solid phase transition rates, pointing to the pronounced impact of the crystal sizes. While in the case of small crystals (average size 1.2 μm) lp–np–lp transition (Figure 3.44(b)) could be detected within 3 s, the same transition of larger crystals (average size 28 μm) required two orders of magnitude longer time to switch from np to lp form (Figure 3.44(a)). These findings indicate the overall switching mechanism to be largely controlled by intracrystalline diffusion and give important insights into the tailoring of dynamics of switchable MOFs.

Miura *et al.* reported on the temporal engineering of guest-induced cp–op response in the series of mixed-metal frameworks DUT-8(M_1,M_2) (M_1 = Ni, M_2 = Co) by tuning of cobalt content.[256] *In situ* time-resolved PXRD investigations with the frame rate of 10 PXRD per second were carried to directly observe the pore opening with a high time resolution after imposing a pressure pulse on the crystal ensemble (Figure 3.45).

The phase transition is clearly detected after a few seconds (DUT-8(Ni)) by the rise of the (110) peak at 0.49 Å^{-1} of the DUT-8(Ni)_op phase, whereas

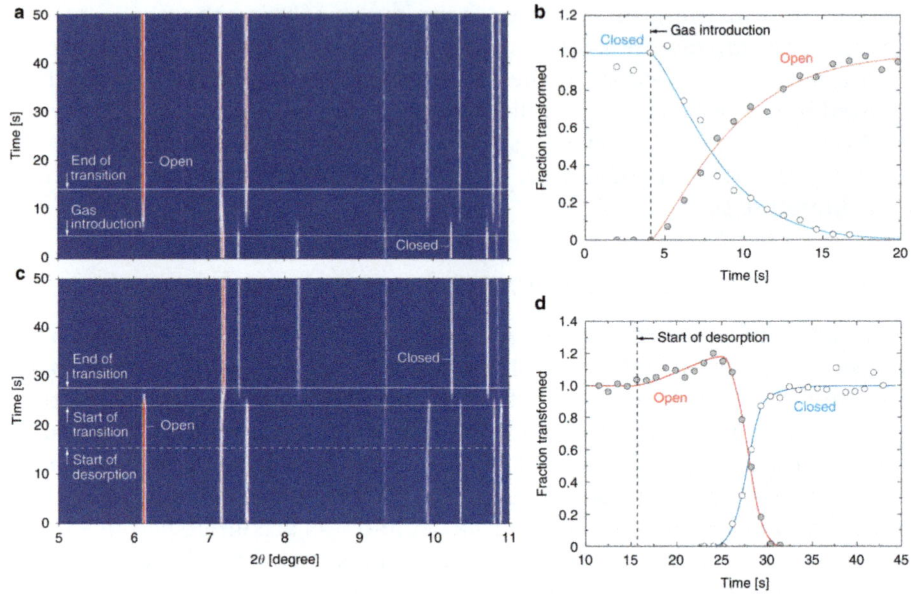

Figure 3.43 (a) Colour map of PXRD patterns for CO_2 adsorption on ELM-11 at 273 K and constant pressure of 40.8 kPa. (b) Normalised peak intensities (fractions of the phase transformed) from the 002 reflection at 6.1° for the open phase and the 020 reflection at 10.2° for the closed phase shown in (a). The two solid lines after 4.15 s were obtained by fitting of the Kolmogorov–Johnson–Mehl–Avrami (KJMA) equation. (c) Colour map of PXRD patterns for CO_2 desorption at 273 K during pressure depression from 100 to 3.8 kPa. (d) Fractions of the phase transformed from the 002 reflection for the open phase and the 020 reflection for the closed phase shown in (c). Reproduced from ref. 255, [http://doi.org/s41467-020-17625-3], under the terms of the CC BY 4.0 license https://creativecommons.org/licenses/by/4.0/.

increasing of cobalt amount leads to longer nucleation time. These observations have been also confirmed by adsorption experiments, performed in the same conditions, and analysing the pressure profile in the adsorption cell after the pressure jump. Moreover, the results have been confirmed by optical microscopy, observing the switching of single particles. The peak intensities of the op phase were analysed on the basis of the KJMA equation (where α is the fraction of op phase at time t) but the index n varied for the different samples which made it difficult to compare the k parameters directly (Table 3.2).

3.4.2 *In Situ* TEM

Because of the high sensitivity of MOFs to the electron beam, HR-TEM imaging of MOFs remains challenging and required both advanced equipment and certain experimental experience of the operator.[257,258] Han *et al.* defined the highest

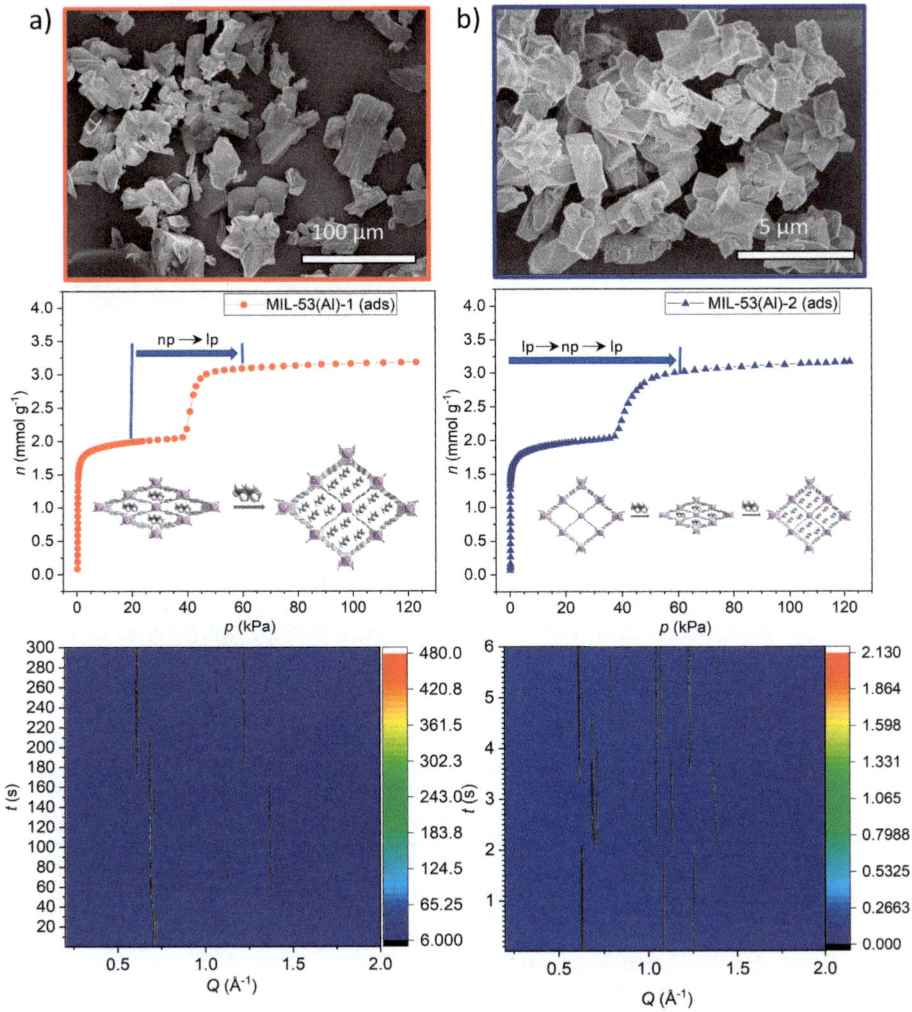

Figure 3.44 SEM images and adsorption isotherms of *n*-butane on MIL-53(Al) at 298 K showing the corresponding pressure step and expected phase transition and time-resolved synchrotron PXRD for MIL-53(Al) samples with average crystallite size of 1.2 μm (a) and 28 μm (b). Adapted from ref. 256 with permission from the Royal Society of Chemistry.

radiation doses of 10–20 $e^- A^{-2}$ for MOFs and COFs.[259] At the moment only a few groups in the world work on MOFs and COFs imaging *via* TEM at high resolution.[257,259,260] Following the changes in the crystal structure by HR-TEM is even more challenging since often environmental sample holders are required.

At the moment, the studies on the phase transition in MOFs, monitored by HR-TEM, are rather rare. Paesani *et al.* studied the switching of MIL-53(Cr) nanocrystals by the combination of environmental TEM and computer simulations.[261] The low-dose TEM beam conditions (~5 $e^- Å^{-2}$ cumulative dose) were

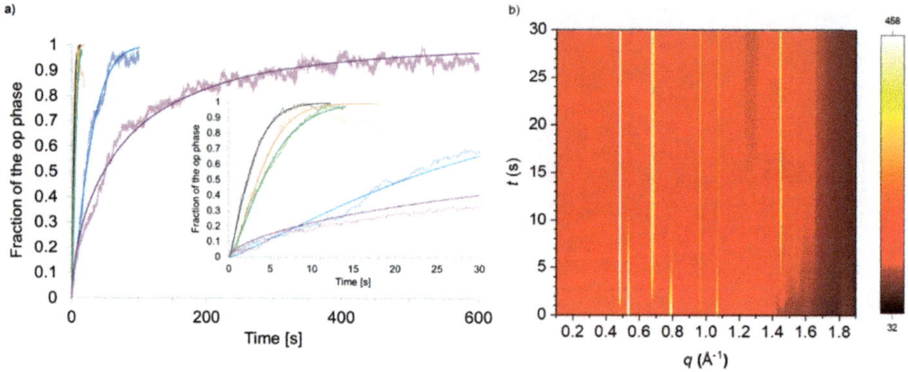

Figure 3.45 (a) Fractions of the open pore phase for DUT-8(Ni) (black), DUT-8($Ni_{0.75}Co_{0.25}$) (orange), DUT-8($Ni_{0.50}Co_{0.50}$) (green), DUT-8($Ni_{0.25}Co_{0.75}$) (blue) and DUT-8(Co) (purple) at room temperature based on *in situ* synchrotron PXRD measurements. Each pale colour jagged symbols correspond to the change in the normalised peaks intensity and the lines with dark colour were obtained by fitting the experimental data to the KJMA equation. (b) Contour plot of time-resolved *in situ* PXRD patterns collected upon DCM adsorption on DUT-8(Ni) at 298 K. Adapted from ref. 269 with permission from John Wiley & Sons, Copyright 2022.

Table 3.2 Experimental and calculated parameters for *in situ* synchrotron PXRD measurements on DUT-8(Ni), DUT-8($Ni_{0.75}Co_{0.25}$), DUT-8($Ni_{0.50}Co_{0.50}$), DUT-8($Ni_{0.25}Co_{0.75}$) and DUT-8(Co) at 298 K. Adapted from ref. 269 with permission from John Wiley & Sons, Copyright 2022.

Sample	n	k	R^2	$P_{ini,Mani}$/kPa	P_{max}/kPa
DUT-8(Ni)	1.31	0.238	0.9995	97	46
$Ni_{0.75}Co_{0.25}$	1.51	0.104	0.9919	97	46
$Ni_{0.50}Co_{0.50}$	1.36	0.103	0.9992	97	52
$Ni_{0.25}Co_{0.75}$	1.20	0.018	0.9983	97	52
DUT-8(Co)	0.64	0.05	0.9988	97	50

used during *in situ* ETEM experiments to mitigate the potential damage of the MOF nanocrystals. In the first step, the empty lp phase with mismatched errors of ~3% in unit cell parameters was confirmed for a MIL-53(Cr) nanocrystal, exposed to an ultra-high vacuum of TEM microscope (Figure 3.46). The dosing of 3 mbar of water in environmental TEM cell has induced a dramatic structural transformation, reflected by the changes in its diffraction pattern, accompanied by lp–np transitions. Further calcination of the nanocrystal at 300 °C led to the structural transition back in lp phase. The repeatability of the transition was confirmed in the second cycle of water adsorption.

Farha *et al.*[262] monitored the phase transition from a microporous Zr-based MOF, *scu*-NU-906, to a mesoporous MOF, *csq*-NU-1008 and the transition was monitored *in situ* by TEM in a liquid cell (Figure 3.47).

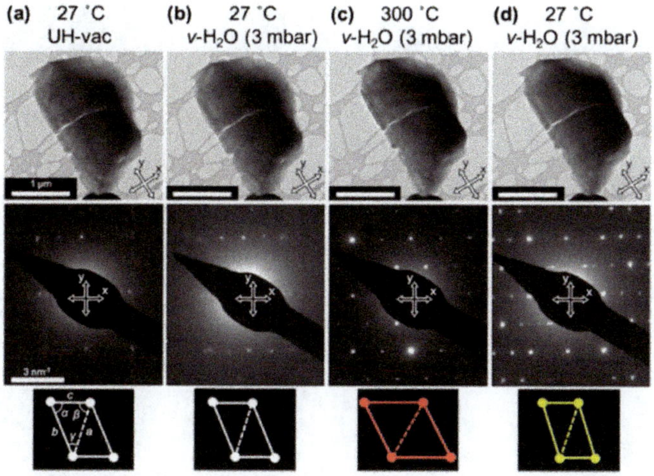

Figure 3.46 *In situ* ETEM images (top row) and diffraction patterns (middle row) of one MIL-53(Cr) nanocrystal at four different environmental conditions during breathing; (a) 27 °C and UH-vacuum, (b) 27 °C and water vapour (pre-calcination), (c) 300 °C and water vapour (calcined), and (d) 27 °C and water vapour (post-calcination). Reproduced from ref. 261 with permission from American Chemical Society, Copyright 2017.

In this experiment, both the shape morphology of the crystallites and lattice constants were changed with time. After 2 h of experiments, the crystallite shape changed from oval-shape in NU-906 to the rod shape morphology in NU-1008 framework. FFT-transformations of the images, obtained from both phases indicates the doubling of the lattice constants upon the spatio-temporal evolution process.

Presently, no further reports on adsorption-induced transitions in flexible MOFs could be found in the literature. This may be connected with subjective and objective factors such as the complexity of the experimental setup and crystallite size dependency in flexible MOFs.

3.5 Outlook

Impressive progress has been made in the development of *in situ* characterisation methods for flexible MOFs. However, many challenges are still ahead. The most intriguing question is the temporal evolution of structural transformations and their mechanisms. In particular dedicated methods with high time resolution in the range of 10^{-3}–10^{-9} s or multiple time scales are not well developed.[10,254] Only few studies at synchrotrons have given first insights into time-resolved framework transformations so far. Moreover, the limited availability of synchrotron beamtime poses limitations for the field. There is a clear need in developing inexpensive methods that are easily applied in the laboratory with laboratory X-ray sources. Advanced high power laboratory

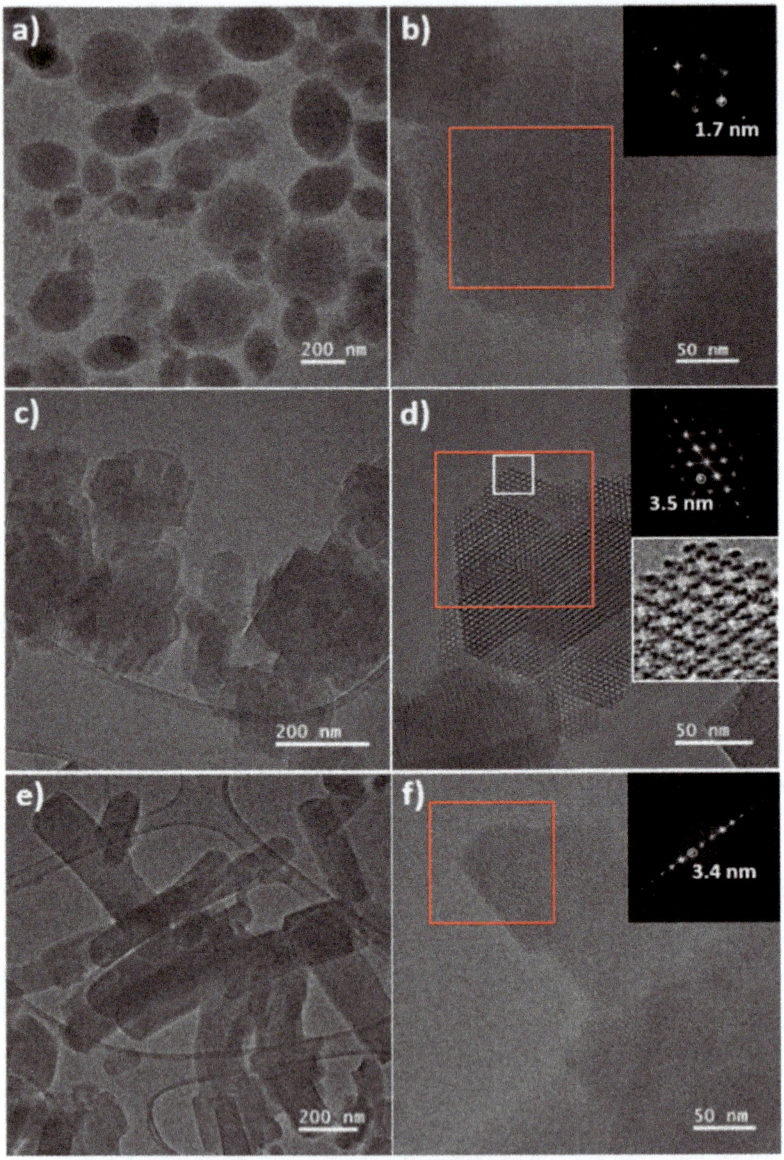

Figure 3.47 TEM images during the reaction illustrating the phase transition over time. Low-magnification images and HRTEM images of (a), (b) NU-906, (c), (d) NU-1008 isolated after 2 h, and (e), (f) NU-1008 isolated after 2 days. Insets: fast Fourier transform (FFT) of the image inside the red square, cropped at the predominant lattice fringes, and enlarged image inside the white square. Reproduced from ref. 262 with permission from American Chemical Society, Copyright 2020.

X-ray sources but also detectors with improved sensitivity may gain acceptance and become affordable helping to propel the field.

Furthermore, an obvious direction is the combination of multiple *in situ* methods. In particular, combining local and global structural information is important. While an integration of optical probes into diffraction instrumentation is within reach or has been even reached in a few cases, the combination of NMR or EPR techniques with diffraction remains challenging. However, thermal methods (calorimetry, TG, DSC) also give important insights and their combination with spectroscopy and adsorption tools may give valuable insights and be operable in standard laboratories at acceptable cost. The field can certainly profit from *operando* techniques used in other fields such as catalysis or battery science. However, often overlooked are the specific demands for measuring the adsorbed amount and the precise temperature control for sample and gas phase that require specific instrumentation, quite different compared with *operando* catalysis setups that often operate by heating or cooling only the solid catalyst and thus creating huge temperature gradients.

Another future direction involves multiple stimuli. A few examples of electrical field induced, or optical switching were discussed. The wider exploration of transformations induced by magnetic fields, radiation, mechanical pressure, *etc.*, *via* parallelised adsorption–diffraction techniques is important for the exploration of new functionalities such as switchable membranes, catalysts, electronic materials, sensors, or microfluidic devices.

The huge amount of data in particular for time-resolved methods will also require novel data deposition infrastructures. The scientific community must actively discuss and shape the future electronic publication schemes giving access to advanced analytical data obeying the FAIR principles.[263] The recently developed AIF format[264] for adsorption data is only a small step in this direction but may serve as a nucleus to develop interoperable data management structures.

List of Abbreviations

2D EXSY	Two-dimensional exchange spectroscopy
AIF	Adsorption information file
APS	Advanced photon source
cp	Closed pore
CPL	Coordination polymer
CSA	Chemical shift anisotropy
CuPyr-I	Bis[1-(4-pyridyl)butane-1,3-dione]copper(II)
Cw	Constant wave
DAC	Diamond anvil cell
DFT	Density functional theory
DMF	*N*,*N*-dimethylformamide
DSC	Differential scanning calorimetry
DUT	Dresden University of Technology

DUT-8(Ni) $Ni_2(ndc)_2dabco$, where ndc = 2,6-napthalenedicarboxylate and dabco = 1,4-diazabicyclo[2.2.2]octane
ELM Elastic layer material
EPR Electron paramagnetic resonance
ESRF European Synchrotron Radiation Facility
ETEM Environmental transmission electron microscopy
FFT Fast Fourier transform
GCMC Grand canonical Monte Carlo
GPa Gigapascal
HKUST Hong Kong University of Technology
HKUST-1 Cu_3(1,3,5-benzenetricarboxylate)$_2$
HR-TEM High-resolution transmission electron microscopy
IR Infrared spectroscopy
IPA Isopropyl alcohol
JT Jann–Teller
JUK Jagiellonian University in Kraków
lp Large pore
MAS Magic-angle spinning
MEW Methanol–ethanol–water
MIL Materials of Institute Lavoisier
MIL-53 $Al(bdc)_2(OH)$ where bdc = benzenedicarboxylate
mIm 2-Methylimidazole
MOF Metal–organic framework
NGA Negative gas adsorption
NIST National Institute of Standards and Technology
NLC Negative linear compressibility
NMR Nuclear magnetic resonance
np Narrow pore
NPD Neutron powder diffraction
NU Northwestern University
op Open pore
PJC Pressure jump cell
PTM Pressure transmitting medium/media
PXRD Powder X-ray diffraction
REI Rare earth ions
SAXS Small-angle X-ray scattering
SCC Sapphire capillary cell
SCD Supercritical CO_2 drying
SHF Sheffield Framework
TEM Transmission electron microscopy
TG Thermogravimetric analysis
TMI Transition metal ions
UiO Universitetet i Oslo
UV-vis Ultraviolet–visible light spectroscopy
WAXS Wide-angle X-ray scattering
ZFS Zero-field splitting

ZIF	Zeolitic Imidazolate Framework
ZIF-8	$Zn(mIm)_2$ where mIm = 2-methyl imidazole
ZIF-65	$Zn(nIm)_2$ where nIm = 2-nitro imidazole
ZIF-90	$Zn(ICA)_2$ where ICA = imidazolate-2-carboxyaldehyde

References

1. A. Schneemann, V. Bon, I. Schwedler, I. Senkovska, S. Kaskel and R. A. Fischer, *Chem. Soc. Rev.*, 2014, **43**(16), 6062.
2. I. Senkovska and V. Bon, *In Situ* X-ray Diffraction and XAS Methods, in *The Chemistry of Metal–Organic Frameworks: Synthesis, Characterization, and Applications*, ed. S. Kaskel, Wiley-VCH, Weinheim, 2016, pp. 691–727.
3. V. Bon, E. Brunner, A. Pöppl and S. Kaskel, *Adv. Funct. Mater.*, 2020, **30**(41), 1907847.
4. F. Bonino, C. Lamberti and S. Bordiga, IR and Raman Spectroscopies Probing MOFs Structure, Defectivity, and Reactivity, in *The Chemistry of Metal–Organic Frameworks: Synthesis, Characterization, and Applications*, ed. S. Kaskel, Wiley-VCH, Weinheim, 2016, pp. 657–690.
5. E. Brunner and M. Rauche, *Chem. Sci.*, 2020, **11**(17), 4297.
6. M. Mendt, M. Šimėnas and A. Pöppl, Electron Paramagnetic Resonance, in *The Chemistry of Metal–Organic Frameworks: Synthesis, Characterization, and Applications*, ed. S. Kaskel, Wiley-VCH, Weinheim, 2016, pp. 629–656.
7. S. I. Brückner, J. Pallmann and E. Brunner, Nuclear Magnetic Resonance of Metal–Organic Frameworks (MOFs), in *The Chemistry of Metal–Organic Frameworks: Synthesis, Characterization, and Applications*, ed. S. Kaskel, Wiley-VCH, Weinheim, 2016, pp. 607–628.
8. H. Petersen and C. Weidenthaler, *Inorg. Chem. Front.*, 2022, **9**(16), 4244.
9. K. I. Hadjiivanov, D. A. Panayotov, M. Y. Mihaylov, E. Z. Ivanova, K. K. Chakarova, S. M. Andonova and N. L. Drenchev, *Chem. Rev.*, 2021, **121**(3), 1286.
10. J. D. Evans, V. Bon, I. Senkovska, H.-C. Lee and S. Kaskel, *Nat. Commun.*, 2020, **11**(1), 2690.
11. S. Kitagawa and M. Kondo, *Bull. Chem. Soc. Jpn.*, 1998, **71**(8), 1739.
12. K. Uemura, R. Matsuda and S. Kitagawa, *J. Solid State Chem.*, 2005, **178**(8), 2420.
13. S. Horike, S. Shimomura and S. Kitagawa, *Nat. Chem.*, 2009, **1**(9), 695.
14. S. Krause, N. Hosono and S. Kitagawa, *Angew. Chem., Int. Ed.*, 2020, **59**(36), 15325.
15. J. E. Mondloch, M. J. Katz, N. Planas, D. Semrouni, L. Gagliardi, J. T. Hupp and O. K. Farha, *Chem. Commun.*, 2014, **50**(64), 8944.
16. A. J. Howarth, A. W. Peters, N. A. Vermeulen, T. C. Wang, J. T. Hupp and O. K. Farha, *Chem. Mater.*, 2017, **29**(1), 26.

17. A. P. Nelson, O. K. Farha, K. L. Mulfort and J. T. Hupp, *J. Am. Chem. Soc.*, 2009, **131**(2), 458.
18. N. Klein, H. C. Hoffmann, A. Cadiau, J. Getzschmann, M. R. Lohe, S. Paasch, T. Heydenreich, K. Adil, I. Senkovska, E. Brunner and S. Kaskel, *J. Mater. Chem.*, 2012, **22**(20), 10303.
19. N. Kavoosi, V. Bon, I. Senkovska, S. Krause, C. Atzori, F. Bonino, J. Pallmann, S. Paasch, E. Brunner and S. Kaskel, *Dalton Trans.*, 2017, **46**(14), 4685.
20. S. Ehrling, I. Senkovska, V. Bon, J. D. Evans, P. Petkov, Y. Krupskaya, V. Kataev, T. Wulf, A. Krylov, A. Vtyurin, S. Krylova, S. Adichtchev, E. Slyusareva, M. S. Weiss, B. Büchner, T. Heine and S. Kaskel, *J. Mater. Chem. A*, 2019, **7**(37), 21459.
21. L. Abylgazina, I. Senkovska, S. Ehrling, V. Bon, P. St. Petkov, J. D. Evans, S. Krylova, A. Krylov and S. Kaskel, *CrystEngComm*, 2021, **23**(3), 538.
22. U. Stoeck, S. Krause, V. Bon, I. Senkovska and S. Kaskel, *Chem. Commun.*, 2012, **48**(88), 10841.
23. S. Krause, V. Bon, I. Senkovska, U. Stoeck, D. Wallacher, D. M. Többens, S. Zander, R. S. Pillai, G. Maurin, F.-X. Coudert and S. Kaskel, *Nature*, 2016, **532**(7599), 348.
24. B. Garai, V. Bon, S. Krause, F. Schwotzer, M. Gerlach, I. Senkovska and S. Kaskel, *Chem. Mater.*, 2020, **32**(2), 889.
25. E. J. Carrington, C. A. McAnally, A. J. Fletcher, S. P. Thompson, M. Warren and L. Brammer, *Nat. Chem.*, 2017, **9**(9), 882.
26. Cavitation of Water Confined in Hydrophobic Nanoporous Materials, in *Proceedings of the 10th International Symposium on Cavitation (CAV2018)*, ed. J. Katz, ASME Press, 2018, pp. 239–243.
27. C. Duan, R. Karnik, M.-C. Lu and A. Majumdar, *Proc. Natl. Acad. Sci. U. S. A.*, 2012, **109**(10), 3688.
28. Y. Sakata, S. Furukawa, M. Kondo, K. Hirai, N. Horike, Y. Takashima, H. Uehara, N. Louvain, M. Meilikhov, T. Tsuruoka, S. Isoda, W. Kosaka, O. Sakata and S. Kitagawa, *Science*, 2013, **339**(6116), 193.
29. Y. Hijikata, S. Horike, D. Tanaka, J. Groll, M. Mizuno, J. Kim, M. Takata and S. Kitagawa, *Chem. Commun.*, 2011, **47**(27), 7632.
30. D. Tanaka, A. Henke, K. Albrecht, M. Moeller, K. Nakagawa, S. Kitagawa and J. Groll, *Nat. Chem.*, 2010, **2**(5), 410.
31. S. Ehrling, H. Miura, I. Senkovska and S. Kaskel, *Trends Chem.*, 2021, **3**(4), 291.
32. L. Abylgazina, I. Senkovska, R. Engemann, S. Ehrling, T. E. Gorelik, N. Kavoosi, U. Kaiser and S. Kaskel, *Front. Chem.*, 2021, **9**, 674566.
33. O. Erhart, P. A. Georgiev and H. Krautscheid, *CrystEngComm*, 2019, **21**(43), 6523.
34. M. Shivanna, Q.-Y. Yang, A. Bajpai, S. Sen, N. Hosono, S. Kusaka, T. Pham, K. A. Forrest, B. Space, S. Kitagawa and M. J. Zaworotko, *Sci. Adv.*, 2018, **4**(4), eaaq1636.
35. E. J. Carrington, R. Pétuya, R. K. Hylton, Y. Yan, D. Antypov, G. R. Darling, M. S. Dyer, N. G. Berry, A. P. Katsoulidis and M. J. Rosseinsky, *Cryst. Growth Des.*, 2019, **19**(10), 5604.

36. S. Ehrling, E. M. Reynolds, V. Bon, I. Senkovska, T. E. Gorelik, J. D. Evans, M. Rauche, M. Mendt, M. S. Weiss, A. Pöppl, E. Brunner, U. Kaiser, A. L. Goodwin and S. Kaskel, *Nat. Chem.*, 2021, **13**(6), 568.
37. P. S. Petkov, V. Bon, C. L. Hobday, A. B. Kuc, P. Melix, S. Kaskel, T. Düren and T. Heine, *Phys. Chem. Chem. Phys.*, 2019, **21**(2), 674.
38. A. De, M. Maliuta, I. Senkovska and S. Kaskel, *Langmuir*, 2022, **38**(46), 14073.
39. L. Abylgazina, I. Senkovska and S. Kaskel, *ChemRxiv*, 2021, DOI: 10.26434/chemrxiv.13618808.v1.
40. C. Gropp, S. Canossa, S. Wuttke, F. Gándara, Q. Li, L. Gagliardi and O. M. Yaghi, *ACS Cent. Sci.*, 2020, **6**(8), 1255.
41. A. D. MacNaught and A. Wilkinson, *Compendium of Chemical Terminology: IUPAC Recommendations*, Blackwell Science, Oxford, IUPAC Chemical Nomenclature Series, 2nd edn, 1997.
42. S. K. Elsaidi, M. H. Mohamed, D. Banerjee and P. K. Thallapally, *Coord. Chem. Rev.*, 2018, **358**, 125.
43. H. Tanaka and M. T. Miyahara, *Curr. Opin. Chem. Eng.*, 2019, **24**, 19.
44. F.-X. Coudert, A. Boutin, M. Jeffroy, C. Mellot-Draznieks and A. H. Fuchs, *ChemPhysChem*, 2011, **12**(2), 247.
45. A. Kondo, H. Noguchi, L. Carlucci, D. M. Proserpio, G. Ciani, H. Kajiro, T. Ohba, H. Kanoh and K. Kaneko, *J. Am. Chem. Soc.*, 2007, **129**(41), 12362.
46. Z. Cao, L. Chen, S. Li, M. Yu, Z. Li, K. Zhou, C. Liu, F. Jiang and M. Hong, *Chem. – Asian J.*, 2019, **14**(20), 3597.
47. J. Albalad, R. A. Peralta, M. T. Huxley, S. Tsoukatos, Z. Shi, Y.-B. Zhang, J. D. Evans, C. J. Sumby and C. J. Doonan, *Chem. Sci.*, 2021, **139**, 15363.
48. N. Klein, C. Herzog, M. Sabo, I. Senkovska, J. Getzschmann, S. Paasch, M. R. Lohe, E. Brunner and S. Kaskel, *Phys. Chem. Chem. Phys.*, 2010, **12**(37), 11778.
49. B. Mu, F. Li, Y. Huang and K. S. Walton, *J. Mater. Chem.*, 2012, **22**(20), 10172.
50. A. Schneemann, R. Rudolf, S. J. Baxter, P. Vervoorts, I. Hante, K. Khaletskaya, S. Henke, G. Kieslich and R. A. Fischer, *Dalton Trans.*, 2019, **48**(19), 6564.
51. S. A. Moggach, T. D. Bennett and A. K. Cheetham, *Angew. Chem., Int. Ed.*, 2009, **48**(38), 7087.
52. J. Seo, R. Matsuda, H. Sakamoto, C. Bonneau and S. Kitagawa, *J. Am. Chem. Soc.*, 2009, **131**(35), 12792.
53. D. Yuan, D. Zhao and H.-C. Zhou, *Inorg. Chem.*, 2011, **50**(21), 10528.
54. C. Serre, F. Millange, C. Thouvenot, M. Noguès, G. Marsolier, D. Louër and G. Férey, *J. Am. Chem. Soc.*, 2002, **124**(45), 13519.
55. T. Loiseau, C. Serre, C. Huguenard, G. Fink, F. Taulelle, M. Henry, T. Bataille and G. Férey, *Chem. – Eur. J.*, 2004, **10**(6), 1373.
56. C. Serre, S. Bourrelly, A. Vimont, N. A. Ramsahye, G. Maurin, P. L. Llewellyn, M. Daturi, Y. Filinchuk, O. Leynaud, P. Barnes and G. Férey, *Adv. Mater.*, 2007, **19**(17), 2246.
57. F. Millange, N. Guillou, R. I. Walton, J.-M. Grenèche, I. Margiolaki and G. Férey, *Chem. Commun.*, 2008, **39**, 4732.

58. M. K. Taylor, T. Runčevski, J. Oktawiec, J. E. Bachman, R. L. Siegelman, H. Jiang, J. A. Mason, J. D. Tarver and J. R. Long, *J. Am. Chem. Soc.*, 2018, **140**(32), 10324.
59. M. K. Taylor, T. Runčevski, J. Oktawiec, M. I. Gonzalez, R. L. Siegelman, J. A. Mason, J. Ye, C. M. Brown and J. R. Long, *J. Am. Chem. Soc.*, 2016, **138**(45), 15019.
60. S. Krause, V. Bon, I. Senkovska, D. M. Többens, D. Wallacher, R. S. Pillai, G. Maurin and S. Kaskel, *Nat. Commun.*, 2018, **9**(1), 1573.
61. S. Krause, J. D. Evans, V. Bon, I. Senkovska, P. Iacomi, F. Kolbe, S. Ehrling, E. Troschke, J. Getzschmann, D. M. Többens, A. Franz, D. Wallacher, P. G. Yot, G. Maurin, E. Brunner, P. L. Llewellyn, F.-X. Coudert and S. Kaskel, *Nat. Commun.*, 2019, **10**(1), 3632.
62. J. Schaber, S. Krause, S. Paasch, I. Senkovska, V. Bon, D. M. Többens, D. Wallacher, S. Kaskel and E. Brunner, *J. Phys. Chem. C*, 2017, **121**(9), 5195.
63. F. Kolbe, S. Krause, V. Bon, I. Senkovska, S. Kaskel and E. Brunner, *Chem. Mater.*, 2019, **31**(16), 6193.
64. M. Brokate, *Hysteresis and Phase Transitions*, Springer New York, New York, NY, Springer eBook Collection Mathematics and Statistics, 1996, vol. 121.
65. C. Serre, C. Mellot-Draznieks, S. Surblé, N. Audebrand, Y. Filinchuk and G. Férey, *Science*, 2007, **315**(5820), 1828.
66. Y.-X. Shi, W.-X. Li, W.-H. Zhang and J.-P. Lang, *Inorg. Chem.*, 2018, **57**(14), 8627.
67. V. Bon, S. Krause, I. Senkovska, N. Grimm, D. Wallacher, D. M. Többens and S. Kaskel, *Angew. Chem.*, 2021, **133**(21), 11841.
68. J. Anwar and D. Zahn, *Adv. Drug Delivery Rev.*, 2017, **117**, 47.
69. T. R. Jensen, T. K. Nielsen, Y. Filinchuk, J.-E. Jørgensen, Y. Cerenius, E. M. Gray and C. J. Webb, *J. Appl. Crystallogr.*, 2010, **43**(6), 1456.
70. E. Salas-Colera, Á. Muñoz-Noval, C. Heyman, C. O. Ania, J. B. Parra, S. García-Granda, S. Calero, J. Rubio-Zuazo and G. R. Castro, *J. Synchrotron Radiat.*, 2015, **22**(1), 42.
71. J. Gandara-Loe, A. Missyul, F. Fauth, L. L. Daemen, Y. Q. Cheng, A. J. Ramirez-Cuesta, P. I. Ravikovitch and J. Silvestre-Albero, *J. Mater. Chem. A*, 2019, **7**(24), 14552.
72. S. Hiraide, H. Tanaka and M. T. Miyahara, *Dalton Trans.*, 2016, **45**(10), 4193.
73. R. Pallach, J. Keupp, K. Terlinden, L. Frentzel-Beyme, M. Kloß, A. Machalica, J. Kotschy, S. K. Vasa, P. A. Chater, C. Sternemann, M. T. Wharmby, R. Linser, R. Schmid and S. Henke, *Nat. Commun.*, 2021, **12**(1), 4097.
74. S. Wannapaiboon, A. Schneemann, I. Hante, M. Tu, K. Epp, A. L. Semrau, C. Sternemann, M. Paulus, S. J. Baxter, G. Kieslich and R. A. Fischer, *Nat. Commun.*, 2019, **10**(1), 346.
75. A. Knebel, B. Geppert, K. Volgmann, D. I. Kolokolov, A. G. Stepanov, J. Twiefel, P. Heitjans, D. Volkmer and J. Caro, *Science*, 2017, **358**(6361), 347.
76. R. Matsuda, R. Kitaura, S. Kitagawa, Y. Kubota, R. V. Belosludov, T. C. Kobayashi, H. Sakamoto, T. Chiba, M. Takata, Y. Kawazoe and Y. Mita, *Nature*, 2005, **436**(7048), 238.

77. T. Devic, F. Salles, S. Bourrelly, B. Moulin, G. Maurin, P. Horcajada, C. Serre, A. Vimont, J.-C. Lavalley, H. Leclerc, G. Clet, M. Daturi, P. L. Llewellyn, Y. Filinchuk and G. Férey, *J. Mater. Chem.*, 2012, **22**(20), 10266.
78. V. Bon, N. Klein, I. Senkovska, A. Heerwig, J. Getzschmann, D. Wallacher, I. Zizak, M. Brzhezinskaya, U. Mueller and S. Kaskel, *Phys. Chem. Chem. Phys.*, 2015, **17**(26), 17471.
79. H. Tanaka, S. Hiraide, A. Kondo and M. T. Miyahara, *J. Phys. Chem. C*, 2015, **119**(21), 11533.
80. S. Krause, J. D. Evans, V. Bon, I. Senkovska, S. Ehrling, P. Iacomi, D. M. Többens, D. Wallacher, M. S. Weiss, B. Zheng, P. G. Yot, G. Maurin, P. L. Llewellyn, F.-X. Coudert and S. Kaskel, *Chem. Sci.*, 2020, **11**(35), 9468.
81. V. Bon, S. Krause, I. Senkovska, N. Grimm, D. Wallacher, D. M. Többens and S. Kaskel, *Angew. Chem., Int. Ed.*, 2021, **60**(21), 11735.
82. L. Li, R.-B. Lin, R. Krishna, X. Wang, B. Li, H. Wu, J. Li, W. Zhou and B. Chen, *J. Mater. Chem. A*, 2017, **5**(36), 18984.
83. L. Li, R.-B. Lin, R. Krishna, X. Wang, B. Li, H. Wu, J. Li, W. Zhou and B. Chen, *J. Am. Chem. Soc.*, 2017, **139**(23), 7733.
84. J. Y. Kim, J. Park, J. Ha, M. Jung, D. Wallacher, A. Franz, R. Balderas-Xicohténcatl, M. Hirscher, S. G. Kang, J. T. Park, I. H. Oh, H. R. Moon and H. Oh, *J. Am. Chem. Soc.*, 2020, **142**(31), 13278.
85. C. R. Groom, I. J. Bruno, M. P. Lightfoot and S. C. Ward, *Acta Crystallogr., Sect. B: Struct. Sci., Cryst. Eng. Mater.*, 2016, **72**(Pt 2), 171.
86. W. N. Cohn, *Phys. Rev.*, 1933, **44**(4), 326.
87. G. J. Piermarini, *J. Res. Natl. Inst. Stand. Technol.*, 2001, **106**(6), 889.
88. C. E. Weir, E. R. Lippincott, A. van Valkenburg and E. N. Bunting, *J. Res. Natl. Bur. Stand., A Phys. Chem.*, 1959, **63A**(1), 55.
89. W. A. Bassett, *High Pressure Res.*, 2009, **29**(2), 163.
90. S. A. Moggach, D. R. Allan, S. Parsons and J. E. Warren, *J. Appl. Crystallogr.*, 2008, **41**(2), 249.
91. C. L. Hobday, C. H. Woodall, M. J. Lennox, M. Frost, K. Kamenev, T. Düren, C. A. Morrison and S. A. Moggach, *Nat. Commun.*, 2018, **9**(1), 1429.
92. J. Sotelo, C. H. Woodall, D. R. Allan, E. Gregoryanz, R. T. Howie, K. V. Kamenev, M. R. Probert, P. A. Wright and S. A. Moggach, *Angew. Chem., Int. Ed.*, 2015, **54**(45), 13332.
93. G. J. Piermarini, S. Block, J. D. Barnett and R. A. Forman, *J. Appl. Phys.*, 1975, **46**(6), 2774.
94. G. Shen, J. S. Smith, C. Kenney-Benson and S. Klotz, *High Pressure Res.*, 2021, **41**(2), 175.
95. A. Dawson, D. R. Allan, S. Parsons and M. Ruf, *J. Appl. Crystallogr.*, 2004, **37**(3), 410.
96. A. Katrusiak, *High-pressure Devices*, International Tables for Crystallography, International Union of Crystallography, 2019.
97. R. Angel and J. Gonzalez-Platas, *J. Appl. Crystallogr.*, 2013, **46**(1), 252.
98. S. Parsons, *ECLIPSE Program for Masking High Pressure Diffraction Images and Conversion between CCD Image Formats*, 2010.

99. K. W. Chapman, G. J. Halder and P. J. Chupas, *J. Am. Chem. Soc.*, 2008, **130**(32), 10524.
100. C. J. McMonagle, D. R. Allan, M. R. Warren, K. V. Kamenev, G. F. Turner and S. A. Moggach, *J. Appl. Crystallogr.*, 2020, **53**(6), 1519.
101. R. Finger, N. Kurtzemann, T. C. Hansen and H. Kohlmann, *J. Appl. Crystallogr.*, 2021, **54**(Pt 3), 839.
102. C. J. McMonagle, *Effect of Pressure on Porous Materials*, Edinburg, 2018.
103. N. J. Brooks, B. L. L. E. Gauthe, N. J. Terrill, S. E. Rogers, R. H. Templer, O. Ces and J. M. Seddon, *Rev. Sci. Instrum.*, 2010, **81**(6), 64103.
104. S. Dissegna, P. Vervoorts, C. L. Hobday, T. Düren, D. Daisenberger, A. J. Smith, R. A. Fischer and G. Kieslich, *J. Am. Chem. Soc.*, 2018, **140**(37), 11581.
105. K. A. S. Usman, J. W. Maina, S. Seyedin, M. T. Conato, L. M. Payawan, L. F. Dumée and J. M. Razal, *NPG Asia Mater.*, 2020, **12**(1), 1.
106. P. Vervoorts, C. L. Hobday, M. G. Ehrenreich, D. Daisenberger and G. Kieslich, *Z. Anorg. Allg. Chem.*, 2019, **645**(15), 970.
107. L. A. Price, C. J. Ridley, C. L. Bull, S. A. Wells and A. Sartbaeva, *CrystEngComm*, 2021, **23**(33), 5615.
108. P. J. Chupas, K. W. Chapman, C. Kurtz, J. C. Hanson, P. L. Lee and C. P. Grey, *J. Appl. Crystallogr.*, 2008, **41**(4), 822.
109. J. Andrieux, C. Chabert, A. Mauro, H. Vitoux, B. Gorges, T. Buslaps and V. Honkimäki, *J. Appl. Crystallogr.*, 2014, **47**(1), 245.
110. P. Vervoorts, J. Stebani, A. S. J. Méndez and G. Kieslich, *ACS Mater. Lett.*, 2021, **3**(12), 1635.
111. A. J. Graham, J.-C. Tan, D. R. Allan and S. A. Moggach, *Chem. Commun.*, 2012, **48**(10), 1535.
112. A. L. Spek, *J. Appl. Crystallogr.*, 2003, **36**(1), 7.
113. C. J. McMonagle, P. Comar, G. S. Nichol, D. R. Allan, J. González, J. A. Barreda-Argüeso, F. Rodríguez, R. Valiente, G. F. Turner, E. K. Brechin and S. A. Moggach, *Chem. Sci.*, 2020, **11**(33), 8793.
114. S. C. McKellar, A. J. Graham, D. R. Allan, M. I. H. Mohideen, R. E. Morris and S. A. Moggach, *Nanoscale*, 2014, **6**(8), 4163.
115. H. Bux, F. Liang, Y. Li, J. Cravillon, M. Wiebcke and J. Caro, *J. Am. Chem. Soc.*, 2009, **131**(44), 16000.
116. D. Fairen-Jimenez, R. Galvelis, A. Torrisi, A. D. Gellan, M. T. Wharmby, P. A. Wright, C. Mellot-Draznieks and T. Düren, *Dalton Trans.*, 2012, **41**(35), 10752.
117. C. L. Hobday, T. D. Bennett, D. Fairen-Jimenez, A. J. Graham, C. A. Morrison, D. R. Allan, T. Düren and S. A. Moggach, *J. Am. Chem. Soc.*, 2018, **140**(1), 382.
118. P. Serra-Crespo, E. Stavitski, F. Kapteijn and J. Gascon, *RSC Adv.*, 2012, **2**(12), 5051.
119. P. Ramaswamy, J. Wieme, E. Alvarez, L. Vanduyfhuys, J.-P. Itié, P. Fabry, V. van Speybroeck, C. Serre, P. G. Yot and G. Maurin, *J. Mater. Chem. A*, 2017, **5**(22), 11047.
120. P. G. Yot, K. Yang, V. Guillerm, F. Ragon, V. Dmitriev, P. Parisiades, E. Elkaïm, T. Devic, P. Horcajada, C. Serre, N. Stock, J. P. S. Mowat, P. A.

Wright, G. Férey and G. Maurin, *Eur. J. Inorg. Chem.*, 2016, **2016**(27), 4424.
121. S. Henke, M. T. Wharmby, G. Kieslich, I. Hante, A. Schneemann, Y. Wu, D. Daisenberger and A. K. Cheetham, *Chem. Sci.*, 2018, **9**(6), 1654.
122. P. Z. Moghadam, A. Li, S. B. Wiggin, A. Tao, A. G. P. Maloney, P. A. Wood, S. C. Ward and D. Fairen-Jimenez, *Chem. Mater.*, 2017, **29**(7), 2618.
123. Z. Su, Y.-R. Miao, G. Zhang, J. T. Miller and K. S. Suslick, *Chem. Sci.*, 2017, **8**(12), 8004.
124. P. Iacomi, J. S. Lee, L. Vanduyfhuys, K. H. Cho, P. Fertey, J. Wieme, D. Granier, G. Maurin, V. van Speybroeck, J.-S. Chang and P. G. Yot, *Chem. Sci.*, 2021, **12**(15), 5682.
125. J. M. Bermúdez-García, M. Sánchez-Andújar, S. Castro-García, J. López-Beceiro, R. Artiaga and M. A. Señarís-Rodríguez, *Nat. Commun.*, 2017, **8**, 15715.
126. N. A. Benedek and C. J. Fennie, *Phys. Rev. Lett.*, 2011, **106**(10), 107204.
127. A. Abragam, *The Principles of Nuclear Magnetism*, Oxford University Press, Oxford, Oxford Science Publications, 2007, vol. 32.
128. R. R. Ernst, G. Bodenhausen and A. Wokaun, *Principles of Nuclear Magnetic Resonance in One and Two Dimensions*, Clarendon Press, Oxford, Oxford science publications, 2004, vol. 14.
129. A. J. Beeler, A. M. Orendt, D. M. Grant, P. W. Cutts, J. Michl, K. W. Zilm, J. W. Downing, J. C. Facelli, M. S. Schindler and W. Kutzelnigg, *J. Am. Chem. Soc.*, 1984, **106**(25), 7672.
130. J. A. Ripmeester and C. I. Ratcliffe, *Energy Fuels*, 1998, **12**(2), 197.
131. *Solid-state NMR Spectroscopy: Principles and Applications*, ed. M. J. Duer, Blackwell Science, Oxford, 2005.
132. E. R. Andrew, A. Bradbury and R. G. Eades, *Nature*, 1959, **183**(4678), 1802.
133. A. Pines, M. G. Gibby and J. S. Waugh, *J. Chem. Phys.*, 2003, **59**(2), 569.
134. J. Schaefer and E. O. Stejskal, *J. Am. Chem. Soc.*, 1976, **98**(4), 1031.
135. M. Rauche, S. Ehrling, S. Krause, I. Senkovska, S. Kaskel and E. Brunner, *Chem. Commun.*, 2019, **55**(62), 9140.
136. H. Hoffmann, M. Debowski, P. Müller, S. Paasch, I. Senkovska, S. Kaskel and E. Brunner, *Materials*, 2012, **5**(12), 2537.
137. B. E. G. Lucier, S. Chen and Y. Huang, *Acc. Chem. Res.*, 2018, **51**(2), 319.
138. Y. T. A. Wong, V. Martins, B. E. G. Lucier and Y. Huang, *Chem. – Eur. J.*, 2019, **25**(8), 1848.
139. V. J. Witherspoon, J. Xu and J. A. Reimer, *Chem. Rev.*, 2018, **118**(20), 10033.
140. M. Haouas, F. Taulelle and C. Martineau, *Prog. Nucl. Magn. Reson. Spectrosc.*, 2016, **94–95**, 11.
141. M. Haouas, *Materials*, 2018, **11**(8), 1416.
142. M. Hunger and T. Horvath, *J. Chem. Soc., Chem. Commun.*, 1995, (14), 1423.
143. W. Zhang, S. Xu, X. Han and X. Bao, *Chem. Soc. Rev.*, 2011, **41**(1), 192.
144. F. Blanc, M. Leskes and C. P. Grey, *Acc. Chem. Res.*, 2013, **46**(9), 1952.
145. J. M. Griffin, A. C. Forse and C. P. Grey, *Solid State Nucl. Magn. Reson.*, 2016, **74–75**, 16.

146. J. B. Richter, C. Eßbach, I. Senkovska, S. Kaskel and E. Brunner, *Chem. Commun.*, 2019, **55**(43), 6042.
147. T. Meissner, S. K Goh, J. Haase, M. Richter, K. Koepernik and H. Eschrig, *J. Phys.: Condens. Matter*, 2013, **26**(1), 15501.
148. M. Benzaqui, R. S. Pillai, A. Sabetghadam, V. Benoit, P. Normand, J. Marrot, N. Menguy, D. Montero, W. Shepard, A. Tissot, C. Martineau-Corcos, C. Sicard, M. Mihaylov, F. Carn, I. Beurroies, P. L. Llewellyn, G. de Weireld, K. Hadjiivanov, J. Gascon, F. Kapteijn, G. Maurin, N. Steunou and C. Serre, *Chem. Mater.*, 2017, **29**(24), 10326.
149. G. Mali, M. Mazaj, I. Arìon, D. Hanel, D. Arion and Z. Jagliç, *J. Phys. Chem. Lett.*, 2019, **10**(7), 1464.
150. A. E. Khudozhitkov, S. S. Arzumanov, D. I. Kolokolov and A. G. Stepanov, *Microporous Mesoporous Mater.*, 2020, **300**, 110155.
151. M. R. Hansen, R. Graf and H. W. Spiess, *Acc. Chem. Res.*, 2013, **46**(9), 1996.
152. Y. Yan, D. I. Kolokolov, I. da Silva, A. G. Stepanov, A. J. Blake, A. Dailly, P. Manuel, C. C. Tang, S. Yang and M. Schröder, *J. Am. Chem. Soc.*, 2017, **139**(38), 13349.
153. N. C. Burtch, A. Torres-Knoop, G. S. Foo, J. Leisen, C. Sievers, B. Ensing, D. Dubbeldam and K. S. Walton, *J. Phys. Chem. Lett.*, 2015, **6**(5), 812.
154. J. H. Lee, S. Jeoung, Y. G. Chung and H. R. Moon, *Coord. Chem. Rev.*, 2019, **389**, 161.
155. Z. Chang, D.-H. Yang, J. Xu, T.-L. Hu and X.-H. Bu, *Adv. Mater.*, 2015, **27**(36), 5432.
156. D. Freude, M. Hunger and H. Pfeifer, *Chem. Phys. Lett.*, 1982, **91**(4), 307.
157. R. Giovine, C. Volkringer, M.-A. Springuel-Huet, A. Nossov, F. Blanc, J. Trébosc, T. Loiseau, J.-P. Amoureux, O. Lafon and F. Pourpoint, *J. Phys. Chem. C*, 2017, **121**(35), 19262.
158. T. Ito and J. Fraissard, *J. Chem. Phys.*, 1998, **76**(11), 5225.
159. J. Fraissard and T. Ito, *Zeolites*, 1988, **8**(5), 350.
160. C. I. Ratcliffe, Xenon NMR, in *Annual Reports on NMR Spectroscopy*, Elsevier, 1998, pp. 123–221.
161. D. Raftery, Xenon NMR Spectroscopy, in *Annual Reports on NMR Spectroscopy*, Elsevier, 2006, pp. 205–270.
162. J. Demarquay and J. Fraissard, *Chem. Phys. Lett.*, 1987, **136**, 314.
163. V. V. Terskikh, I. L. Moudrakovski, S. R. Breeze, S. Lang, C. I. Ratcliffe, J. A. Ripmeester and A. Sayari, *Langmuir*, 2002, **18**(15), 5653.
164. A. K. Jameson, C. J. Jameson and H. S. Gutowsky, *J. Chem. Phys.*, 2003, **53**(6), 2310.
165. C. J. Jameson, A. K. Jameson and S. M. Cohen, *J. Chem. Phys.*, 2003, **59**(8), 4540.
166. I. Moudrakovski, D. V. Soldatov, J. A. Ripmeester, D. N. Sears and C. J. Jameson, *Proc. Natl. Acad. Sci. U. S. A.*, 2004, **101**(52), 17924.
167. M.-A. Springuel-Huet, A. Nossov, Z. Adem, F. Guenneau, C. Volkringer, T. Loiseau, G. Férey and A. Gédéon, *J. Am. Chem. Soc.*, 2010, **132**(33), 11599.

168. H. C. Hoffmann, B. Assfour, F. Epperlein, N. Klein, S. Paasch, I. Senkovska, S. Kaskel, G. Seifert and E. Brunner, *J. Am. Chem. Soc.*, 2011, **133**(22), 8681.
169. X. Kong, E. Scott, W. Ding, J. A. Mason, J. R. Long and J. A. Reimer, *J. Am. Chem. Soc.*, 2012, **134**(35), 14341.
170. V. Bon, J. Pallmann, E. Eisbein, H. C. Hoffmann, I. Senkovska, I. Schwedler, A. Schneemann, S. Henke, D. Wallacher, R. A. Fischer, G. Seifert, E. Brunner and S. Kaskel, *Microporous Mesoporous Mater.*, 2015, **216**, 64.
171. R. M. Marti, J. D. Howe, C. R. Morelock, M. S. Conradi, K. S. Walton, D. S. Sholl and S. E. Hayes, *J. Phys. Chem. C*, 2017, **121**(46), 25778.
172. Y. Lu, B. E. G. Lucier, Y. Zhang, P. Ren, A. Zheng and Y. Huang, *Phys. Chem. Chem. Phys.*, 2017, **19**(8), 6130.
173. A. C. Forse, M. I. Gonzalez, R. L. Siegelman, V. J. Witherspoon, S. Jawahery, R. Mercado, P. J. Milner, J. D. Martell, B. Smit, B. Blümich, J. R. Long and J. A. Reimer, *J. Am. Chem. Soc.*, 2018, **140**(5), 1663.
174. P. Melix and T. Heine, *ChemPhysChem*, 2021, **22**(22), 2336.
175. M. Sin, N. Kavoosi, M. Rauche, J. Pallmann, S. Paasch, I. Senkovska, S. Kaskel and E. Brunner, *Langmuir*, 2019, **35**(8), 3162.
176. K. Roztocki, M. Rauche, V. Bon, S. Kaskel, E. Brunner and D. Matoga, *ACS Appl. Mater. Interfaces*, 2021, **13**(24), 28503.
177. J. A. Weil and J. R. Bolton, *Electron Paramagnetic Resonance: Elementary, Theory and Practical Applications*, Wiley-Interscience; A John Wiley & Sons, Hoboken (New Jersey), 2nd edn, 2007.
178. V. Chechik, E. Carter and D. Murphy, *Electron Pharmagnetic Resonance*, Oxford University Press, Oxford, Oxford Chemistry Primers, 2016.
179. S. Kaskel, *Chemistry of Metal–Organic Frameworks*, Wiley-VCH, Weinheim, Germany, 2016.
180. N. M. Atherton, *Principles of Electron Spin Resonance*, Horwood, Chichester, Ellis Horwood PTR Prentice Hall Physical Chemistry Series, 1993.
181. J. R. Pilbrow, *Transition Ion Electron Paramagnetic Resonance*, Clarendon Press, Oxford, 1990.
182. F. Neese, *Coord. Chem. Rev.*, 2009, **253**(5), 526.
183. A. Brückner, *Chem. Ing. Tech.*, 2014, **86**(11), 1871.
184. A. Brückner, *Chem. Soc. Rev.*, 2010, **39**(12), 4673.
185. A. Brückner, *Phys. Chem. Chem. Phys.*, 2003, **5**(20), 4461.
186. M. Labanowska, *ChemPhysChem*, 2001, **2**(12), 712.
187. G. Férey and C. Serre, *Chem. Soc. Rev.*, 2009, **38**(5), 1380.
188. G. Férey, *Dalton Trans.*, 2016, **45**(10), 4073.
189. A. Schneemann, V. Bon, I. Schwedler, I. Senkovska, S. Kaskel and R. A. Fischer, *Chem. Soc. Rev.*, 2014, **43**(16), 6062.
190. E. Borfecchia, S. Maurelli, D. Gianolio, E. Groppo, M. Chiesa, F. Bonino and C. Lamberti, *J. Phys. Chem. C*, 2012, **116**(37), 19839.
191. I. Nevjestić, H. Depauw, P. Gast, P. Tack, D. Deduytsche, K. Leus, M. van Landeghem, E. Goovaerts, L. Vincze, C. Detavernier, P. van der Voort, F. Callens and H. Vrielinck, *Phys. Chem. Chem. Phys.*, 2017, **19**(36), 24545.

192. M. Mendt, P. Vervoorts, A. Schneemann, R. A. Fischer and A. Pöppl, *J. Phys. Chem. C*, 2019, **123**(5), 2940.
193. D. M. Polyukhov, S. Krause, V. Bon, A. S. Poryvaev, S. Kaskel and M. V. Fedin, *J. Phys. Chem. Lett.*, 2020, **11**(15), 5856.
194. M. Brustolon, *Electron Paramagnetic Spectroscopy: A Practitioner's Toolkit*, Wiley, Hoboken, NJ, 2009.
195. V. Bon, N. Klein, I. Senkovska, A. Heerwig, J. Getzschmann, D. Wallacher, I. Zizak, M. Brzhezinskaya, U. Mueller and S. Kaskel, *Phys. Chem. Chem. Phys.*, 2015, **17**(26), 17471.
196. S. Henke, D. C. Florian Wieland, M. Meilikhov, M. Paulus, C. Sternemann, K. Yusenko and R. A. Fischer, *CrystEngComm*, 2011, **13**(21), 6399.
197. I. Nevjestić, H. Depauw, K. Leus, G. Rampelberg, C. A. Murray, C. Detavernier, P. van der Voort, F. Callens and H. Vrielinck, *J. Phys. Chem. C*, 2016, **120**(31), 17400.
198. O. Kozachuk, M. Meilikhov, K. Yusenko, A. Schneemann, B. Jee, A. V. Kuttatheyil, M. Bertmer, C. Sternemann, A. Pöppl and R. A. Fischer, *Eur. J. Inorg. Chem.*, 2013, **2013**(26), 4546–4557.
199. I. Nevjestić, H. Depauw, K. Leus, V. Kalendra, I. Caretti, G. Jeschke, S. van Doorslaer, F. Callens, P. van der Voort and H. Vrielinck, *ChemPhysChem*, 2015, **16**(14), 2968.
200. S. Ehrling, M. Mendt, I. Senkovska, J. D. Evans, V. Bon, P. Petkov, C. Ehrling, F. Walenszus, A. Pöppl and S. Kaskel, *Chem. Mater.*, 2020, **32**(13), 5670.
201. A. Pöppl, S. Kunz, D. Himsl and M. Hartmann, *J. Phys. Chem. C*, 2008, **112**(7), 2678.
202. K. Hadjiivanov, M. Mihaylov, D. Panayotov, E. Ivanova and K. Chakarova, Chapter 2. Isotopes in the FTIR investigations of solid surfaces, in *Spectroscopic Properties of Inorganic and Organometallic Compounds*, ed. R. Douthwaite, S. Duckett, and J. Yarwood, Royal Society of Chemistry, Cambridge, 2014, pp. 43–78.
203. Y. Chen, J. Zhang, J. Li and J. V. Lockard, *J. Phys. Chem. C*, 2013, **117**(39), 20068.
204. D. Matoga, K. Roztocki, M. Wilke, F. Emmerling, M. Oszajca, M. Fitta and M. Bałanda, *CrystEngComm*, 2017, **19**(22), 2987.
205. K. Roztocki, M. Szufla, V. Bon, I. Senkovska, S. Kaskel and D. Matoga, *Inorg. Chem.*, 2020, **59**(15), 10717.
206. A. Sławek, J. M. Vicent-Luna, B. Marszałek, B. Gil, R. E. Morris, W. Makowski and S. Calero, *Chem. Mater.*, 2018, **30**(15), 5116.
207. P. Kanoo, R. Matsuda, H. Sato, L. Li, N. Hosono and S. Kitagawa, *Chem. - Eur. J.*, 2020, **26**(10), 2148.
208. F. Millange, C. Serre and G. Férey, *Chem. Commun.*, 2002, (8), 822.
209. A. Vimont, A. Travert, P. Bazin, J.-C. Lavalley, M. Daturi, C. Serre, G. Férey, S. Bourrelly and P. L. Llewellyn, *Chem. Commun.*, 2007, (31), 3291.
210. L. Hamon, P. L. Llewellyn, T. Devic, A. Ghoufi, G. Clet, V. Guillerm, G. D. Pirngruber, G. Maurin, C. Serre, G. Driver, W. van Beek, E. Jolimaître,

A. Vimont, M. Daturi and G. Férey, *J. Am. Chem. Soc.*, 2009, **131**(47), 17490.
211. L. Hamon, H. Leclerc, A. Ghoufi, L. Oliviero, A. Travert, J.-C. Lavalley, T. Devic, C. Serre, G. Férey, G. de Weireld, A. Vimont and G. Maurin, *J. Phys. Chem. C*, 2011, **115**(5), 2047.
212. S. Bourrelly, B. Moulin, A. Rivera, G. Maurin, S. Devautour-Vinot, C. Serre, T. Devic, P. Horcajada, A. Vimont, G. Clet, M. Daturi, J.-C. Lavalley, S. Loera-Serna, R. Denoyel, P. L. Llewellyn and G. Férey, *J. Am. Chem. Soc.*, 2010, **132**(27), 9488.
213. S. Andonova, E. Ivanova, J. Yang and K. Hadjiivanov, *J. Phys. Chem. C*, 2017, **121**(34), 18665.
214. M. Mihaylov, K. Chakarova, S. Andonova, N. Drenchev, E. Ivanova, E. A. Pidko, A. Sabetghadam, B. Seoane, J. Gascon, F. Kapteijn and K. Hadjiivanov, *Chem. Commun.*, 2016, **52**(7), 1494.
215. H. Leclerc, T. Devic, S. Devautour-Vinot, P. Bazin, N. Audebrand, G. Férey, M. Daturi, A. Vimont and G. Clet, *J. Phys. Chem. C*, 2011, **115**(40), 19828.
216. G. Kumari, K. Jayaramulu, T. K. Maji and C. Narayana, *J. Phys. Chem. A*, 2013, **117**(43), 11006.
217. M. R. Ryder, B. Civalleri, T. D. Bennett, S. Henke, S. Rudić, G. Cinque, F. Fernandez-Alonso and J.-C. Tan, *Phys. Rev. Lett.*, 2014, **113**(21), 215502.
218. A. Krylov, I. Senkovska, S. Ehrling, M. Maliuta, S. Krylova, E. Slyusareva, A. Vtyurin and S. Kaskel, *Chem. Commun.*, 2020, **56**(59), 8269.
219. A. Krylov, I. Yushina, E. Slyusareva, S. Krylova, A. Vtyurin, S. Kaskel and I. Senkovska, *Phys. Chem. Chem. Phys.*, 2022, **24**(6), 3788.
220. J. Seo, C. Bonneau, R. Matsuda, M. Takata and S. Kitagawa, *J. Am. Chem. Soc.*, 2011, **133**(23), 9005.
221. K. Roztocki, F. Formalik, A. Krawczuk, I. Senkovska, B. Kuchta, S. Kaskel and D. Matoga, *Angew. Chem., Int. Ed.*, 2020, **59**(11), 4491.
222. A.-L. Lavoisier, Elements of Chemistry, in *A New Systematic Order, Containing All the Modern Discoveries Illustrated With Thirteen Copperplates*, Gale ECCO, 2018.
223. F. Rouquerol, J. Rouquerol, K. S. W. Sing, P. L. Llewellyn and G. Maurin, *Adsorption by Powders and Porous Solids: Principles, Methodology and Applications*, Elsevier/Academic Press, Amsterdam, 2nd edn, 2014.
224. *Calorimetry and Thermal Methods in Catalysis*, ed. A. Auroux, Springer Berlin Heidelberg, Berlin, Heidelberg, New York, Dordrecht, London, Springer Series in Materials Science, 2013, vol. 154.
225. *Handbook of Thermal Analysis and Calorimetry*, ed. M. E. Brown, and P. K. Gallagher, Elsevier, Amsterdam, 2nd edn, 2003.
226. E. Calvet, H. Prat and H. A. Skinner, *Recent Progress in Microcalorimetry*, Elsevier Science, Burlington, 1963.
227. P. Wollmann, M. Leistner, W. Grählert, O. Throl, F. Dreisbach and S. Kaskel, *Microporous Mesoporous Mater.*, 2012, **149**(1), 86.
228. P. L. Llewellyn and G. Maurin, *C. R. Chim.*, 2005, **8**(3), 283.
229. J. Rouquerol, S. Partyka and F. Rouquerol, *J. Chem. Soc., Faraday Trans. 1*, 1977, **73**, 306.

230. M. Rubeš, A. D. Wiersum, P. L. Llewellyn, L. Grajciar, O. Bludský and P. Nachtigall, *J. Phys. Chem. C*, 2013, **117**(21), 11159.
231. L. Grajciar, A. D. Wiersum, P. L. Llewellyn, J.-S. Chang and P. Nachtigall, *J. Phys. Chem. C*, 2011, **115**(36), 17925.
232. M. Rubeš, L. Grajciar, O. Bludský, A. D. Wiersum, P. L. Llewellyn and P. Nachtigall, *ChemPhysChem*, 2012, **13**(2), 488.
233. J. W. Yoon, Y.-K. Seo, Y. K. Hwang, J.-S. Chang, H. Leclerc, S. Wuttke, P. Bazin, A. Vimont, M. Daturi, E. Bloch, P. L. Llewellyn, C. Serre, P. Horcajada, J.-M. Grenèche, A. E. Rodrigues and G. Férey, *Angew. Chem., Int. Ed.*, 2010, **49**(34), 5949.
234. Q. Yang, S. Vaesen, F. Ragon, A. D. Wiersum, D. Wu, A. Lago, T. Devic, C. Martineau, F. Taulelle, P. L. Llewellyn, H. Jobic, C. Zhong, C. Serre, G. de Weireld and G. Maurin, *Angew. Chem., Int. Ed.*, 2013, **52**(39), 10316.
235. F. Ragon, B. Campo, Q. Yang, C. Martineau, A. D. Wiersum, A. Lago, V. Guillerm, C. Hemsley, J. F. Eubank, M. Vishnuvarthan, F. Taulelle, P. Horcajada, A. Vimont, P. L. Llewellyn, M. Daturi, S. Devautour-Vinot, G. Maurin, C. Serre, T. Devic and G. Clet, *J. Mater. Chem. A*, 2015, **3**(7), 3294.
236. G. Férey, *Chem. Soc. Rev.*, 2008, **37**(1), 191.
237. S. Kitagawa, R. Kitaura and S. Noro, *Angew. Chem., Int. Ed.*, 2004, **43**(18), 2334.
238. S. Aguado, G. Bergeret, M. P. Titus, V. Moizan, C. Nieto-Draghi, N. Bats and D. Farrusseng, *New J. Chem.*, 2011, **35**(3), 546.
239. F. Salles, G. Maurin, C. Serre, P. L. Llewellyn, C. Knöfel, H. J. Choi, Y. Filinchuk, L. Oliviero, A. Vimont, J. R. Long and G. Férey, *J. Am. Chem. Soc.*, 2010, **132**(39), 13782.
240. J. A. Mason, J. Oktawiec, M. K. Taylor, M. R. Hudson, J. Rodriguez, J. E. Bachman, M. I. Gonzalez, A. Cervellino, A. Guagliardi, C. M. Brown, P. L. Llewellyn, N. Masciocchi and J. R. Long, *Nature*, 2015, **527**(7578), 357.
241. P. Iacomi, B. Zheng, S. Krause, S. Kaskel, G. Maurin and P. L. Llewellyn, *Chem. Mater.*, 2020, **32**(8), 3489.
242. Y. Du, B. Wooler, M. Nines, P. Kortunov, C. S. Paur, J. Zengel, S. C. Weston and P. I. Ravikovitch, *J. Am. Chem. Soc.*, 2015, **137**(42), 13603.
243. S. Krause, J. D. Evans, V. Bon, I. Senkovska, S. Ehrling, U. Stoeck, P. G. Yot, P. Iacomi, P. Llewellyn, G. Maurin, F.-X. Coudert and S. Kaskel, *J. Phys. Chem. C*, 2018, **122**(33), 19171.
244. N. A. Ramsahye, G. Maurin, S. Bourrelly, P. L. Llewellyn, T. Devic, C. Serre, T. Loiseau and G. Ferey, *Adsorption*, 2007, **13**(5), 461.
245. C. Cuadrado-Collados, C. K. Rojas-Mayorga, B. Saavedra, M. Martinez-Escandell, J. M. Osiński, P. Z. Moghadam, D. Fairen-Jimenez and J. Silvestre-Albero, *J. Phys. Chem. C*, 2019, **123**(18), 11699.
246. C. Cuadrado-Collados, J. Fernández-Català, F. Fauth, Y. Q. Cheng, L. L. Daemen, A. J. Ramirez-Cuesta and J. Silvestre-Albero, *J. Mater. Chem. A*, 2017, **5**(39), 20938.
247. S. Devautour-Vinot, G. Maurin, F. Henn, C. Serre, T. Devic and G. Férey, *Chem. Commun.*, 2009, (19), 2733.

248. T. D. Bennett, J.-C. Tan, Y. Yue, E. Baxter, C. Ducati, N. J. Terrill, H. H.-M. Yeung, Z. Zhou, W. Chen, S. Henke, A. K. Cheetham and G. N. Greaves, *Nat. Commun.*, 2015, **6**, 8079.
249. M. T. Wharmby, S. Henke, T. D. Bennett, S. R. Bajpe, I. Schwedler, S. P. Thompson, F. Gozzo, P. Simoncic, C. Mellot-Draznieks, H. Tao, Y. Yue and A. K. Cheetham, *Angew. Chem., Int. Ed.*, 2015, **54**(22), 6447.
250. J. M. Chourot, A. Le Bail and D. Chevalier, *High Pressure Res.*, 2000, **19**(1–6), 191.
251. Y. Grosu, V. Eroshenko, J. M. Nedelec and J. P. E. Grolier, *Phys. Chem. Chem. Phys.*, 2015, **17**(3), 1572.
252. J. Rodriguez, I. Beurroies, T. Loiseau, R. Denoyel and P. L. Llewellyn, *Angew. Chem., Int. Ed.*, 2015, **54**(15), 4626.
253. M. Chorążewski, P. Zajdel, T. Feng, D. Luo, A. R. Lowe, C. M. Brown, J. B. Leão, M. Li, M. Bleuel, G. Jensen, D. Li, A. Faik and Y. Grosu, *ACS Nano*, 2021, **15**(5), 9048.
254. D. J. Cerasale, D. C. Ward and T. L. Easun, *Nat. Rev. Chem.*, 2022, **6**, 9–30.
255. S. Hiraide, Y. Sakanaka, H. Kajiro, S. Kawaguchi, M. T. Miyahara and H. Tanaka, *Nat. Commun.*, 2020, **11**(1), 3867.
256. V. Bon, N. Busov, I. Senkovska, N. Bönisch, L. Abylgazina, A. Khadiev, D. Novikov and S. Kaskel, *Chem. Commun.*, 2022, **58**(75), 10492.
257. C. Wiktor, M. Meledina, S. Turner, O. I. Lebedev and R. A. Fischer, *J. Mater. Chem. A*, 2017, **5**(29), 14969.
258. Y. Zhu, J. Ciston, B. Zheng, X. Miao, C. Czarnik, Y. Pan, R. Sougrat, Z. Lai, C.-E. Hsiung, K. Yao, I. Pinnau, M. Pan and Y. Han, *Nat. Mater.*, 2017, **16**(5), 532.
259. D. Zhang, Y. Zhu, L. Liu, X. Ying, C.-E. Hsiung, R. Sougrat, K. Li and Y. Han, *Science*, 2018, **359**(6376), 675.
260. Y. Liu, Y. Ma, Y. Zhao, X. Sun, F. Gándara, H. Furukawa, Z. Liu, H. Zhu, C. Zhu, K. Suenaga, P. Oleynikov, A. S. Alshammari, X. Zhang, O. Terasaki and O. M. Yaghi, *Science*, 2016, **351**(6271), 365.
261. L. R. Parent, C. H. Pham, J. P. Patterson, M. S. Denny, S. M. Cohen, N. C. Gianneschi and F. Paesani, *J. Am. Chem. Soc.*, 2017, **139**(40), 13973.
262. J. Lyu, X. Gong, S.-J. Lee, K. Gnanasekaran, X. Zhang, M. C. Wasson, X. Wang, P. Bai, X. Guo, N. C. Gianneschi and O. K. Farha, *J. Am. Chem. Soc.*, 2020, **142**(10), 4609.
263. M. D. Wilkinson, M. Dumontier, I. J. J. Aalbersberg, G. Appleton, M. Axton, A. Baak, N. Blomberg, J.-W. Boiten, L. B. da Silva Santos, P. E. Bourne, J. Bouwman, A. J. Brookes, T. Clark, M. Crosas, I. Dillo, O. Dumon, S. Edmunds, C. T. Evelo, R. Finkers, A. Gonzalez-Beltran, A. J. G. Gray, P. Groth, C. Goble, J. S. Grethe, J. Heringa, P. A. C. 't Hoen, R. Hooft, T. Kuhn, R. Kok, J. Kok, S. J. Lusher, M. E. Martone, A. Mons, A. L. Packer, B. Persson, P. Rocca-Serra, M. Roos, R. van Schaik, S.-A. Sansone, E. Schultes, T. Sengstag, T. Slater, G. Strawn, M. A. Swertz, M. Thompson, J. van der Lei, E. van Mulligen, J. Velterop, A. Waagmeester, P. Wittenburg, K. Wolstencroft, J. Zhao and B. Mons, *Sci. Data*, 2016, **3**, 160018.

264. J. D. Evans, V. Bon, I. Senkovska and S. Kaskel, *Langmuir*, 2021, **37**(14), 4222.
265. R. Numaguchi, H. Tanaka, S. Watanabe and M. T. Miyahara, *J. Chem. Phys.*, 2013, **138**(5), 54708.
266. Y. Kubota, M. Takata, R. Matsuda, R. Kitaura, S. Kitagawa and T. C. Kobayashi, *Angew. Chem., Int. Ed.*, 2006, **45**(30), 4932.
267. S. Klotz, J.-C. Chervin, P. Munsch and G. Le Marchand, *J. Phys. D: Appl. Phys.*, 2009, **42**(7), 75413.
268. T. Lesage, C. Verrier, P. Bazin, J. Saussey and M. Daturi, *Phys. Chem. Chem. Phys.*, 2003, **5**(20), 4435.
269. H. Miura, V. Bon, I. Senkovska, S. Ehrling, N. Bönisch, G. Mäder, S. Grünzner, A. Khadiev, D. Novikov, K. Maity, A. Richter and S. Kaskel, *Adv. Mater.*, 2022, e2207741.

CHAPTER 4

Theoretical Understanding and Insights on Flexible Metal–Organic Frameworks

L. VANDUYFHUYS[a], M. WAHIDUZZAMAN[b], S. M. J. ROGGE[a], G. MAURIN[b] AND V. VAN SPEYBROECK[a*]

[a]Center for Molecular Modeling, Ghent University, Technologiepark 46, Zwijnaarde B-9052, Belgium; [b]ICGM, Univ. Montpellier, CNRS, ENSCM, Montpellier 34095, France
*E-mail: Veronique.VanSpeybroeck@UGent.be

4.1 Introduction

Molecular simulations, which embody all the methods able to model a system at the electronic and/or atomic level, have substantially contributed to the understanding of the most intriguing physical properties of metal-organic frameworks (MOFs) emerging from their unique flexibility, which is tuneable upon the application of various stimuli. Over the last decade, the computational chemistry, and physics communities have developed a wide array of novel and far-reaching methods—quantum, classical, statistical—and/or adapted existing techniques to afford larger time and length scales to tackle this challenging topic.

This chapter begins by presenting an updated and generalised classification of flexible MOFs to account for the most recently discovered compliant phenomena, including MOF melting and counterintuitive negative

Inorganic Materials Series No. 13
Flexible Metal–Organic Frameworks: Structural Design, Synthesis and Properties
Edited by Susumu Kitagawa

Published by the Royal Society of Chemistry, www.rsc.org

gas adsorption (NGA). The proposed spatiotemporal terminology distinguishes between stimuli-induced displacive and reconstructive structure transformation, and ranges from localised to highly correlated modes over which the flexibility operates. We briefly introduce the appropriate computational methods required to model distinctively each category of flexible modes. We aimed to cover mostly quantum-chemical and force field modelling techniques and give some outlook to the most recent machine learning developments.

The second section provides an overview of the computational challenges one might encounter in modelling flexible MOFs, from the construction of reliable structural models to the determination of accurate potential energy surfaces and their effective sampling. These key successive steps provide the underpinning elements to enable the prediction of the properties of flexible MOFs, and beyond, to gain microscopic insight into the origin of their intriguing behaviours. Herein, our objective is to give the reader an overview of the available computational techniques to address the complexity of those different stages alongside a basic understanding of the underlying theory. Readers are referred to specialised books and detailed reviews to obtain more technical details. The second section ends by compiling the diverse properties that can be simulated for a flexible MOF.

The third section discusses a list of representative modelling studies realised on prototypical flexible MOFs. This section is by no means an exhaustive review of existing modelling works in the field. The objective here is rather to highlight the invaluable contributions of computational approaches to understand the diverse flexible modes triggered by the application of various stimuli, often in tandem with the deployment of sophisticated experimental tools.

Finally, we discuss the prospect of uncovering new compliant behaviours by adapting existing theoretical methodologies or developing new ones to characterise complex phenomena and cover multiple scales in both space and time.

4.2 Classifying Flexibility in MOFs and Associated Computational Challenges

MOFs form a highly versatile class of nanostructured materials that are composed of a large variety of organic and inorganic building blocks. The interplay between these different building blocks as well as with guest molecules or other external stimuli is mediated through a range of interactions, from weak dispersive interactions over hydrogen bonds, coordination bonds, and strong covalent bonds. Due to this versatility in building blocks and interaction strengths, MOFs are prone to an enormous diversity of flexibility modes, a term that will be adopted in a broad sense in this chapter to cover all dynamic modes in which the framework atoms move from their equilibrium properties. As a result, no single computational approach can be expected to adequately cover all different types of flexibility.[1,2] In general, multiple approaches, both computational and experimental in nature, should be combined to completely interrogate flexibility in MOFs and to understand how this flexibility can be harnessed for functional applications, from shock absorption to gas sensing, as outlined in detail in this chapter.

Flexible MOFs—also interchangeably named stimuli-responsive MOFs, switchable MOFs, soft porous crystals, smart MOFs, or, when focussing on guest-induced flexibility, third-generation porous coordination polymers—have been discussed extensively before and nomenclature to name these modes of flexibility has grown accordingly in recent decades.[3–20] To properly guide the discussion in this chapter, we opted to build upon this earlier introduced terminology while simultaneously accounting for newer flexible phenomena in MOFs, such as MOF melting, leading us to propose the classification visualised in Figure 4.1.[†] A first axis in Figure 4.1 focusses on the spatial range over which the mode of flexibility impacts the MOF structure or, conversely, on the degree of correlation of this mode of flexibility across the material. The two extremes on this continuous axis are completely localised modes of flexibility, such as the lability of water, protons, and hydroxide ions at the inorganic nodes of defective UiO-66 samples,[21,22] and completely concerted modes of flexibility, such as swelling in MIL-88.[23–25] A second, discrete axis focusses on the bond network of the material. If the mode of flexibility retains the bond network, it is defined as displacive. In that case, bonds are only distorted in this process, similar to displacive phase transformations.[26] In contrast, reconstructive flexibility modes are accompanied by a reorganisation of the bond network. For instance, while the single-crystal-to-single-crystal breathing transformation in the MIL-53(Al) family is a highly correlated displacive transformation,[27–29] the single-crystal-to-amorphous transformations in ZIF-4 and ZIF-67 (ZIF = zeolitic imidazolate framework) are reconstructive in nature as bond breaking takes place.[30,31] For displacive phase transformations, the initial and final phases are constrained as they should exhibit a group/subgroup relationship; no such requirement holds for reconstructive transitions.[26]

Depending on where the mode of flexibility under investigation is located along the two axes in Figure 4.1, different computational approaches should be adopted. This also implies that, when different modes of flexibility occur in the same material, a different set of computational tools may be needed to investigate each of these modes. For instance, while the displacive transformations from ZIF-4 to either ZIF-4-I, ZIF-4-cp-I, ZIF-4-cp-II, or ZIF-4-cp-III could all be described by force fields, the reconstructive nature of the phase transition from ZIF-4 to the amorphous high-pressure or high-temperature

---

[†]While rather general terms such as network flexibility and shape-responsive fitting are not taken up explicitly in Figure 4.1, they are present under more specific names such as gate opening and swelling.

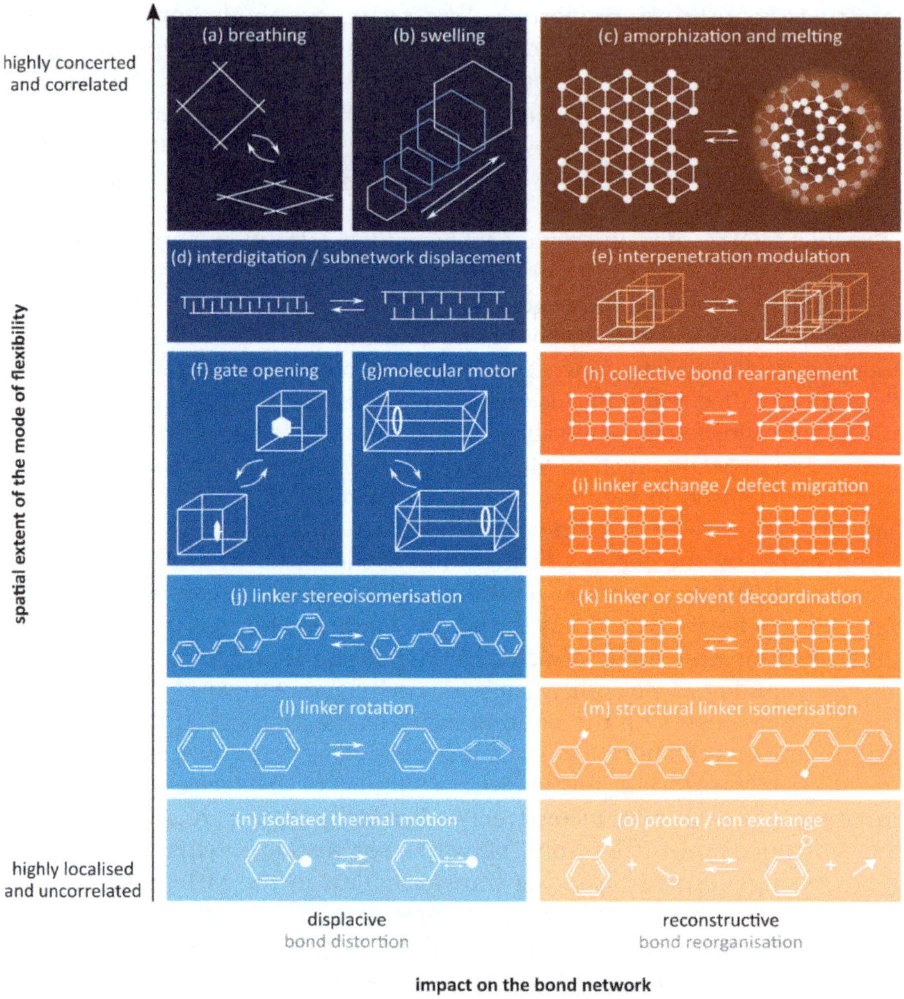

Figure 4.1 Classification of the different modes of flexibility in MOFs based on the spatial extent of the flexibility (or, conversely, its degree of correlation) on the one hand and on whether the flexibility induces bond reorganisation (reconstructive) or not (displacive) on the other hand. Depending on where the flexible phenomenon at play is situated in this diagram, different simulation approaches are necessary to adequately characterise and understand the flexibility. The different modes of flexibility classified above are not necessarily mutually exclusive, as for instance mesopore formation requires the rearrangement of bonds and migration of defects.

ZIF-4 phases precludes this classical technique.[31] Therefore, Figure 4.1 focusses on the different modes of flexibility rather than on the material in which it occurs. This also implies that the classification adopted here is equally valid for 3D MOF crystallites, 2D MOF sheets, and 1D MOF chains, as well as for other materials.

Next to the two main axes of interest in Figure 4.1, which will be discussed in detail below, also the type of external stimulus that triggers the flexibility affects which computational methods can be adopted. For instance, when studying adsorption-induced flexibility, different hybrid modelling schemes exist that account both for the inherent flexibility of the material as well as for the variation in the amount of adsorbates in the material.[32–39] Together with the structural complexities discussed in Section 4.3.1, this will require one to further select the appropriate computational tools to describe that trigger, as we will illustrate in Section 4.4 with various case studies.

4.2.1 Local *vs.* Concerted Modes of Flexibility: The Need for Larger Models

As shown in Figure 4.1, flexibility can occur very locally, in which the MOF framework only undergoes changes in a nanometre-sized region, but can also impact a much larger spatial region, up to the point at which the whole crystal may be affected. As discussed below, this spatial extent will determine which computational tools become accessible. For computational studies, it is important to distinguish between this spatial extent that we will focus on here and the amplitude with which the mode is excited. This can be well illustrated by the gate-opening phenomenon in ZIF-8.[40] The amplitude of this flexibility can be tuned by altering the type and concentration of guest molecules, whereas the spatial extent of each gate-opening event is fixed and remains limited to the direct environment of each of the six-membered ring apertures (although strong correlations between adjacent apertures exist).[40] Similarly, while the single-crystal-to-single-crystal phase transformations in MIL-53(Cr) and DMOF-1(Zn) show a different amplitude in volume contraction (*ca.* 32% for MIL-53(Cr) *vs. ca.* 48% for DMOF-1(Zn)[41,42]), the phase transformation affects the whole crystal of both materials and can hence be described by similar computational approaches.

In general, the larger the extent of the flexibility mode, the larger the structural model should be in order for it to fully describe this flexibility. For instance, finite cluster models of about 1 nanometre in diameter were adopted to describe the flexibility of the proton topology on the nodes of UiO-66 and NU-1000,[43] whereas a bigger, complete UiO-66 unit cell was necessarily used to understand the flexibility associated with the introduction of a hemilabile linker in the material[44] or when modelling the material's amorphisation.[45] However, increasing the size of the structural model also leads to a rise in the computational resources needed to perform these simulations,[2] as illustrated for instance by recent GPU-intensive simulations on a one-million-atom model of MIL-53(Al).[46] This is especially true when the timescale associated with the flexible mode under investigation becomes larger, hence, when the mode becomes slower. In that case, dedicated enhanced techniques may be necessary to fully describe the spatiotemporal evolution of the material and to understand how its spatial and temporal behaviour are intertwined, as discussed more extensively in Section 4.3.3.3.[2]

To limit the size of the structural model while mimicking the true behaviour of the bulk material more closely than cluster models, researchers ubiquitously resort to periodic boundary conditions. Under periodic boundary conditions, three unit cell vectors are defined that make up the cell matrix. By construction, any translation along a combination of these three unit cell vectors transforms the model into itself. As a result, one can predict the behaviour of a large material—in principle, assumed to be infinitely large—based on knowledge of the behaviour of the much smaller unit cell, as outlined in more detail in Section 4.3.1. While periodic boundary conditions are frequently adopted for this reason, they limit the modes of flexibility that can be expressed to those that fully fit within the simulation cell while simultaneously requiring that the mode of flexibility affects adjacent simulation cells in exactly the same way. Periodic boundary conditions are therefore known to overestimate a material's stability (in case the weakest mode of flexibility does not fit in the simulation cell and is artificially suppressed) as well as to overestimate the cooperative nature of phase transitions and the associated transition barriers.[46–50] In addition, periodic boundary conditions exclude the study of surface-dominated phenomena or size effects, which are crucial for several modes of flexibility, especially in thin films.

As discussed more extensively in Section 4.3.2, two main classes of techniques have traditionally been adopted to model the interactions in MOFs. Quantum-mechanical methods, and especially density-functional theory (DFT), are attractive approaches to extract information with (near) quantum-mechanical accuracy. However, their computational cost limits the time and length scales of the models under study. In contrast, force fields (FFs)—or interatomic potentials—allow one to study much larger systems, as the detailed electronic structure of the material is discarded in favour of predefined interatomic potentials. Despite the tremendous amount of MOF works based on these two classes of methods, three main challenges can be defined for future work.

1. Flexible MOF systems with strong electron correlations, such as DMOF-1(Cu), MIL-47(V), and MIL-53(Fe), are difficult—if not impossible—to describe within the DFT framework as DFT fails to account for strong electron correlation. As a result, more accurate QM-based methods are necessary, which need to be sufficiently inexpensive to be used for the different types of flexibility occurring in MOFs.
2. While classical force fields can describe larger models of flexibility, they are not suitable to describe reactive events and even sometimes specific adsorptive events with sufficient accuracy. Reactive force fields, such as ReaxFF, partially meet this challenge by explicitly including classical interaction potentials describing the dissociation and reassociation of bonds through a bond order formalism, although the fixed form of this dissociation term limits the accuracy of the approach.[51,52] Therefore, other methods need to be developed to obtain near-QM accuracy at near-FF length scales. In this respect, current research on

machine-learning potentials for MOFs seems to start filling this gap, although the field is only at its infancy.[53,54]

3. While the length scale attainable by force fields is larger than for QM-based methods, the largest flexible MOFs described by force fields to date are a few tens of nanometres in size. Given that many modes of flexibility greatly exceed this length scale, additional techniques are necessary to describe systems at 100 nm size and above. In this respect, coarse-graining techniques may help tackling this challenge, although also here the field remains at its infancy.[55–58]

4.2.2 Displacive *vs.* Reconstructive Modes of Flexibility: The Need for Disordered Structure Prediction and Reactive Interatomic Potentials

As mentioned above, a second major question that should be answered before choosing an appropriate computational tool to describe a given mode of flexibility is whether the mode is displacive or reconstructive. For reconstructive modes, computational models capable of describing bond breaking and formation should be adopted (which most often means DFT methods or, less frequently, reactive force fields are used), whereas displacive modes can be described by classical methods such as non-reactive force fields (referred to below as force fields). Given the larger computational cost of DFT approaches compared to force field methods, reconstructive modes of flexibility with a given spatial range are more expensive to simulate than displacive modes with the same range. For instance, the melting of ZIF-4 was studied through DFT molecular dynamics (MD) simulations, following the evolution of a 4.38 nm^3 unit cell over a time span of 65 to 225 ps, reaching an important milestone.[59] To put this into perspective, water intrusion in ZIF-8, which induces flexibility on a similar length scale as the ZIF-4 melting, could be studied on a ~21.9 nm^3 unit cell over a timespan of several nanoseconds as classical force fields were used to model the interatomic interactions.[60] To estimate the computational feasibility of studying a certain mode of flexibility, it is hence of paramount importance to distinguish between displacive and reconstructive modes of flexibility.

However, making this distinction is not always as unambiguous, not even from an experimental point-of-view. While it is experimentally rather straightforward to distinguish between amorphous and crystalline end products, or between elastic or inelastic transitions, for instance through X-ray diffraction and diverse advanced spectroscopies, respectively, this is far from being the case for the discrimination between displacive *vs.* reconstructive modes of flexibility that we are interested in here. For instance, while single-crystal-to-amorphous transitions are always reconstructive as bonds are broken during amorphisation, single-crystal-to-single-crystal transitions can, depending on the pathway followed during the phase transformation, be either displacive (as in the case of breathing in MIL-53 [28]) or reconstructive (as for instance in the case of the linker coiling in ZAG-6, which is accompanied by bond rearrangement throughout the crystal[61]). In addition, the MIL-53 breathing is a prime example of a mode that is inelastic in nature, although it retains the bond arrangement. Therefore, choosing an appropriate computational method to describe flexibility critically depends on the pathway carved between the initial and the final state of the material. Also here, a close interplay between experimental and computational tools could help to understand the elusive transition pathways between different metastable states, for instance to explain the observed 'partial amorphisation' in several frameworks.[42,62]

Based on this discussion, two major challenges can be defined to model flexibility along this axis.

1. While flexibility is often regarded as occurring between a given initial and final state, computationally interrogating a material's flexibility requires approaches that faithfully capture not only these two states, but also the full transition pathway. This is extremely challenging, since MOFs can typically follow multiple transition pathways in phase space, especially in materials for which the building blocks allow for so-called combinatorial or adaptive flexibility between multiple disordered states[63,64] In a standard MD simulation, only one transition is realised between the two states, while a proper characterisation of the transition requires a procedure to average over all accessible pathways at the given thermodynamic conditions and to understand which pathway is preferred under which conditions. In this regard, pressure *vs.* volume equations of state to model phase transitions in which the cell volume plays a critical role during the transition have been developed[65,66] but extensions to other triggers and collective variables are necessary.
2. As methods to describe reconstructive modes of flexibility are computationally more expensive, these modes are characterised to a lesser degree than displacive modes of flexibility—an exemplar of this is given by comparing the number of theoretical studies on MOF melting *versus* MOF breathing. As a result, computational methods that are cheaper than DFT approaches but more accurate than force fields—to an extent that they can be used to describe reconstructive events—are critically needed. Similar to the second challenge in Section 4.2.1, MLPs could play a pivotal role in this respect.

4.3 Theoretical Aspects on How to Model Flexible MOFs

In this section, a brief review is given on the various levels of complexity one is confronted with when modelling flexible MOFs starting from the atomic scale based on underlying physical principles. This section is written in a bottom-up approach to give a clear overview of the full sequence of a molecular modelling exercise starting from the construction of the molecular structure model towards derivation of macroscopic properties. Obviously, the field is very broad and depending on the experimental question to be tackled, other techniques need to be used. In what follows crucial elements of the modelling workflow are highlighted. To facilitate readers interested in a specific topic to easily navigate this section, we include an overview of the topics covered in this section in Table 4.1. We conclude this section with a brief overview (Section 4.3.5) on which modelling techniques might be used to derive experimental properties which specifically relate to MOF flexibility.

To set the scene, we first introduce 'the sequence of a molecular modelling exercise', which is illustrated in Figure 4.2.[67] The first step, which is especially important for the modelling of flexible MOFs is the construction of reliable

Table 4.1 Overview of all theoretical aspects and topics covered in the various subsections of Section 4.3

Section	Topics covered
4.3.1	Structural model, periodic boundary conditions, finite size effects, defects, disorder, phase coexistence, transition mechanism and mesoscale simulation
4.3.2	Born–Oppenheimer approximation, density functional theory (DFT), classical force fields (FF), machine learning potentials (MLP) and coarse graining (CG)
4.3.3	Statistical physics, thermodynamic conditions, ensemble and probability distribution
4.3.3.1	Geometry optimisation and normal mode analysis
4.3.3.2	Molecular dynamics (MD), Monte Carlo (MC) and ensemble averages
4.3.3.3	Enhanced sampling, collective variable (CV), umbrella sampling and metadynamics
4.3.4.1	Thermodynamic response properties, thermal expansion, bulk modulus, transition pressures, adsorption isotherm, osmotic ensemble and hybrid MC/MD
4.3.4.2	Thermodynamic potentials and ensembles, legendre transformation, (order of) phase transition, pressure- and adsorption-induced transition, osmotic potential and theoretical models for thermodynamic potentials
4.3.4.3	Diffusion, gate opening and transition rate

Figure 4.2 The sequence of a modelling exercise where the basic quantity is the multidimensional potential energy surface (PES) for a well-chosen molecular structure. This PES needs to be sampled in the interesting regions after which the target properties at operating conditions can be derived. Figure reproduced from ref. 67 with permission from The Royal Society Publishing, Copyright 2023.

structural models to represent the material in a realistic way. In principle, real MOFs are finite size crystals with a certain morphology and always have a certain degree of defects at various length scales (see Figure 4.3). Recent literature has clearly given evidence that spatial disorder at various length scales, may have a large impact on the flexible behaviour of MOFs. Therefore, the importance of this first step may not be underestimated when modelling flexible behaviour in MOFs.

Once a proper structural model has been defined, it is necessary to obtain an accurate representation of the potential energy surface (PES), which is a multidimensional energy surface in terms of the various nuclear coordinates. For complex materials like MOFs, the PES is a highly dimensional surface, making it rather challenging to explore. The situation can be compared to a topographical map used when hiking in mountains, which is illustrated for a hypothetical free energy map in Figure 4.9. For clarity, the PES is not the same as the free energy surface (FES). The FES is obtained by accounting for appropriate contributions rising from the partition function

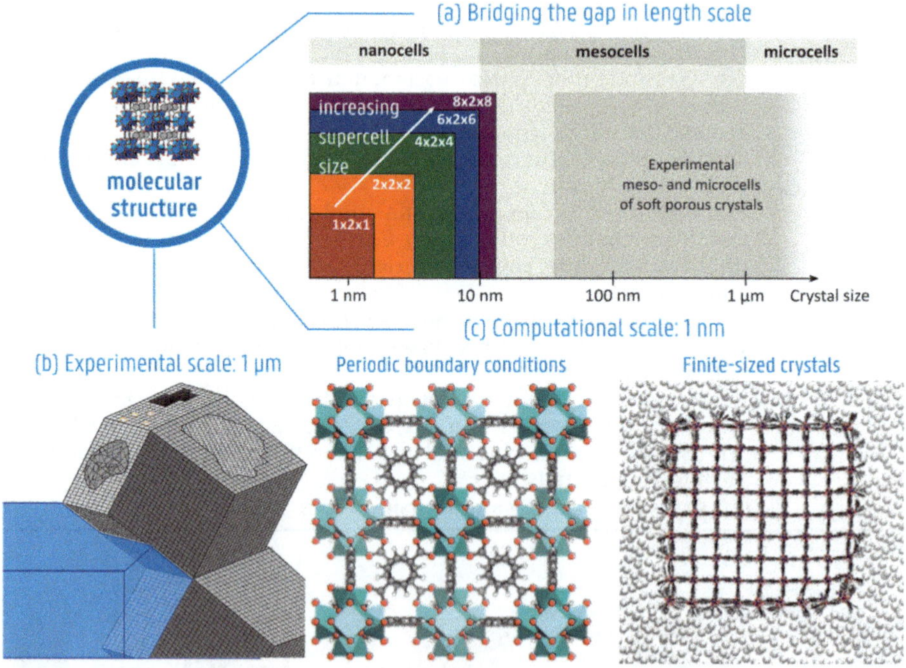

Figure 4.3 MOF molecular structure represented either using periodic boundary conditions or by means of terminated finite-size crystals. In both cases, the major challenge is to bridge the gap in length scales between feasible computational simulations (~1 nm) and experiment (~1 μm). Pane (a) is adapted from ref. 47 with permission from Springer Nature, Copyright 2019. Pane (b) is adapted from ref. 2 with permission from Elsevier, Copyright 2021. Pane (c) 'finite-sized crystals' is adapted from ref. 50 with permission from Frontiers Media S.A., Copyright 2021.

in the appropriate thermodynamics ensemble (more details are given in Sections 4.3.3 and 4.3.4.2). In principle, when interested in describing phenomena mimicking experimental conditions one should focus on the FES to find interesting paths leading the material from one state to the other, however the PES is a key ingredient in the construction of the FES (*vide infra*). When interested in going from one stable minimum on the FES to another, one needs to find the minimum free energy paths leading the system from one state to the other. For this purpose, enhanced sampling methods are required which allow one to also scan with greater efficiency lower probability regions such as activated states separating two metastable minima. For flexible MOFs such metastable states are of major importance as their existence and associated energy determine to a large extent the kinetics of phase transformations. Finally, the last step of the modelling sequence consists in extracting and interpreting the interesting properties under the right conditions. Ideally, properties are determined at conditions that may be compared to experimental findings. To link the atomistic information to the macroscopic properties, one needs thermodynamic and statistical routines. For each step of the modelling exercise, the choice of computational approaches determines to a large extent the reliability and accuracy of the computationally predicted properties. In any case, this results into a compromise between accuracy and feasible computational time given the available computational resources (*i.e.* number of central processing unit (CPU) nodes/cores, number of graphics processing unit (GPU) cores/nodes, memory, *etc.*).

Previous steps are in essence necessary for any modelling tasks in material science. As all previous aspects have a major impact on the final modelling results, we briefly discuss some key points on each of the mentioned modelling choices both in the sequence of the modelling exercise and the structural representation of the material.

4.3.1 Structural Models for Realistic MOFs

MOFs are far from perfectly ordered periodic materials and they are characterised by spatial disorder at various length scales. First of all, MOFs have a finite size varying from orders of nanometres to the micrometre scale and they exhibit a certain morphology. Furthermore, for practical usage, crystal particles need to be associated with binders or deposited to diverse supports to shape them into macroscopically stable bodies of different size. Second, at the level of the crystal particle, a broad range of spatial gradients may be present that vary from the sub-nanometre scale to several tens of micrometres. As such, one can distinguish between short-range, mid-range and long-range heterogeneities. Typical examples of short-range disorder are point defects, due to missing linkers or undercoordinated metal sites. Mid-range disorder may occur when various point defects coalescence into correlated nanoregions or when mesopores are present. Finally, long-range disorder would typically occur when various parts of the crystal have undergone a phase transformation whereas others remain in the original phase.

So far, most modelling efforts have considered rather idealistic model representations of the MOF, where the material is represented as one unit cell or a supercell, comprising various unit cells, with periodic boundary conditions. The number of atoms per conventional unit cell fluctuates to a large extent from one MOF to another, *e.g.*, 76, 456, 540 and 1728 atoms are present in the unit cells of MIL-53(Al), UiO-66, MOF-808, and DUT-49, respectively, and dictates the attainable level of theory to describe the potential energy surface, as will be explained in the next section.

Usage of periodically extended structural models has large implications for the observable effects with respect to flexible behaviour of MOFs. A quite extensive body of simulations was performed on the prototypical flexible MIL-53(Al) material. From a thermodynamic point-of-view, one can characterise flexibility by determining the Helmholtz free energy profile in terms of important collective variables, which undergo large changes during the transformation.[68] For MIL-53 materials having topological flexibility, the volume takes the role of such collective variable and many Helmholtz free energy profiles were determined for MIL-53 type materials using force field-based modelling techniques.[69–72] Simulations performed on relatively small periodically extended systems systematically predicted a collective mechanism for the transformation from one phase to the other. In this case the entire system cooperatively transforms between the two phases. However, when larger periodically extended supercells were used having sizes in the mesoscale range and having *ca.* 10 000 atoms in their supercells, it was observed that phase transformations did not occur cooperatively anymore.[47] Instead, parts of the system undergo a large pore to a contracted pore transition whereas other parts remain temporarily in the original large pore phase. In this case the dimensions of the system are sufficiently large to allow a sort of spatial disorder within one supercell. These observations were confirmed and extended with a consideration of a million number of atoms simulation with the MIL-53(Al) system exploiting the massive parallelism of GPUs using the OpenMM software package and by implementing a new pressure control algorithm to allow for fully anisotropic unit cell fluctuations.[46] When subjecting the system to external pressures, the behaviour shown in Figure 4.4 was observed. In the lower pressure regime, the system transforms in a layer-by-layer mechanism, whereas at higher pressures the transition initiates at discrete nucleation points, with domain formation. This case study clearly emphasised the importance of the structural model on the simulation results regarding flexible behaviour.

In principle, one does not need to assume periodic boundary conditions, but instead finite sized nanocrystallites could be considered within a simulation. Such approach was applied to model nanocrystallites of DMOF-1(Zn) by setting up a crystallite model containing *ca.* 250 000 atoms, as illustrated in Figure 4.3(c). In this case the effects of the external surface are explicitly included in the calculations.[48] These simulations showed that the transition between two phases nucleates near the external surface and then propagates further through the crystal until a full phase transformation has taken place.

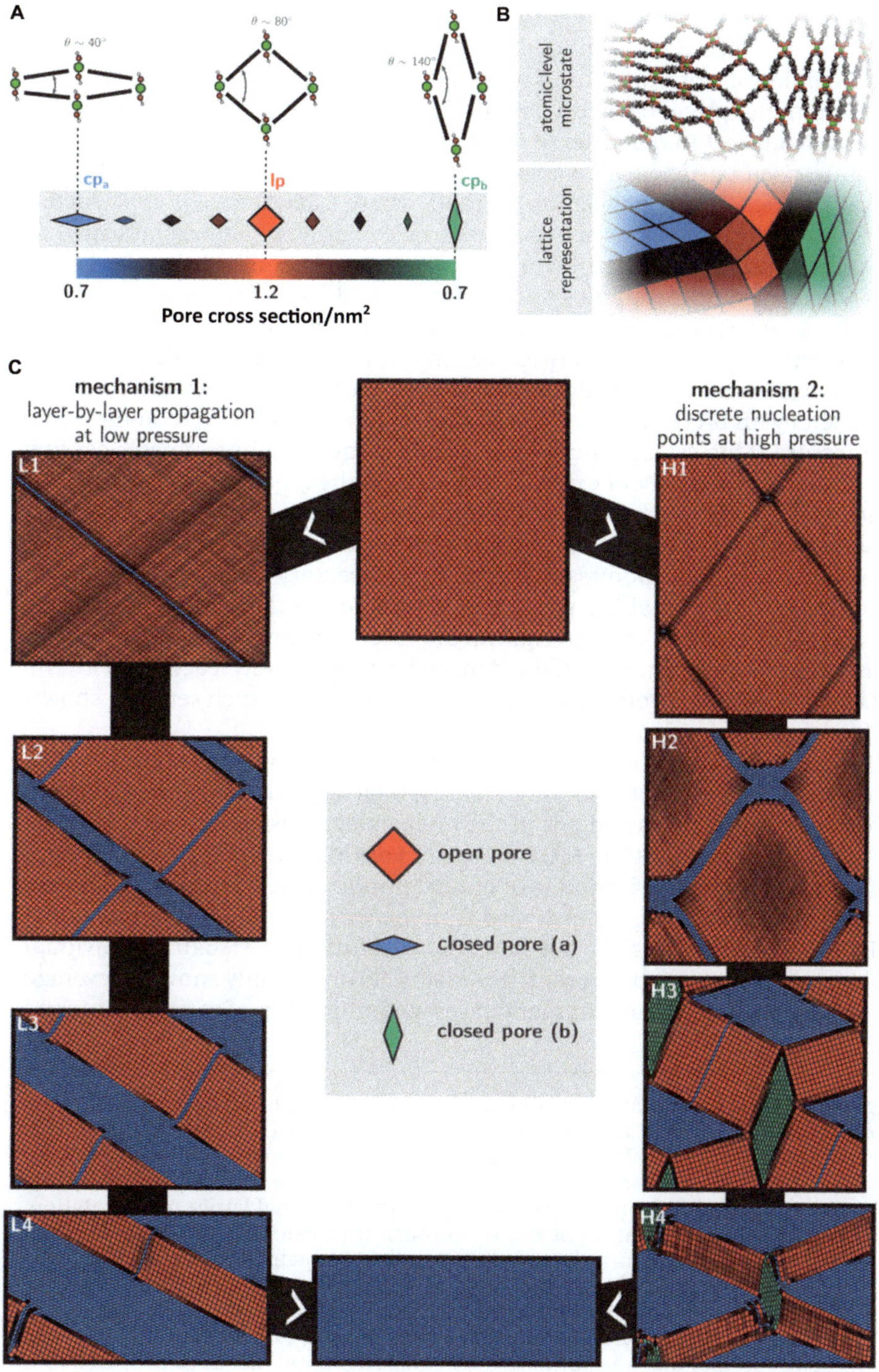
A
θ ~ 40°
θ ~ 80°
θ ~ 140°
cp_a
lp
cp_b
0.7
1.2
0.7
Pore cross section/nm²
B
atomic-level microstate
lattice representation
C
mechanism 1:
layer-by-layer propagation at low pressure
mechanism 2:
discrete nucleation points at high pressure
L1
L2
L3
L4
H1
H2
H3
H4
open pore
closed pore (a)
closed pore (b)

(continued)

Using such an approach is very interesting, but immediately poses the question on how to configure initial structural models for such finite sized crystals. Simulations on finite sized nanocrystallites are very scarce in the MOF field.[48] Smaller metal particles have been simulated with DFT, but their sizes are much smaller in the order of 5 nm.[73,74]

Finite sized particle with a certain morphology can be constructed using the atomistic Wulff construction method, where a 3D surface is constructed by determining crystal faces with low surface energy, which can be scaled to any crystal size.[73,75,76] To truly mimic experimental set-ups finite sized crystals with a specific morphology need to be embedded in an external medium which can exert in a more natural way an external pressure or simulate a heat bath. Such an approach was recently implemented with the use of an external pressure by embedding the crystal particle in a periodic box surrounded by an inert medium,[48] which could in reality resemble noble gases or mercury. This set up would largely mimic mercury porosimetry set-ups used to study the effect of external pressure on the mechanical behaviour of materials.[77] When setting up such system, one also needs to define force field parameters that mimic the interaction between the external medium and the crystal particle. Some points of attention are necessary here, *e.g.*, they should be of those kind to enable a phase transformation of the crystal particle but on the other hand as the external medium is inert and does not enter the pore system, they should be defined in such a way to avoid reactivity with the MOF crystallite. An overall schematic representation of such set up is shown in Figure 4.3(c).

Previous structural models, rely on representations of perfect crystals, where no defects are introduced. In reality, MOF crystals have different types of disorder and defects and one of the main challenges is to construct realistic models to represent the MOF material. For this the link with experimental structural characterisation is of crucial importance. The characterisation of defects is a challenging task also from experimental point-of-view. Most microscopy techniques do not have the resolution to visualise individual point defects in these beam sensitive materials or can only show the average of defects on a mesoscale. In many cases a combination of various advanced

---

Figure 4.4 *(continued)* (A) Quadrilateral representation of individual one-dimensional channels within the winerack topology of MIL-53(Al). The colour of each quadrilateral is determined based on the instantaneous cross-sectional area of the corresponding channel (B) Example configuration of the framework and corresponding two-dimensional lattice representation. (C) Visualisation of the two observed transition mechanisms for the lp–cp transformation in MIL-53(Al), as explained in Section 4.2.2. For lower pressures, a layer-by-layer mechanism is preferred in which the rhombus shape of the pores is approximately preserved throughout the transition (left). For higher pressures, we observe discrete nucleation points and the formation of concerted-pore and large-pore domains (right). Reproduced from ref. 46 with permission from Frontiers Media S.A., Copyright 2021.

experimental techniques is required to fully characterise the material and to give valuable input for the modeller to generate initial plausible models.

The synergy between experimental characterisation and modelling efforts can nicely be illustrated for the UiO-66 material, where besides the standard nuclear magnetic resonance and thermogravimetric analysis techniques, advanced structural characterisations as well as high resolution imaging techniques were deployed, including pair distribution function analysis on high quality X-ray diffraction (XRD) patterns[78] and scanning electron diffraction in particular to characterise defective nanodomains.[79] Starting from such experimental information, molecular models for these defective materials can be generated. An elegant way for the *in silico* construction of MOF structure models is to maximally exploit the topology of this family of materials. Many groups constructed a large variety of hypothetical MOF structure databases starting with nets available from the Reticular Chemistry Structure Resource (RCSR) and mapping onto these nets diverse building blocks.[80,81] From the resulting ideal structure model, one needs to introduce structure defects in line with the experimental evidences in most of the MOFs. UiO-66(Zr), built from 12-connected Zr-bricks, linked together in the **fcu** topology is regarded as the prototypical defective MOF. To create the missing node defect model, one needs to remove one Zr-brick leading to the **reo** topology (Figure 4.5(B)). Recent extensive characterisations gave also evidence for the existence of **bcu** and **scu** topologies in defective materials. **Bcu** topologies are obtained by 8-connected missing linker defects, **reo** by 8-connected missing cluster defects whereas **scu** by a combination of 4,8 connected missing-cluster defects. These defects can be introduced in a systematic way through dedicated experimental procedure,[82] as illustrated in Figure 4.5(A).

Previous elaborations clearly pointed out that modelling realistic MOF crystals having dimensions comparable to experimental crystal sizes in the mesoscale, necessitates the development of advanced techniques to build realistic structural models for the MOF crystals, having finite size and with inclusion of spatial disorder at various length scales. In principle this should

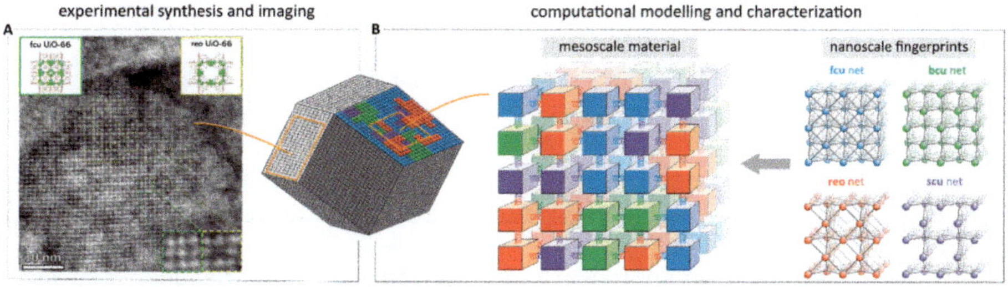

Figure 4.5 Definition of mesoscale and nanoscale cells. (A) Comparison with experimental transmission electron microscopy (TEM). (B) Schematic illustration of various procedures to build mesoscale systems containing several nanoscale fingerprints. Pane A reproduced from ref. 82 with permission from the American Chemical Society, Copyright 2021.

be done in a close feedback loop with experimental work, where the material can be characterised at various length scales and with various resolutions to give validation for the proposed structural model.

4.3.2 Potential Energy Surface

Once a proper structural model has been constructed, one needs to select a method to model the interactions between the framework atoms, for which various techniques are available as illustrated in Figure 4.6. In most

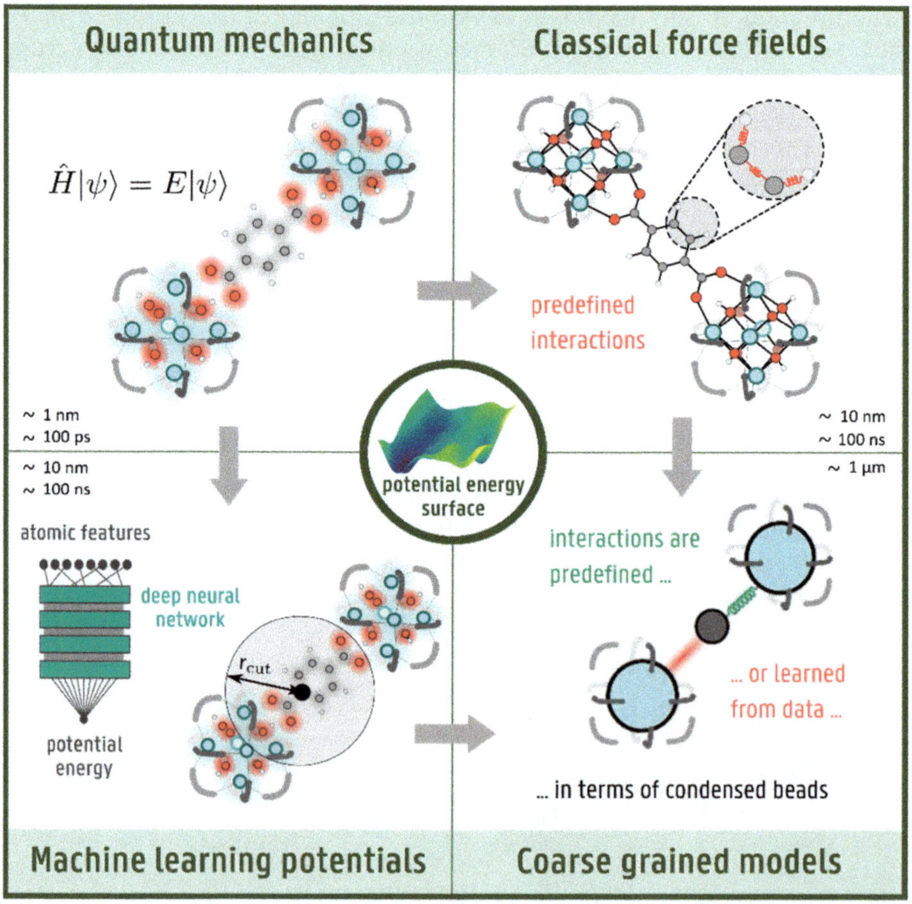

Figure 4.6 Four different approaches for describing the potential energy surface: quantum mechanical approach involving solving the Schrödinger equation, classical force fields involving parameterised springs between atoms, machine learning potentials trained to mimic certain input data, and coarse-grained models defined as interactions between condensed beads. Figure reproduced from ref. 67 with permission of The Royal Society Publishing, Copyright 2023.

simulations, one assumes the Born–Oppenheimer approximation where the dynamics of the electronic degrees of freedom is separated from the nuclear motion. The underlying principle behind this approximation is that the motion of the electrons is much faster than the motion of the nuclei, given the large mass difference. Within the Born–Oppenheimer approximation, one constructs the PES by solving the electronic Schrödinger equation for fixed nuclear coordinates. To obtain the complete PES, one in principle needs to calculate the energies and forces along all nuclear degrees of freedom, which is a computationally very expensive task given the multidimensional nature of the PES for complex materials like MOFs.

As mentioned before, one in principle needs to use quantum mechanical based methods to determine the interatomic forces and construct the PES. DFT is by far the most popular method for this purpose thanks to its attractive trade-off between accuracy and speed. However, the accuracy of DFT methods critically depends on the exchange-correlation functional[83–86] and the description of long-range dispersion interactions.[87–91] The attainable length and time scales of current simulation methods are dictated by the level of theory used to describe the PES. When residing to DFT-based methods, the attainable length scales are in principle limited to the nanometer and MD runs extend to a few hundreds of picoseconds. In a recent contribution, the terminology spatiotemporal phenomena was coined, referring to the correlation between spatial gradients and the observed dynamics of the system.[92] When using DFT-based methods, attainable spatiotemporal windows are much too small to mimic experimental flexible behaviour or realistic MOF crystals. In an attempt to increase length and time scales of realistic systems, classical force fields have been extensively used, where the interactions between the atoms are approximated by simple analytical functions, neglecting the quantum description of the electrons (see upper right panel of Figure 4.6).[93–96] Within the field of MOFs, classical force fields have been extensively used to simulate flexible behaviour of MOFs.[70,95,97–102] In many cases such force fields were used to determine Helmholtz free energy surfaces along some collective variable which is representative for the phase transformation, some representative examples will be given in Section 4.4. Many types of force fields have been suggested in literature, ranging from force fields derived specifically for a single MOF[70,97–99] to force fields applicable to a much wider range of MOFs.[95,100–102] While most force fields start from a fixed molecular topology and have fixed atomic charges, some are able to describe polarisation[103,104] or even chemical bond breaking and generally referred to as reactive force fields.[51,105] Several groups have developed a general methodology to derive force fields from *ab initio* input.[93,94,96,106–109] The main drawback of classical force fields is their lack in accuracy compared to QM methods, lack of transferability, *e.g.*, force fields derived under certain thermodynamic conditions are not necessarily applicable in other operating windows, the inability of most force fields to simulate bond formation/breakage, and in some specific cases, like for MOFs incorporating open metal sites, they fail to describe the host–guest interactions.

An interesting direction which is in full exploration is the development of machine learning potentials (MLPs), where a numerical potential is generated with on a non-linear regression method based on underlying quantum mechanical training data (see lower left panel of Figure 4.6). MLPs hold the promise to calculate much more efficiently the PES with an accurate matching of the underlying QM data, however, within the field of MOFs there are surprisingly few research reports on MLPs developed for MOFs. For MOF-5 a Behler–Parrinello neural network was developed based on *ca.* 20 000 reference QM configurations.[53] There are various challenges to tackle when trying to extend MLPs for flexible MOFs. One major challenge consists in finding systematic ways to generate an extensive and representative QM dataset that enables the description of MOFs close to but also far from equilibrium. The latter is necessary to describe phase transformations and other phenomena with large deformations. When compared to small molecules and other solids, MOFs are inherently large periodic systems with often a few hundred of atoms in their unit cell, so that generating the QM dataset is computationally the heaviest step when deriving MLPs for MOFs. Interestingly, the field of MLP development is in full exploration and very efficient new mathematical frameworks are being devised, which allow to drastically reduce the number of underlying QM data points. A very recent work succeeded in generating an accurate and transferable MLP for both rigid and flexible framework materials, requiring at least two orders of magnitude less input data with respect to the state of the art.[54] The methodology is built around the Neural Equivariant interatomic potential (NequIP), which is a message passing neural network that achieves remarkable data efficiency due to its use of rotationally equivariant features in the characterisation of atomic environments.[110] Figure 4.7 illustrates the proof-of-concept of this methodology with the prototypical UiO-66, showing that with NequIP, only a few hundreds of reference configurations are necessary to obtain a mean average error on the forces below 20 meV Å^{-1}, which is regarded as highly accurate in the field. While being very promising, the extension of those concepts to MOFs with defects

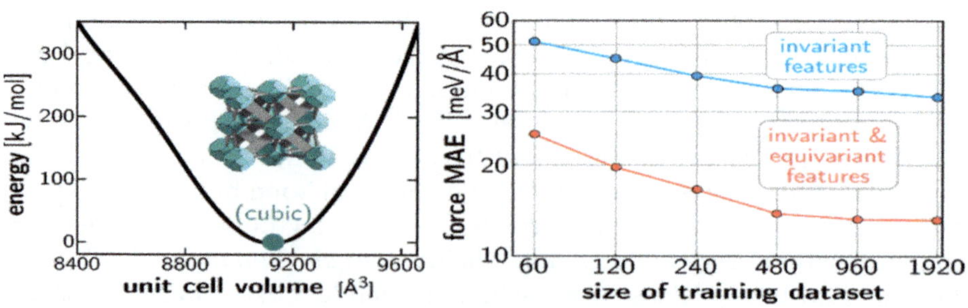

Figure 4.7 (Left) Pressure *vs.* volume curve for UiO-66(Zr) from which training data are extracted to train MLPs. (Right) Test error as a function of the size of the training data set with rotationally invariant (blue) and rotationally equivariant models (red). Figure reproduced from ref. 67 with permission of the Royal Society Publishing, Copyright 2023.

and including guest molecules necessitates further method development. To account for localised defects, one could still use a cluster-based approach, as was done by Eckhoff and Behler in generation of their MOF-5 MLP force field.[53] It would be necessary to include sufficient defective clusters in the underlying QM data set.

Another major point of attention is how to describe the host–guest interactions and more generally long-range interactions. Treatment of long range interactions is a major challenge within the field of MLPs.[111] The first papers describing guests@MOFs with MLPs are now appearing.[53,112,113] In this case it will be essential to include representative structures in the training data set that enable a proper host–guest interaction description. The problem is rather complex as the particular interactions will depend on the degree of loading within the confined environment. Efficient MLP algorithms will have to be set up where also concepts such as active learning will be adopted to train MLPs in an iterative way and to include more training data on the fly.[114–116] Finally, when interested in developing MLPs for MOFs having mesopores, it will probably be necessary to complement MLPs which in essence describe short-range interactions with physical models to describe longer range interactions such as electrostatics. Treatment of long-range interactions is still a major challenge within the field of MLPs and it will be necessary for researchers interested in simulating MOFs to closely follow recent development on the more methodological side within the MLP community.[117]

Finally, when interested in describing phenomena on even larger length and time scales than the ones accessible with classical force fields or MLPs, one can reside to coarse-grained (CG) methods. In this case, one abandons the atomic resolution and groups a set of atoms in effective particles—also called CG beads—which interact with each other by means of effective interactions (see lower right panel of Figure 4.6). As one greatly reduces the number of degrees of freedom and orders of magnitude in accessible scales can be bridged. CG methods within the field of nanostructured materials are nearly unexplored so far. Typically, a CG force field for HKUST-1 was developed to describe the stability of mesopores in the material.[57,118] Depending on the CG resolution, semi-quantitative predictions were obtained for mechanical and structural properties, however, macroscopic effects depending on the local deformations such negative thermal expansion could not be described properly. To apply CG methods to flexible materials, one would need methods to systematically map atomistic representations to CG beads, where a balance is found between the sparseness of the CG representation and the attainable resolution in CG space. One major problem with the usage of CG methods, is the loss of physically relevant interpretation of the dynamics of the system. As one removes the highest frequencies from the system, CG particles experience less friction and are therefore artificially accelerated. From a conceptual point-of-view, methods are available that could be explored in the MOF field, for example one could introduce some additional fictitious particles that interact with the CG beads.[119,120] However, those methods were applied on much simpler systems than MOFs and their applicability within the MOF field needs to be tested.

4.3.3 Phase Space Sampling Protocol

Once the molecular model and level of theory have been selected, the remaining ingredient required for a molecular simulation is a protocol to generate molecular configurations representative for the material at hand under a given set of conditions. To understand the physical significance of these configurations, we need to apply statistical mechanics which reveals that thermodynamic properties are expressed as statistical averages of their corresponding microscopic quantities. The central concept herein is the generation of a statistical ensemble made of a large number of replicas of the system at hand (see Figure 4.8). Each replica satisfies identical thermodynamic boundary conditions (*e.g.*, control of temperature pressure, *etc.*) but exhibits fluctuations on its microscopic properties (*i.e.*, energy level in a quantum mechanical treatment or atomic coordinates and momenta in a classical treatment, *vide infra*). These fluctuations are mathematically described through a probability distribution function whose form depends on the specific ensemble, *i.e.*, on the specific set of thermodynamic boundary conditions. The most widely used ensembles are graphically depicted in Figure 4.8.

Considering the *NVT* ensemble as an example, also denoted as the canonical ensemble, where each replica has identical number of particles N, occupy a same volume V and can exchange heat with a heat bath at the same temperature T. However, two replicas from this *NVT* ensemble usually have a different internal energy. The fluctuations on the energy within this canonical ensemble is described by the so-called Boltzmann probability distribution:

$$p_r^{NVT} = \frac{g_r}{Z} e^{-\beta E_r} \text{ with } Z^{NVT} = \sum_r g_r e^{-\beta E_r} \tag{4.1}$$

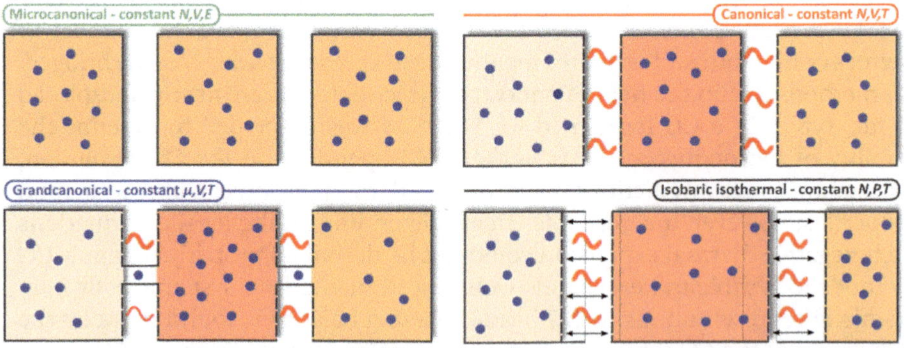

Figure 4.8 Graphical depiction of the most widely applied ensembles as defined in statistical physics. For each ensemble, three replicas of the system are depicted with a visual indication of their internal energy (background colour), volume (black edge) and number of particles (blue dots) as well as exchange of heat (red waves) work (black arrows) and particles (canals between replicas). The $NP\sigma_a = 0T$ ensemble,[65] which is not depicted in the figure, is an extension of the isobaric isothermal ensemble to solids and non-ideal liquids, in which not only the unit cell volume, but also its shape is allowed to fluctuate.

Herein, we assumed that the dynamics of the system is described quantum mechanically resulting in a set of allowed energy levels $E_r(N,V)$ that depend on the number of particles N and volume V. Furthermore, g_r is the degeneracy of the quantum state E_r, $\beta = \frac{1}{k_B T}$ is a measure for the temperature, and Z is the canonical partition function. In the following paragraphs, we outline how various techniques can be applied to generate molecular configurations to construct such a partition function as well as how to use it to compute thermodynamic observables.

4.3.3.1 Static Approach

In the static approach, one considers a single molecular configuration corresponding to the global minimum of the PES and, hence, the ground state of the system. In principle, this ground state energy corresponds to the energy at a temperature of absolute zero and is a crude approximation at elevated temperatures. As a possible second step, however, the Hessian matrix containing the second-order derivatives of the PES towards the nuclear coordinates evaluated at the global minimum is computed and diagonalised to extract the normal modes and frequencies. These frequencies allow one to approximate the energy levels in the partition function (eqn (4.1)) by those of uncoupled harmonic oscillators and estimate the corresponding free energy, entropy, and any other thermodynamic property at elevated temperature. Unfortunately, in many cases such a static approach cannot adequately capture the behaviour of flexible MOFs at finite temperature. On the one hand, flexible MOFs usually have a highly non-harmonic PES implying that the harmonic oscillator approximation cannot be applied beyond a vanishingly small region around the minimum. On the other hand, the PES of flexible MOFs might also feature many local minima with comparable energies instead of a single well-defined global minimum, resulting in multiple phases whose relative stability needs to be considered. To properly account for such dynamic flexibility, one can apply dynamic sampling techniques.

4.3.3.2 Dynamic Sampling

To overcome the issues related to a static approach, one resorts to dynamic sampling of the potential energy surface using techniques such as MD and Monte Carlo (MC) as illustrated in the top panel of Figure 4.9. In both techniques one usually switches from a quantum mechanical description of the nuclear degrees of freedom to a classical treatment.[‡]

[‡]Whenever the MOF contains hydrogen atoms whose behaviour is crucial for the process under investigation, such classical approach is no longer plausible and one needs to apply more involved techniques such as path integral molecular dynamics.

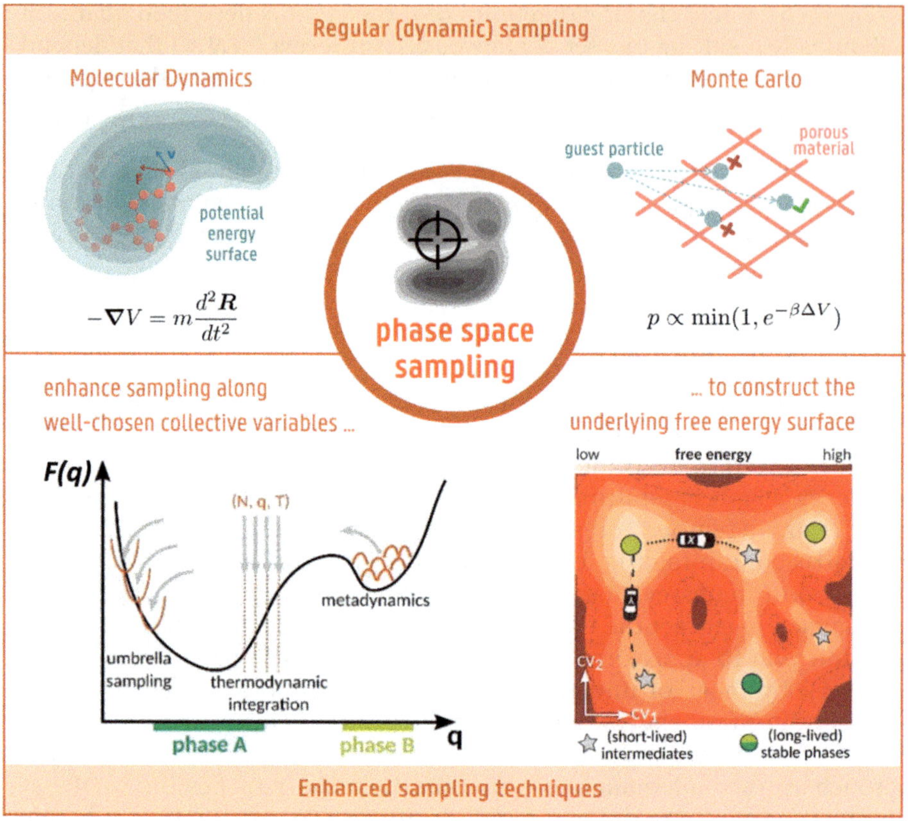

Figure 4.9 Illustration of various techniques to sample the potential energy surface. Molecular dynamics tracking the systems time evolution through integration of Newton's equation of motion (upper left). Monte Carlo in which random moves are proposed and accepted/rejected through importance sampling (upper right). Advanced techniques to enhance sampling along well-chosen reaction coordinates (*e.g.*, umbrella sampling, metadynamics and thermodynamic integration) (bottom left) which in turn allow to construct the free energy surface and identify the (meta)stable states as well as pathways connecting these states (bottom right). Bottom panes of the figure adapted from ref. 2 with permission from Elsevier, Copyright 2021.

Herein, the potential energy is expressed as function of the positions of all nuclei (denoted as $\boldsymbol{R}$) and as a result, the probability distribution function can be expressed as a continuous function over nuclear configurations[§] $\boldsymbol{R}$:

$$p^{NVT}(\boldsymbol{R}) = C \cdot e^{-\beta V(\boldsymbol{R})} \tag{4.2}$$

[§]In principle, the probability distribution is also a function of nuclear momenta. However, whenever one is interested in properties that do not depend explicitly on the momenta, they can be integrated out. Therefore, we will ignore the momenta here.

where C represents a constant related to the partition function and normalises the probability distribution function. If we now consider a physical property whose microscopic value $A(\boldsymbol{R})$ depends on the nuclear configuration $\boldsymbol{R}$, we can express the thermodynamic ensemble average $\langle A \rangle$ as:

$$\langle A \rangle = \int A(\boldsymbol{R})\, p(\boldsymbol{R})\mathrm{d}\boldsymbol{R} \tag{4.3}$$

Unfortunately, this expression is not very practical, as the integral requires to scan all configurations and compute their corresponding probability $p(\boldsymbol{R})$ and $A(\boldsymbol{R})$ value, which is highly unfeasible given the dimensionality of the configurational space.¶ Furthermore, many (if not most) of these configurations $\boldsymbol{R}$ will correspond to unrealistic situations (*e.g.*, with overlapping atoms) with a very high energy and hence a negligible probability $p(\boldsymbol{R}) \approx 0$. As such, they do not contribute to the thermodynamic average in eqn (4.3). MD and MC simulations overcome this issue by generating a set of configurations $\boldsymbol{R}_i$ which represent probable samples taken from the probability distribution function $p(\boldsymbol{R})$ of the ensemble of choice. These samples are generated subsequently, in which each configuration follows from the previous one either by integrating Newton's equation of motion (MD) or by proposing and accepting/rejecting random moves according to importance sampling (MC). For more details on the MD and MC algorithms, the readers are referred to dedicated textbooks.[121] Finally, the thermodynamic ensemble average from eqn (4.3) can be rewritten by applying the so-called ergodic hypothesis, stating that ensemble averages over replicas are identical to dynamical averages of a single replica. Hence, the series of molecular configurations $\boldsymbol{R}_i$ generated by MD or MC of a single replica can be used to approximate the thermodynamic ensemble average as a sample average:

$$\langle A \rangle \approx \frac{1}{M}\sum_{i=1}^{M} A(\boldsymbol{R}_i) \tag{4.4}$$

where M represents the total number of configurational samples. In the limit for an infinite number of samples, this sample average reduces to the exact thermodynamic average whenever the ensemble and its sampling protocol are ergodic. The regular MD and MC simulations outlined in this section are adequate techniques for sampling the potential energy surface of systems that remain in the close neighbourhood of a local minimum and extract thermodynamic properties. As such, one can compute properties including the thermal expansion coefficient and bulk modulus among others as will be further outlined in Section 4.3.4.1.

¶To illustrate, consider a molecule consisting of 10 atoms and assume you construct a 3D cartesian grid for each atom consisting of 10 × 10 × 10 points (which is extremely coarse). The integral in eqn (4.3) would then already require 1000^{10} evaluations of the energy, which is completely unfeasible.

4.3.3.3 Enhanced Sampling Techniques

In many cases flexible MOFs undergo transitions from one local minimum to another under influence of various triggers such as temperature or mechanical pressure as discussed in Section 4.2. These transitions are activated processes with a very low probability to occur in standard MD or MC simulations because they require crossing an intermediate state with very high energy and hence very low probability. As a result, adequately simulating such processes requires to enhance the configurational sampling along the direction of the transition. Various so-called enhanced sampling techniques exist, and they can be categorised into two main groups based on whether or not they require the definition of a reaction coordinate (RC). In current literature, the terminology collective variable, order parameter and reaction coordinate are often used interchangeably. In this respect, the convention of Peters *et al.*[122] defines a collective variable as any function of the molecular configuration $\boldsymbol{R}$, an order parameter as a collective variable that is able to distinguish between various (meta)stable states and a reaction coordinate as an order parameter able to describe the dynamical progress along the process of interest. However, to avoid unwillingly emphasising chemical reactions, we will here refer to the relevant degree of freedom that can describe the dynamical progress as collective variable, as is also commonly done in literature. For some forms of flexibility, it is straightforward to find an adequate collective variable to describe the transition. For example, for pressure-induced breathing in MIL-53(Al) from an open pore to a contracted pore, it has been reported that the unit cell volume is a good collective variable and the free energy profile $F(V)$ can be constructed using the enhanced sampling technique of choice.[65,72] However, for more complex forms of flexibility such as the linker-flip of CAU-13, illustrated in Figure 4.10, it is much less trivial to identify proper collective variables.[123] Ongoing research investigates the use of machine learning techniques to automatically determine appropriate collective variables.[123] Once such collective variable is known, enhanced sampling techniques such as metadynamics,[124,125] umbrella sampling[126] or thermodynamic integration[127] can be applied to construct the free energy profile $F(q)$ in terms of the collective variable q as illustrated in Figure 4.9.

Metadynamics accomplishes this by means of a non-equilibrium simulation in which bias potentials are added to the PES on the fly during the simulation (indicated by orange hills in the bottom left pane of Figure 4.9) to stimulate the system to visit other states. After convergence, *i.e.*, when each state in terms of the collective variable is visited with equal probability, the collection of bias potential can be inverted to obtain the unbiased free energy profile. Umbrella sampling consists of a series of independent equilibrium simulations with predetermined fixed bias potentials (indicated by the orange parabolas in the bottom left pane of Figure 4.9) which can be combined afterwards to obtain the unbiased free energy profile by means of, *e.g.*, the weighted histogram analysis method (WHAM).[128] Thermodynamic integration also consists of a series of independent equilibrium simulations,

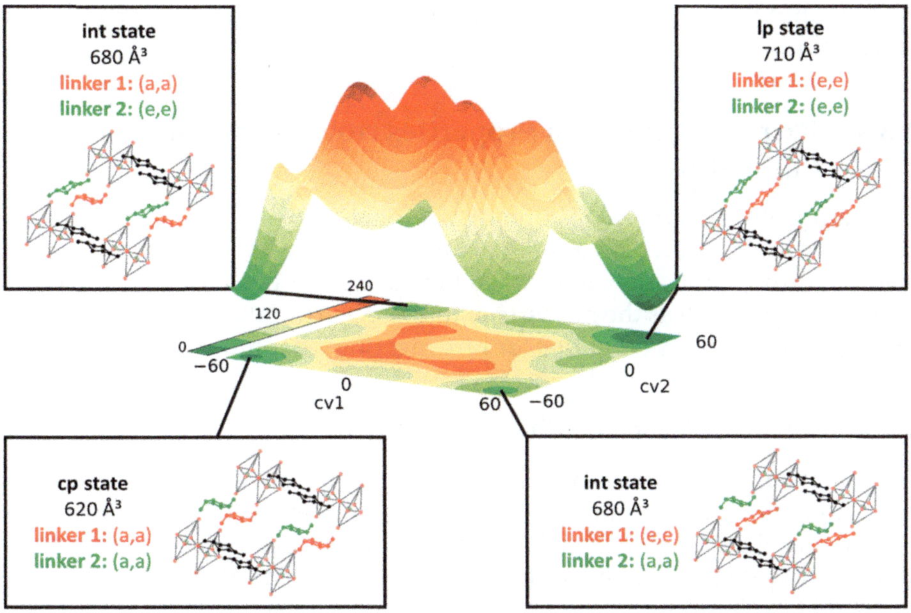

Figure 4.10 Free energy surface (in kJ mol^{-1} and obtained with variationally enhanced sampling, an enhanced sampling method) of CAU-13 at 300 K as a function of collective variables corresponding to the ring inversion in CAU-13. The insets schematically depict the folding pathways of CAU-13 as well as well as the corresponding unit cell volumes. Adapted from ref. 123 with permission from the American Chemical Society, Copyright 2018.

but instead of biasing the system, it is constrained to be at a specific value of the collective variable (indicated by the dashed vertical lines in the bottom panel of Figure 4.9) and afterwards the derivative of the free energy towards this collective variable is integrated over the simulated range. Even though many other techniques exist, such as variationally enhanced sampling,[129] which have furthermore been tested for efficiency and accuracy in determining the free energy profile of breathing MOFs,[72] we do not aim at giving a detailed description of these methods here.

Once the free energy profile is known, all relevant thermodynamic properties can be extracted as will be discussed in Section 4.3.4.2. Finally, alternative techniques exist that do not require any prior definition of a collective variable. One such example is replica exchange,[130] in which multiple simulations are run in parallel but at different thermodynamic conditions and according to an appropriate Boltzmann factor, the state of the systems are interchanged between the various replicas. Another example is transition path sampling,[131,132] where an ensemble of transition paths is generated by means of shooting moves from an initial pathway connecting the reactant and product region of the transition. Although such methods are attractive from a conceptual point of view, they are computationally very expensive and therefore not always feasible.

4.3.4 Extracting Physical Properties from Molecular Simulations

Estimating physical properties from the output of molecular simulations can be either as straightforward as taking a simple sample average or much more involved through the application of statistical physics depending on the type of simulation as well as the property of interest. In the following subsections, we discuss the various scenarios for investigating flexibility in MOFs.

4.3.4.1 Direct Response of Flexible MOFs

Some of the most straightforward physical properties to extract from molecular simulations that relate to flexibility are the unit cell parameters/unit cell volume changes alongside local geometric features variations (bending and dihedral angles, *etc.*), bulk modulus and (volumetric) thermal expansion coefficient as they reflect the direct structural response of the MOF towards variations in applied pressure or temperature. To estimate these properties (as illustrated on Figure 4.11), it is sufficient to perform MD simulations in the $NP\sigma_a = 0T$ ensemble, which allows the MOF to adapt its unit cell volume V and shape towards the applied temperature T and pressure P, and extract the equilibrium volume as the average of the instantaneous volume.

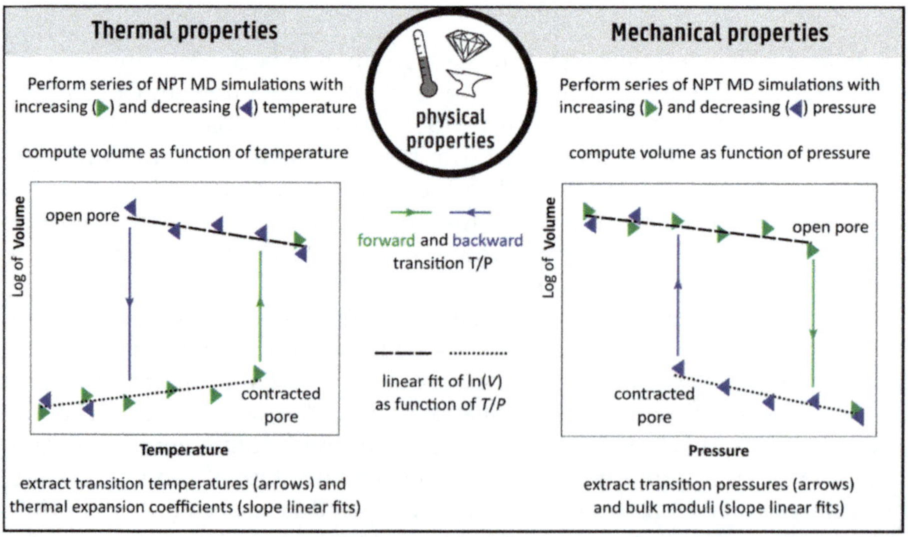

Figure 4.11 Examples on how to extract thermal and mechanical properties directly from molecular simulations.

By performing a series of MD simulations over a range of pressures and/or temperatures and extracting the equilibrium volume, one gains access to the $V(P)$ and $V(T)$ reponses. The bulk modulus B and thermal expansion coefficient α_V can then be obtained by means of the following thermodynamic relations:

$$B = -V\left(\frac{\partial P}{\partial V}\right)_{N,T} \tag{4.5}$$

$$\alpha_V = \frac{1}{V}\left(\frac{\partial V}{\partial T}\right)_{N,P} \tag{4.6}$$

Hence, by plotting the logarithm of volume ($\ln V$) as function of pressure/temperature, fitting a straight line (as indicated in Figure 4.11 by the dashed and dotted lines for open and contracted pores respectively) and computing the slope, one can determine the bulk modulus and thermal expansion coefficient. Using this approach, molecular simulations can be used to reveal that flexible MOFs are typically characterised by low bulk moduli while some others exhibit the counterintuitive phenomenon of negative thermal expansion (*i.e.*, a material with $\alpha_V < 0$ which shrinks in volume upon heating).[133–135] The same approach can also be applied to investigate pressure- or temperature-induced breathing transitions by sweeping the applied pressure or temperature over larger ranges and plotting the $V(P)$ or $V(T)$ profile to detect sudden decrease/increase in volume (indicated by the arrows on Figure 4.11). However, considering the pressure-induced transitions in particular, one has to be careful with the nature of the observed transitions as they can originate not only from the thermodynamic driving force (*i.e.*, the equilibrium pressure) but also from large fluctuations in the instantaneous pressure.[65] Such large instantaneous fluctuations originate from an (unrealistically) small unit cell size of the system. In order to obtain more reliable estimates of the breathing transitions, as well as to gain access to the energetics of the transition, one should construct the underlying free energy profile as outlined in Section 4.3.4.2.

As discussed previously, various other breathing triggers exist apart from mechanical pressure and temperature. One of the most often investigated triggers is guest adsorption. Unfortunately, simulating adsorption-induced breathing is more complicated as it requires an ensemble in which both the number of particles N as well as the unit cell volume and shape fluctuates, *i.e.*, the osmotic or μPT[l] ensemble, and it is highly non-trivial to have a fluctuating number of particles in MD or a dynamically vibrating framework in MC. Two approaches are possible to deal with these issues. The first is combining MD and MC in a so-called hybrid MC/MD simulation,[38,39] the second approach first parameterises[69,136] or constructs[68] free energy profiles in the canonical (NVT) or grand canonical (μVT) ensemble and then translates to the desired osmotic (μPT) ensemble by means of the

[l]More accurately we should denote it as the $N_{\text{cell}}\mu_{\text{guest}}PT$ ensemble, as we need at least one control variable that fixes the system size (which is the number of MOF unit cells N_{cell} here).

Legendre transformation. We will briefly discuss the first method here, for the second method we refer to Section 4.3.4.2. Hybrid MC/MD simulations combine the sampling of fluctuating number of adsorbed guest molecules from MC with the dynamic sampling of atomic vibrations and framework flexibility from MD. Let us first consider the response of flexible materials in terms of guest loading (commonly denoted as the adsorption isotherm) or unit cell volume, which can be computed as first order derivatives of the osmotic potential X:

$$N(\mu,P,T) = -\left(\frac{\partial X}{\partial \mu}\right)_{P,T} \tag{4.7}$$

$$V(\mu,P,T) = \left(\frac{\partial X}{\partial P}\right)_{\mu,T} \tag{4.8}$$

In a first approach to such hybrid simulations, one directly samples the osmotic ensemble.[38] As such, the number of guest molecules, the atomic vibrations of guests and host, the host unit cell volume and shape are all sampled dynamically. As a result, one can directly compute the equilibrium number of guests and unit cell volume from as a sample average given by eqn (4.4). Such direct sampling has been applied to qualitatively predict structural transitions as well as adsorption isotherms for adsorption of CO_2 in MIL-53(Cr).[38] However, this approach is expected to fail whenever large free energy barriers separate adsorption-induced metastable states. Therefore, a different hybrid approach was developed in which hybrid MC/MD simulations are performed in the grand canonical ensemble, in which atoms can vibrate and the unit cell shape can fluctuate, but the unit cell volume is kept fixed.[39] This achieved through application of MD in the $NV\underline{\sigma}_a = \overline{\overline{0}}T$ ensemble.[65] After running such hybrid simulations for a range of fixed unit cell volumes, the grand potential, given by an integration of the adsorption isotherm, is Legendre transformed towards the osmotic ensemble and osmotic adsorption isotherms and volume response can be computed through eqn (4.7) and (4.8). More details will be given in Section 4.3.4.2.

For other types of flexibility, a slightly different approach has to be followed. Consider the example of linker rotation as indicated on Figure 4.1 to illustrate this. As a first step, we define a collective variable that unambiguously determines the rotational state of the linker (*e.g.*, an appropriate dihedral angle). Next, we choose an appropriate thermodynamic ensemble to mimic the conditions of interest. Then we compute the instantaneous value of that parameter along the simulation trajectory. Finally, we compute the average value of the collective variable and repeat the procedure for different thermodynamic control variables to investigate the evolution. Alternatively, one can construct the free energy profile along the reaction coordinate to identify possible multiple (meta)stable states. The construction of such free energy profiles is discussed in Section 4.3.4.2.

4.3.4.2 Free Energy Surfaces

All equilibrium properties can be derived once the thermodynamic potential is known as function of its thermodynamic control variables. This thermodynamic potential is commonly denoted as the free energy. However, various choices for this potential are possible, each corresponding with an associated statistical ensemble and all are connected to each other by means of Legendre transformations. An example is the Helmholtz free energy $F(N,V,T)$ as function of the thermodynamic control parameters N, V and T. Whenever the various (meta)stable states accessible to the flexible MOF can be uniquely distinguished by the unit cell volume, this Helmholtz free energy can be used to describe pressure-induced, temperature-induced and adsorption-induced breathing of flexible MOFs.[68] However, in other cases the various (meta)stable states are not well distinguished by the unit cell volume, which reveals the need for another collective variable q. In this case, one can further partition the free energy into contributions associated with a specific value of this collective variable and construct a free energy profile $F(q)$ as function of q. Such an approach is required to investigate for example linker rotation.

Let us first consider the case of pressure-induced breathing in Al-fumarate as illustrated in Figure 4.12. By means of the enhanced sampling techniques discussed in Section 4.3.3.3 the Helmholtz free energy $F(V;N_0,T_0)$ can be constructed as function of unit cell volume V at fixed number of particles N_0** and temperature T_0 (Figure 4.12(a)). Through the following Legendre transformation:

$$\begin{gathered} F(N,V,T) \rightarrow G(N,P,T) = \min_V\left[F(N,V,T) + PV\right] \\ V = \left(\frac{\partial G}{\partial P}\right)_{N\,T} \text{ is the solution of} \left(\frac{\partial F}{\partial V}\right)_{N\,T} = -P \end{gathered} \tag{4.9}$$

One can construct the Gibbs free enthalpy $G(P;N_0,T_0)$ (Figure 4.12(c)), which is the thermodynamic potential in the *NPT* ensemble. Finally, the corresponding $V(P;N_0,T_0)$ is then readily computed as a simple derivative (Figure 4.12(d)).

This $V(P;N_0,T_0)$ profile represents the MOF unit cell volume response towards the applied mechanical pressure P. As such, it also allows to extract the bulk modulus through application of eqn (4.5). However, one can also detect a forward pressure-induced transition from the open pore to the contracted pore upon increase of the applied pressure as a discontinuous decrease in volume (indicated with a green arrow in Figure 4.12(d)), similarly a backward transition from the contracted pore back to the open pore can also be identified (indicated with a blue arrow in Figure 4.12(d)).

From an experimental point-of-view, one can measure such $V(P)$ profiles through mercury porosimetry measurements[137] and, hence, the simulated transition pressures can be compared with corresponding experimental

**Fixed number of particles here indicates a fixed number of MOF unit cells as well as no adsorbing guest molecules.

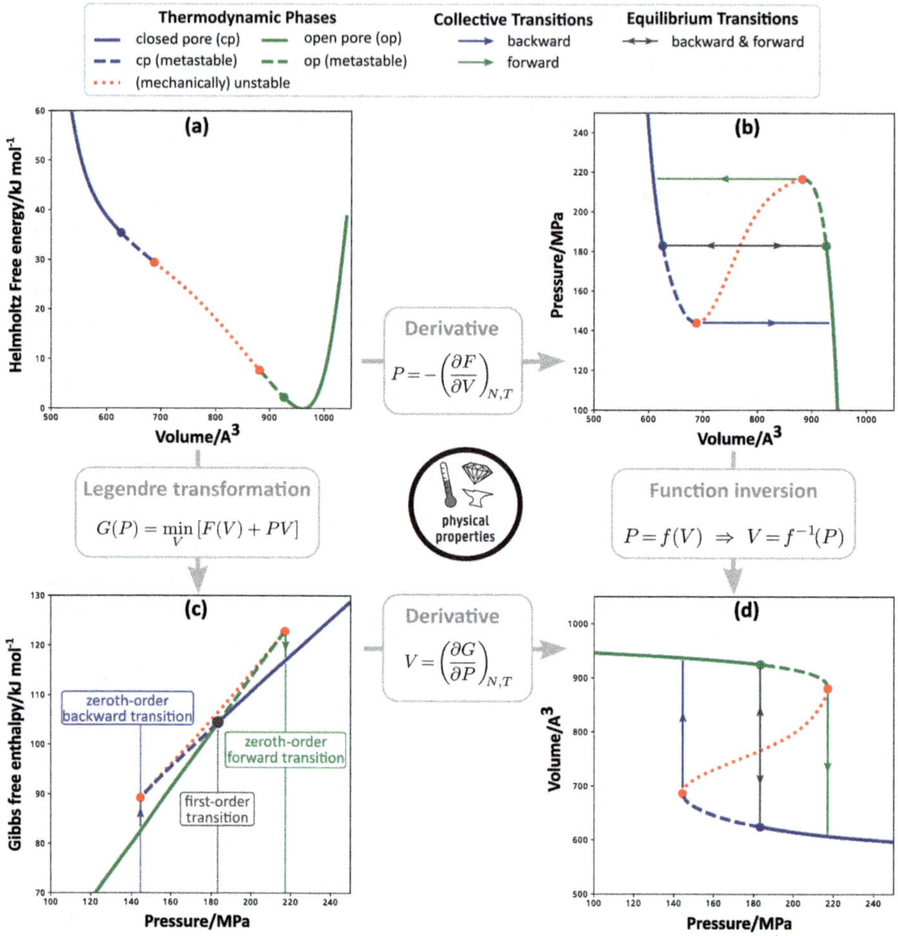

Figure 4.12 Illustration of thermodynamic potentials and derived properties for Al-fumarate at fixed temperature of 300 K (a) Helmholtz free energy as function of unite cell volume computed from enhanced molecular simulation.[68] (b) Pressure *vs.* volume profile resulting from the (negative) function derivative of the Helmholtz free energy. (c) Gibbs free enthalpy as a Legendre transformation of the Helmholtz free energy and (d) volume *vs.* pressure profile resulting from the function derivative of the Gibbs free enthalpy.

data (see Section 4.4 for illustrations). However, such experiments do not provide access to the full free energy profile revealing the added-value of modelling as various additional insights can be obtained from the free energy profile. First, one can identify breathing transitions as either zeroth or first order phase transitions, for which multiple classifications exist. In this text, the Ehrenfest classification[138] is used, which states that a zeroth order pressure-induced phase transition is characterised by a discontinuous change in Gibbs free enthalpy as function of the pressure, while a first order transition would be characterised by a continuous Gibbs free enthalpy but a discontinuous change in its first derivative towards the pressure, *i.e.*, the unit cell volume.[††] Whether or not the transition is of zeroth or first order depends on the transition mechanism. On the one hand, collective behaviour imposes that all unit cells simultaneously undergo the transition.[69,139] In this mechanism, a forward transition from the open pore to the contracted pore will only happen when the open pore ceases to be thermodynamically (meta)stable, resulting in an irreversible transition to the closed pore associated with a discontinuous change in Gibbs free enthalpy as illustrated by the green arrow at a pressure slightly lower than 220 MPa in Figure 4.12(c), and hence a zeroth order transition. On the other hand, if each unit cell could transition independently allowing for phase coexistence, the MOF would always seek to be in global thermodynamic equilibrium, *i.e.*, in the state of lowest Gibbs free enthalpy, resulting in a reversible transition with continuous changes in free enthalpy as illustrated by the grey dot at a pressure slightly higher than 180 MPa in Figure 4.12(c), and hence a first order phase transition. The question now remains, what is the correct transition mechanism? In this respect, it has been observed in molecular simulations that small unit cells transition in a collective fashion, while larger unit cells can transition in a layer-by-layer fashion and exhibit phase coexistence.[47] The fact that the cell size has a determining role in the transition mechanism indicates that large super cells are required to adequately model the transitions. Furthermore, simulations on very large unit cells revealed that also the magnitude of the mechanical pressure influences the transition mechanism: in the lower pressure regime the system transforms in a layer-by-layer mechanism,[140] whereas at higher pressures the transition initiates at discrete nucleation points.[46]

A second aspect where access to the full free energy profile provides additional insight is with respect to the mechanical work associated with the phase transition.[68] According to the first and second law of thermodynamics, the total work performed on or by a flexible MOF during an isothermal pressure-induced phase transition is given by:

$$W_{\text{MOF}} = \Delta F_{\text{MOF}} + T\Delta S_{\text{universe}} \geq \Delta F_{\text{MOF}} \tag{4.10}$$

For reversible transitions, *e.g.*, those corresponding to a first order transition, the work performed on or by the MOF is equal to the corresponding change in Helmholtz free energy. However, whenever the transition is of zeroth order and hence thermodynamically irreversible, the work on or by the MOF is increased by a term related to the entropy increase of the universe. Such considerations are crucial when investigating MOFs for mechanical energy storage applications.[141]

††More generally, an nth order transition is characterised by a discontinuous change in an nth order derivative of the thermodynamic potential, while all lower order derivatives remain continuous.

Until now, free energy profiles are discussed in terms of unit cell volume for the investigation of pressure-induced breathing. However, as has been mentioned previously, MOFs can exhibit various forms of flexibility that are not always adequately described by the unit cell volume but require more involved collective variables. Furthermore, the transitions can be triggered by various stimuli other than pressure such as temperature, adsorption of guest molecules, electric fields and more. Therefore, the use of molecular simulations to construct an appropriate thermodynamic potential and investigate the flexibility requires three aspects: (1) a proper collective variable q to distinguish the various (meta)stable states between which the material transitions; (2) an adequate sampling scheme to visit all these relevant states in the simulation and; (3) an appropriate way to apply the relevant trigger X. The first two aspects were already discussed in Section 4.3.3.3. The third aspect can be achieved in two ways, either by directly applying the trigger in the appropriate ensemble (*e.g.*, applying an electric field during the molecular simulation) or by means of the Legendre transformation (as was outlined in the previous paragraph for pressure-induced breathing). As a result, one gains access to the thermodynamic potential $F(q;X)$ in terms of a well-chosen collective variable q for a given range of the appropriate trigger X. Such profile then allows to trace the evolution as function of the applied trigger X of the (meta)stable states as (local) minima of the thermodynamic potential as well as the pathways connecting these various states. The insight gained from this allows to determine the thermodynamic conditions required for triggering the transitions.

As was mentioned before, guest adsorption is one of the most commonly investigated triggers for framework transformations. Therefore, it is interesting to elaborate on how to model such transitions in more details. Unfortunately, it is far from trivial to directly calculate the material response of the number of adsorbed guests and host unit cell volume by means of *e.g.*, hybrid MC/MD simulations as discussed in Section 4.3.4.1. Therefore, various alternative approaches aim to construct the osmotic potential X of a guest-loaded MOF as function of the chemical potential μ (or equivalently the vapour pressure p). Most of these approaches first construct or parameterise the thermodynamic potential in a different ensemble (*i.e.*, the canonical or grand-canonical ensemble) and afterwards transform it to the osmotic ensemble through application of the Legendre transform:

From canonical ensemble to osmotic ensemble:

$$F(N,V,T) \rightarrow X(\mu,P,T) = \min_{N,V}\left[F(N,V,T) + PV - \mu N\right]$$

$$\left.\begin{array}{l} N = -\left(\dfrac{\partial X}{\partial \mu}\right)_{P,T} \\ V = \left(\dfrac{\partial X}{\partial P}\right)_{\mu,T} \end{array}\right\} \text{is the solution of} \left\{ \begin{array}{l} \left(\dfrac{\partial F}{\partial N}\right)_{V,T} = \mu \\ \left(\dfrac{\partial F}{\partial V}\right)_{N,T} = -P \end{array}\right. \tag{4.11}$$

From grand canonical ensemble to osmotic ensemble:

$$\Omega(\mu,V,T) \to X(\mu,P,T) = \min_V\left[\Omega(\mu,V,T) + PV\right]$$
$$V = \left(\frac{\partial X}{\partial P}\right)_{\mu,T} \text{ is the solution of} \left(\frac{\partial \Omega}{\partial V}\right)_{\mu,T} = -P \tag{4.12}$$

In a first approach, Coudert *et al.* provided a thermodynamic framework based on the osmotic potential of a discrete number of (meta)stable phases, each phase i with a fixed volume V^i, to provide a qualitative understanding of guest-induced structural transitions.[36] First, the grand potential of each phase i is expressed by means of thermodynamic integration of the adsorption isotherm $N_{\text{ads}}^{(i)}$ of that phase:[‡‡]

$$\Omega^{(i)}(p,T) = F_{\text{host}}^{(i)}(T) - \int_0^p N_{\text{ads}}^{(i)}(T,\tilde{p})V_{\text{m}}(\tilde{p},T)\text{d}\tilde{p} \tag{4.13}$$

The first term corresponds to the grand potential at vapour pressure $p = 0$ (or equivalently a chemical potential $\mu = -\infty$) in phase i and is equal to the Helmholtz free energy of the empty host in phase i. Next, the osmotic potential of phase i can be computed through application of the Legendre transformation (eqn (4.12)), which in the current case of discrete phases results in the addition of a term $pV^{(i)}$ with $V^{(i)}$ the volume of phase i. By imposing thermodynamic equilibrium between two phases i and j, *i.e.*, $X^{(i)}(p_{\text{tr}},T) = X^{(j)}(p_{\text{tr}},T)$, one can finally compute the equilibrium vapour pressure p_{tr} of the transition from i to j. Coudert *et al.* applied this model by fitting fictitious rigid single-phase adsorption isotherms to different regions of experimental flexible framework isotherm and computed differences in host free energy $F_{\text{host}}^{(j)} - F_{\text{host}}^{(i)}$ by equating the model-predicted transition pressure from state i to j with the experimentally observed transition pressure.[36] Through combination with Ideal Adsorption Solution Theory (IAST), the model was later extended to compute adsorption selectivity of mixtures, which was denoted as Osmotic Framework Adsorption Solution Theory (OFAST).[136] In a second approach, a mean-field was applied to derive the Helmholtz free energy in the canonical ensemble of a guest-loaded flexible framework as function of the number of adsorbed guests N as well as the unit cell volume V:[69,139]

$$F(N,V,T) = F_{\text{host}}(V,T) + F_{\text{guest}}\left(N,V_{\text{pore}}(V),T\right) + N\cdot\Delta U_{\text{int}}(V,T) \tag{4.14}$$

Herein $F_{\text{host}}(V)$ again represents the Helmholtz free energy of the empty host, but now as a continuous function of the unit cell volume. F_{guest} is the free energy of N guest molecules confined within the pore volume V_{pore} without interacting with the framework, as described by an *a priori* chosen equation of state such as the van der Waals equation of state. Finally, ΔU_{int} represents the interaction energy of a single molecule with a rigid host framework at unit cell volume V. All three terms require input (F_{host}, V_{pore}, ΔU and parameters

‡‡In principle, the thermodynamic integration is done in terms of the chemical potential. However, by means of substitution for which the molar volume V_{m} is the jacobian, the integral is performed in terms of the vapour pressure p.

required for the guest equation of state F_{guest}) that can be extracted from molecular simulations and/or experiments. By applying the Legendre transformation given in eqn (4.11), the osmotic potential is obtained, which then allows to compute the material response properties according to eqn (4.7) and (4.8). This model was used to reproduce the experimentally observed CO_2-induced breathing of MIL-53(Al) as well as the lack of CH_4-induced breathing. Furthermore, the model has been extended towards (1) the additional application of mechanical pressure and rationalise the thermodynamic conditions required for observing negative gas adsorption[142] as well as (2) towards multicomponent adsorption with additional mechanical pressure to illustrate increased selectivity of MIL-53 upon application of mechanical pressure.[143] Finally, a third approach for the construction of osmotic potentials starts from hybrid MC/MD simulations in the grand canonical ensemble as was discussed previously in Section 4.3.4.1. As a result of these simulations, adsorption isotherms $N(\mu,V)$ (which account for vibrating atoms and flexible unit cell shape) were computed as function of chemical potential and unit cell volume. By means of thermodynamic integration, one can construct the corresponding grand potential:

$$\Omega(\mu,V,T) = F_{host}(V,T) - \int_{-\infty}^{\mu} N(\tilde{\mu},V,T)\mathrm{d}\tilde{\mu} \tag{4.15}$$

which can then easily be transformed to the osmotic potential using the Legendre transform given in eqn (4.12). This protocol was demonstrated to be successful in accurately describing breathing transitions in MIL-53(Al) triggered by Ne, Ar, Xe, CH_4 and CO_2.[39]

4.3.4.3 Kinetic Properties

Although classical thermodynamics allows the extraction of thermodynamic properties, it does not allow computation of dynamic properties such as diffusion constants or reaction rates, for that we need statistical physics. However, at this point the investigation of kinetic properties has not resulted yet into substantial increased insight into the flexibility of MOFs. One can compute for example the self or transport diffusion coefficients of guest molecules diffusing through the nanopores of the MOF using Einstein's relation.[121] A myriad of computational works has been reported to explore the impact of the MOF flexibility on the dynamics of the confined guest molecules.[144,145] The consideration of a flexible force field for the MOF framework was demonstrated to be crucial to accurately describe the guest diffusion in ultra-small channel MOFs (typically the contracted version of MIL-53) as well as cage-like MOFs (like ZIF-8 or derivatives) implying the passage of molecules throughout small windows that is controlled by flip, pore gating or local structural relaxations. Free energy differences associated with gate opening can be computing through application of the methods discussed in Section 4.3.4.2. As such, one can apply the model of Coudert *et al.* (*i.e.*, eqn (4.13) in combination with the Legendre transformation eqn (4.12)) to relate the gate-opening pressure to the host deformation free energy as well as the Langmuir adsorption characteristics (*i.e.*, the Henry constant and saturation uptake) of the porous 'open' phase.[146] Tanaka *et al.* computed grand potential profiles $\Omega(q;\mu,V,T)$ for Ar adsorption in ZIF-8 as function of the methyl-imidazole linker orientation (*i.e.*, a dihedral angle we here denote as q), which is responsible for gate opening, and applied Legendre transformations to construct to corresponding osmotic potential.[147] As a result, they concluded that upon steady increase of the Ar vapour pressure, the linker initially gradually reorients followed with a sudden transition at higher pressures that could be identified as the gate opening transition. For more details on gate opening, the readers are referred to specialised books and detailed reviews.[144,145,148]

On the other hand, one could also investigate the kinetics of the structural transition itself through application of the Bennet–Chandler expression for reaction rates[121] or through application of transition state theory.[149,150] However, to the best of our knowledge, such investigation has not yet been done for MOFs in literature.

4.3.5 Overview of Modelling Techniques to Derive Properties Related to MOF Flexibility

To summarise various aspects of the modelling workflow that have been discussed in this section, we give a brief overview on modelling techniques that can be used to derive properties related to MOF flexibility and that can be related to experimental measurements. Figure 4.13 gives an overview of experimentally derived properties related to flexibility together with various theoretical models and their accessible length and time scales. Given the broadness of the field and the diversity of experimental measurements that can be performed to obtain insight into flexibility of MOFs, this section is not intended to be comprehensive but focus on the most commonly used techniques. Furthermore, given the fact that the molecular modelling field evolves incredibly fast, both in terms of attainable accuracy, length and time scales and accessible computational resources, the overview given hereafter necessarily needs to be seen in the framework of what is possible today with current computational techniques.[151]

The first set of properties to be derived from simulations are structural and textural properties of stable or (meta)stable phases. Depending on the required level of accuracy one can calculate these from static simulations, *i.e.*, by performing a geometry optimisation and analysing the textural properties such as pore size distribution, void fraction, *etc.* The correctness of the proposed structure can be cross-validated with XRD patterns, capturing the atomic structure of the material. For these purposes, classical force fields are mostly appropriate certainly when one is interested in performing a quick screening of a lot of materials and their textural and geometric properties. Some care needs to be taken to capture also dynamic rearrangements of the framework, either global or local. In this case, either more structures need

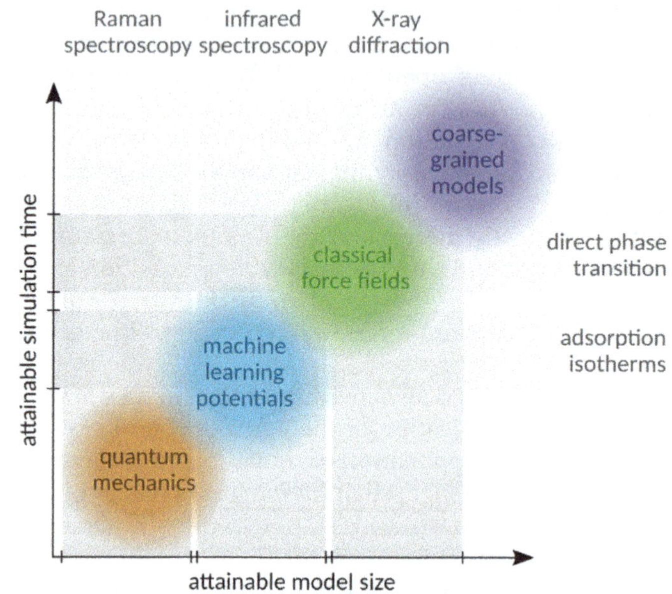

Figure 4.13 Overview of experimentally derived properties related to flexibility together with various theoretical models and their accessible length and time scales.

to be taken into account or one needs to switch to molecular dynamics techniques to average over a broader set of structures.[152]

Secondly, spectroscopic characterisation is a very strong method to track various phases of flexible materials. The field of spectroscopy is very broad, and it is not the intention to give a full overview here, for this we refer to dedicated reviews.[153,154] In the context of flexibility, vibrational spectroscopy is often used to distinguish the various phases of soft porous frameworks. Especially the modes in the low vibrational spectrum or the so called 'terahertz' vibrations have proven to be very indicative to recognise various phases of flexible MOFs.[155,156] Various vibrational spectra can be calculated each having their own selection rules. Within infrared spectroscopy, the magnitude the observed peaks upon interaction with electromagnetic radiation depends on the magnitude of the induced dipole moment, whereas within Raman spectroscopy one probes the induced polarisability. In principle we can calculate the infrared spectrum both by classical force fields as with quantum mechanical based methods, as a method to extract from the calculations the dipole moment or the atomic charges is required. In Raman spectroscopy, we need to rely on a method that gives the polarisability of the electron distribution and thus here only quantum mechanical based methods, where we have access to the electron density, are valid.

Thirdly, we can use various 'direct' techniques to access the phase transition in flexible MOFs experimentally. Mercury porosimetry, for instance, has been extensively adopted to characterise the change in cell volume upon pressurisation for prototypical pressure-responsive MOFs such as MIL-53(Cr),[41] DUT-49(Cu)[157] and CUK-1(Mn),[158] among others. Recently, Pallach *et al.* showed that the structural transitions in functionalised MOF-5 derivatives could also be followed directly *via* optical microscopy, since the sample turns from transparent to dull and cracked.[159] Complementing this information from a simulation viewpoint is not always straightforward. Besides requiring *in silico* access to the experimental probe, directly modelling the phase transformation also demands for sufficiently large simulation models and long simulation times, because these transformation are often size-dependent activated processes.[160] As a result, it forms an impetus to develop new simulation procedures to efficiently construct the potential energy surface and sample the available phase space.

Lastly to probe stimuli-responsive behaviour in terms of guest adsorption, measuring adsorption isotherms yields information whether the material shows various forms of switchability. Typical isotherms which may be found in case of gating, multistep adsorption, breathing and negative gas adsorption are shown in Figure 4.14. As discussed in Section 4.3.4, adsorption isotherms of flexible materials can be calculated using a variety of techniques. In current literature, adsorption isotherms are mostly calculated using GCMC simulations based on classical force fields as a huge amount of energy evaluations are necessary. When the aim is to quickly screen a variety of materials in terms of their potential flexible behaviour, this methodology is certainly fine, however, when one is interested in accurate values of the Henry constants or adsorption enthalpies a strong dependence on the applied force field may exist.[161,162] Also in case specific interactions with the framework become important, which are not well represented by the classical force fields, care must be taken in using classical force fields. Other methods for example based on importance sampling may be used in this case.[163] Furthermore, it is expected that in the near future also methodologies will be developed that rely on a MLP to describe the PES for constructing adsorption isotherms.

The previous section gives a brief overview of commonly used experimental techniques to probe MOF flexibility and the relation with modelling techniques on how to simulate these properties. As mentioned earlier, this section is certainly not comprehensive but gives information on some of the most commonly used techniques.

4.4 Case Studies

This section highlights through typical illustrations, the contribution of the diverse computational approaches introduced in the previous sections for the prediction and *in-depth* understanding of stimuli-induced flexible behaviour exhibited by prototypical MOFs.

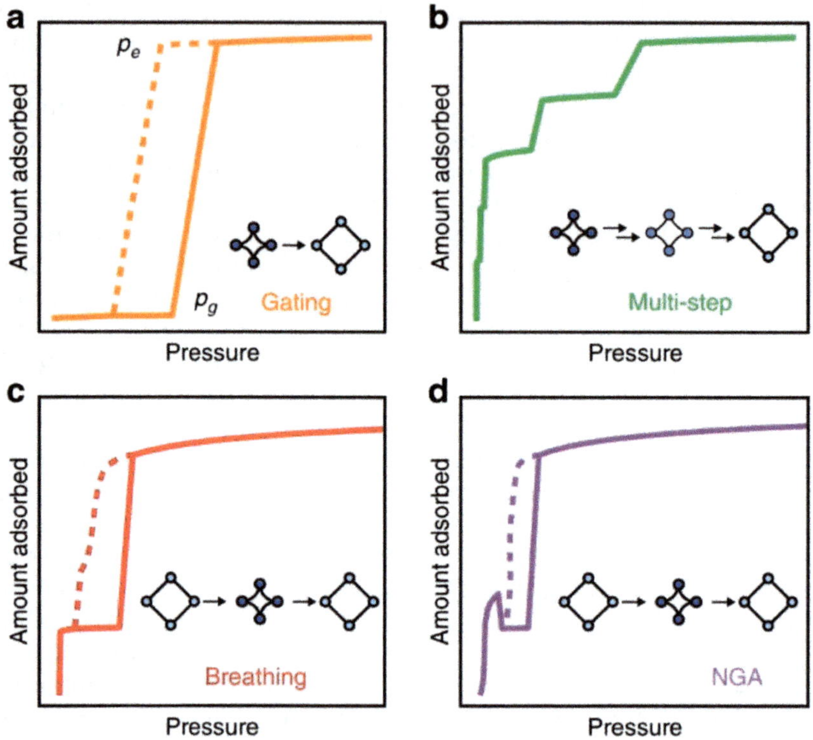

Figure 4.14 Dynamic adsorption phenomena in MOFs. (a) Gate opening. (b) Multistep adsorption. (c) Breathing. (d) Negative gas adsorption (NGA). Adsorption is depicted by solid lines and desorption by dashed lines. Reproduced from ref. 164 with permission from Springer Nature, Copyright 2020.

4.4.1 Gate Opening

Gate opening, classified as a displacive transition in Figure 4.1, is the phenomenon in which functionalised ligands can rotate (or in some cases movements of functional groups in the inorganic nodes) as a response to various external stimuli, thereby changing the pore environment of the material. As no bond reformation is associated with this type of flexibility, it can be modelled both by DFT and classical methods, as illustrated by the case studies below.

4.4.1.1 Temperature Induced

As a typical example, DFT calculations demonstrated that a temperature controlled dynamic opening of the MFU-4 pore aperture (which is made up of four chlorine atoms belong to the inorganic nodes) of only 2.52 Å enables to control the adsorption of D_2 and H_2 by exploiting the kinetic nature of

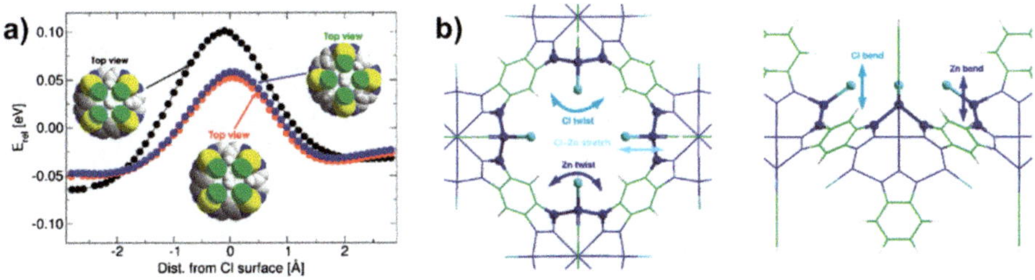

Figure 4.15 DFT calculations. (a) Potential energy surface of hydrogen penetrating through the gate from the large to the small cavity of MFU-4. Energies are with respect to free H_2. (b) Low-energy optical phonons in MFU-4, with phonon energies corresponding to 45–85 K. Adapted from ref. 165 with permission from John Wiley & Sons, Copyright 2013.

the quantum-sieving phenomenon.[165] This narrow pore aperture acts as a barrier for H_2 to pass from one cavity to the other with a corresponding calculated energy barrier of *ca.* 100 meV on the potential energy surface for a perfect and stiff network (see Figure 4.15(a)). Since the interaction of H_2 with the gate is governed by Pauli repulsion, this energy barrier changes exponentially according to the H_2–Cl separating distances. As such, small structural changes of the MOF are assumed to cause large variations in the energy barrier. Indeed, the DFT calculated lowest-energy optical phonons in MFU-4 occur at energies corresponding to 45 K–85 K (see Figure 4.15(b)). These phonons are associated with the vibrational modes involving the four Cl atoms that define the aperture size between the small and large cavities of the MOF, and hence they were shown to control the only pathway for hydrogen to diffuse through the framework.

4.4.1.2 *Adsorption Induced*

ZIFs are prototypical MOFs that manifest the guest adsorption-induced gate opening–contraction phenomenon. This phenomenon was shown to be associated with a structural switching between contracted- and open-gate structures owing to the flexibility and dynamic behaviour of the linkers accompanied by a substantial reorganisation of the adsorbate molecules in the MOF cages.[166] In general, gate opening leads to substantially wider apertures to access the cages that allow larger molecules to pass through these windows, and even enhance the diffusion rate of adsorbed molecules through the MOF's porosity. In a typical example, the CO_2 adsorption-triggered gate opening in ZIF-8 was explored by MD simulations implementing generic flexible force fields for the MOF framework.[167] A gradual contraction of the gate was predicted at low CO_2 pressure, followed by a sudden opening at larger uptakes due to the conformational change of the linkers. These MD simulations revealed that the aperture-size of the six-membered rings decrease from 3.375 Å to 3.135 Å for a CO_2 concentration of 0.5 to 20 molecules

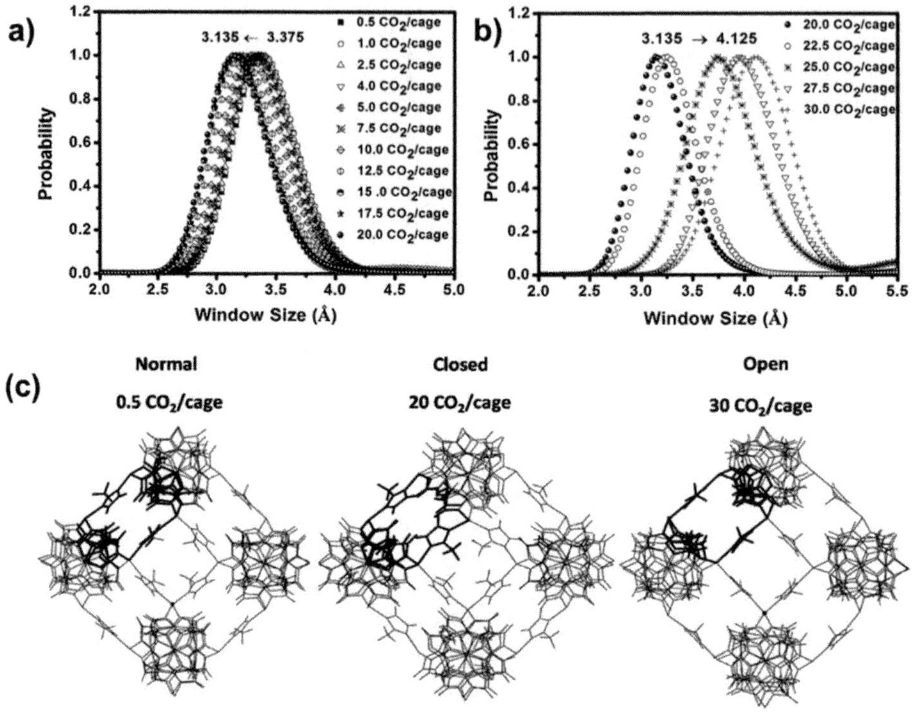

Figure 4.16 Flexible force field molecular dynamics. Window size distribution (six-membered rings) of ZIF-8 upon adsorption of CO_2 at 300 K (a) from 0.5 to 20 CO_2 per cage and (b) from 20 to 30 CO_2 per cage. (c) Illustration of the different guest-induced gate forms of ZIF-8. Adapted from ref. 167 with permission from Elsevier, Copyright 2016.

per cage. However, for a CO_2 loading of 20 to 30 molecules per cage the same gate expands drastically from 3.135 Å to 4.125 Å (see Figure 4.16).

The guest-induced open–contraction gate switching of ZIF-8 was also explored by *ab initio* MD (AIMD) calculations.[168] It was revealed that N_2 induces a continuous evolution of the initial contracted gate structure towards a wider open-gate form as the N_2 loading increases. In contrast to the previously assumed first-order structural transition,[40,169] here the gate opening was revealed to be associated to a gradual change in the swing angle distribution upon increasing N_2 concentration as depicted in Figure 4.17.

Hybrid Gibbs ensemble Monte Carlo (GEMC)/MD simulations were also performed to explore the adsorption-induced structural transition of ZIF-8 using a fully flexible MOF framework.[169] These calculations first confirmed that at a high N_2 loading (*ca.* 50 molecules per unit cell), ZIF-8 transforms from the contracted pore to the open pore structure. More interestingly, the multi-step experimental N_2 adsorption isotherm at 77 K was accurately captured by these simulations, which enabled to gain unprecedented microscopic insight into the adsorption phenomena with the identification of

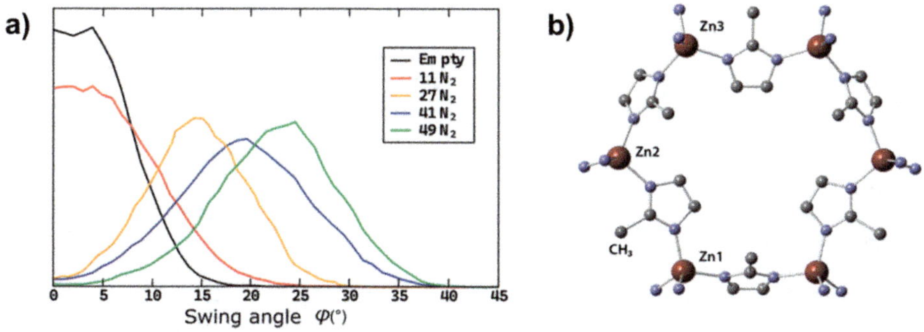

Figure 4.17 *Ab initio* MD simulations. (a) Distribution of the imidazolate linkers' swing angle φ of ZIF-8 upon variable N_2 loadings at 77 K. (b) Linker swing angle φ corresponds to the Zn3–Zn2–Zn1–CH_3 dihedral angle of the six-ring windows. Adapted from ref. 168 with permission from John Wiley & Sons, Copyright 2017.

three distinct regions corresponding to a pore filling of the ZIF-8 contracted structure below 0.005 bar, followed by a second region where a structural transition occurs from the contracted structure to the open form at a pressure between 0.005 and 0.01 bar, and above this pressure, the ZIF-8 structure remains fully open (Figure 4.18(a)). The corresponding unit cell size of ZIF-8 decreases slightly with increasing N_2 pressure and then slowly increases until a critical pressure of 0.01 bar, at which a structural transition occurs, and the cell size increases sharply as shown in Figure 4.18(b). The orientation of the imidazolate rings was shown to undergo a drastic change upon this structural transition while the guest-induced initial contraction of the cell was attributed to the internal stress exerted by the adsorbed N_2 molecules in the pore.

Other computational strategies including the osmotic framework adsorbed solution theory[170] and the Gauge cell Monte Carlo[147] approaches were equally considered to follow the guest-induced structural transition of ZIF-8. Interestingly, flexible force field MD simulations conducted at 300 K revealed that for some other guest molecules like H_2O, there is a counter-intuitive closing of the gate as evidenced in the reduction of the imidazolate linkers' swing angles (Figure 4.19).[60]

Adsorption-stimulated gate-opening of ZIF-8 was also described by free energy analysis using thermodynamic integration of adsorption isotherms obtained by grand canonical and gauge cell Monte Carlo simulations. The gate opening and closing mechanisms of ZIF-8 (*i.e.*, reversible reorientation of 2-methylimidazolate (MeIM) linkers) were probed by evaluating the change of the osmotic free energy of the system upon Ar adsorption–desorption.[147] In doing so, individual adsorption isotherms were calculated by GCMC simulations for a wide range of rotational angle value, θ_{IM} of the MeIM linkers. Furthermore, the change of the Helmholtz free energy of the ZIF-8 was evaluated for variable linker rotations so that the osmotic free energy change

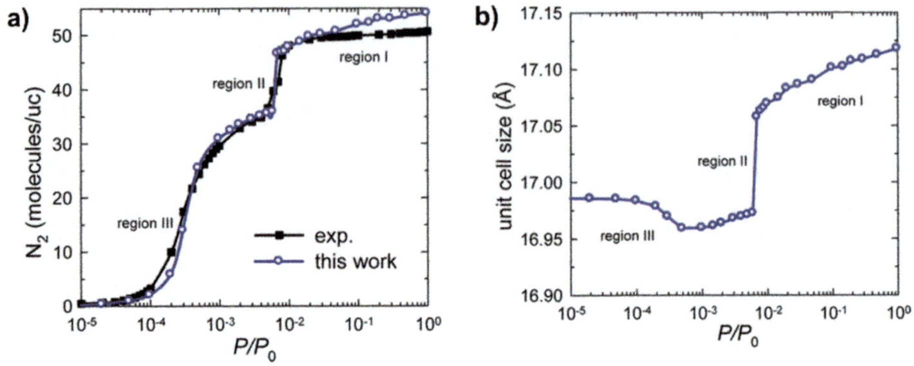

Figure 4.18 Hybrid GEMC/MD simulations. (a) N_2 simulated and experimental adsorption isotherm of ZIF-8 at 77 K. (b) Evolution of the corresponding simulated unit cell dimension of ZIF-8 upon N_2 adsorption. Adapted from ref. 169 with permission from American Chemical Society, Copyright 2013.

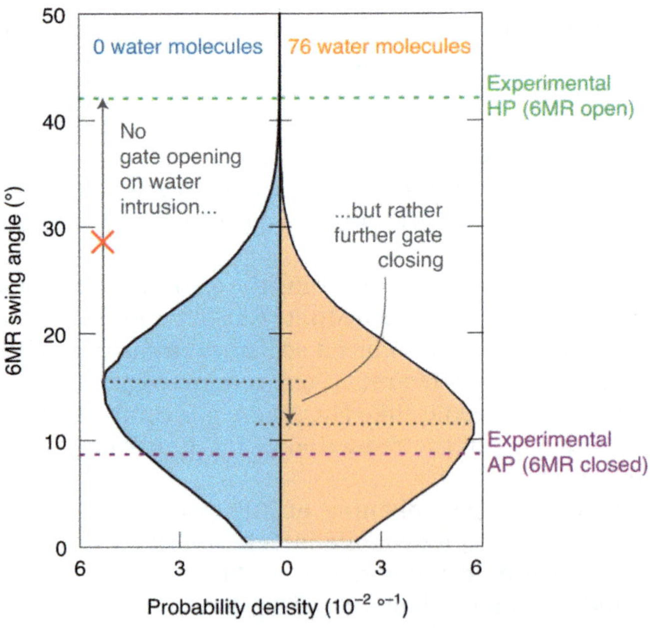

Figure 4.19 Flexible force field MD simulations. Probability density of the 6MR dihedral swing angle of the ZIF-8 linkers, during a 5 ns MD run at 300 K and 0 MPa with either 0 or 76 water molecules per unit cell. Experimental ambient- (AP) and high-pressure (HP) phase linker swing angle are indicated in purple and green, respectively. Adapted from ref. 60 with permission from Springer Nature, Copyright 2021.

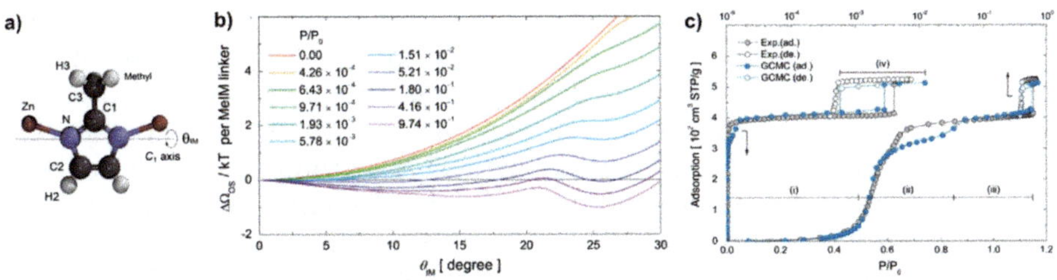

Figure 4.20 (a) Definition of rotational axis of the MeIM linker in ZIF-8, (b) osmotic free energy change calculated from the GCMC adsorption isotherms at 87 K, and (c) comparison between adsorption isotherms measured experimentally and predicted from the free energy analysis with the aid of the GCMC isotherms at 91 K. Adapted from ref. 147 with permission from American Chemical Society, Copyright 2014.

could adequately correlate with the presence of the hysteresis loop observed in the experimental adsorption isotherms. It was thus demonstrated that ZIF-8 linkers undergo a gradual reorientation with increasing Ar pressure followed by a spontaneous structural transition at a given Ar pressure implying a large rotational angle change. The osmotic free energy profile obtained at $T = 87$ K (see Figure 4.20(b)) shows that, during the adsorption process, the global minimum of the free energy profile shifts toward a larger θ_{IM}, which is accompanied with a gradual expansion of the pore window size. A further increase in the relative Ar pressure results in a local minimum at around $\theta_{IM} = 25.5^{\circ}$. Spontaneous structural transition then occurred at this rotational angle as the activation energy (metastable states to global minima) becomes smaller than the energy fluctuation of the system (0.48 kT per MeIM linker).

Interestingly, during the desorption process the system remains at a local minimum at $\theta_{IM} = 25.5^{\circ}$ regardless of the decrease of Ar pressure until a structural transition takes place with the opposite reorientation of the MeIM linkers at a smaller θ_{IM}. In this case, the activation energy (0.58 kT per MeIM linker) is found to be larger than the energy fluctuation of the system (0.48 kT per MeIM linker) obtained for the spontaneous structural transition upon adsorption. This also explains the experimentally observed hysteresis present in the desorption branch. Finally, based on the free energy analysis, GCMC calculated adsorption isotherm at $T = 91$ K (Figure 4.20(c)) shows a very good overall agreement with the experimental data over the entire pressure range.

Recently, the osmotic free energy based thermodynamic model was also applied to elucidate the gate opening behaviour of the doubly catenated—doubly interpenetrated, or two-fold interlocked—Zn-CAT MOF [$Zn_2(bdc)_2$(4,4′-bpy)] where bdc = benzene-1,4-dicarboxylic acid and 4,4′-bpy = 4,4′-bipyridine (Zn-CAT).[146] The experimentally observed sigmoidal shape acetylene (C_2H_2) adsorption isotherm is believed to be afforded by mutual movement of the two identical framework motifs of the Zn-CAT MOF at certain gas pressure. The gate-opening (*i.e.*, creation of extra voids during the adsorption) and gate-closing pressures depend on the guest induced free energy change of the system and corresponding activation energy required for a structural transition to happen. This deformation energy is attributed to the van der Waals interactions between two interpenetrating structural motifs, and therefore, provides an excellent platform to tune the adsorption properties by changing the steric and chemical (electronic) properties of the linkers. As such, it was demonstrated that by introducing different ratios of functionalised bdc–X linkers (X = $-NO_2$ or $-NH_2$), it is possible to tune the gating behaviour of the MOF and thereby to optimise the desired shape of the acetylene isotherm for different operating conditions.

Note that adsorption induced gate opening of an interpenetrated system implies a first order transition from a non-porous to open pore phases accessible to guest molecules. Therefore, a simple two-phase model was used to describe the gating behaviour of the Zn-CAT MOF. Here, the free energy calculation was based on the thermodynamic integration of the continuous isotherm (rigid-host approximation) obtained by Langmuir-type fit of the experimental isotherms for the open phase. Indeed, the free energy profiles calculated from the experimental adsorption and desorption isotherms at T = 273 K show a free energy difference (ΔF) of only 5.2 kJ mol^{-1} (per unit cell) between open and close gate states of the Zn-CAT system. This relatively small ΔF value clearly indicates highly flexible nature of the MOF, and an adsorption induced gate-opening can be achieved at a low-to-moderate gas pressure. It was also shown that the introduction of the $-NO_2$ and $-NH_2$ in the linkers resulted in further decrease in the ΔF values as shown in Table 4.2.

Free energy based calculations were also applied to understand the experimentally observed gating behaviour of a mutually interpenetrating jungle gym (JG) framework.[171] Here, the free energy calculation was based on the thermodynamic integration of the solvation force obtained by GCMC simulations. This method is however quite difficult to apply to actual flexible MOFs with complex structures. Thereby each jungle gym framework motif was replicated with uniform structureless rods made of benzene rings, and LJ-methane was used as adsorbate/probe molecule. The base point of the free energy is set to the central configuration (*i.e.*, intersection of a JG is located

Table 4.2 Calculated gate opening–closing pressures and free energy differences (ΔF) of acetylene (C_2H_2) adsorption at 273 K in Zn-CAT, and its derivatives. Adapted from ref. 146 with permission from Springer Nature, Copyright 2022.

	P_{go}/kPa	P_{gc}/kPa	P_{trans}/kPa	ΔF/kJ mol^{-1}
Zn-CAT	45.4	16.7	27.5	5.2
Zn-CAT-$(NH_2)_{10}$	51.7	18.2	30.7	5.1
Zn-CAT-$(NH_2)_{20}$	60.9	21.2	35.9	5.1
Zn-CAT-$(NH_2)_{40}$	83.2	22.6	43.4	4.4
Zn-CAT-$(NO_2)_{20}$	12.1	10.2	11.1	2.6
Zn-CAT-$(NO_2)_{50}$	6.5	7.0	6.7	1.8
Zn-CAT-$(NO_2)_{100}$	1.0	0.9	0.96	0.3

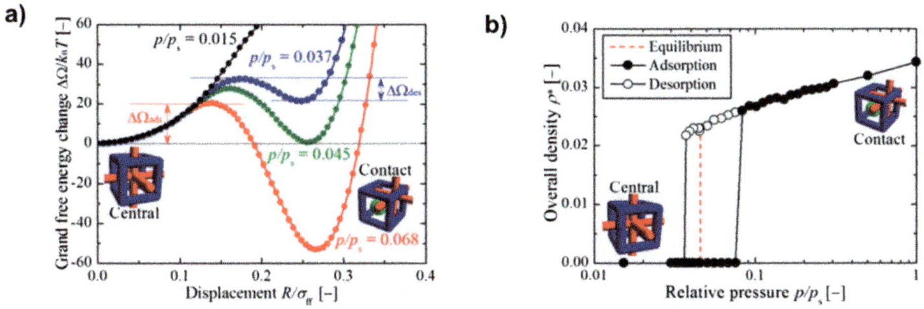

Figure 4.21 (a) Grand free energy profiles of the interpenetrated framework calculated for different relative displacement of JG motifs at different relative pressures. (b) Simulated adsorption isotherms obtained for the benzene rod model system. Adapted from ref. 171 with permission from American Institute of Physics, Copyright 2009.

at the centre of other JG cube) at which the JG system is considered closed, and no adsorption could occur (Figure 4.21). The free energy increases with increasing gas pressure as JG framework approaches to a close contact configuration, *i.e.*, the intersection of one JG moves diagonally from the central position of other JG. For a higher gas pressure at $P/P_0 = 0.037$, a local minimum appears around the displacement of 0.25 σ (LJ sigma of the methane molecule), and methane molecules able to enter into the pores of a 'contact configuration' as if the gate is open. A further increase in the gas pressure increases the adsorbed amount, and JG contact-configuration becomes more stable. At a relative pressure of $P/P_0 = 0.045$, the free energy profile becomes bistable and above this pressure the global minimum shifts into the central configuration in place of the contact configuration. Note that, the switch of the global minimum in the free energy profile naturally yields a step wise adsorption isotherm.

4.4.1.3 Synergistic Pressure-adsorption Induced

The application of an external mechanical pressure can control the gate size of MOFs. However, this is often accompanied by amorphisation of the structure, as observed for, *e.g.*, ZIF-8 (see Section 4.4.4). However, the presence of guest molecules enhances the resistance of the architecture, which in turn enables the use of a higher pressure to modulate the gate size/shape of the MOFs. Flexible force field MD simulations demonstrated that the gate size of propane and propylene loaded ZIF-8 gradually decreases when the mechanical pressure increases. Notably, a ~0.3 Å reduction being achieved at 1 GPa (see Figure 4.22).[172] This mechanically controlled size of the ZIF-8 gate was shown to create a more pronounced decrease of the diffusivity of the bulkiest propane compared to propylene. This synergistic combination of mechanical-control pore architecture and adsorption in flexible MOFs was further predicted to boost the kinetic propylene/propane selectivity by about one

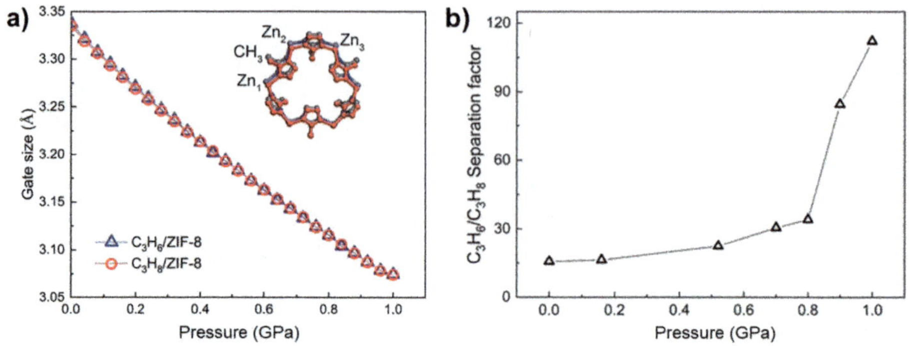

Figure 4.22 Flexible force field MD. (a) Evolution of the gate size for the hydrocarbon-loaded ZIF-8 averaged over the MD trajectories. (b) Predicted evolution of the propylene/propane diffusion selectivity as a function of the applied mechanical pressure at 300 K. Adapted from ref. 172 with permission from John Wiley & Sons, Copyright 2019.

order of magnitude leading to a value of ~110 under the application of an external pressure of 1 GPa *vs.* 18 GPa without the application of mechanical pressure (see Figure 4.22(b)).

4.4.2 Swelling

When compared to first order structural transitions, the second order transition denoted as the swelling behaviour of MOFs—associated with a gradual but highly concerted expansion of their unit cell dimensions upon guest-adsorption (see Figure 4.1)—was theoretically explored to a much lesser extent. We are aware of only a single computational study on the prototypical swelling of MIL-88 MOF[25] with the implementation of MD simulations using a newly derived set of MOF-FF flexible force field parameters.[173] The authors first calculated free energy *vs.* volume profiles of MIL-88B(Fe)-Cl as a function of methanol loading and temperature (Figure 4.23). These predictions revealed an initial gradual swelling of the contracted form at the initial stage of adsorption, followed by an abrupt transition from a semi-contracted to an open pore form at a given methanol loading. Microscopic analysis of these MD data highlighted that the MOF pore opening is associated with distribution changes of the counterions (Cl^-) around the Fe_3O oxo-trimer inorganic units and of the solvent coordination with those counterions. These observations demonstrated that the solvation of the counterions play a major role in the swelling behaviour of the MIL-88B. From an energetic perspective, it was found that the volume changes of MIL-88B(Fe)-Cl are governed by a subtle interplay between the internal energy and entropy of the MOF framework, the solvation of the counter ions and the incipient cohesion (mainly through hydrogen bonds) of the liquid methanol in the cavities.

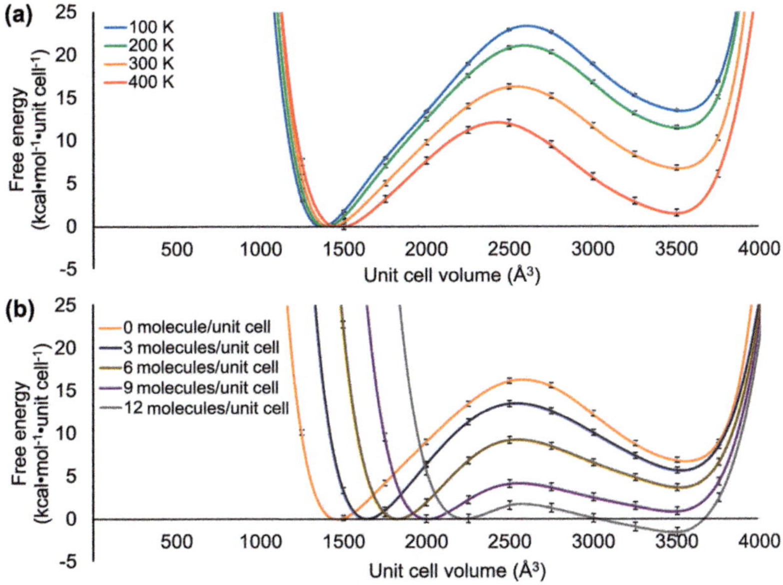

Figure 4.23 Flexible force field MD. Free-energy landscapes obtained for MIL-88B(Fe)-Cl: (a) empty MOF at temperatures ranging from 100 to 400 K and (b) methanol loaded MOF at concentration ranging from 0 to 12 molecules per unit cell at T = 300 K. Adapted from ref. 173 with permission from American Chemical Society, Copyright 2021.

4.4.3 Breathing

While breathing is, similar to the swelling phenomenon discussed in Section 4.4.2, a concerted mode of flexibility, it contrasts with this swelling behaviour as the breathing transition does not occur gradually but rather abruptly between well-defined phases, as illustrated schematically in Figure 4.1. In what follows, case studies on the various stimuli that can induce breathing are introduced.

4.4.3.1 Temperature Induced

Flexible force-field MD simulations implementing a modified version of the Amber force field were considered to explore the temperature-driven reversible transition of ZIF-7 between its contracted pore and large pore phases (Figure 4.24).[174] These calculations revealed that the crystal density of the initial ZIF-7 contracted pore phase decreases with increasing temperature, followed by step like changes corresponding to the formation of the open pore phase at about 600 K. Upon cooling, a step-change from the open pore to the contracted pore phases was also evidenced with only a small hysteresis (Figure 4.24). These calculations suggest that the formation of the ZIF-7 contracted pore phase is controlled by incremental nonbonded interactions that drives the structure to 'fold' into a high-density, low-symmetry state.

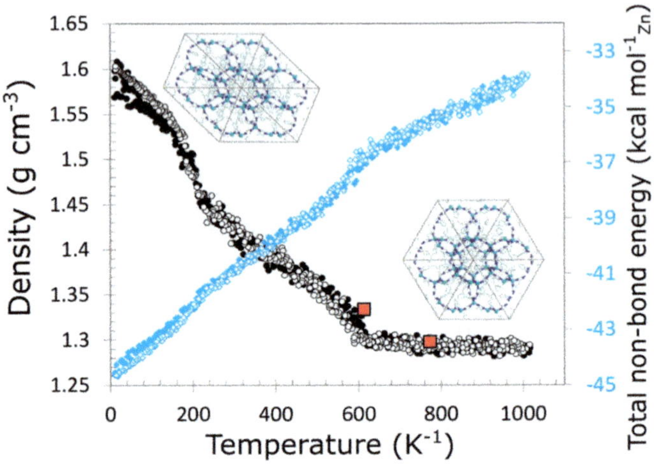

Figure 4.24 Flexible force-field MD simulations. Density changes of ZIF-7 as a function of temperature associated to the contracted pore to open pore phase transitions. Closed and open symbols correspond to heating and cooling, respectively. Red squares show XRD measured ZIF-7 crystal densities at 613 K and 773 K. Adapted from ref. 174 with permission from American Chemical Society, Copyright 2015.

4.4.3.2 *Adsorption Induced*

MIL-53 is one of the most representative examples of breathing MOFs, showing a cell expansion of its unit cell volumes up to almost 35% upon guest adsorption.[6] Flexible force field MD simulations in the $NP\sigma_a = 0T$ ensemble[§§] initially unravelled the mechanism at the origin of CO_2-triggered reversible two-step structural switching between the open pore (orthorhombic symmetry, space group *Imcm*) and the contracted pore (monoclinic symmetry, space group *C*2/*c*) forms of this MOF. The structural change associated with a drastic reduction of the channel aperture from 12.8 to 7.9 Å as determined by *in situ* powder XRD (see Figure 4.25).[99] It was demonstrated that once the adsorption stimulus reaches a critical value, a soft mode appears and triggers the structural switching of the MOF.[38,175,176] Combining MD and GCMC techniques in a hybrid osmotic Monte Carlo (HOMC) scheme further enabled to quantitatively (transition pressure and amount adsorbed) capture the unusual shape of the experimental CO_2 adsorption isotherm for this breathing MOF in the whole range of gas pressure (see Figure 4.25(b)).

The exploration of the guest assisted-structural transitions of diverse MIL-53(Al, Cr, Ga, Sc) was also performed using DFT and AIMD calculations.[177,178] Thermodynamic and analytical models have also been proposed to rationalise the dynamic behaviours of this family of materials.[36,139,143]

---

§§This is an extension of the constant pressure ensemble and allows to control each component of the stress tensor in addition to the hydrostatic pressure, which is applied isotropically, see the caption of Figure 4.8.

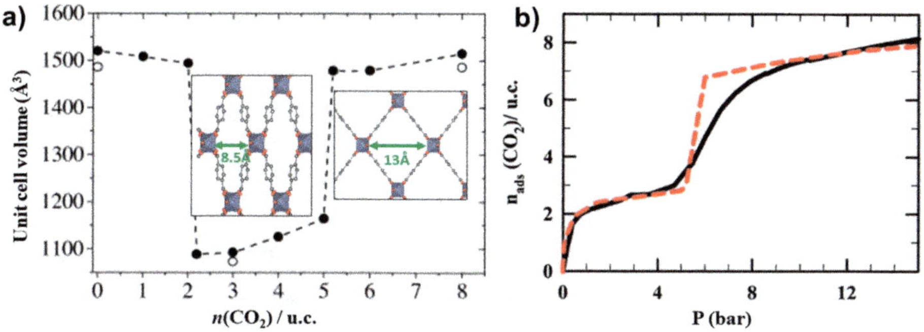

Figure 4.25 Flexible force field MD and HOMC simulations. (a) Evolution of the simulated unit cell volume for MIL-53(Cr) as a function of the CO_2 concentration (solid symbols correspond to *in situ* XRD data) at 303 K. Reproduced with permission from John Wiley & Sons, copyright 2008. (b) HOMC simulated (dashed line) and experimental (solid line) CO_2 adsorption isotherms at 303 K. Reproduced with permission from the American Chemical Society, Copyright 2012.

4.4.3.3 Pressure Induced

Flexible force field MD simulations performed in the $NP\overline{\overline{\sigma}}_a = \overline{\overline{0}}T$ ensemble in tandem with high-pressure XRD and Hg-porosimetry, were able to explore the breathing behaviour of both MIL-53(Cr)[176] and MIL-47(V)[77] MOFs under mechanical pressure with the application of an external pressure ranging from 50–137 MPa respectively. A generalised thermodynamic approach[65,68] has been developed and applied to this family of pressure-induced breathing MOFs including MIL-53, DMOF-1, MIL-47, CUK-1 and DUT-49, among others,[46,68,71,142,158,179] as well as to breathing induced by temperature and gas adsorption. Interestingly, the Helmholtz free energy profiles and the pressure *vs.* volume equations-of-state constructed from flexible force-field MD simulations were able to categorise the pressure-induced structural behaviour of breathing MOFs into three classes: shock absorbers, triggered dispersers or lossy springs as reported in Figure 4.26(a). For example, the open pore phase of MIL-53(Al) is metastable and behaves as a triggered disperser (see Figure 4.26(b)), in line with a fully irreversible pressure-induced structural contraction which would only return to the initial open pore upon application of high enough tension.

4.4.3.4 Electric Field Induced

The electric field induced breathing behaviour of MIL-53(Cr) was computationally explored by means of flexible force field MD simulations in the $NP\sigma_a = 0T$ ensemble.[180] An external electric field with the same magnitude along the x, y and z directions was applied on each atom of the MOF framework such as $F_i = -q_i E$ where F_i is the force on each atom i, E is the electric

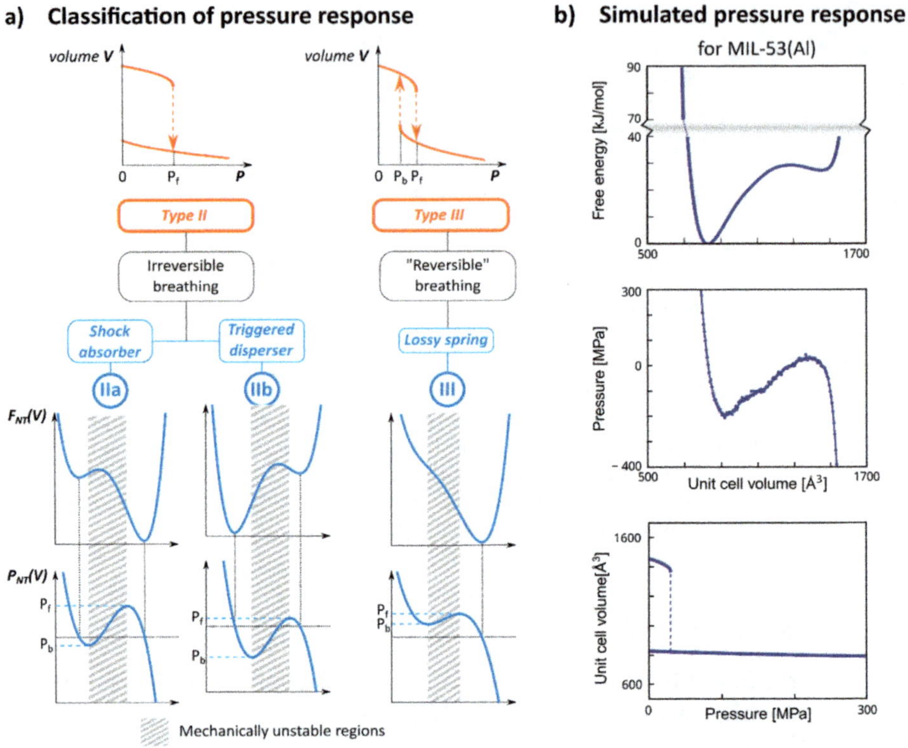

Figure 4.26 Flexible force-field MD simulations and thermodynamics approach. (a) Illustration of the possible experimental responses of breathing MOFs upon the application of a mechanical pressure (top panel), hypothetical Helmholtz free energy $F_{NT}(V)$ and pressure *vs.* volume equation-of-states $P_{NT}(V)$ with N number of particles, T temperature and V volume (bottom panel). (b) Illustration of the free energy profile (upper row), mechanical equations-of-state (middle row) and volume *vs.* external pressure profile (bottom row) for MIL-53(Al). Adapted from ref. 68 with permission from Springer Nature, Copyright 2018.

field, and q_i is the partial charge on atom i. These calculations revealed that the application of an electric field of 1.75 V nm^{-1} induces a structure polarisation change accompanied with a switching from the initial open pore form to a contracted pore version (Figure 4.27(a))—similar to what was observed upon guest adsorption, thermal and mechanical stimuli—resulting from a collective motion in such a way that the shear and compression processes occur simultaneously. This is reflected by the concomitant drop of the unit cell parameters b and β once the MIL-53(Cr) structure switches under the application of an electric field of 1.75 V nm^{-1} (Figure 4.27(b)). Interestingly, further HOMC simulations revealed that the application of an external electric field would force the structure to remain in the contracted pore form during the adsorption process, leading to a complete separation of CO_2 over CH_4 *via* a size exclusion effect.

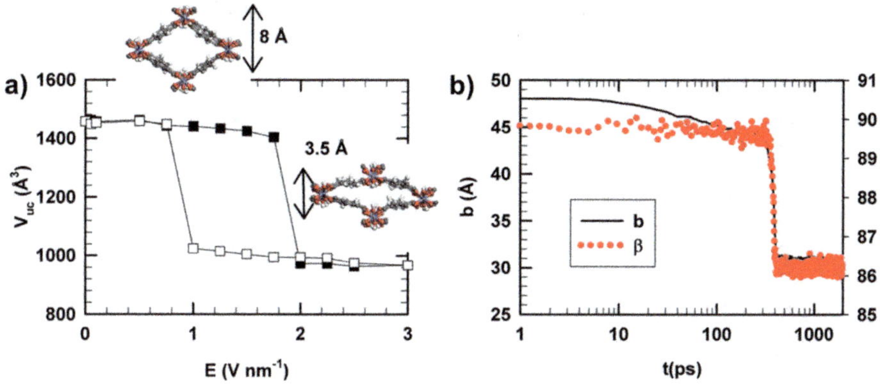

Figure 4.27 Flexible force-field MD. (a) Evolution of the unit cell volume of MIL-53(Cr) as a function of external electric field (E). Full and empty symbols correspond to increasing and decreasing electrical field range, respectively, at 300 K and 1 bar. (b) Evolution of the unit cell parameters b and β of MIL-53(Cr) as a function of time at E = 1.75 V nm^{-1} at 300 K and 1 bar. Adapted from ref. 180 with permission from American Chemical Society, Copyright 2017.

In a relevant note, DFT calculations have recently evidenced that the application of an electric field decreases the charge transfer between the CO_2 molecules and the adsorption sites of MIL-53(Al) which results in reduced binding energy and consequently lower CO_2 adsorption capacity.[181]

4.4.4 Crystalline to Amorphous Phase Changes

The application of a mechanical constraint can provoke a structural instability of a series of MOFs that can be predicted as well as understood computationally.[20,166] In contrast to the aforementioned modes of flexibility, these crystalline to amorphous phase transitions are reconstructive in nature (see Figure 4.1), making DFT the method of choice to computationally model it. DFT calculations revealed that the pressure-driven mechanical instability of ZIF-8 frameworks to shear deformations occurs at P > 0.3 GPa.[182] Both the simulated C_{11} and C_{12} elastic constants appeared to slightly increase up to 0.15 GPa and then decrease above 0.2 GPa. In contrast, the calculated C44 shear modulus was found to remain almost constant up to 0.1 GPa and then linearly decreased (Figure 4.28, top panel) consistent with force-field MD simulations reported in another work.[183] The bulk modulus (K) was predicted to initially increase up to 0.15 GPa and decrease above 0.2 GPa (Figure 4.28, bottom panel), in line with an initial stiffening of the structure followed by a structural softening at higher mechanical pressure. Analysis of the harmonic lattice dynamics evidenced that C_{11} and C_{12} are predominantly affected by lattice vibrations involving the symmetric 'gate-opening' motion of imidazolate linkers, where the four imidazolate linkers of the same squared channel swing perfectly in-phase, while C_{44} is related to the asymmetric 'gate-opening'

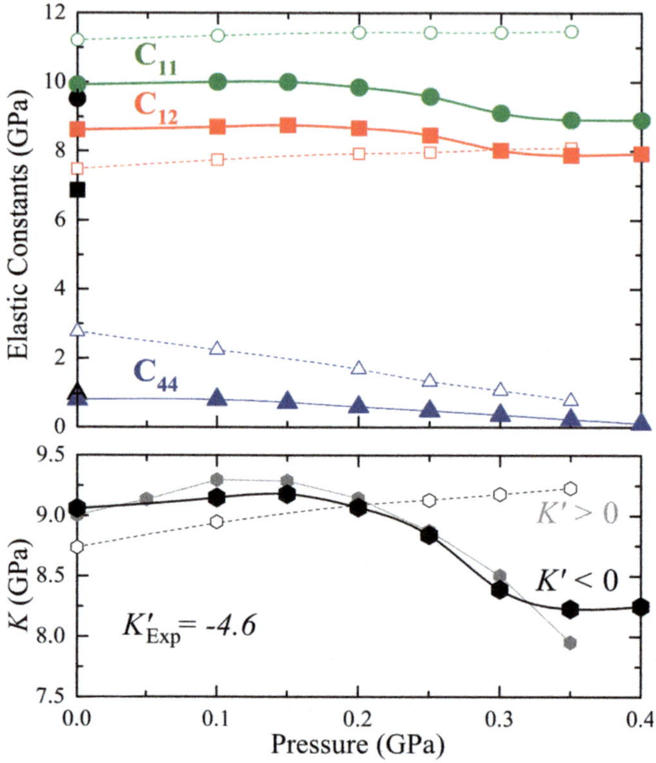

Figure 4.28 DFT and flexible force-field MD. Upper panel depicts elastic constants of ZIF-8 as a function of pressure, with filled symbols (DFT calculations)[2] and open symbols (flexible force field MD simulations).[183] In the lower panel, the bulk modulus as a function of pressure is determined from computed elastic constants ($K = 1/3(C_{11} + 2C_{12})$; black symbols) and from the pressure *vs.* volume equation-of-state ($K = -V\partial P/\partial V$; grey symbols). Adapted from ref. 182 with permission from American Physical Society, Copyright 2019.

motion, where the two pairs of imidazolate linkers on the same squared channel swing out-of-phase.

The determination of structural models for amorphous MOFs is, however, highly challenging and therefore calls for the deployment of sophisticated simulation techniques as recently reviewed,[184] including Reverse Monte Carlo (RMC) methods, Continuous Random Networks (CRN), classical and *ab initio* molecular dynamics, reactive force fields, and simulated assembly/polymerisation algorithms. Typically, the RMC approach is coupled with neutron- and XRD data to construct amorphous/glass-like structure models. This dual approach has been applied to develop an amorphous model of ZIF-4 starting with a continuous random network of amorphous SiO_2 and replacing Si by Zn and O by an imidazolate linker in the appropriate orientation.[185] This RMC strategy was further exploited to deliver the first melt-quenched MOF

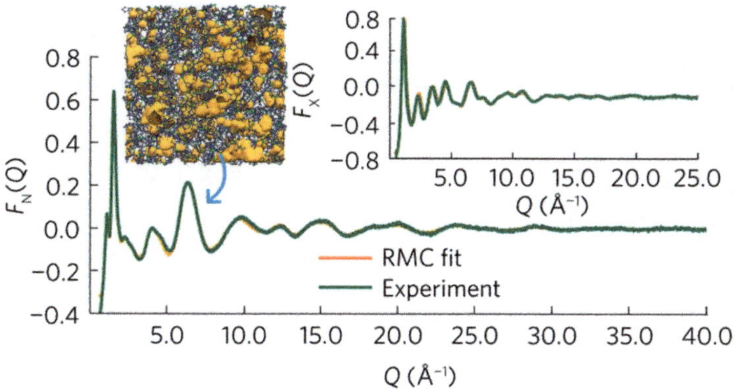

Figure 4.29 Reverse MC method. Construction of an amorphous model of ZIF-4 tuned on pair distribution functions of neutron total scattering data. Adapted from ref. 59 with permission from Springer Nature, Copyright 2017.

glass structure model[59] starting with the amorphous model of ZIF-4 model and experimental neutron pair distribution function analysis (Figure 4.29).

4.4.5 Reversible Loss of Long-range Order

While the crystalline to amorphous transitions discussed in Section 4.4.4 are assumed to occur irreversibly, a novel concept of 'frustrated flexibility' in MOFs was recently introduced in which this loss of long-range order can occur reversibly.[159] Frustrated flexibility, which falls in the top right corner of Figure 4.1, given the breaking and reformation of bonds, arises from the incompatibility of intra-framework dispersion forces with the geometrical constraints of the inorganic building units. Flexible force-field MD simulations in tandem with a number of experimental techniques demonstrated the phenomenon of frustrated flexibility in a series of functionalised MOF-5 derivatives, which feature additional intra-framework dispersion forces competing against the strong and directional coordination bonds responsible for structural rigidity. The functionalisation of the benzenedicarboxylate linkers by alkoxy groups of different lengths denoted as MOF-5-CX enabled to identify this new type of stimuli-responsive structural flexibility of the MOF framework characterised by an unprecedented reversible loss and recovery of crystalline order under full retention of the MOF framework connectivity and topology.

These MD simulations evidenced that MOF-5-C3 under the application of a mechanical pressure switches from a homogeneous and narrow cell volume distribution to a broad one with volumes only weakly correlated between neighbouring cells while the different cell angle types are organised and form layers (see Figure 4.30). Interestingly, upon mechanical pressure

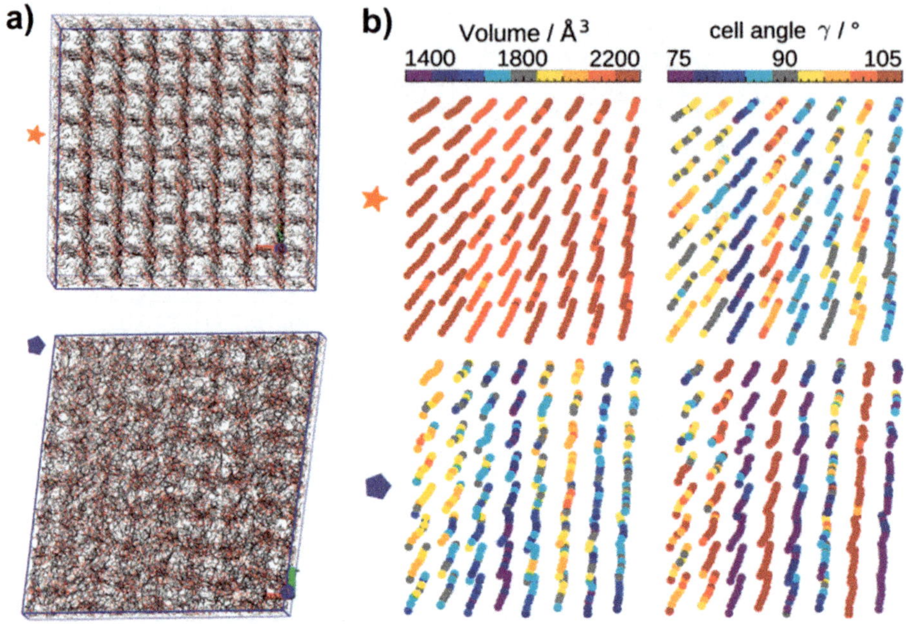

Figure 4.30 Flexible force-field MD. (a) Side view of an alkoxy-functionalised MOF-5-C3 before (top panel; red star) and after (bottom panel; blue pentagon) the pressure-induced phase transformation. (b) Individual cell volumes and angles γ of MOF-5-C3 associated to these two structures. Adapted from ref. 159 with permission from Springer Nature, Copyright 2021.

release, the framework connectivity of MOF-5-C3 partially preserves loss of long-range order driven by intra-framework dispersion forces.

4.4.6 Transition Towards More Complex Metastable States

DFT and flexible force field MD simulations have been deployed to gain microscopic insight into the NGA behaviour of DUT-49 observed upon adsorption of a series of guest molecules, which can be classified as a specific form of breathing in Figure 4.1.[142,186,187] It was demonstrated that this abnormal phenomenon originates from a guest-triggered concerted rotation of the metal–organic polyhedra leading to a severe deformation of the linkers while preserving the connectivity of the MOF framework (Figure 4.31(a) and (b)). In particular, it was shown that a bistable porous material in which the empty open pore phase is thermodynamically more stable but the contracted pore phase is metastable under ambient conditions can indeed give rise to NGA.[187] Free-energy profiles as a function of the unit cell volume were constructed from several constant-volume (N, V, T) flexible force-field MD simulations inserting a variable number of methane molecules (matching with the corresponding experimental adsorption uptakes at different pressure points)

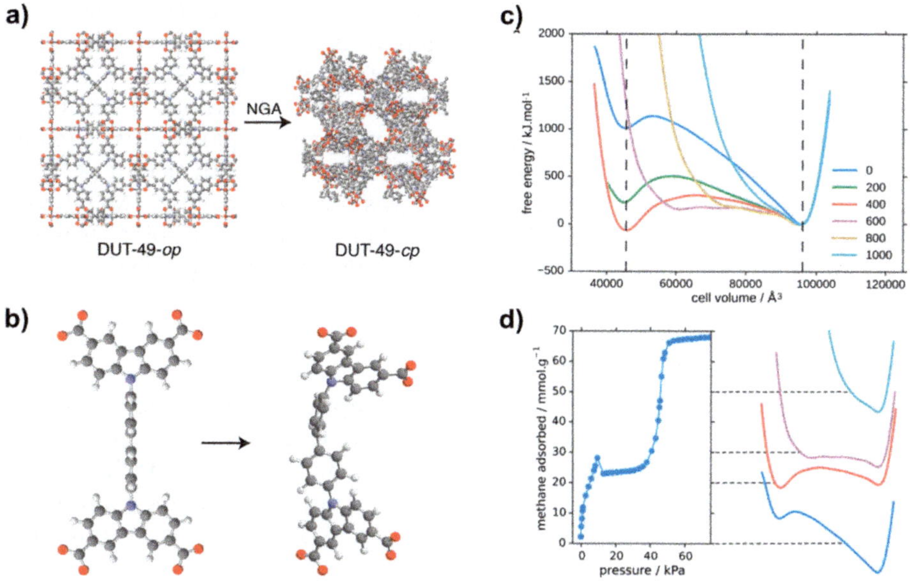

Figure 4.31 Flexible force-field MD simulations and thermodynamic approach. (a) Illustration of DUT-49-op (open pore) and DUT-49-cp (contracted pore). (b) Buckling-like deformation of the ligand demonstrated during the NGA process. (c) Free-energy profiles obtained from (*N*, *V*, *T*) simulations as a function of unit cell volume at 120 K with increasing amounts of methane molecules per unit cell. Volumes corresponding to the open pore and contracted pore phases are displayed as dashed lines. (d) The methane isotherm for DUT-49 at 111 K is shown with the corresponding free-energy profile for a given methane loading. Adapted from ref. 187 with permission from Cell Press, Copyright 2016.

in the simulation cell at 120 K (Figure 4.31(c) and (d)). It was revealed that for a methane loading of 400 molecules per unit cell (*i.e.*, 20.2 mmol g^{-1}) the contracted phase of DUT-49 becomes thermodynamically more stable due to enhanced host–guest interactions than the initial open pore form. However, despite the shift of global minima in the free energy landscape, an energy barrier of ~300 kJ mol^{-1} still persists for the open to contracted phase transition, which significantly reduced to 180 kJ mol^{-1} as the methane loading increased to 600 molecules per unit cell (*i.e.*, 30.4 mmol g^{-1}) allowing the system to adopt higher density forms. Notably, this 30.4 mmol g^{-1} methane loading does not induce a stable contracted form in the free energy landscape. It was argued that it is likely that by releasing methane molecules from the pore, DUT-49 structure transforms from a metastable state at 600 guest molecules per unit cell to a stable contracted phase at 400 guest molecules per unit cell. The system then recovers the open pore form at higher methane loadings due to increasing guest–guest and host–guest short-range repulsion forces. Finally, in an alternative approach to investigate NGA, a semi-analytical mean field model was applied to construct Helmholtz free energy profiles of guest loaded MOFs which were then translated towards the osmotic ensemble using the Legendre transformation (see Section 4.3.4.2) with additional application of mechanical pressure.[142] This approach showed the sensitivity of NGA towards various thermodynamic parameters such as empty framework flexibility, guest–host interaction strength as well as external mechanical pressure. Furthermore, it was revealed that even MOFs that do not exhibit NGA in absence of mechanical pressure can exhibit NGA when an appropriate mechanical pressure (or possibly tension) is applied.

4.4.7 Linker Dynamics

While linker dynamics is a rather broad concept that can include both displacive (*e.g.*, gate opening or molecular motors) or reconstructive (*e.g.*, bond rearrangement) modes of flexibility, they occur in a more local fashion than breathing, swelling, or amorphisation. As a result, more detailed and computationally expensive methods can be adopted to model these phenomena, as illustrated below.

4.4.7.1 Adsorption Induced

The guest-induced ligand dynamics of MOFs is expected to affect their adsorption and separation properties.[188,189] For example, HOMC simulations performed on the ultra-small channel-like Zr-MOF MIL-140B demonstrated that the adsorption of hexane isomers induces a significant tilting of the MOF organic linkers which varies in the range of 7.8–10.2° depending on the vapour pressure of the guest molecules.[190] Interestingly, the consideration of this guest-triggered ligand flexibility enabled to reproduce fairly the single component adsorption isotherms for all hexane isomers as well as their adsorption hierarchy in quinary mixture (Figure 4.32).

4.4.7.2 Electric Field Induced

Flexible force-field MD simulations revealed that the application of an external electric field can control the rotation of organic linkers (rotor-like behaviour) in the isoreticular metal–organic frameworks (IRMOFs) with the objective to tune the transport of methane in their pores.[191] In this work, the organic linkers of IRMOF-1 and IRMOF-7 were functionalised by different halogens to create permanent dipole moments on the linkers making them responsive to electric field.

The Cl-functionalised version of IRMOF-7 with the highest dipole moment (*i.e.*, highest torque on a ligand) was shown to undergo a 25° linker rotation under the application of an electric field of 8 V nm^{-1} (see Figure 4.33(a)). It was evidenced that the functionalisation of IRMOF-7 with two Cl atoms enables to lower the electric-field strength (as low as 0.5 V nm^{-1}) to rotate

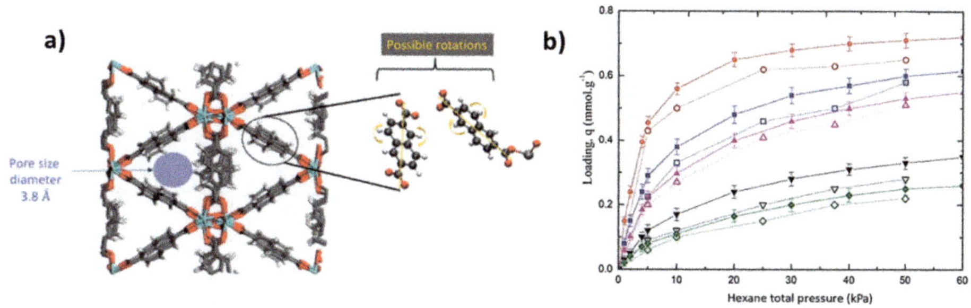

Figure 4.32 HOMC simulations. (a) Illustration of the MIL-140B structure (viewed along the channel) and possible organic linker rotations. (b) Single component adsorption isotherms for all hexane isomers compared to the corresponding experimental data. HOMC simulations: *n*-C6 (red -●-), 2MP (blue -■-), 3MP (pink -▲-), 23DMB (black -▼-) and 22DMB (green -◆-) and experimental data for *n*-C6 (red -○-), 2MP (blue -□-), 3MP (pink -△-), 23DMB (black -▽-) and 22DMB (green -◇-). Adapted from ref. 190 with permission from American Chemical Society, Copyright 2022.

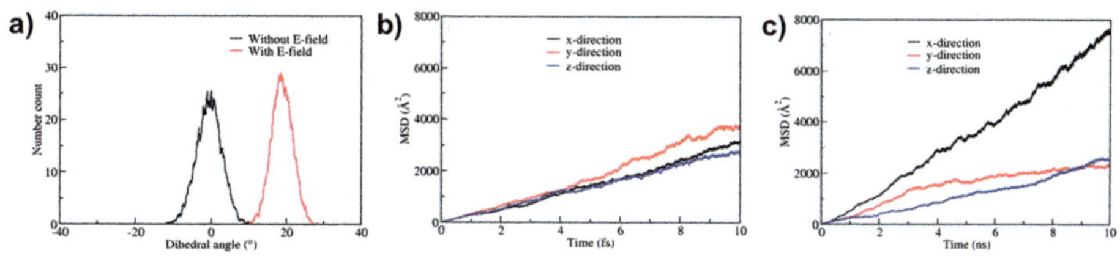

Figure 4.33 Flexible force-field MD simulations (a) Distribution of the dihedral angles of ligands in IRMOF-1-Cl at 100 K with and without the application of an electric field (E = 8 V nm^{-1}). (b) and (c) MSD profiles for methane in CR0–CR0p substituted IRMOF-7-Cl_2 without electric field and with electric field (2 V nm^{-1}). Adapted from ref. 191 with permission from The Royal Society of Chemistry.

the ligands. Interestingly, this electric-field induced rotation of the ligands was demonstrated to boost the methane diffusion (Figure 4.33(b) and (c)) that can be also reversibly controlled by changing the direction of the electric field applied.

4.4.8 Metal–Linker Bond Dynamics

A reversible guest-induced metal–linker bond rearrangement, classified as a reconstructive mode of flexibility with an intermediate spatial extent in Figure 4.1, can also occur in the MOF framework. Climbing image nudged elastic band (CI-NEB) and AIMD simulations were reported for the MOF MFM-300(Sc)[192] and demonstrated that this MOF framework undergoes substantial Sc–carboxylate bond dynamics upon ammonia adsorption that

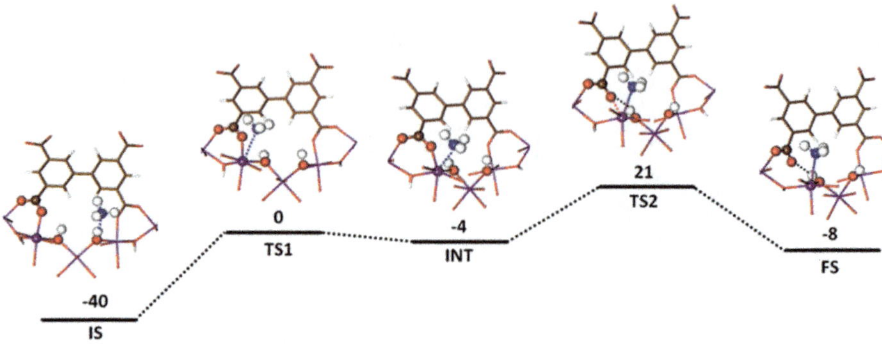

Figure 4.34 DFT-based climbing image nudged elastic band simulations. Potential energy profile for the NH_3-triggered Sc-carboxylate dynamic bonding in MFM-300(Sc) and illustrations of the initial state (IS), transition states (TS1 and TS2), intermediate state (INT), and final state (FS). Colour code: carbon (brown), oxygen (red), nitrogen (blue), hydrogen (white), and scandium (purple). Potential energies are reported in kJ mol^{-1}. Adapted from ref. 192 with permission from American Chemical Society, Copyright 2021.

enables a strong NH_3 binding toward the newly created Sc open metal sites, a key feature to ensure an efficient capture of this toxic molecule. Figure 4.34 illustrates a possible reaction path including Sc–O(carboxylate linker) bond breaking, formation of new O(carboxylate)···OH(μ-OH) hydrogen bond, which enables the NH_3 to coordinate to the Sc *via* its N-atom. AIMD simulations further evidence that the resulting Sc–NH_3 coordinative mode remains stable over time.

The flexibility of a MOF can also be probed by evaluating the metal–linker coordination strength, on the basis that weak coordination ability of a linker to the metal might lead to a high dynamics of the bond prone to induce a flexibility of the MOF framework, and *vice versa*. For example, PCP-N and PCP-C (Figure 4.35) are two isostructural Fe-based MOFs realised with Hppt (3-(2-pyrazinyl)-5-(4-pyridyl)-1,2,4-triazole) and Hdpt (3-(2-pyridyl)-5-(4-pyridyl)-1,2,4-triazole) ligands, respectively.[193] PCP-C is constructed by simply replacing a nitrogen atom from the pyrazinyl ligand of the PCP-N with a carbon atom. It was demonstrated that upon CO_2 and C_2H_2 adsorption, PCP-N undergoes significant structural changes involving dimensional change of the pore to trap additional gas molecules within the porosity while PCP-C is found to be quite rigid. Interestingly, the distinct structural dynamics—due to a subtle atom exchange in the ligand—can be traced back to different coordination ability of nitrogen atoms of pyrazine ($pK_b = 0.65$) and pyridine ($pK_b = 8.8$). To confirm this hypothesis, DFT calculations were performed to evaluate the ligand dissociation energies of the coordination bond between Fe and the two ligands. In this approach, representative discrete model complexes for PCP-N and PCP-C were constructed with one Fe^{2+} ion coordinated to four ppt or dpt ligands, respectively. The ligand dissociation energies

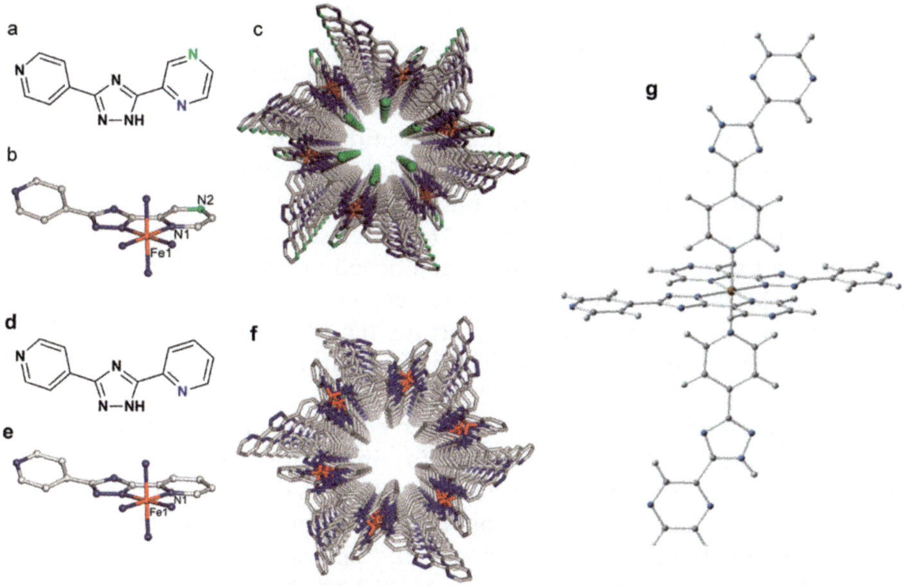

Figure 4.35 Structural representation of PCP-N and PCP-C. (a) and (d) Chemical structures of Hppt and Hdpt ligands, respectively. (b) and (e) Coordination environment of the Fe(II) centre in PCP-N and PCP-C, respectively. (c) and (f) Framework structures for PCP-N and PCP-C viewed along the *c*-direction, respectively. (g) DFT optimised finite metal-complex cleaved from the PCP-N, which was used for the ligand dissociation energy calculations. Adapted from ref. 193 with permission from American Chemical Society, Copyright 2015.

were then deduced by removing one ligand from the equatorial position. It appears that calculated ligand dissociation energy for the for the dpt ligand in PCP-C is 1.3 kcal mol^{-1} higher than the same for a ppt ligand of the PCP-N. As such, the authors argued that ppt ligands in the PCP-N are weakly coordinated to the Fe centre as compared to the dpt ligands in PCP-C, and thereby induces the flexibility in PCP-N framework while PCP-C is rigid.

4.5 Perspectives

Within this chapter, a comprehensive overview was given on the various types of flexible behaviour that exist within MOFs ranging from very localised flexible modes in the framework to more collective and spatially extended phenomena. Flexibility is induced by external stimuli such as temperature, pressure, guest adsorption, external fields, *etc.* Upon the application of these stimuli, the MOF framework may retain its bond integrity or bonds may be broken. Experimentally, one can recognise flexible behaviour by monitoring the response of the material to the external stimuli, for example by following the structural changes of the material by *in situ* XRD or indirectly by mercury porosimetry or by measuring adsorption isotherms. However, it is very difficult experimentally to follow the kinetics of these structural changes since they mostly occur in a timescale hardly attainable by the standard experimental techniques. Therefore, theoretical insights and computer simulations have become indispensable in understanding the flexible behaviour of MOFs. However, modelling flexibility in all its forms of complexity starting from the atomic scale and reaching out to the macroscopic scale, entails a formidable challenge for the modeller. Hereafter, a few challenges, and perspectives are listed that we have identified based on the ambition to model flexibility in MOFs in a more realistic way accounting for all forms of complexity present in a real material at various length and time scales, while being exposed to various stimuli which may potentially be exerted in a simultaneous way.

The various challenges may be brought in context to the four steps identified in 'the sequence of a modelling exercise' in Section 4.3.

One important component in modelling MOFs is the construction of the PES, *i.e.*, the second step in the modelling sequence, and more in particular the level of theory used to perform this task. Ideally the interactions are modelled at the quantum level, where DFT is nowadays the method of choice for modelling MOFs at the quantum level. Within relying on DFT calculations one can rely on very expensive MD simulations as was extensively detailed in this chapter, however in some cases one can already reveal useful insights into the potential flexibility by evaluating the metal–linker coordination strength as was shown in one of the case studies (Section 4.4.8). Irrespective of the successes obtained with DFT, it is important to emphasise that DFT in combination with currently available exchange-correlation functionals may on itself be not accurate enough to model some challenging MOFs having complicated electronic structures. This is for example the case for MOFs with complicated spin structures or MOFs, which exhibit strong electron correlation. Experiments revealed that for MIL-53 with metals like Fe^{3+} or Sc^{3+}, having complicated spin structures, the relative phase stabilities depend on the particular spin state.[194] For paddlewheel MOFs like DMOF-1 and DUT-49, extensive research has been performed on their stability in terms of the specific metal included in the paddlewheel. More in particular open metal sites may occur and various spin configurations are possible. Strong electron correlation is expected within the paddlewheel units when containing metals such as copper, cobalt or nickel.[195] To mediate the problem of describing electron correlations in a more accurate way, one can resort to rather expensive methods like the random phase approximation (RPA)[196–199] and the so-called GW approach,[200,201] which can be formulated using Green's function determined by perturbation theory. There are a few examples where such methods were used in the context of flexibility of MOFs.[196] However, these methods are very expensive and for the simulation of exotic phenomena that are governed by strong electron correlations, RPA and the GW approach also fall short, despite their increased complexity, due to their perturbative nature.

Other approaches have been devised that are nonperturbative. The dynamical mean-field theory (DMFT) also targets Green's function but restricts its spatial coordinates to a small cluster. Green's function in this cluster is then determined *via* a self-consistent set of equations, which still requires solving a many-body problem for the cluster (which is often referred to as the impurity problem).[202–204] A closely related approach that only uses the equal-time components of Green's function, still restricted to a cluster, is the recent Density Matrix Embedding Theory (DMET) proposal.[205–208] While both DMFT and DMET can be very successful, the approximations they incorporate are less well controlled and can only be systematically improved by increasing the size of the clusters and this leads to an exponential increase in complexity. While major advances have been made the last few years to apply such methods on more realistic systems,[208–210] their application to MOFs has so far, to the best of our knowledge, not been conducted. There are also direct wave-function based methods to deal with interactions, requiring a choice of orbitals in which the wavefunction is expressed, *e.g.*, the Hartree–Fock orbitals. At an exponential cost in the number of orbitals, the configuration interaction (CI) method is basically exact.[211] While being attractive from conceptual point of view, their applicability for large systems as MOFs is so far not feasible.[211] Within the field of many body quantum physics and quantum information theory, dedicated techniques have been devised to solve strongly correlated systems such as tensor network states,[212–214] but so far they were applied to systems which are much simpler than the MOFs of interest here. In summary, the treatment of MOFs with strong electron correlations, poses a major challenge and important methodological advances would be necessary. To advance in this area a strong interaction between various communities is necessary, namely the fundamental quantum physics and more application oriented MOF communities.

Another important challenge in modelling flexible MOFs, is to devise structural models for MOFs which describe them in a more realistic way, *i.e.*, having various forms of spatial disorder at various length scales. As explained in Section 4.3.1, disorder may extend from the sub-nanometre to the micrometre scale. Even if computational models would allow to simulate systems reaching to the micrometre scale, a major challenge is the construction of representative structural model for the MOF. Many experimental imaging and spectroscopic techniques exist to characterise the MOF structure, encompassing spectroscopic, diffraction, and microscopy methods, each with its own space-time window. They enable to characterise *in situ* either local or spatially more extended effects.[153,215] However, despite major advances in the field of *in situ* and *operando* spectroscopy, it remains very challenging to obtain a one-to-one correspondence between experimentally measured fingerprints and atomic scale information necessary to build structural models. Most techniques do not have the resolution to visualise individual point defects in these beam-sensitive materials or can only show the average of defects on a mesoscale. In many cases one needs to combine information from various spectroscopic techniques to validate the proposed structural model for the MOF. For example, NMR probes the local chemical environment, making it suitable to investigate materials lacking long-range order. Techniques such as ^{129}Xe and ^{2}H NMR have proven indispensable to characterise switching transitions and linker flipping in MOFs.[216] Vibrational spectroscopies such as Raman and IR can help to reveal structural information on the particular phase. More in particular modes in the terahertz region were found to be characteristic fingerprints for flexible materials.[217–219] Highly promising is the usage of TEM to characterise MOFs with atomic-level resolution. However, TEM is challenging, as most MOFs are very beam-sensitive materials. To progress in this field and to build more realistic structural models of MOFs a close synergistic action is required between theoretical modelling groups and experimental communities having access to the most advanced spectroscopic, microscopic, and imaging tools. From a modelling point-of-view, one needs to make many choices when building larger structural models having spatial disorder, and cross-coupling with experiment is necessary to select among the plethora of models, structures that are representative of the real material.

An important future direction within the MOF field, is to develop methods to describe the interatomic interactions with quantum accuracy on systems having sizes comparable to experimental scales, *i.e.*, reaching out to the micrometre scale. The increase in accessible length and time scales in modelling is certainly one of the biggest challenges in the field. As discussed in Section 4.3.2, a very promising direction is the development of MLPs for MOFs. While the field of MLPs has evolved substantially from a methodological point-of-view, it is only recently that their application and development within the MOF community appeared. The relatively slow development of MLPs within the MOF community can probably partly be ascribed to the complexity of generating representative training data to fit the MLPs for MOFs. Some recent contributions used a building block approach where training data were designed based on building blocks of the MOF to reduce the number of configurations that need to be simulated with quantum mechanical data.[53] However, new MLP algorithms are being developed which allow to substantially reduce the number of training data. These are very promising, however, the success of MLP application to realistic MOFs having sizes comparable to experimental samples, necessitates training data which may represent the broad variety in spatial heterogeneities found in MOFs. This is a highly non-trivial task, given the complexity of the structural models itself—see previous paragraph. As flexibility in MOFs is furthermore an activated process, one needs also to be able to generate training data in an efficient way, also for less probable regions of phase space. For the last aspect one can reside to enhanced sampling MD techniques to generate data in these lower probability regions of phase space. In any case, to fully explore the potential of MLPs within the MOF field, a close interaction should be established between researchers of the MLP community and the MOF community. Furthermore, to allow simulations of systems having millions of atoms, a close connection would also be necessary with computer scientists, to parallelise and optimise the algorithms to achieve the highest possible efficiency on currently available computer architectures being composes both of CPU and GPU parts.

As has become clear from this book chapter, modelling has already contributed substantially to the understanding of flexibility in MOFs, however, most of our insights so far yield thermodynamic information on the flexibility. For example, under given conditions one can construct thermodynamic potential like Helmholtz free energy curves along important collective variables. Probably one of the most challenging topics specifically within the field of flexible MOFs is how to obtain kinetics on the phase transformation. Both for experimentalists as for theoreticians, this is a very challenging task, as it depends on the occurrence of metastable states and activation barriers which are hard to assess both experimentally and theoretically. From a theoretical point-of-view, one needs proper sampling techniques for structurally extended models and ideally one could build so-called transition mechanism diagrams which indicate which transition mechanism would occur for various crystal sizes, spatial heterogeneities and external stimuli and predict for each of these mechanisms the kinetics of the transformation.[46] As mentioned in Section 4.3.3, it is not always trivial to define proper collective variables to sample configuration space. An interesting direction would be to explore also transition path sampling methods, where no prior information is necessary on the collective variables is necessary. Within transition path sampling (TPS), an ensemble of transition paths is created on the basis of an initial trajectory, which may lead to new reaction paths.[220–222] Within the field of crystal nucleation, TPS has successfully been used to unravel nucleation mechanisms but path sampling methods have hardly been applied for nanoporous materials due to the computational expense of generating many paths.[221,222] Usage of MLPs opens perspectives to use TPS for the materials of interest and discover new transition mechanisms. From the dynamics of the generated ensemble of transition paths, a reaction rate coefficient can in principle be determined, using methods like transition interface sampling or others.[223–226] To connect with experimental data, one should ideally have access to experimental kinetic information, which is so far still scarce in the field. Currently, experimental endeavours are undertaken to build dedicated *in situ* cells to unravel the spatiotemporal response of MOFs under operating conditions.

Finally, so far most of the theoretical modelling efforts, have focused on the effect of applying one trigger to the MOF and following the response of the material. It would be highly interesting to apply advanced thermodynamic and kinetic models to unravel the response of MOFs upon exposure to multiple triggers simultaneously and reveal how correlated the response to the various triggers is. As such, one could unravel how to tune the simultaneous application of various triggers in such a way to obtain a very specific desired response for particular applications. Some examples in this direction are already given by the investigation of the impact of applying mechanical pressure on negative gas adsorption[142] as well as how to tune the separation of gas mixtures by adjusting the mechanical pressure.[143]

In conclusion, this chapter has given an intriguing insight into the complex field of flexibility within MOFs and how theory and simulations can contribute to the understanding and ultimately the prediction of flexibility. We formulated a few challenges, which are by no sense comprehensive but show the hurdles ahead both from theory and experiment. Furthermore, the challenges show that further advances in the field, will require a close collaboration among theoreticians and experimentalists in the MOF field but also collaborations with other communities in the fields of fundamental quantum physics, computer science, spectroscopy and imaging, data science should be embraced. In any case inspiration from various fields will have to be merged and a synergistic approach between various communities will be necessary.

Acknowledgements

L.V. and V.V.S acknowledge the Research Board of Ghent University (BOF) for funding. S.M.J.R. acknowledges support by the Fund for Scientific Research Flanders (FWO) through a senior postdoctoral fellowship (grant no. 12T3522N) and the Research Board of Ghent University (BOF).

References

1. G. Fraux, S. Chibani and F.-X. Coudert, *Philos. Trans. R. Soc., A*, 2019, **377**, 20180220.
2. V. Van Speybroeck, S. Vandenhaute, A. E. J. Hoffman and S. M. J. Rogge, *Trends Chem.*, 2021, **3**, 605.
3. S. Kitagawa and K. Uemura, *Chem. Soc. Rev.*, 2005, **34**, 109.
4. S. Bureekaew, S. Shimomura and S. Kitagawa, *Sci. Technol. Adv. Mater.*, 2008, **9**, 014108.
5. S. Horike, S. Shimomura and S. Kitagawa, *Nat. Chem.*, 2009, **1**, 695.
6. G. Férey and C. Serre, *Chem. Soc. Rev.*, 2009, **38**, 1380.
7. A. Schneemann, V. Bon, I. Schwedler, I. Senkovska, S. Kaskel and R. A. Fischer, *Chem. Soc. Rev.*, 2014, **43**, 6062.
8. S. S. Nagarkar, A. V. Desai and S. K. Ghosh, *Chem. – Asian J.*, 2014, **9**, 2358.
9. Z. Chang, D.-H. Yang, J. Xu, T.-L. Hu and X.-H. Bu, *Adv. Mater.*, 2015, **27**, 5432.
10. R. E. Morris and L. Brammer, *Chem. Soc. Rev.*, 2017, **46**, 5444.
11. J.-P. Zhang, H.-L. Zhou, D.-D. Zhou, P.-Q. Liao and X.-M. Chen, *Natl. Sci. Rev.*, 2018, **5**, 907.
12. W. Danowski, T. van Leeuwen, S. Abdolahzadeh, D. Roke, W. R. Browne, S. J. Wezenberg and B. L. Feringa, *Nat. Nanotechnol.*, 2019, **14**, 488.
13. J. H. Lee, S. Jeoung, Y. G. Chung and H. R. Moon, *Coord. Chem. Rev.*, 2019, **389**, 161.
14. S. Krause, N. Hosono and S. Kitagawa, *Angew. Chem.*, 2020, **59**, 15325.

15. F. Bigdeli, C. T. Lollar, A. Morsali and H.-C. Zhou, *Angew. Chem.*, 2020, **59**, 4652.
16. B. H. Wilson, C. S. Vojvodin, G. Gholami, L. M. Abdulla, C. A. O'Keefe, R. W. Schurko and S. J. Loeb, *Chem*, 2021, **7**, 202.
17. P. Zhao, S. C. E. Tsang and D. Fairen-Jimenez, *Cell Rep. Phys. Sci.*, 2021, **2**, 100544.
18. C. L. Hobday and G. Kieslich, *Dalton Trans.*, 2021, **50**, 3759.
19. S. Seth and S. Jhulki, *Mater. Horiz.*, 2021, **8**, 700.
20. F.-X. Coudert, *Chem. Mater.*, 2015, **27**, 1905.
21. C. A. Trickett, K. J. Gagnon, S. Lee, F. Gándara, H.-B. Bürgi and O. M. Yaghi, *Angew. Chem., Int. Ed.*, 2015, **54**, 11162.
22. C. Caratelli, J. Hajek, F. G. Cirujano, M. Waroquier, F. X. Llabrés i Xamena and V. Van Speybroeck, *J. Catal.*, 2017, **352**, 401.
23. C. Mellot-Draznieks, C. Serre, S. Surblé, N. Audebrand and G. Férey, *J. Am. Chem. Soc.*, 2005, **127**, 16273.
24. S. Surblé, C. Serre, C. Mellot-Draznieks, F. Millange and G. Férey, *Chem. Commun.*, 2006, 284.
25. P. Horcajada, F. Salles, S. Wuttke, T. Devic, D. Heurtaux, G. Maurin, A. Vimont, M. Daturi, O. David, E. Magnier, N. Stock, Y. Filinchuk, D. Popov, C. Riekel, G. Férey and C. Serre, *J. Am. Chem. Soc.*, 2011, **133**, 17839.
26. U. Müller, *Symmetry Relationships Between Crystal Structures: Applications of Crystallographic Group Theory in Crystal Chemistry*, Oxford University Press, Oxford, 1st edn, 2013.
27. F. Millange, C. Serre and G. Férey, *Chem. Commun.*, 2002, 822.
28. C. Serre, F. Millange, C. Thouvenot, M. Noguès, G. Marsolier, D. Louër and G. Férey, *J. Am. Chem. Soc.*, 2002, **124**, 13519.
29. T. Loiseau, C. Serre, C. Huguenard, G. Fink, F. Taulelle, M. Henry, T. Bataille and G. Férey, *Chem. – Eur. J.*, 2004, **10**, 1373.
30. R. N. Widmer, G. I. Lampronti, S. Anzellini, R. Gaillac, S. Farsang, C. Zhou, A. M. Belenguer, C. W. Wilson, H. Palmer, A. K. Kleppe, M. T. Wharmby, X. Yu, S. M. Cohen, S. G. Telfer, S. A. T. Redfern, F.-X. Coudert, S. G. MacLeod and T. D. Bennett, *Nat. Mater.*, 2019, **18**, 370.
31. R. N. Widmer, G. I. Lampronti, S. Chibani, C. W. Wilson, S. Anzellini, S. Farsang, A. K. Kleppe, N. P. M. Casati, S. G. MacLeod, S. A. T. Redfern, F.-X. Coudert and T. D. Bennett, *J. Am. Chem. Soc.*, 2019, **141**, 9330.
32. S. Duane, A. D. Kennedy, B. J. Pendleton and D. Roweth, *Phys. Lett. B*, 1987, **195**, 216.
33. B. Mehlig, D. W. Heermann and B. M. Forrest, *Phys. Rev. B: Condens. Matter Mater. Phys.*, 1992, **45**, 679.
34. R. Faller and J. J. de Pablo, *J. Chem. Phys.*, 2002, **116**, 55.
35. M. Jeffroy, A. H. Fuchs and A. Boutin, *Chem. Commun.*, 2008, 3275.
36. F.-X. Coudert, M. Jeffroy, A. H. Fuchs, A. Boutin and C. Mellot-Draznieks, *J. Am. Chem. Soc.*, 2008, **130**, 14294.
37. J. A. Greathouse, T. L. Kinnibrugh and M. D. Allendorf, *Ind. Eng. Chem. Res.*, 2009, **48**, 3425.

38. A. Ghoufi and G. Maurin, *J. Phys. Chem. C*, 2010, **114**, 6496.
39. S. M. J. Rogge, R. Goeminne, R. Demuynck, J. J. Gutiérrez-Sevillano, S. Vandenbrande, L. Vanduyfhuys, M. Waroquier, T. Verstraelen and V. Van Speybroeck, *Adv. Theory Simul.*, 2019, **2**, 1.
40. D. Fairen-Jimenez, S. A. Moggach, M. T. Wharmby, P. A. Wright, S. Parsons and T. Düren, *J. Am. Chem. Soc.*, 2011, **133**, 8900.
41. I. Beurroies, M. Boulhout, P. L. Llewellyn, B. Kuchta, G. Férey, C. Serre and R. Denoyel, *Angew. Chem., Int. Ed.*, 2010, **49**, 7526.
42. J. Wieme, S. M. J. Rogge, P. G. Yot, L. Vanduyfhuys, S.-K. Lee, J.-S. Chang, M. Waroquier, G. Maurin and V. Van Speybroeck, *J. Mater. Chem. A*, 2019, **7**, 22663.
43. D. Yang, V. Bernales, T. Islamoglu, O. K. Farha, J. T. Hupp, C. J. Cramer, L. Gagliardi and B. C. Gates, *J. Am. Chem. Soc.*, 2016, **138**, 15189.
44. X. Feng, J. Hajek, H. S. Jena, G. Wang, S. K. P. Veerapandian, R. Morent, N. De Geyter, K. Leyssens, A. E. J. Hoffman, V. Meynen, C. Marquez, D. E. De Vos, V. Van Speybroeck, K. Leus and P. Van Der Voort, *J. Am. Chem. Soc.*, 2020, **142**, 3174.
45. S. M. J. Rogge, J. Wieme, L. Vanduyfhuys, S. Vandenbrande, G. Maurin, T. Verstraelen, M. Waroquier and V. Van Speybroeck, *Chem. Mater.*, 2016, **28**, 5721.
46. S. Vandenhaute, S. M. J. Rogge and V. Van Speybroeck, *Front. Chem.*, 2021, **9**, 718920.
47. S. M. J. Rogge, M. Waroquier and V. Van Speybroeck, *Nat. Commun.*, 2019, **10**, 4842.
48. J. Keupp and R. Schmid, *Adv. Theory Simul.*, 2019, **2**, 1900117.
49. J. Keupp, J. P. Dürholt and R. Schmid, *Faraday Discuss.*, 2021, **225**, 324.
50. L. Schaper, J. Keupp and R. Schmid, *Front. Chem.*, 2021, **9**, 757680.
51. A. C. T. van Duin, S. Dasgupta, F. Lorant and W. A. Goddard, *J. Phys. Chem. A*, 2001, **105**, 9396.
52. T. P. Senftle, S. Hong, M. M. Islam, S. B. Kylasa, Y. Zheng, Y. K. Shin, C. Junkermeier, R. Engel-Herbert, M. J. Janik, H. M. Aktulga, T. Verstraelen, A. Grama and A. C. T. van Duin, *npj Comput. Mater.*, 2016, **2**, 1.
53. M. Eckhoff and J. Behler, *J. Chem. Theory Comput.*, 2019, **15**, 3793.
54. S. Vandenhaute, M. Cools-Ceuppens, T. Verstraelen and V. Van Speybroeck, *npj Comput. Mater.*, 2023, **9**, 19.
55. J. P. Dürholt, R. Galvelis and R. Schmid, *Dalton Trans.*, 2016, **45**, 4370.
56. J. P. Durholt, J. Keupp and R. Schmid, *Eur. J. Inorg. Chem.*, 2016, 4517.
57. R. Semino, J. P. Dürholt, R. Schmid and G. Maurin, *J. Phys. Chem. C*, 2017, **121**, 21491.
58. S. M. J. Rogge, *Faraday Discuss.*, 2021, **225**, 271.
59. R. Gaillac, P. Pullumbi, K. A. Beyer, K. W. Chapman, D. A. Keen, T. D. Bennett and F.-X. Coudert, *Nat. Mater.*, 2017, **16**, 1149.
60. Y. Sun, S. M. J. Rogge, A. Lamaire, S. Vandenbrande, J. Wieme, C. R. Siviour, V. Van Speybroeck and J.-C. Tan, *Nat. Mater.*, 2021, **20**, 1015.
61. A. U. Ortiz, A. Boutin, K. J. Gagnon, A. Clearfield and F.-X. Coudert, *J. Am. Chem. Soc.*, 2014, **136**, 11540.

62. V. Nozari, C. Calahoo, J. M. Tuffnell, P. Adelhelm, K. Wondraczek, S. E. Dutton, T. D. Bennett and L. Wondraczek, *Sci. Rep.*, 2020, **10**, 3532.
63. A. Simonov and A. L. Goodwin, *Nat. Rev. Chem*, 2020, **4**, 657.
64. E. M. Reynolds, E. H. Wolpert, A. R. Overy, L. Mizzi, A. Simonov, J. N. Grima, S. Kaskel and A. L. Goodwin, *Faraday Discuss.*, 2021, **225**, 241.
65. S. M. J. Rogge, L. Vanduyfhuys, A. Ghysels, M. Waroquier, T. Verstraelen, G. Maurin and V. Van Speybroeck, *J. Chem. Theory Comput.*, 2015, **11**, 5583.
66. S. M. J. Rogge, S. Caroes, R. Demuynck, M. Waroquier, V. Van Speybroeck and A. Ghysels, *J. Chem. Theory Comput.*, 2018, **14**, 1186.
67. V. Van Speybroeck, *Philos. Trans. R. Soc., A*, 2023, **381**, 20220239.
68. L. Vanduyfhuys, S. M. J. Rogge, J. Wieme, S. Vandenbrande, G. Maurin, M. Waroquier and V. Van Speybroeck, *Nat. Commun.*, 2018, **9**, 204.
69. L. Vanduyfhuys, A. Ghysels, S. M. J. Rogge, R. Demuynck and V. Van Speybroeck, *Mol. Simul.*, 2015, **41**, 1311.
70. L. Vanduyfhuys, T. Verstraelen, M. Vandichel, M. Waroquier and V. Van Speybroeck, *J. Chem. Theory Comput.*, 2012, **8**, 3217.
71. J. Wieme, L. Vanduyfhuys, S. M. J. Rogge, M. Waroquier and V. Van Speybroeck, *J. Phys. Chem. C*, 2016, **120**, 14934.
72. R. Demuynck, S. M. J. Rogge, L. Vanduyfhuys, J. Wieme, M. Waroquier and V. Van Speybroeck, *J. Chem. Theory Comput.*, 2017, **13**, 5861.
73. G. D. Barmparis, Z. Lodziana, N. Lopez and I. N. Remediakis, *Beilstein J. Nanotechnol.*, 2015, **6**, 361.
74. S. Ali, V. S. Myasnichenko and E. C. Neyts, *Phys. Chem. Chem. Phys.*, 2016, **18**, 792.
75. W. H. Sun and G. Ceder, *Surf. Sci.*, 2013, **617**, 53.
76. R. Tran, Z. Xu, B. Radhakrishnan, D. Winston, W. Sun, K. A. Persson and S. P. Ong, *Sci. Data*, 2016, **3**, 160080.
77. P. G. Yot, Q. Ma, J. Haines, Q. Yang, A. Ghoufi, T. Devic, C. Serre, V. Dmitriev, G. Férey, C. Zhong and G. Maurin, *Chem. Sci.*, 2012, **3**, 1100.
78. M. J. Cliffe, W. Wan, X. D. Zou, P. A. Chater, A. K. Kleppe, M. G. Tucker, H. Wilhelm, N. P. Funnell, F. X. Coudert and A. L. Goodwin, *Nat. Commun.*, 2014, **5**, 8.
79. D. N. Johnstone, F. C. N. Firth, C. P. Grey, P. A. Midgley, M. J. Cliffe and S. M. Collins, *J. Am. Chem. Soc.*, 2020, **142**, 13081.
80. S. Bureekaew, V. Balwani, S. Amirjalayer and R. Schmid, *CrystEngComm*, 2015, **17**, 344.
81. Y. J. Colon, D. A. Gomez-Gualdron and R. Q. Snurr, *Cryst. Growth Des.*, 2017, **17**, 5801.
82. X. Feng, H. S. Jena, C. Krishnaraj, D. Arenas-Esteban, K. Leus, G. Wang, J. Sun, M. Rüscher, J. Timoshenko, B. Roldan Cuenya, S. Bals and P. V. D. Voort, *J. Am. Chem. Soc.*, 2021, **143**, 21511.
83. K. Burke, *J. Chem. Phys.*, 2012, **136**, 150901.
84. A. J. Cohen, P. Mori-Sánchez and W. Yang, *Chem. Rev.*, 2012, **112**, 289.
85. N. Mardirossian and M. Head-Gordon, *Mol. Phys.*, 2017, **115**, 2315.

86. M. G. Medvedev, I. S. Bushmarinov, J. Sun, J. P. Perdew and K. A. Lyssenko, *Science*, 2017, **355**, 49.
87. S. Grimme, J. Antony, S. Ehrlich and H. Krieg, *J. Chem. Phys.*, 2010, **132**, 1.
88. R. J. Maurer, C. Freysoldt, A. M. Reilly, J. G. Brandenburg, O. T. Hofmann, T. Björkman, S. Lebègue and A. Tkatchenko, *Annu. Rev. Mater. Res.*, 2019, **49**, 1.
89. M. Stöhr, T. Van Voorhis and A. Tkatchenko, *Chem. Soc. Rev.*, 2019, **48**, 4118.
90. J. Hermann, R. A. DiStasio and A. Tkatchenko, *Chem. Rev.*, 2017, **117**, 4714.
91. S. Grimme, A. Hansen, J. G. Brandenburg and C. Bannwarth, *Chem. Rev.*, 2016, **116**, 5105.
92. Van V. Speybroeck, S. Vandenhaute, A. E. J. Hoffman and S. M. J. Rogge, *Trends Chem.*, 2021, **3**, 605.
93. L. Vanduyfhuys, S. Vandenbrande, T. Verstraelen, R. Schmid, M. Waroquier and V. Van Speybroeck, *J. Comput. Chem.*, 2015, **36**, 1015.
94. S. Bureekaew, S. Amirjalayer, M. Tafipolsky, C. Spickermann, T. K. Roy and R. Schmid, *Phys. Status Solidi B*, 2013, **250**, 1128.
95. M. A. Addicoat, N. Vankova, I. F. Akter and T. Heine, *J. Chem. Theory Comput.*, 2014, **10**, 880.
96. L. Vanduyfhuys, S. Vandenbrande, J. Wieme, M. Waroquier, T. Verstraelen and V. Van Speybroeck, *J. Comput. Chem.*, 2018, **39**, 999.
97. M. Tafipolsky, S. Amirjalayer and R. Schmid, *J. Comput. Chem.*, 2007, **28**, 1169.
98. L. Zhao, Q. Yang, Q. Ma, C. Zhong, J. Mi and D. Liu, *J. Mol. Model.*, 2011, **17**, 227.
99. F. Salles, A. Ghoufi, G. Maurin, R. G. Bell, C. Mellot-Draznieks and G. Férey, *Angew. Chem., Int. Ed.*, 2008, **47**, 8487.
100. J. K. Bristow, D. Tiana and A. Walsh, *J. Chem. Theory Comput.*, 2014, **10**, 4644.
101. J. K. Bristow, J. M. Skelton, K. L. Svane, A. Walsh and J. D. Gale, *Phys. Chem. Chem. Phys.*, 2016, **18**, 29316.
102. D. E. Coupry, M. A. Addicoat and T. Heine, *J. Chem. Theory Comput.*, 2016, **12**, 5215.
103. T. Verstraelen, S. Vandenbrande and P. W. Ayers, *J. Chem. Phys.*, 2014, **141**, 194114.
104. T. Verstraelen, P. W. Ayers, V. Van Speybroeck and M. Waroquier, *J. Chem. Phys.*, 2013, **138**, 074108.
105. K. Chenoweth, A. C. T. van Duin and W. A. Goddard, *J. Phys. Chem. A*, 2008, **112**, 1040.
106. V. Barone, I. Cacelli, N. De Mitri, D. Licari, S. Monti and G. Prampolini, *Phys. Chem. Chem. Phys.*, 2013, **15**, 3736.
107. S. Vandenbrande, M. Waroquier, V. Van Speybroeck and T. Verstraelen, *J. Chem. Theory Comput.*, 2017, **13**, 161.
108. S. Grimme, *J. Chem. Theory Comput.*, 2014, **10**, 4497.

109. S. K. Burger, M. Lacasse, T. Verstraelen, J. Drewry, P. Gunning and P. W. Ayers, *J. Chem. Theory Comput.*, 2012, **8**, 554.
110. S. Batzner, A. Musaelian, L. Sun, M. Geiger, J. P. Mailoa, M. Kornbluth, N. Molinari, T. E. Smidt and B. Kozinsky, *Nat. Commun.*, 2022, **13**, 2453.
111. A. Grisafi and M. Ceriotti, *J. Chem. Phys.*, 2019, **151**, 204105.
112. Y. Yu, W. Zhang and D. Mei, *J. Phys. Chem. C*, 2022, **126**, 1204.
113. O. Tayfuroglu, A. Kocak and Y. Zorlu, *Phys. Chem. Chem. Phys.*, 2022, **24**, 11882.
114. C. Schran, F. L. Thiemann, P. Rowe, E. A. Müller, O. Marsalek and A. Michaelides, *Proc. Natl. Acad. Sci. U. S. A.*, 2021, **118**, e2110077118.
115. G. Sivaraman, A. N. Krishnamoorthy, M. Baur, C. Holm, M. Stan, G. Csányi, C. Benmore and Á. Vázquez-Mayagoitia, *npj Comput. Mater.*, 2020, **6**, 1.
116. W. Wang, T. Yang, W. H. Harris and R. Gómez-Bombarelli, *Chem. Commun.*, 2020, **56**, 8920.
117. J. Behler and G. Csányi, *Eur. Phys. J. B*, 2021, **94**, 142.
118. J. P. Durholt, R. Galvelis and R. Schmid, *Dalton Trans.*, 2016, **45**, 4370.
119. A. Davtyan, J. F. Dama, G. A. Voth and H. C. Andersen, *J. Chem. Phys.*, 2015, **142**, 154104.
120. A. Davtyan, G. A. Voth and H. C. Andersen, *J. Chem. Phys.*, 2016, **145**, 224107.
121. D. Frenkel and B. Smit, *Understanding Molecular Simulation: From Algorithms to Applications*, Academic Press, San Diego, 2002.
122. B. Peters, *Reaction Rate Theory and Rare Events*, 1st edn, 2017, https://www.elsevier.com/books/reaction-rate-theory-and-rare-events/peters/978-0-444-56349-1, accessed 3 May 2022.
123. R. Demuynck, J. Wieme, S. M. J. Rogge, K. D. Dedecker, L. Vanduyfhuys, M. Waroquier and V. Van Speybroeck, *J. Chem. Theory Comput.*, 2018, **14**, 5511.
124. A. Laio and F. L. Gervasio, *Rep. Prog. Phys.*, 2008, **71**, 126601.
125. O. Valsson, P. Tiwary and M. Parrinello, *Annu. Rev. Phys. Chem.*, 2016, **67**, 159.
126. G. M. Torrie and J. P. Valleau, *J. Comput. Phys.*, 1977, **23**, 187.
127. J. G. Kirkwood, *J. Chem. Phys.*, 1935, **3**, 300.
128. S. Kumar, D. Bouzida, R. H. Swendsen, P. A. Kollman and J. M. Rosenberg, *J. Comput. Chem.*, 1992, **13**, 1011.
129. O. Valsson and M. Parrinello, *Phys. Rev. Lett.*, 2014, **113**, 1.
130. Y. Sugita and Y. Okamoto, *Chem. Phys. Lett.*, 1999, **314**, 141.
131. C. Dellago, P. G. Bolhuis, F. S. Csajka and D. Chandler, *J. Chem. Phys.*, 1998, **108**, 1964.
132. P. G. Bolhuis, D. Chandler, C. Dellago and P. L. Geissler, *Annu. Rev. Phys. Chem.*, 2002, **53**, 291.
133. J. C. Tan and A. K. Cheetham, *Chem. Soc. Rev.*, 2011, **40**, 1059.
134. F.-X. Coudert and A. H. Fuchs, *Coord. Chem. Rev.*, 2016, **307**, 211.

135. S. E. Tallentire, F. Child, I. Fall, L. Vella-Zarb, I. R. Evans, M. G. Tucker, D. A. Keen, C. Wilson and J. S. O. Evans, *J. Am. Chem. Soc.*, 2013, **135**, 12849.
136. F.-X. Coudert, *Phys. Chem. Chem. Phys.*, 2010, **12**, 10904.
137. P. G. Yot, Z. Boudene, J. Macia, D. Granier, L. Vanduyfhuys, T. Verstraelen, V. Van Speybroeck, T. Devic, C. Serre, G. Férey, N. Stock and G. Maurin, *Chem. Commun.*, 2014, **50**, 9462.
138. G. Jaeger, *Arch. Hist. Exact Sci.*, 1998, **53**, 51.
139. A. Ghysels, L. Vanduyfhuys, M. Vandichel, M. Waroquier, V. Van Speybroeck and B. Smit, *J. Phys. Chem. C*, 2013, **117**, 11540.
140. C. Triguero, F.-X. Coudert, A. Boutin, A. H. Fuchs and A. V. Neimark, *J. Phys. Chem. Lett.*, 2011, **2**, 2033.
141. P. G. Yot, L. Vanduyfhuys, E. Alvarez, J. Rodriguez, J.-P. Itié, P. Fabry, N. Guillou, T. Devic, I. Beurroies, P. L. Llewellyn, V. Van Speybroeck, C. Serre and G. Maurin, *Chem. Sci.*, 2016, **7**, 446.
142. L. Vanduyfhuys and V. Van Speybroeck, *Commun. Phys.*, 2019, **2**, 1.
143. L. Vanduyfhuys and G. Maurin, *Advanced Theory and Simulations*, 2019, p. 1900124.
144. N. A. Ramsahye and G. Maurin, in *Modelling and Simulation in the Science of Micro- and Meso-Porous Materials*, ed. C. R. A. Catlow, V. Van Speybroeck, and R. A. van Santen, Elsevier, 2018, pp. 63–97.
145. B. C. Bukowski, F. J. Keil, P. I. Ravikovitch, G. Sastre, R. Q. Snurr and M.-O. Coppens, *Adsorption*, 2021, **27**, 683.
146. M. Bonneau, C. Lavenn, J.-J. Zheng, A. Legrand, T. Ogawa, K. Sugimoto, F.-X. Coudert, R. Reau, S. Sakaki, K. Otake and S. Kitagawa, *Nat. Chem.*, 2022, **14**, 816.
147. H. Tanaka, S. Ohsaki, S. Hiraide, D. Yamamoto, S. Watanabe and M. T. Miyahara, *J. Phys. Chem. C*, 2014, **118**, 8445.
148. H. Tanaka and M. T. Miyahara, *Curr. Opin. Chem. Eng.*, 2019, **24**, 19.
149. H. Eyring, *J. Chem. Phys.*, 1935, **3**, 107.
150. P. Hänggi, P. Talkner and M. Borkovec, *Rev. Mod. Phys.*, 1990, **62**, 251.
151. C. Chang, V. L. Deringer, K. S. Katti, V. Van Speybroeck and C. M. Wolverton, *Nat. Rev. Mater.*, 2023, 1.
152. S. Borgmans, S. M. J. Rogge, J. S. De Vos, C. V. Stevens, P. Van Der Voort and V. Van Speybroeck, *Angew. Chem.*, 2021, **60**, 8913.
153. V. Bon, E. Brunner, A. Poppl and S. Kaskel, *Adv. Funct. Mater.*, 2020, **30**, 32.
154. K. I. Hadjiivanov, D. A. Panayotov, M. Y. Mihaylov, E. Z. Ivanova, K. K. Chakarova, S. M. Andonova and N. L. Drenchev, *Chem. Rev.*, 2021, **121**, 1286.
155. A. E. J. Hoffman, L. Vanduyfhuys, I. Nevjestić, J. Wieme, S. M. J. Rogge, H. Depauw, P. Van Der Voort, H. Vrielinck and V. Van Speybroeck, *J. Phys. Chem. C*, 2018, **122**, 2734.
156. A. E. J. Hoffman, I. Senkovska, J. Wieme, A. Krylov, S. Kaskel and V. V. Speybroeck, *J. Mater. Chem. A*, 2022, **10**, 17254.

157. S. Krause, J. D. Evans, V. Bon, I. Senkovska, P. Iacomi, F. Kolbe, S. Ehrling, E. Troschke, J. Getzschmann, D. M. Többens, A. Franz, D. Wallacher, P. G. Yot, G. Maurin, E. Brunner, P. L. Llewellyn, F.-X. Coudert and S. Kaskel, *Nat. Commun.*, 2019, **10**, 3632.
158. P. Iacomi, J. S. Lee, L. Vanduyfhuys, K. H. Cho, P. Fertey, J. Wieme, D. Granier, G. Maurin, V. Van Speybroeck, J.-S. Chang and P. G. Yot, *Chem. Sci.*, 2021, **12**, 5682.
159. R. Pallach, J. Keupp, K. Terlinden, L. Frentzel-Beyme, M. Kloß, A. Machalica, J. Kotschy, S. K. Vasa, P. A. Chater, C. Sternemann, M. T. Wharmby, R. Linser, R. Schmid and S. Henke, *Nat. Commun.*, 2021, **12**, 4097.
160. C. L. Hobday, S. Krause, S. M. J. Rogge, J. D. Evans and H. Bunzen, *Front. Chem.*, 2022, **9**, 829906.
161. H. Fang, H. Demir, P. Kamakoti and D. S. Sholl, *J. Mater. Chem. A*, 2014, **2**, 274.
162. J. Yu, L.-H. Xie, J.-R. Li, Y. Ma, J. M. Seminario and P. B. Balbuena, *Chem. Rev.*, 2017, **117**, 9674.
163. S. Vandenbrande, M. Waroquier, V. Van Speybroeck and T. Verstraelen, *J. Chem. Theory Comput.*, 2018, **14**, 6359.
164. J. D. Evans, V. Bon, I. Senkovska, H.-C. Lee and S. Kaskel, *Nat. Commun.*, 2020, **11**, 2690.
165. J. Teufel, H. Oh, M. Hirscher, M. Wahiduzzaman, L. Zhechkov, A. Kuc, T. Heine, D. Denysenko and D. Volkmer, *Adv. Mater.*, 2013, **25**, 635.
166. P. Iacomi and G. Maurin, *ACS Appl. Mater. Interfaces*, 2021, **13**, 50602.
167. T. Chokbunpiam, S. Fritzsche, C. Chmelik, J. Caro, W. Janke and S. Hannongbua, *Chem. Phys. Lett.*, 2016, **648**, 178.
168. F. Coudert, *ChemPhysChem*, 2017, **18**, 2732.
169. L. Zhang, Z. Hu and J. Jiang, *J. Am. Chem. Soc.*, 2013, **135**, 3722.
170. C. Zhang, J. A. Gee, D. S. Sholl and R. P. Lively, *J. Phys. Chem. C*, 2014, **118**, 20727.
171. S. Watanabe, H. Sugiyama, H. Adachi, H. Tanaka and M. T. Miyahara, *J. Chem. Phys.*, 2009, **130**, 164707.
172. B. Zheng and G. Maurin, *Angew. Chem.*, 2019, **131**, 13872.
173. S. Siwaipram, P. A. Bopp, J. Keupp, L. Pukdeejorhor, J.-C. Soetens, S. Bureekaew and R. Schmid, *J. Phys. Chem. C*, 2021, **125**, 12837.
174. Y. Du, B. Wooler, M. Nines, P. Kortunov, C. S. Paur, J. Zengel, S. C. Weston and P. I. Ravikovitch, *J. Am. Chem. Soc.*, 2015, **137**, 13603.
175. A. Ghoufi, G. Maurin and G. Férey, *J. Phys. Chem. Lett.*, 2010, **1**, 2810.
176. A. Ghoufi, A. Subercaze, Q. Ma, P. G. Yot, Y. Ke, I. Puente-Orench, T. Devic, V. Guillerm, C. Zhong, C. Serre, G. Férey and G. Maurin, *J. Phys. Chem. C*, 2012, **116**, 13289.
177. V. Haigis, F.-X. Coudert, R. Vuilleumier and A. Boutin, *Phys. Chem. Chem. Phys.*, 2013, **15**, 19049.
178. A. M. Walker, B. Civalleri, B. Slater, C. Mellot-Draznieks, F. Corà, C. M. Zicovich-Wilson, G. Román-Pérez, J. M. Soler and J. D. Gale, *Angew. Chem.*, 2010, **122**, 7663.

179. J. Wieme, S. M. J. Rogge, P. G. Yot, L. Vanduyfhuys, S.-K. Lee, J.-S. Chang, M. Waroquier, G. Maurin and V. Van Speybroeck, *J. Mater. Chem. A*, 2019, 7, 22663.
180. A. Ghoufi, K. Benhamed, L. Boukli-Hacene and G. Maurin, *ACS Cent. Sci.*, 2017, 3, 394.
181. K. Chen, R. Singh, J. Guo, Y. Guo, A. Zavabeti, Q. Gu, R. Q. Snurr, P. A. Webley and G. K. Li, *ACS Appl. Mater. Interfaces*, 2022, **14**, 13904.
182. J. Maul, M. R. Ryder, M. T. Ruggiero and A. Erba, *Phys. Rev. B: Condens. Matter Mater. Phys.*, 2019, **99**, 014102.
183. A. U. Ortiz, A. Boutin, A. H. Fuchs and F.-X. Coudert, *J. Phys. Chem. Lett.*, 2013, **4**, 1861.
184. N. Castel and F.-X. Coudert, *J. Phys. Chem. C*, 2022, **126**, 6905.
185. T. D. Bennett, A. L. Goodwin, M. T. Dove, D. A. Keen, M. G. Tucker, E. R. Barney, A. K. Soper, E. G. Bithell, J. C. Tan and A. K. Cheetham, *Phys. Rev. Lett.*, 2010, **104**, 115503.
186. S. Krause, V. Bon, I. Senkovska, U. Stoeck, D. Wallacher, D. M. Többens, S. Zander, R. S. Pillai, G. Maurin, F.-X. Coudert and S. Kaskel, *Nature*, 2016, **532**, 348.
187. J. D. Evans, L. Bocquet and F.-X. Coudert, *Chem*, 2016, **1**, 873.
188. R. J. Verploegh, S. Nair and D. S. Sholl, *J. Am. Chem. Soc.*, 2015, **137**, 15760.
189. K. Zhang, R. P. Lively, C. Zhang, R. R. Chance, W. J. Koros, D. S. Sholl and S. Nair, *J. Phys. Chem. Lett.*, 2013, **4**, 3618.
190. H. Zhao, G. Maurin and A. Ghoufi, *J. Phys. Chem. C*, 2022, **126**, 2905.
191. S. Namsani and A. O. Yazaydin, *Mol. Syst. Des. Eng.*, 2018, **3**, 951.
192. P. Lyu, A. M. Wright, A. López-Olvera, P. G. M. Mileo, J. A. Zárate, E. Martínez-Ahumada, V. Martis, D. R. Williams, M. Dincă, I. A. Ibarra, G. Maurin, M. Dincă, I. A. Ibarra and G. Maurin, *Chem. Mater.*, 2021, **33**, 6186.
193. Y. Ma, R. Matsuda, H. Sato, Y. Hijikata, L. Li, S. Kusaka, M. Foo, F. Xue, G. Akiyama, R. Yuan and S. Kitagawa, *J. Am. Chem. Soc.*, 2015, **137**, 15825.
194. P. G. Yot, K. Yang, V. Guillerm, F. Ragon, V. Dmitriev, P. Parisiades, E. Elkaïm, T. Devic, P. Horcajada, C. Serre, N. Stock, J. P. S. Mowat, P. A. Wright, G. Férey and G. Maurin, *Eur. J. Inorg. Chem.*, 2016, **2016**, 4424.
195. H. Ha, Y. Kim, D. Kim, J. Lee, Y. Song, S. Kim, M. H. Park, Y. Kim, H. Kim, M. Yoon and M. Kim, *Chem. – Eur. J.*, 2019, **25**, 14414.
196. J. Wieme, K. Lejaeghere, G. Kresse and V. Van Speybroeck, *Nat. Commun.*, 2018, **9**, 4899.
197. H. Eshuis, J. E. Bates and F. Furche, *Theor. Chem. Acc.*, 2012, **131**, 1084.
198. J. Harl and G. Kresse, *Phys. Rev. Lett.*, 2009, **103**, 056401.
199. X. Ren, P. Rinke, C. Joas and M. Scheffler, *J. Mater. Sci.*, 2012, **47**, 7447.
200. M. Shishkin, M. Marsman and G. Kresse, *Phys. Rev. Lett.*, 2007, **99**, 4.
201. L. Hedin, *Phys. Rev.*, 1965, **139**, A796.
202. A. Georges, G. Kotliar, W. Krauth and M. J. Rozenberg, *Rev. Mod. Phys.*, 1996, **68**, 13.

203. G. Kotliar, S. Y. Savrasov, K. Haule, V. S. Oudovenko, O. Parcollet and C. A. Marianetti, *Rev. Mod. Phys.*, 2006, **78**, 865.
204. D. Vollhardt, *Ann. Phys.*, 2012, **524**, 1.
205. G. Knizia and G. K.-L. Chan, *J. Chem. Theory Comput.*, 2013, **9**, 1428.
206. S. Wouters, C. A. Jiménez-Hoyos, Q. Sun and G. K. L. Chan, *J. Chem. Theory Comput.*, 2016, **12**, 2706.
207. Z.-H. Cui, T. Zhu and G. K.-L. Chan, *J. Chem. Theory Comput.*, 2020, **16**, 119.
208. H. Q. Pham, M. R. Hermes and L. Gagliardi, *J. Chem. Theory Comput.*, 2020, **16**, 130.
209. L. Gagliardi, D. G. Truhlar, G. Li Manni, R. K. Carlson, C. E. Hoyer and J. W. L. Bao, *Acc. Chem. Res.*, 2017, **50**, 66.
210. T. Y. Zhu and G. K. L. Chan, *Phys. Rev. X*, 2021, **11**, 13.
211. S. O. Odoh, C. J. Cramer, D. G. Truhlar and L. Gagliardi, *Chem. Rev.*, 2015, **115**, 6051.
212. S. R. White, *Phys. Rev. Lett.*, 1992, **69**, 2863.
213. F. Verstraete, V. Murg and J. I. Cirac, *Adv. Phys.*, 2008, **57**, 143.
214. U. Schollwöck, *Ann. Phys.*, 2011, **326**, 96.
215. L. Hanna and J. V. Lockard, *J. Phys.: Condens.Matter*, 2019, **31**, 483001.
216. F. Kolbe, S. Krause, V. Bon, I. Senkovska, S. Kaskel and E. Brunner, *Chem. Mater.*, 2019, **31**, 6193.
217. A. Krylov, A. Vtyurin, P. Petkov, I. Senkovska, M. Maliuta, V. Bon, T. Heine, S. Kaskel and E. Slyusareva, *Phys. Chem. Chem. Phys.*, 2017, **19**, 32099.
218. M. R. Ryder and J. C. Tan, *Dalton Trans.*, 2016, **45**, 4154.
219. M. R. Ryder, B. Civalleri, T. Bennett, S. Henke, S. Rudić, G. Cinque, F. Fernandez-Alonso and J. C. Tan, *Phys. Rev. Lett.*, 2014, **113**, 1.
220. C. Dellago, P. G. Bolhuis and P. L. Geissler, *Adv. Chem. Phys.*, 2002, **123**, 1.
221. J. Anwar and D. Zahn, *Angew. Chem., Int. Ed.*, 2011, **50**, 1996.
222. G. T. Beckham, B. Peters and B. L. Trout, *J. Phys. Chem. B*, 2008, **112**, 7460.
223. T. S. van Erp and P. G. Bolhuis, *J. Comput. Phys.*, 2005, **205**, 157.
224. T. S. van Erp, D. Moroni and P. G. Bolhuis, *J. Chem. Phys.*, 2003, **118**, 7762.
225. D. Moroni, P. G. Bolhuis and T. S. van Erp, *J. Chem. Phys.*, 2004, **120**, 4055.
226. R. Cabriolu, K. M. S. Refsnes, P. G. Bolhuis and T. S. van Erp, *J. Chem. Phys.*, 2017, **147**, 17.

CHAPTER 5

New Dimensions of Flexible MOFs: Toward Complex Systems and Devices

RUI-BIAO LIN[a], XIAO-MING CHEN[a], SUSUMU KITAGAWA[b] AND BANGLIN CHEN[c*]

[a]MOE Key Laboratory of Bioinorganic and Synthetic Chemistry, GBRCE for Functional Molecular Engineering, School of Chemistry, IGCME, Sun Yat-Sen University, Guangzhou, 510275, China; [b]Institute for Integrated Cell-Material Sciences, Kyoto University, Yoshida, Sakyo-ku, Kyoto, 606-8501, Japan; [c]Fujian Provincial Key Laboratory of Polymer Materials, College of Chemistry & Materials Science, Fujian Normal University, Fuzhou, 350007, China

*E-mail: banglin.chen@fjnu.edu.cn

5.1 Introduction

Porous solids are of fundamental importance for their recognition of molecules and broad applications in many processes in modern human society, such as purification, filtration, extraction, catalysis, drying and cooling.[1] Numerous metal–organic frameworks (MOFs) or porous coordination polymers (PCPs) have been developed to serve as novel porous solids, assembled by inorganic species and organic linkers through strong coordination bonds.[2–4] The hybridisation of inorganic and organic moieties endows single-entity MOFs to integrate robust coordination geometry from metal centres and flexibility from the responsive structural dynamics of organic ligands,

Inorganic Materials Series No. 13
Flexible Metal–Organic Frameworks: Structural Design, Synthesis and Properties
Edited by Susumu Kitagawa

Published by the Royal Society of Chemistry, www.rsc.org

leading to the emergence of flexible MOFs. In the past 25 years, flexible MOFs have emerged and are well-known for their diverse (global or local) structural responses upon exposure to external stimuli, such as different guest species, temperature change, mechanical forces and light irradiation.[5–10]

The diverse structural dynamics not only can arise locally from the rotations and vibrations of organic moieties or bending and rotations of the bonding interactions, but also from global network displacements like sliding, breathing and shrinking in layered or polycatenated structures.[11–14] In some cases, coordination bonds can even break and reform with dramatic structural transformations. One major change during those transitions of flexible MOFs is in their pore structure leading to different pore sizes, volumes, and shapes, that would affect the uptake capacity or adsorption selectivity for guest molecules. Another feature of flexible MOFs is their multiple intermediate structures, including thermodynamically stable phases that show gradual structural distortions in contrast to each other. Those bistable or multi-stable flexible MOFs can exhibit stepwise (sometimes with hysteresis) stimuli-responsive behaviours. Under the stimulations of different physical and chemical factors, structural transitions in flexible MOFs would be accompanied by corresponding changes in their photo-/electric-/magnetic-response, differentiating themselves from rigid MOFs on application performance.[15] For instance, flexible MOFs with adsorption-induced structural transitions show sigmoidal adsorption isotherms with relatively large working capacity because of their drastic changes in the adsorbed amounts at a small pressure swing. Another potential advantage is that the adsorption heat of the exothermic sorption in flexible MOFs can be consumed by accompanied endothermic deformation of framework structures, facilitating heat management. For those mild stimulations, such as photon irradiation upon light, variation of spin states in magnetic fields, or changes in charge–dipole moments, the collaborative framework flexibility would amplify the responsive behaviour of flexible MOFs.

The study on the properties and functions of flexible MOFs can be dated back over two decades. There are a variety of dynamic modes of flexibility that have been investigated. For two-dimensional (2D) coordination networks, the sliding motion during the adsorption of guest molecules can form an open structure, known as gate-opening behaviour, which results in stepwise adsorption isotherms, as exemplified by $[Cu(bpy)(BF_4)_2(H_2O)_2]\cdot(bpy)$ (ELM-11) and $[Cu_2(dhbc)_2(bpy)]$ (bpy = 4,4′-bipyridine and dhbc = 2,5-dihydroxybenzoate).[16,17] For some 3D MOFs with flexible coordination bonding, minor change in local coordination geometry can lead to significant pore expansion and contraction, as illustrated by [Cr(OH)(bdc)] (MIL-53(Cr)) (bdc = benzene-1,4-dicarboxylate),[6] [Co(bdp)] (bdp = 1,4-benzenedipyrazolate),[18] MIL-88[19,20] and MCF-18.[21] Another mode is represented by the simple glide motion of frameworks in interpenetrated systems without any local distortion or bond cleavage. For example, a pillar-layered 3D framework $[M_2(bdc)_2(bpy)]$ (M = Zn (MOF-508), M = Cu) was revealed to switch between a non-porous structure (guest-free, closed phase) and a porous framework with a pore diameter of

4×4 Å^2 (upon guest incorporation, open phase).[17,22] This dynamic framework of MOF-508, with the bimodal pore system, has been applied for the separation of linear and branched isomers of pentane and hexane molecules through a gas chromatographic column. The dynamic behaviour can also be derived from the dangling substituent groups. For example, [Cu(etz)] (MAF-2) (etz = 3,5-diethyl-1,2,4-triazolate) exhibits small pore apertures blocked by flexible ethyl groups that have a temperature-dependent dynamic motion, which results in abnormal N_2 adsorption at 195 K rather than at 77 K.[23] Overall, flexible MOFs have been extensively studied and practiced for their unique properties and various functions.[23–26] After development over two decades, new flexible phenomena and functions of flexible MOFs have been continuously emerging. In this chapter, we highlight the major developments of representative flexible MOFs for adsorption-based guest storage and separation, chemical sensing, catalytic conversion, proton conducting and other emerging functions in this very important field (Scheme 5.1).

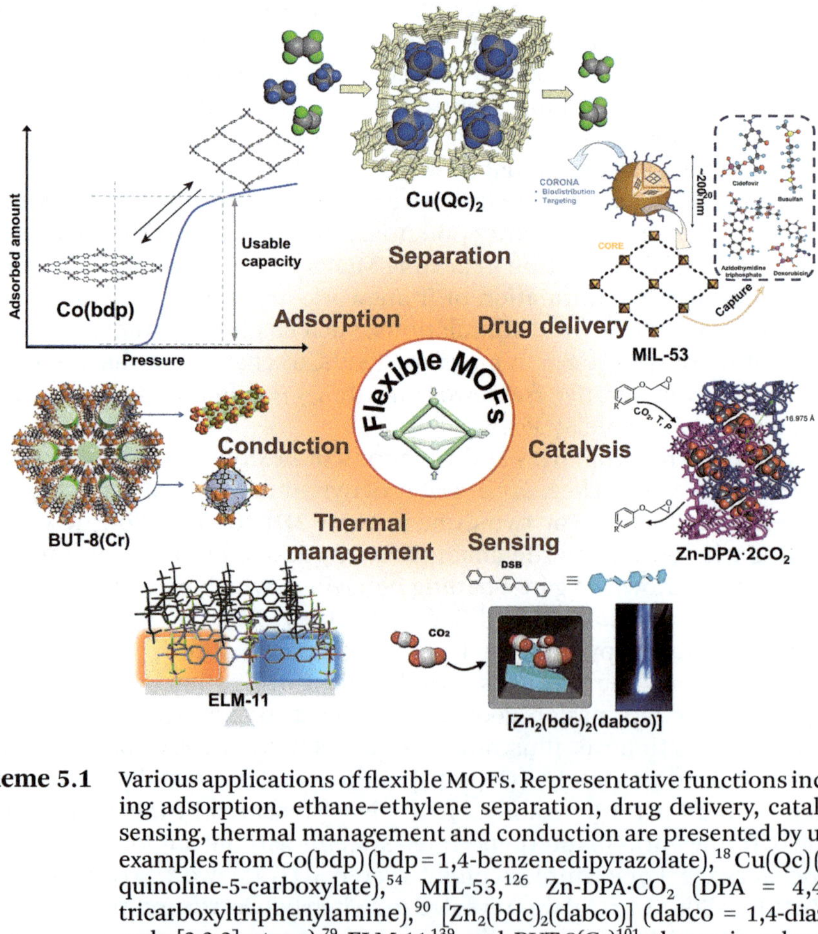

Scheme 5.1 Various applications of flexible MOFs. Representative functions including adsorption, ethane–ethylene separation, drug delivery, catalysis, sensing, thermal management and conduction are presented by using examples from Co(bdp) (bdp = 1,4-benzenedipyrazolate),[18] Cu(Qc) (Qc = quinoline-5-carboxylate),[54] MIL-53,[126] Zn-DPA·CO_2 (DPA = 4,4′,4″-tricarboxyltriphenylamine),[90] [$Zn_2(bdc)_2(dabco)$] (dabco = 1,4-diazabicyclo-[2.2.2]octane),[79] ELM-11[139] and BUT-8(Cr)[101] shown in schematic diagrams, respectively.

5.2 Adsorption

Flexible MOFs exhibit unique adsorption properties due to their structural dynamics that show different types and degrees of response to various external stimuli, resulting in different types of adsorption isotherms, such as stepped or S-shaped adsorption isotherms. Notably, flexible MOFs that exhibit sudden transitions from a non-porous (closed) desolvated phase to a porous (open) phase can rapidly adsorb substantial amounts of gas at a specific threshold pressure while refraining from adsorbing gases at low pressures.[5,16] This feature is greatly conducive their working capacity for gas storage. Furthermore, if the pore size of the open-pore phase is further enlarged, multi-step adsorption behaviour can be displayed, resulting in higher working capacity.[27,28] Flexible MOFs that show such phenomena are also known as fourth-generation flexible metal–organic materials[29] and hold great potential for storing and separating gases.

Adsorbate-dependent gate-opening in flexible MOFs is currently well-known in this field, owing to the different strength of intermolecular interactions. The application of high-pressure sorption facilitates the discovery of such behaviour. In an early example, a flexible framework [Cu(dhbc)$_2$(4,4′-bpy)]·H_2O that prepared from Cu^{II}, 4,4′-bipyridine (4,4′-bpy), and 2,5-dihydroxybenzoic acid (Hdhbc) (Figure 5.1) was reported to exhibit guest-dependent sorption properties based on high-pressure sorption.[17] At ambient temperature, MOF [Cu(dhbc)$_2$(4,4′-bpy)] shows stepwise isotherms for CO_2, CH_4, O_2 and N_2 with different gate-opening threshold pressures. The results of powder X-ray diffraction (PXRD) imply the reversible gate-opening in MOF [Cu(dhbc)$_2$(4,4′-bpy)]. This example was one of the very first studies to report guest-dependent gate-opening.

Given that stepwise sorption behaviour might facilitate the working capacity for adsorbed gas, two flexible MOFs, Co(bdp) and Fe(bdp) (bdp = 1,4-benzenedipyrazolate), were reported to exhibit a high methane adsorption

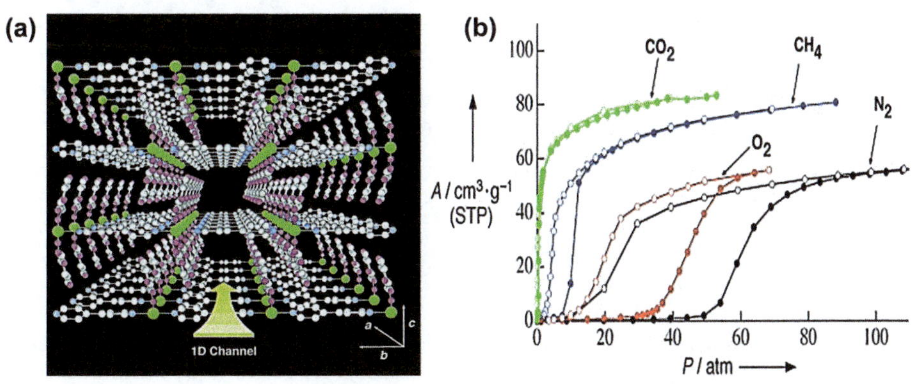

Figure 5.1 (a) Crystal structure of [Cu(dhbc)$_2$(4,4′-bpy)]·H_2O. (b) Single-component isotherms of N_2, CH_4, CO_2 and O_2 at 298 K. Adapted from ref. 17 with permission from John Wiley & Sons, Copyright 2003.

capacity.[18,30,31] Both Fe(bdp) and Co(bdp) exhibit gate-opening adsorption behaviour upon increasing on gas pressure and a stepwise CH_4 isotherm. They undergo gate opening for CH_4 at the threshold pressure, providing the opportunity for boosting the working capacity of methane storage (Figure 5.2). According to high-pressure CH_4 sorption experiments, negligible CH_4 uptake can be observed in Co(bdp) at low-pressure region followed by a sharp increase at threshold pressure, giving a deliverable capacity of 197 cm^3 cm^{-3} when operation in the pressure range of 5.8–65 bar. Similar adsorption behaviour has been observed for both Co(bdp) and Fe(bdp) with only a difference in their gate-opening pressures (16 bar for Co(bdp) and 24 bar for Fe(bdp)).

Acetylene is an important industrial feedstock gas, the safe storage and transport of which is a significant challenge due to its high reactivity and explosive nature. A flexible adsorbent $Cu_2(pzdc)_2(pyz)$ (pzdc = pyrazine-2,3-dicarboxylate; pyz = pyrazine) that is composed of open 1D channels with a 4–6 Å^2 cross-section, was synthesised and applied for highly efficient C_2H_2 adsorption (Figure 5.3(a)).[32] Gas adsorption studies revealed a high and

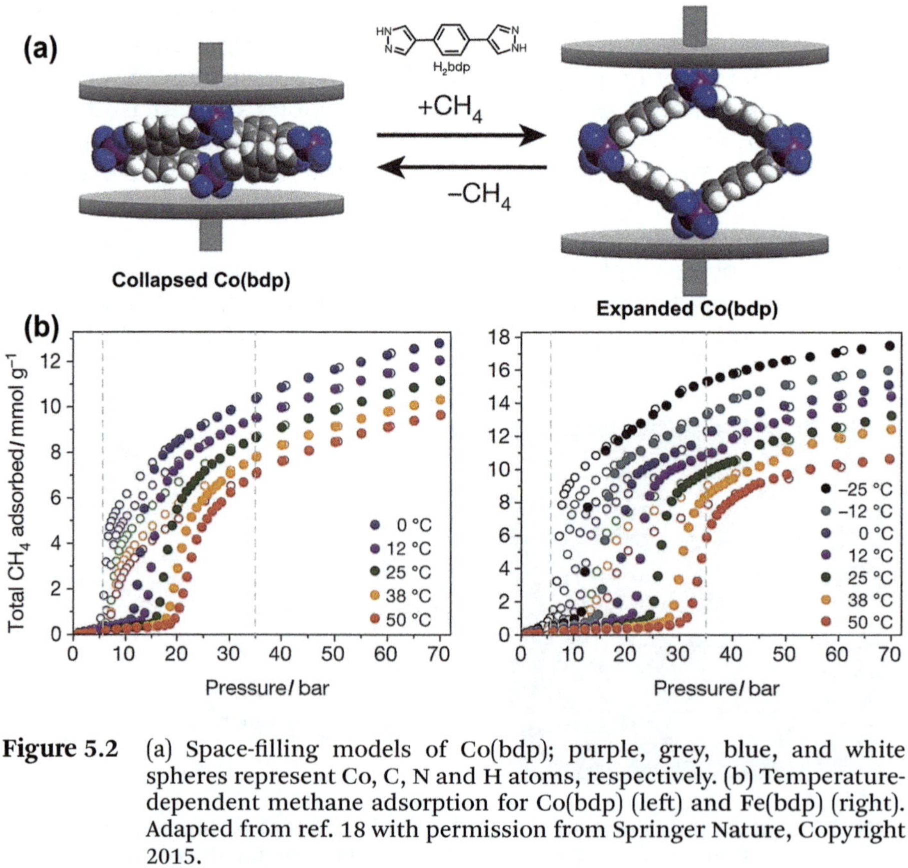

Figure 5.2 (a) Space-filling models of Co(bdp); purple, grey, blue, and white spheres represent Co, C, N and H atoms, respectively. (b) Temperature-dependent methane adsorption for Co(bdp) (left) and Fe(bdp) (right). Adapted from ref. 18 with permission from Springer Nature, Copyright 2015.

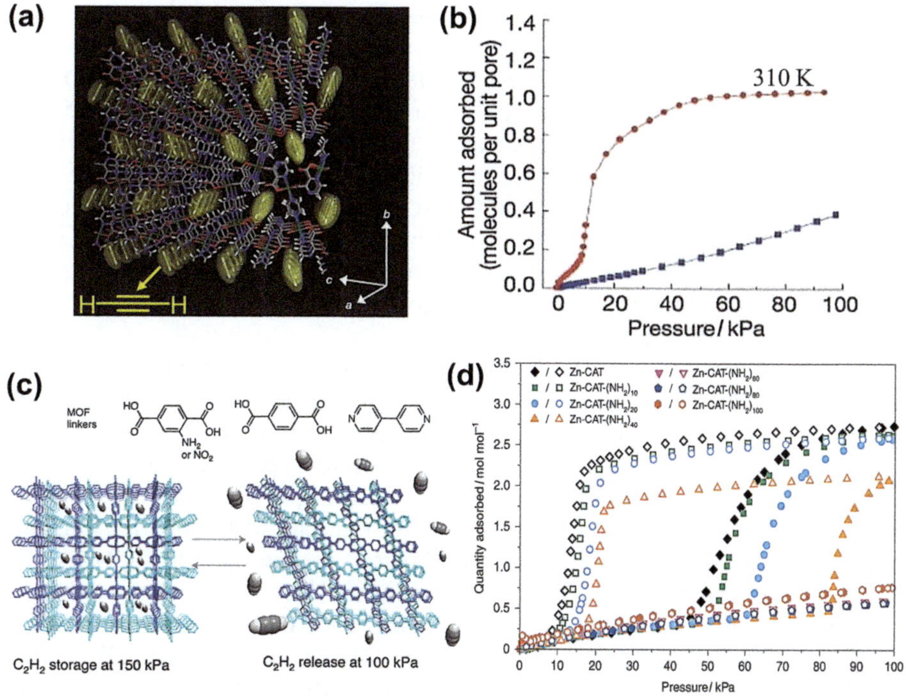

Figure 5.3 (a) Crystal structure of $Cu_2(pzdc)_2(pyz)$ with C_2H_2 at 170 K from MEM/Rietveld analysis. (b) Adsorption isotherms for C_2H_2 (red circles) and CO_2 (blue squares) on $Cu_2(pzdc)_2(pyz)$ at 310 K. Adapted from ref. 32 with permission from Springer Nature, Copyright 2005. (c) Reversible structural transformation of Zn-CAT or Zn-CAT-(X)*n* (X represents the functional moiety $-NH_2$ or $-NO_2$ and *n* represents the molar percent of 0–100%) during the adsorption and desorption of C_2H_2. (d) Acetylene (C_2H_2) adsorption isotherms for all Zn-CAT-$(NH_2)n$ at 273 K. Adapted from ref. 33 with permission from Springer Nature, Copyright 2022.

steep C_2H_2 uptake of 42 cm^3 g^{-1}, whereas gradual increase for CO_2 uptake was observed. The C_2H_2 molecules were indicated to be linearly arranged inside the pores, where hydrogen bonding interactions with two non-coordinated oxygen atoms facilitate such accommodation of C_2H_2 molecules. This study initiated the stable and safe storage of high-density C_2H_2, in contrast to the compression limit of gaseous acetylene at ambient temperature. Additionally, achieving higher useable storage capacity for pure acetylene at room temperature is critical due to the narrow storage pressure (P) and temperature range ($100 < P < 200$ kPa at 298 K) for industrial requirements. Recently, a two-fold interlocked flexible MOF Zn-CAT ($[Zn_2(bdc)_2(4,4'\text{-}bpy)]$, bdc = benzene-1,4-dicarboxylate, 4,4′-bpy = 4,4′-bipyridine) has been reported to show tunable gate-opening and gate-closing pressures (P_{go} and P_{gc}) by replacing bdc^{2-} ligands with functionalised bdc–X ligands in varying ratios ($X = -NO_2$ or $-NH_2$), and exhibits a high useable capacity for acetylene at room temperature (Figure 5.3(b)).[33] *In situ* gas sorption studies and theoretical calculations have revealed that increasing the amount of bdc–NH_2 in Zn-CAT leads to an exponential growth of P_{go}, while the incorporation of bdc-NO_2 has the opposite effect. This is because the former ligand can enhance the interactions between the two interpenetrating frameworks whereas the latter one leads to more repulsion, resulting in a significant change in the structural deformation energy.[34] Thus, the MOF Zn-CAT-$(NH_2)_{100}$ that is constructed all by bdc-NH_2 exhibits acetylene storage with a very high useable capacity of up to ~90 v/v (77% of the overall capacity) under practical pressure range (100–150 kPa) at 298 K. This example well demonstrates the significant advantage of flexible MOF adsorbents for gas storage.

A highly porous MOF DUT-49 was reported to exhibit a unique adsorption phenomenon known as negative gas adsorption (NGA) at elevated gas pressure (Figure 5.4).[35–39] It features large tetrahedral and octahedral cages with large void space (10 Å, 17 Å and 24 Å). For CH_4 adsorption in DUT-49 at 111 K and >10 kPa, structural transitions corresponding to pore contraction arising from the bending of organic linkers have been identified, which can further open back to its original states at elevated pressures. During NGA, the highly porous framework undergoes a remarkable contraction as indicated by its crystal unit cell volume from $V = 100\,072$ Å^3 (calculated density ρ_{calc} = 0.287 g cm^{-3}) to a contracted intermediate DUT-49cp with a unit cell volume of only 47 282 Å^3 (ρ_{calc} = 0.661 g cm^{-3}), corresponding to a unit cell change in volume of more than 50 000 Å^3. And the pore diameter of tetrahedral voids

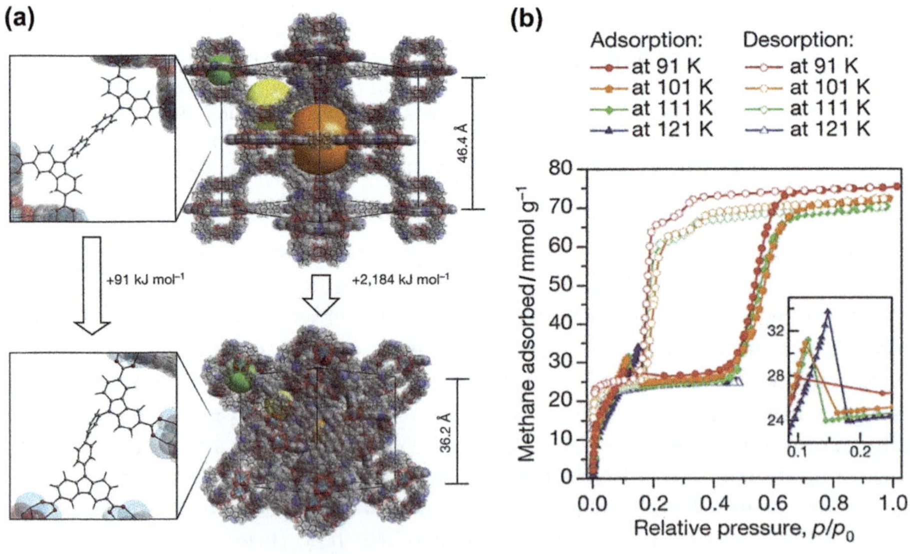

Figure 5.4 (a) Crystal structure of DUT-49op and DUT-49cp showing the conformation of the ligand in change during DUT-49op to DUT-49cp transformation. (b) Temperature-dependent methane adsorption. Adapted from ref. 35 with permission from Springer Nature, Copyright 2016.

was reduced from 16.9 Å to 10.6 Å while the mesoporous octahedral voids contracted from 23.9 Å to 7.8 Å. This structural transformation is reversible and can expand back to the original DUT-49op. A similar sorption phenomenon was observed when DUT-49 was exposed to *n*-butane.

5.3 Adsorptive Separation

Flexible MOFs show unique adsorption behaviours along with structural responses at different levels (global or local transformations) such as pore-opening and gate-opening effects upon the stimulation of gas molecules. Indicated by gas sorption isotherms for different gases, flexible MOFs theoretically can be utilised for gas separation and purification. Given that the gas separation performance of flexible MOFs for gas mixtures can be hardly predicted from their single-component sorption isotherms, they have rarely been applied for the separation of gas mixtures in contrast to those robust MOFs. Co-adsorption of impurity gas could happen simultaneously once the pore structures of flexible MOFs are opened by targeted gas molecules, leading to poor separation performance. Flexible MOFs usually show negligible gas uptake under gas pressures below the threshold of structural transformations, which is unfavourable for the purification of gases with trace impurities. On the other hand, the flexibility of guest gas molecules might also occur upon interactions with flexible host structures, collaboratively increasing the complexity of their sorption behaviours. It is thus quite challenging for flexible MOFs to be applied for gas separation. However, with more research on flexible MOFs, strategies such as rational flexible control through external stimuli, the introduction of functional sites in the structure, and designing flexible and robust MOFs have enabled efficient separation for molecules with different sizes, including but not limited to hydrogen isotopes, alkanes/alkenes, aromatic isomers, even polymers, exhibiting excellent molecular sieving performance.

Hydrogen is a small, low-polarity molecule. The separation of its isotopic gases such as deuterium (D_2) and tritium (T_2) gas is a very challenging but important research topic. Recently, flexible MOFs with tunable channels have been shown to exhibit great potential for separating hydrogen isotopes based on quantum effects.[40] MIL-53(Al) was the first flexible MOF that showed efficient separation of hydrogen gas isotopes through dynamic pore change from narrow pore (np) to large pore (lp) (Figure 5.5).[41] Single-component adsorption isotherms demonstrated that the breathing behaviour of MIL-53(Al) is highly dependent on gas pressures and temperatures. Therefore, optimal quantum effects can be achieved by rationally tuning the temperature and gas pressure. Eventually, MIL-53(Al) showed the highest selectivity of 13.6 at 40 K and 10 mbar. More interestingly, as for the sorption of H_2 and D_2, only D_2 can induce the secondary breathing of MIL-53(Al) to form a very large pore (vlp), which was further demonstrated by *in situ* neutron diffraction experiments.[42] This novel phenomenon can be used to achieve selective adsorption of D_2 with high purity. In addition, the diffusivity of hydrogen isotopes in

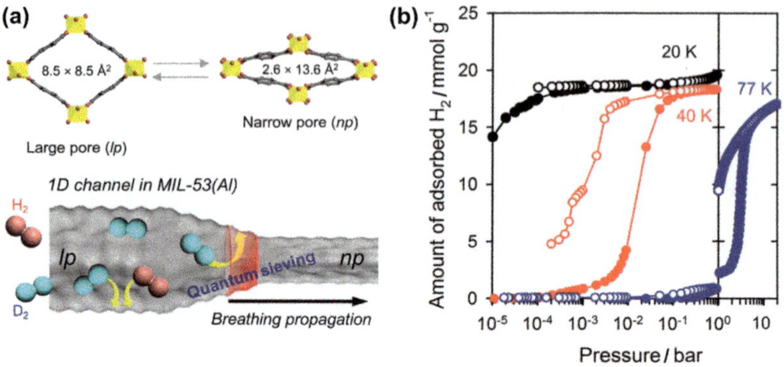

Figure 5.5 (a) The structural transformation diagram between a large pore (lp) and a narrow pore (np) in MIL-53(Al) (top) and schematic diagram of separating hydrogen isotopes by dynamic 1D channel in MIL-53(Al) (bottom). (b) Adsorption isotherms for H_2 in MIL-53(Al) were measured at different temperatures. Adapted from ref. 41 with permission from American Chemical Society, Copyright 2017.

MIL-53(Al) was further elucidated by quasi-elastic neutron scattering.[43] The results showed that the diffusion between isotopes differs significantly with varying temperatures, suggesting that flexible MOFs are a promising material to realise efficient isotope separation in relatively high operating temperatures. In addition, a partially fluorinated Cu-based flexible MOF named FMOFCu was later reported to show the efficient separation of D_2 from isotope mixture with the selectivity of 14 at 25 K.[44] In this flexible MOF, there is a 3D two-fold interpenetrated structure with three cavities of various sizes that are connected by apertures of 3.6 and 2.5 Å. This flexible MOF can still maintain a moderate selectivity of four at a relatively high temperature (77 K).

For carbon monoxide, it can readily coordinate with metal ions *via* π interactions. The molecular shape and properties of CO are similar to N_2, which makes their separation very challenging. Flexible MOFs with open metal sites have demonstrated their capability to separate CO from nitrogen showing high CO adsorption capacity. One such MOF is a three-dimensional structure ([Cu(aip)], aip = 5-azidoisophthalate) with uncoordinated Cu(II) on its pore surface that exhibits highly efficient separation of CO over N_2 due to the reversible self-accelerating CO adsorption.[45] The activated structure was obtained by removing water molecules from the Cu(II) paddle-wheel units in the as-synthesised MOF and showed narrower channels compared to the as-synthesised MOF (Figure 5.6). Comprehensive analyses gave crystallographic evidence that CO molecules can coordinate with Cu(II), leading to a global structural change with the expansion of diffusion paths for further adsorption and thus affording a relatively large space for accommodating extra CO molecules. Such dynamic adsorption behaviour is reminiscent of O_2 trapping on haemoglobin.

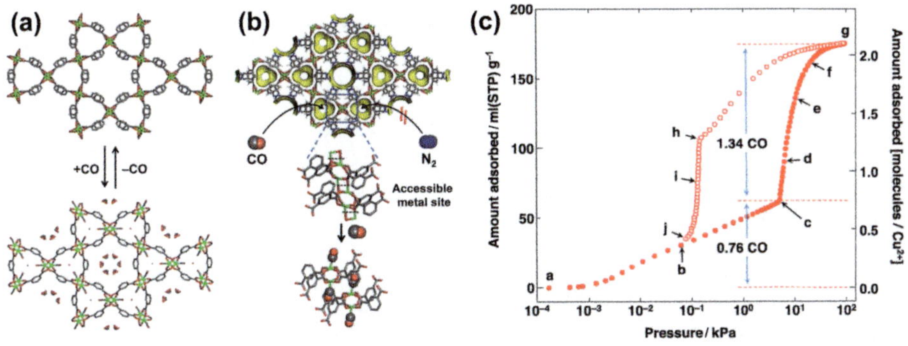

Figure 5.6 (a) Reversible structural transformation of activated MOF during the adsorption and desorption of CO. (b) Schematic images of the access of adsorbates into the pores in activated MOF. (c) CO sorption isotherms on activated MOF at 120 K. Adapted from ref. 45 with permission from American Association for the Advancement of Science, Copyright 2014.

Gas pressure can regulate the pore size of flexible MOFs. Separation of two gases using the flexible MOF can be achieved when there is a significant difference in the pressure required for two gases to gate-opening the MOF. A flexible MOF, [Co(bdp)] with multi-step gate-opening behaviour has been reported to have a high CH_4 capacity.[18] This flexible MOF have also been investigated for CO_2–CH_4 separation under high-pressure (Figure 5.7).[46] Single-component gas adsorption isotherms for CO_2 and CH_4 at 298 K indicated that both gases can trigger gate-opening but occur at distinctly different pressures of 2 and 18 bar, respectively. Multicomponent isotherms revealed that [Co(bdp)] can selectively adsorb CO_2 while almost no CH_4 adsorption under 298 K and at 6.7, 13.9 and 25.3 bar, although CO_2 has triggered the gate-opening. In addition, even upon exposure to a CH_4-rich atmosphere (6 : 94 molar ratio of CO_2–CH_4), [Co(bdp)] can still retain selective adsorption for CO_2 although its CO_2–CH_4 selectivity has dropped to 61. These results revealed that [Co(bdp)] will be a promising material for separating CO_2–CH_4 mixtures.

C_2H_2 and CO_2 have identical molecular sizes and similar physical properties, so it is quite difficult to achieve their separation. Increasing the density of coordination of unsaturated metal sites in MOFs could facilitate the selective binding on unsaturated hydrocarbons. A microporous flexible MOF JNU-1 ([$Zn_3(OH)_2(btca)_2$], H_2btca = benzotriazole-5-carboxylic acid), boasts a dense open metal site (density 3.2 nm^{-3} per unit cell volume) arranged on the 1D diamond-shaped channels surface. It exhibits high C_2H_2–CO_2 selectivity of 285.6, and a C_2H_2 adsorption capacity of 60 mL g^{-1} as well as a C_2H_2-capture amount of 140 mL g^{-1} in a breakthrough experiment (Figure 5.8).[47] Single-crystal XRD analysis and theoretical modelling revealed that C_2H_2 can easily adsorb onto two open Zn(II) metal sites affording higher binding energy of 47.6 kJ mol^{-1}. Interestingly, the absorption of C_2H_2 could show an induced-fit behaviour, which can further enhance binding affinity. However,

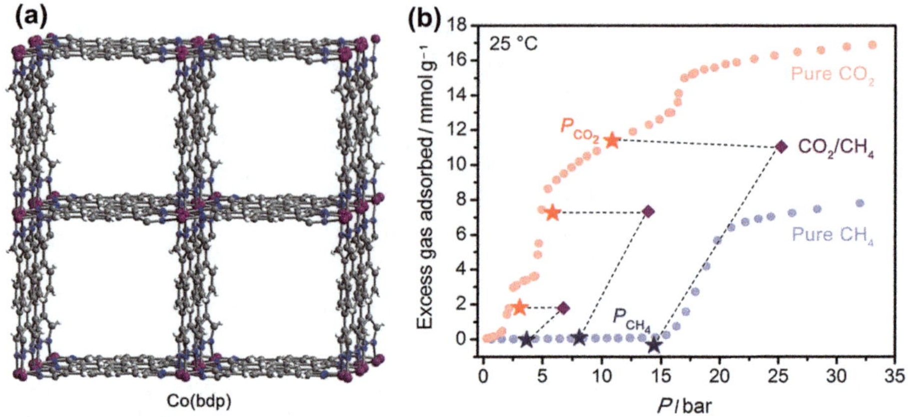

Figure 5.7 (a) Crystal structures of [Co(bdp)]. (b) Multicomponent adsorption experiments were performed at 6.7, 13.9 and 25.3 bar, corresponding to equilibrium CO_2–CH_4 molar ratios of 46 : 54, 42 : 58 and 43 : 57, respectively. Red circles and blue circles represent single-component isotherms for CO_2, CH_4 adsorption in Co(bdp) at 298 K, respectively. Purple diamonds represent the overall amount of gas adsorbed by Co(bdp) (*y*-axis) from a CO_2–CH_4 mixture at a given equilibrium pressure (*x*-axis). Each purple diamond is paired with a corresponding red and blue star: red stars represent the CO_2 adsorbed from the mixture (*y*-axis) at the equilibrium partial pressure of CO_2 (*x*-axis), and blue stars represent the CH_4 adsorbed from the mixture (*y*-axis) at the equilibrium partial pressure of CH_4 (*x*-axis). Adapted from ref. 46 with permission from the American Chemical Society, Copyright 2018.

adsorption–desorption experiments showed that the desorption of C_2H_2 needs a very high temperature (393 K), which may cause energy wastage.

Studying the flexible behaviour of isoreticular flexible MOFs functionalised with different ligands can lead to a deeper understanding of their flexibility mechanism and the development of efficient flexible MOF molecular sieving materials. Two isostructural flexible MOFs, $[Zn_3(OH)_2(pzdc)(tz)]$·DMA and $[Zn_3(OH)_2(pzdc)(atz)]$·DMA (H_3pzdc = 3,5-pyrazoledicarboxylic acid, Htz = 1,2,4-triazole, Hatz = 3-amino-1,2,4-triazole, DMA = *N*,*N*-dimethylacetamide), possess quasi-discrete/gating pores with varying gating energy barriers due to their differences on intra-framework attraction (Figure 5.9).[48] Crystal structure analyses showed that the pore cavities and apertures of $[Zn_3(OH)_2(pzdc)(atz)]$ dramatically shrink upon removal of guests, whereas those in $[Zn_3(OH)_2(pzdc)(tz)]$ only showed slight changes. This suggests that $[Zn_3(OH)_2(pzdc)(atz)]$ has stronger intra-framework attraction and could result in higher gating energy barriers. Therefore, $[Zn_3(OH)_2(pzdc)(atz)]$ completely blocks N_2 and CH_4, which are larger than the pore size, leading to record high selectivity of CO_2–N_2 and CO_2–CH_4 separation. This work revealed that the controlled gated barrier is expected to achieve the ideal molecular sieve effect.

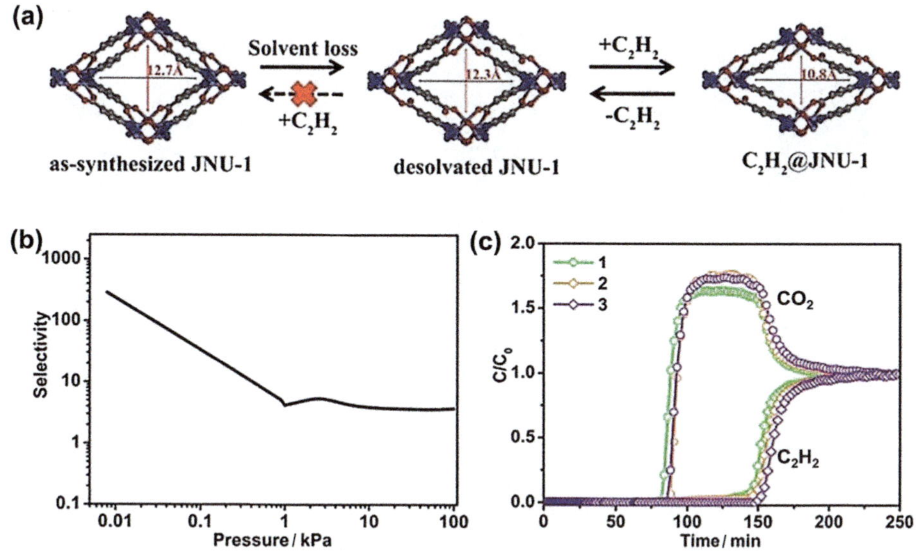

Figure 5.8 (a) Schematic diagram of induced-fit effect in JNU-1. (b) IAST selectivity (calculated based on ideal adsorbed solution theory (IAST)) of JNU-1 for C_2H_2–CO_2 (50 : 50, v/v) at 298 K. (c) Three cycles of experimental column breakthrough curves for JNU-1 in a C_2H_2–CO_2 (50 : 50, v/v) mixture at 298 K. Adapted from ref. 47 with permission from John Wiley & Sons, Copyright 2019.

The flexible MOFs with structural features including suitable pore aperture size and specific functional sites could collaboratively facilitate specific recognition to realise highly efficient gas separation. UTSA-300 ([Zn(dps)$_2$(SiF$_6$)], dps = 4,4′-dipyridylsulfide) is a unique flexible ultramicroporous MOF with a 2D sql network, showing pore structure with narrow pore size (3.3 Å) and multiple fluoride functional sites that allows for selective adsorption of C_2H_2 while completely excludes C_2H_4 and CO_2 under ambient conditions (Figure 5.10).[49] This structure would readily transform into a closed-pore phase UTSA-300a after desolvation due to the existence of strong C–H···F and π–π stacking interactions. Single-component adsorption isotherms revealed that UTSA-300a has a high adsorption capacity of C_2H_2 (76.4 cm^3 g^{-1} at 273 K) whereas rarely adsorbs C_2H_4 and CO_2 under the same conditions. Neutron powder diffraction and molecular modelling studies displayed that the head-on binding of a C_2H_2 molecule to two hexafluorosilicate F atoms expands the closed-pore structure to the pore-opened phase, but that change cannot be triggered by CO_2 and C_2H_4. Therefore, this material could facilely separate a mixture of C_2H_2–C_2H_4 and C_2H_2–CO_2. To further improve the adsorption capacity of C_2H_2, an isoreticular structure of UTSA-300, [Cu(dps)$_2$(SiF$_6$)] (termed as NCU-100 or UTSA-300-Cu), was synthesised by substituting the zinc centres with copper ones (Figure 5.10).[50] The resultant guest-free [Cu(dps)$_2$(SiF$_6$)] has relatively longer metal–F bonds (2.27 Å) causing a larger pore window size (3.6

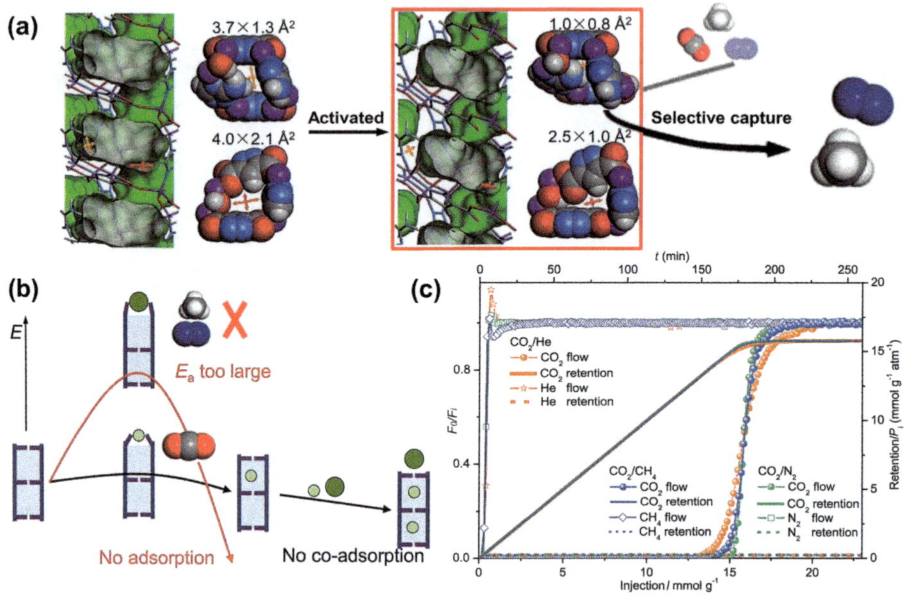

Figure 5.9 (a) Crystal structures of $[Zn_3(OH)_2(pzdc)(atz)]$·DMA (left), $[Zn_3(OH)_2(pzdc)(atz)]$ (right), and schematic diagram of separation performance toward CO_2, CH_4, N_2 in $[Zn_3(OH)_2(pzdc)(atz)]$. (b) Schematic diagram of structure responses and molecular sieving behaviours of the gating action. (c) Column breakthrough curves for $[Zn_3(OH)_2(pzdc)(atz)]$ in 10 : 90 CO_2–N_2, 10 : 90 CO_2–CH_4 and 10 : 90 CO_2–He mixtures. Reproduced from ref. 48 with permission from Cell Press, Copyright 2021.

× 4.3 × 4.2 Å^3), which leads to high low-pressure uptake for C_2H_2 (0.73 mmol g^{-1} at 0.01 bar and 298 K) and gains over two orders of magnitude jump in IAST (ideal adsorbed solution theory) selectivity for C_2H_2–C_2H_4 (1 : 99, v/v). Subsequently, this series of isoreticular flexible MOFs has been also systematically investigated for C_2H_2–CO_2 separation.[51] Among those four analogs (SIFSIX-dps-Zn, SIFSIX-dps-Cu, GeFSIX-dps-Cu and NbOFFIVE-dps-Cu), SIFSIX-dps-Cu shows the highest C_2H_2 uptake capacity (4.57 mmol g^{-1}) and C_2H_2–CO_2 IAST selectivity (1787) at room temperature.

For important gaseous chemical feedstock such as ethylene and propylene, important progresses have also been made by flexible MOFs. A Co-based MOF $[Co(VTTF)]_n$ (VTTF = 2,2′-[1,2-bis(4-benzoic acid)-1,2-ethanediylidene] bis-1,3-benzodithiole) was reported to show a two-step gate-opening process upon exposure to ethylene in contrast to closed-pore phase upon ethane (Figure 5.11).[52] The guest-free MOF is a nonporous 3D framework that was formed by connecting all the skeletal 1D chains *via* Co–S bonds where the Co–S bond can easily split by ethylene molecules because it can form weak d–π interactions with Co atoms in the framework. The intermediate phases during the gate-opening process were confirmed by single-crystal XRD (Figure 5.11). DFT and Monte Carlo calculations demonstrated the

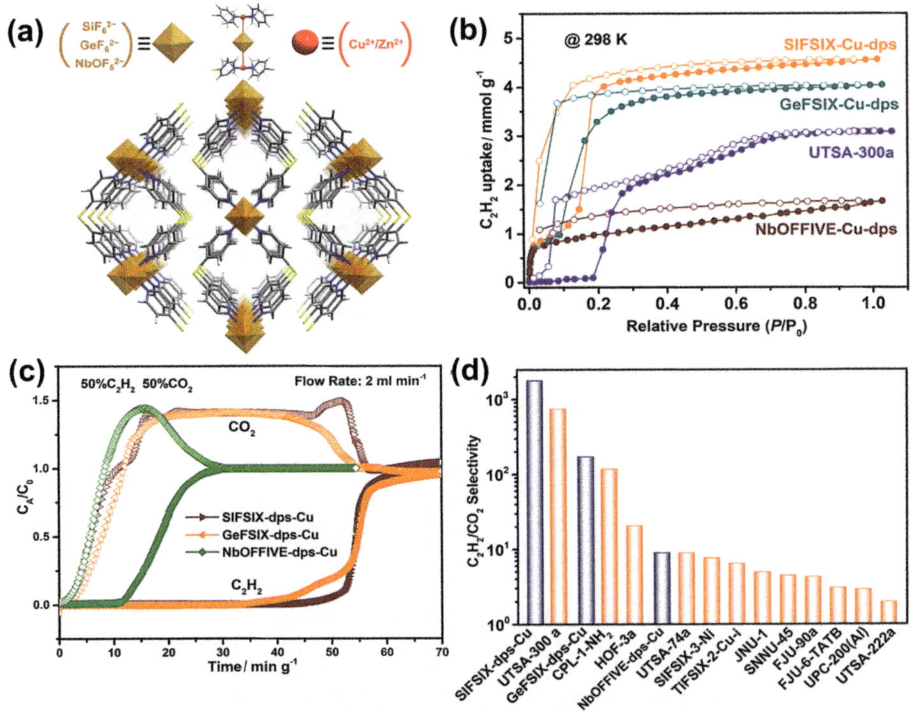

Figure 5.10 (a) Isoreticular 2D MOF structures are constructed by varying anionic linkers and metal nodes. (b) Adsorption isotherms of C_2H_2 in four isoreticular MOFs. (c) Breakthrough curves for 50 : 50 C_2H_2–CO_2 at 298 K. (d) Comparison of IAST selectivity of representative MOFs for 50 : 50 C_2H_2–CO_2. Reproduced from ref. 51 with permission from Springer Nature, Copyright 2022.

possible structural transformation process upon ethylene adsorption. Single-component gas adsorption experiments confirmed the unique gate-opening adsorption for ethylene over ethane that can maintain even at high-pressure (1 MPa) and high temperature (303 K).

A systematic study on gas diffusion in microporous MOFs would facilitate the design of sorbents for important gas separation. Those diffusion processes are typically performed kinetically and thermodynamically, making the diffusion study a very tough task. Such a study was performed by one flexible MOF constructed from butterfly-type ligand OPTz-ipa (5-(5,5-dioxido-10*H*-phenothiazin-10-yl)isophthalic acid) and Cu(II). The resultant MOF [Cu(OPTz-ipa)] features a cage-like cavity and temperature-regulated dynamic bottleneck gates that allow efficient separations of O_2–Ar and ethylene–ethane with high selectivity of ~350 and ~75, respectively (Figure 5.12).[53] Interestingly, the as-synthesised MOF is a 2D layer structure that can readily transform into a 3D structure after activation, namely $[Cu(OPTz\text{-}ipa)]_n$. Meanwhile, OPTz units, a dynamic moiety in the ligands, can be readily modulated

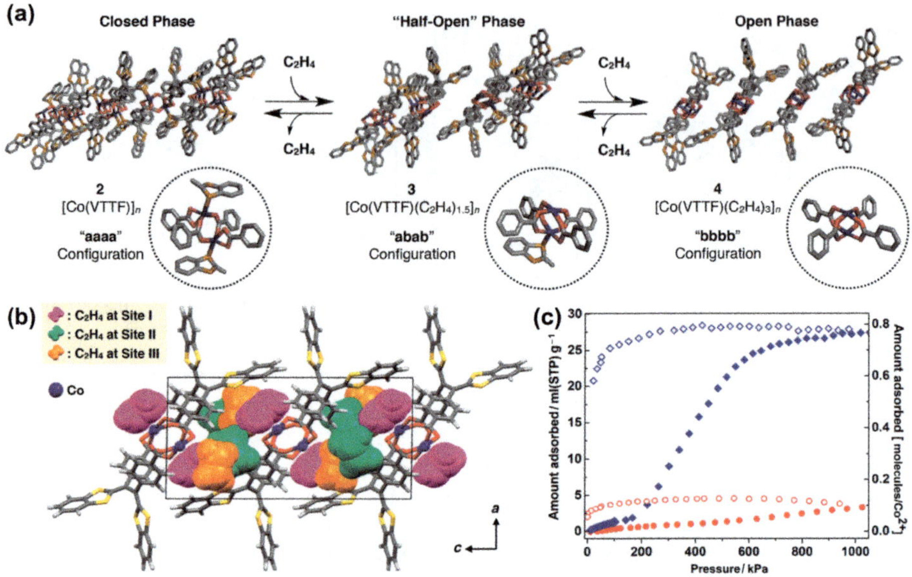

Figure 5.11 (a) The transition between different phases of the Co(VTTF) MOF. (b) Modelled structures of open phase with three different binding sites for ethylene molecules. (c) Adsorption–desorption isotherms of closed phase with ethylene (blue) and ethane (red) at 283 K and 1 MPa. Adapted from ref. 52 with permission from American Chemical Society, Copyright 2017.

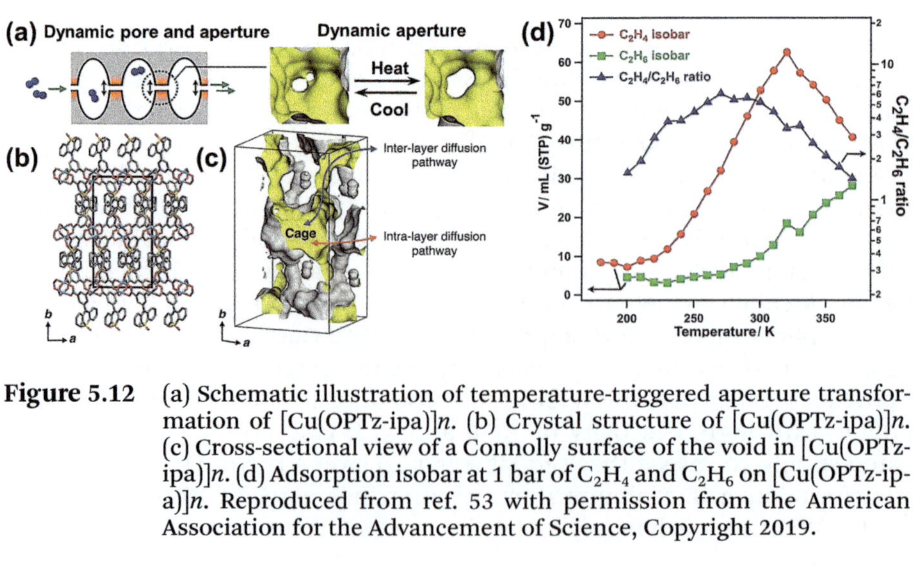

Figure 5.12 (a) Schematic illustration of temperature-triggered aperture transformation of [Cu(OPTz-ipa)]*n*. (b) Crystal structure of [Cu(OPTz-ipa)]*n*. (c) Cross-sectional view of a Connolly surface of the void in [Cu(OPTz-ipa)]*n*. (d) Adsorption isobar at 1 bar of C_2H_4 and C_2H_6 on [Cu(OPTz-ipa)]*n*. Reproduced from ref. 53 with permission from the American Association for the Advancement of Science, Copyright 2019.

by temperature and thus further dominate the pore size of the bottleneck gates to regulate gas diffusion. Adsorption isobars of [Cu(OPTz-ipa)]$_n$ showed that different gases such as H_2, O_2, Ar, N_2, CO, CH_4, C_2H_4 and C_2H_6 are barely adsorbed at low temperatures, while their adsorption amounts gradually become larger with increasing temperature and reaching a maximum value at the respective optimum temperature (T_{max}). Among them, C_2H_4 preferentially adsorbed than C_2H_6 in [Cu(OPTz-ipa)]$_n$ at 200 to 370 K range, and the C_2H_4–C_2H_6 uptake ratio reached a maximum of 6 at 270 K. In addition, dynamic separation experiments suggested that C_2H_4 can be efficiently separated from C_2H_4–C_2H_6 mixtures at 270 K. This work well demonstrates that kinetic separation of ethylene–ethane can be achieved by temperature-dependent local structural motions.

C_2H_4-selective adsorbents typically require multiple adsorption and desorption cycles to achieve high-purity ethylene, whereas C_2H_6-selective adsorbents can directly give pure ethylene by one separation operation. The adaptive properties of flexible MOFs towards guest molecules can also enhance the interactions between adsorbate and adsorbent, facilitating preferential binding of gas molecules. Two isoreticular ultramicroporous MOFs, [Cu(ina)$_2$] (Hina = isonicotinic acid) and [Cu(Qc)$_2$] (HQc = quinoline-5-carboxylic acid), with different pore sizes (~4.1 and ~3.3 Å, respectively) and pore surface contact areas were applied for separation of C_2H_6/C_2H_4 (Figure 5.13).[54] Among them, [Cu(Qc)$_2$] exhibited a higher C_2H_6–C_2H_4 selectivity of 3.4 and a C_2H_6–C_2H_4 uptake ratio of 237% (60.0, 25.3 $cm^3\ cm^{-3}$). Neutron powder diffraction and modelling studies have revealed that [Cu(Qc)$_2$] exhibits self-adaptive adsorption behaviour for C_2H_6, exhibiting multiple weak van der Waals interactions with ethane that results in higher binding affinity in contrast to ethylene.

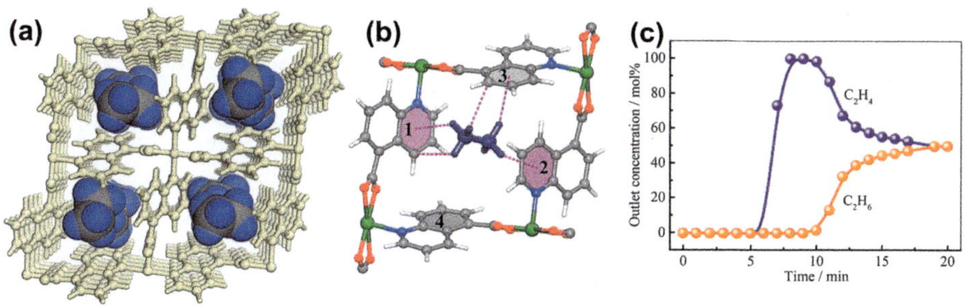

Figure 5.13 (a) Neutron diffraction crystal structures of [Cu(Qc)$_2$]·0.41C_2D_6. (b) Preferential binding sites for C_2D_6 molecules and the close van der Waals contacts within the rhombic cavity of aromatic rings (numbered as 1–4), highlighted C–H···π interactions in pink dashed bonds. (c) Experimental column breakthrough curves for equimolar C_2H_6–C_2H_4 mixture (298 K, 1 bar) in an adsorber bed packed with Cu(Qc)$_2$. Adapted from ref. 54 with permission from American Chemical Society, Copyright 2018.

To obtain pure gas of high grade, the removal of trace impurities from the gas mixture is important but quite challenging, especially for flexible MOFs at pressures below their threshold pressures. MOFs that can maintain a stable open pore space after the removal of solvent molecules and show further opening to large-pore phase at elevated pressures, so-called flexible-robust MOFs, exhibit significant gas adsorption at relatively-low pressures by using their small-pore phrase, and thus applicable for removal of trace impurities without any concern on their co-adsorption from pore opening. Such flexible–robust behaviour can also be observed in other porous framework materials.[55–58] A flexible–robust MOF, $[Cu(bpy)_2(OTf)_2]$ (ELM-12, bpy = 4,4′-bipyridine, OTf^- = trifluoromethanesulfonate), consisting of a rigid square-grid copper bipyridine scaffold with dynamic dangling OTf^- groups (Figure 5.14), was applied to remove trace propyne from propylene.[59] Desolvated ELM-12 contains two types of cavities with the size of 6.1 Å × 4.3 Å × 4.3 Å and 6.8 Å × 4.0 Å × 4.2 Å, respectively, which match well with the molecular size and shape of C_3H_4 (6.2 Å × 3.8 Å × 3.8 Å). Single-component sorption isotherms showed that ELM-12

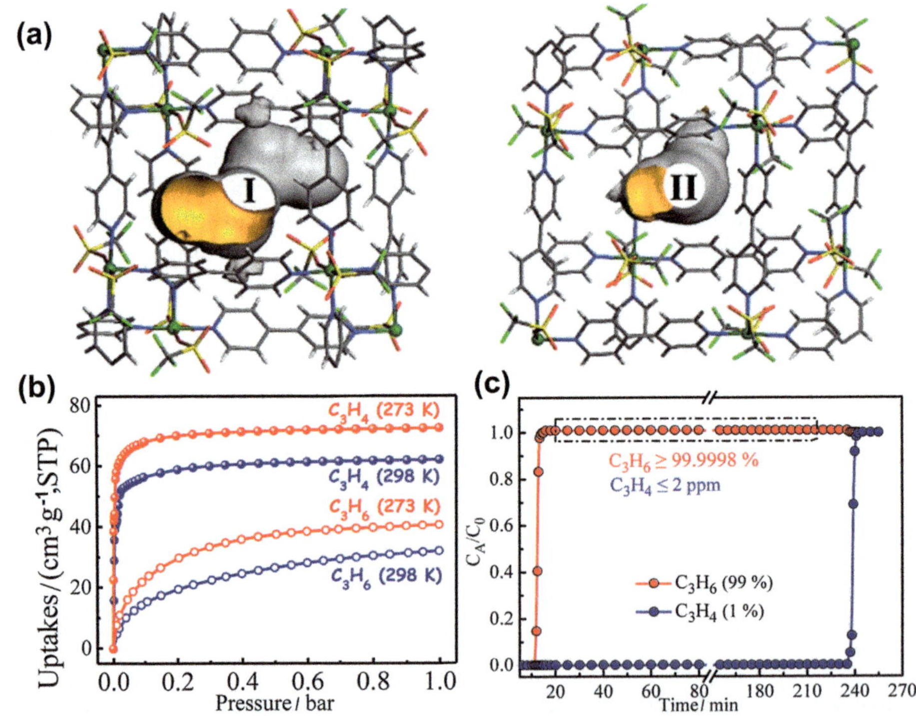

Figure 5.14 (a) Schematic diagrams of the type I and type II cavities in ELM-12. (b) C_3H_4 and C_3H_6 adsorption isotherms of ELM-12. (c) Experimental column breakthrough curves for a C_3H_4–C_3H_6 mixture containing 1% C_3H_4 (298 K, 1.01 bar) in an absorber bed packed with ELM-12. Adapted from ref. 59 with permission from American Chemical Society, Copyright 2017.

exhibits high low-pressure uptake for C_3H_6 (2.55 mmol g^{-1} at 0.1 bar and 298 K), whereas the C_3H_6 adsorption capacity is only 0.67 mmol g^{-1}. The selectivity of ELM-12 for a 1:99 C_3H_4–C_3H_6 mixture was calculated to be 84 based on the IAST approach. High-resolution neutron powder diffraction of the C_3D_4-loaded sample suggested that propyne interacts simultaneously with two oxygen atoms of OTf^- groups from different nets through cooperative C–H···O hydrogen bonds. The efficient separation of ELM-12 for the C_3H_4–C_3H_6 mixture has also been confirmed by breakthrough experiments.

Remarkable separation performance is possible even though there are certain structural transitions that happen at elevated pressures. For example, a two-fold interpenetrated MOF UTSA-200 ([Cu(azpy)$_2$(SiF$_6$)], azpy = 4,4′-azopyridine) with a small aperture size of 3.4 Å and functional SiF_6^{2-} sites was reported to show high sieving effect for separation of C_2H_2–C_2H_4 mixtures.[60] Interestingly, this MOF were also demonstrated exceptional separation capability for the removal of trace C_3H_4 from C_3H_4–C_3H_6 mixtures when comparing to a series of MOFs with similar pore chemistry (Figure 5.15).[61] Single-component gas adsorption studies revealed that UTSA-200 shows high C_3H_4 uptake of 95 cm^3 cm^{-3} but with negligible C_3H_6 adsorption at 0.01

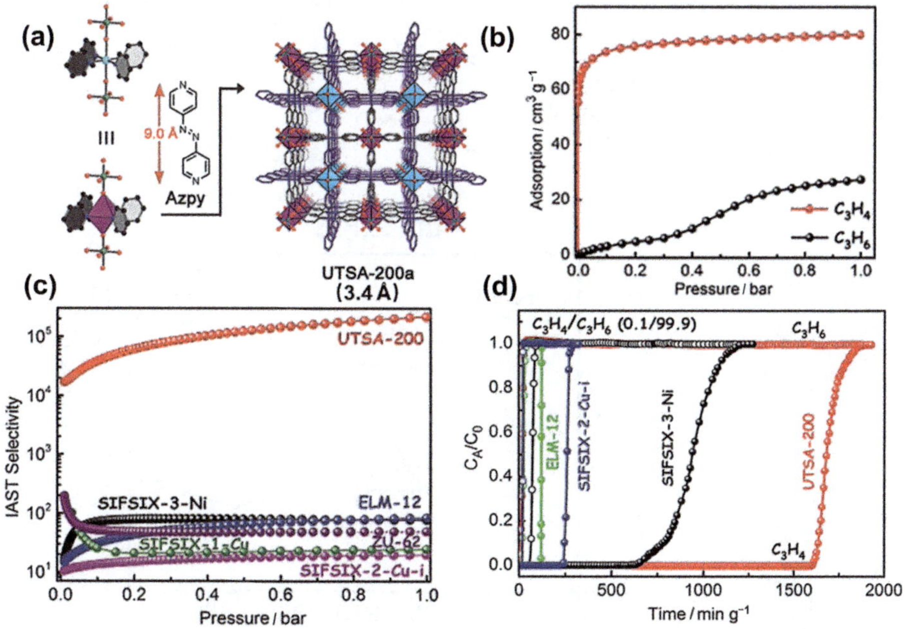

Figure 5.15 (a) Crystal structure of UTSA-200. Adapted from ref. 60 with permission from John Wiley & Sons, Copyright 2017. (b) C_3H_4 and C_3H_6 adsorption isotherms of UTSA-200 at 298 K. (c) IAST selectivity of the indicated MOFs for separating 1:99 (v/v) C_3H_4–C_3H_6 mixture. (d) Experimental breakthrough curves of relevant MOFs under a flow of 2.0 cm^3 min^{-1} at 298 K and 1.01 bar. Adapted from ref. 61 with permission from John Wiley & Sons, Copyright 2018.

bar. The IAST selectivity of UTSA-200 for 1 : 99 C_3H_4–C_3H_6 was calculated to be >20 000. Breakthrough experiments confirmed that trace C_3H_4 in 1 : 99 and 0.1 : 99.9 (v/v) mixtures can be readily removed by UTSA-200, giving C_3H_6 with a purity of 99.9999%. Neutron powder diffraction studies and computational calculations revealed that both compact pore space and fluoride sites contribute jointly to the strong adsorption of C_3H_4, thus realizing highly efficient separation of C_3H_4–C_3H_6.

ZIF-8 (also MAF-4) is one typical MOF that shows robust global network geometry and flexible local structure with dynamic pore apertures and thus is capable of the diffusion of molecules with relatively larger sizes although it only has a static aperture size of 3.4 Å.[62] Interestingly, upon exposure to external electric field (E-field), the ZIF-8 lattice can be polarised into a more rigid one, which changes the size selectivity and gas transportation of ZIF-8 membrane, thus increasing the selectivity for C_3H_6–C_3H_8 mixture (Figure 5.16).[63] XRD and *ab initio* calculations revealed that the ZIF-8 lattice can be reversibly transformed from cubic to monoclinic and triclinic polymorphs upon polarisation under an E-field, among which the *Cm* phase with a large diameter

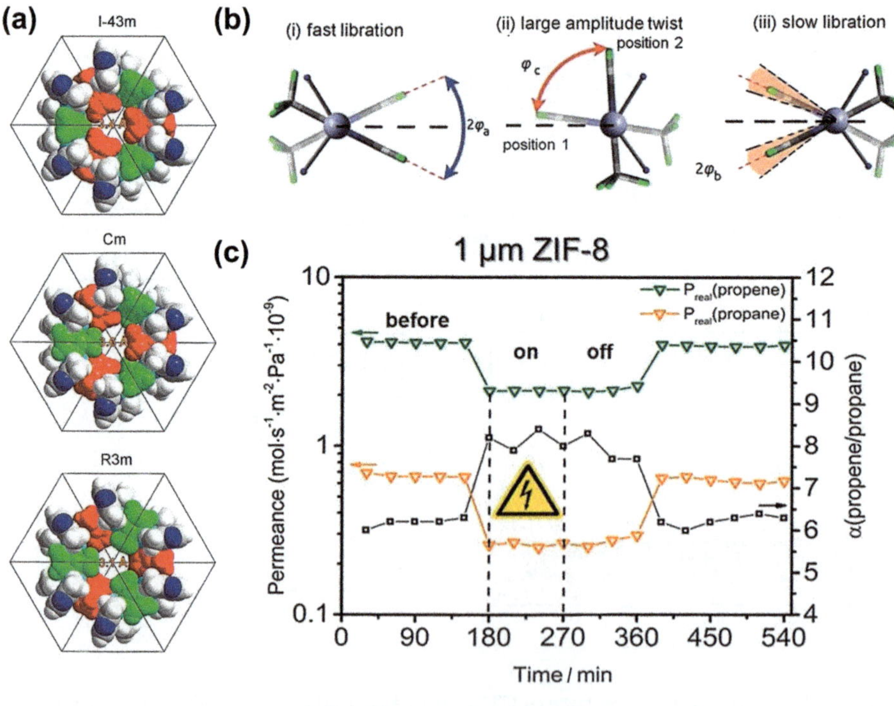

Figure 5.16 (a) Different simulated polymorphs of ZIF-8 and showing the different pore diameters. (b) Schematic diagrams of the possible motions in ZIF-8. (c) Propene–propane mixtures permeation in ZIF-8 and showing the real separation factor α. Adapted from ref. 63 with permission from the American Association for the Advancement of Science, Copyright 2017.

(3.6 Å) is the most dominant polarisable phase. In addition, dielectric spectroscopy, polarisation, and deuterium nuclear magnetic resonance studies showed that low-frequency resonances of ZIF-8 were associated with lattice flexibility and linker movement. The latter was found critical for gas separation. The defibrillation of lattice motion in ZIF-8 crystals here shows improved molecular sieving performance. This work demonstrates well how structural distortions controlled in a non-destructive way can modulate gas transportation in MOFs.

The gate-opening sorption behaviour in flexible MOFs can be temperature-dependent or adsorbate-dependent when the energy barrier of gate-opening might be overcome by relatively strong host–guest interactions of specific gas molecules under certain temperatures. In this context, the separation performance can be adjusted to the maximum by varying sorption temperatures or gas pressures. MOF JNU-3a, containing a one-dimensional (1D) pore channel orthogonally inter-connected by dynamic molecular pockets (pore aperture of 3.7 Å) (Figure 5.17), shows dynamic molecular sieving of propylene from propane.[64] The dynamic pockets were reported to undergo gate-opening upon exposure to propane or propylene at different partial pressures. However, the slim pore channels (4.5 Å × 5.3 Å) facilitate fast

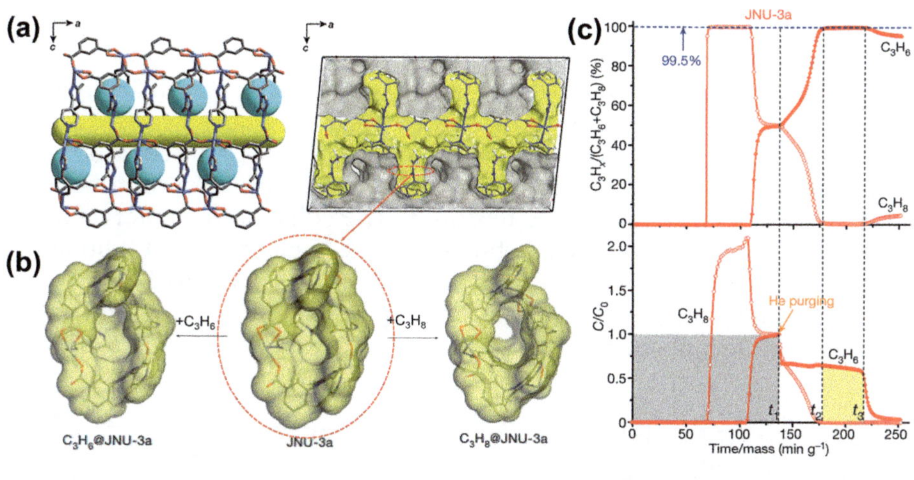

Figure 5.17 (a) The pore structure of JNU-3 viewed along the *b*-axis showing molecular pockets (turquoise) and the 1D channel (yellow). (b) Connolly surfaces comparison to illustrate the opening up of the gourd-shaped aperture along with the adsorption of C_3H_6 or C_3H_8. (c) Breakthrough curves (starting at t = 0) of an equimolar mixture of C_3H_6–C_3H_8 (1.0 cm^3 min^{-1}) on JNU-3a, followed by desorption curves (starting at $t = t_1$) under helium gas (10.0 cm^3 min^{-1}) sweeping at 303 K. C and C_0 are the concentrations of each gas at the outlet and inlet, respectively. Grey area, mixed gas input; yellow area, C_3H_6 gas output; t_1, the beginning of desorption; t_2, the start point of collecting C_3H_6; t_3, the endpoint of collecting C_3H_6. Reproduced from ref. 64 with permission from Springer Nature, Copyright 2021.

adsorption–desorption kinetics. At 303 K, JNU-3a shows C_3H_6 uptake of 58.6 $cm^3\ g^{-1}$ (2.6 mmol g^{-1}) and an IAST C_3H_6–C_3H_8 selectivity (513) for a 50:50 C_3H_6–C_3H_8 mixture. The breakthrough studies indicate that JNU-3a can yield C_3H_6 of polymer grade with productivity of 34.2 L kg^{-1} (99.5%+ purity) in its desorption process after helium purge. Structure and modelling studies indicate both propylene and propane molecules are preferentially located in the pockets, accompanied by a slight deformation of the organic linkers, which supports the dynamic gate-opening in JNU-3a.

For multicomponent mixtures, the temperature-dependent or adsorbate-dependent gate-opening of every gas will be more complex. The control of their gate-opening sorption behaviour for these mixtures to realise separation would be more challenging. An interesting example is represented by one flexible MOF [Cu(L)$_2$(SiF$_6$)] (NTU-65, L = 1,4-di(1*H*-imidazol-1-yl)benzene, Figure 5.18) that features different stepwise sorption isotherms for C_2H_2, CO_2 and C_2H_4, respectively.[65] However, *in situ* sorption PXRD studies revealed there are several intermediate structures during sorption. At elevated temperatures, C_2H_4 and CO_2 triggered gate-opening, and thus show negligible adsorption at 263 and 298 K, respectively. That means a C_2H_4–C_2H_2–CO_2 ternary mixture can be separated well by this MOF *via* altering the sorption temperature. The breakthrough experimental results of NTU-65 for C_2H_4–C_2H_2–CO_2 mixture (90:1:9, v/v/v) confirmed the preferential capture of C_2H_2 and CO_2, giving C_2H_4 with a purity of 99.99% from single breakthrough separation. Density functional theory (DFT) calculations revealed that C_2H_2 and CO_2 also have a strong binding affinity with the framework, which facilitates both gases to open the flexible pore structure in NTU-65.

In contrast to the host framework, the flexibility of guest molecules is usually overlooked. For C_4 hydrocarbon mixtures, more carbon–carbon bonds on their isomers make them ease flexibility. The collaborative interactions of both host- and guest-flexibility thus make the separation of 1,3-butadiene (C_4H_6) for polymer-grade purity (>99.5%) a great challenge. A hydrophilic MOF [Zn$_2$(btm)$_2$] (MAF-23, H$_2$btm = bis(5-methyl-1*H*-1,2,4-triazol-3-yl) methane), with flexible quasi-discrete pore structure, was reported to show

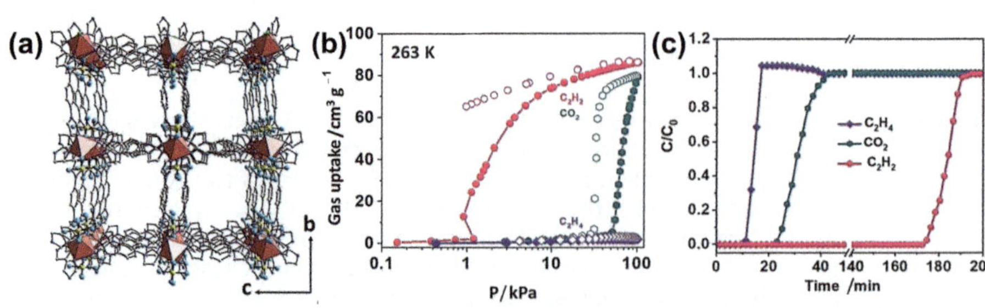

Figure 5.18 (a) Crystal structure of NTU-65 view along crystallographic *a* axis. (b) Adsorption isotherms of NTU-65 for C_2H_2, C_2H_4 and CO_2 at 263 K. (c) Ternary separation of C_2H_4–C_2H_2–CO_2 mixture. Adapted from ref. 65 with permission from John Wiley & Sons, Copyright 2020.

conformation preference on a *cis* isomer of $n\text{-}C_4H_8$, whereas preferring *trans* isomers of C_4H_6 and C_4H_{10}. The calculated adsorption heat of MAF-23 for C_4H_6 is 72 kJ mol^{-1}, which is lower than those for C_4H_{10} (79 kJ mol^{-1}), $n\text{-}C_4H_8$ (87 kJ mol^{-1}), and $i\text{-}C_4H_8$ (101 kJ mol^{-1}), indicating the interactions between C_4H_6 and host framework were weakened. Crystallographic studies on the gas-loaded MAF-23 revealed that stronger C–H···N interactions form for $n\text{-}C_4H_8$, $i\text{-}C_4H_8$, and C_4H_{10} in this MOF in contrast to those for C_4H_6. These results imply that pure C_4H_6 (purity of ≥99.5%) can be directly obtained by adsorption separation of MAF-23 through one single breakthrough operation (Figure 5.19).[66] Column breakthrough experiments for C_4H_6–$n\text{-}C_4H_8$–$i\text{-}C_4H_8$–C_4H_{10} mixture (1 : 1 : 1 : 1 and 5 : 2 : 2 : 1, v/v/v/v) confirmed that C_4H_6 was firstly eluted from the column, whereas C_4H_{10}, $n\text{-}C_4H_8$, and $i\text{-}C_4H_8$ were adsorbed until saturated. This study showed that the host framework of MOFs can induce conformational changes in flexible guest molecules that collaboratively affects their separation performance.

The substitution of the host framework by flexible terminal groups can further increase the complexity of collaborative host–guest interactions. Alkane isomer separation is an important means to improve the quality of gasoline. Generally, more branched alkanes (higher octane number) make gasoline with favourable anti-knock properties. Some reports have already shown the flexible MOF has potential applications in C_6 isomer separation.[67–70] A flexible MOF $[Zn_2(bdc\text{-}NO_2)_2(bpy)]$ (2,5-bis[(2-nitrobenzyl)oxy]-1,4-benzenedicarboxylic acid) incorporated with local flexible substituents was explored for the separation of C_6 alkane isomers (*n*-hexane (HEX), mono-branched 3-methylpentane (3MP), and di-branched 2,2-dimethylbutane (2,2-DMB)), and exhibited a distinct gate-opening behaviour (Figure 5.20).[70] This flexible MOF has a non-interpenetrated 3D pillared-layer structure with a void

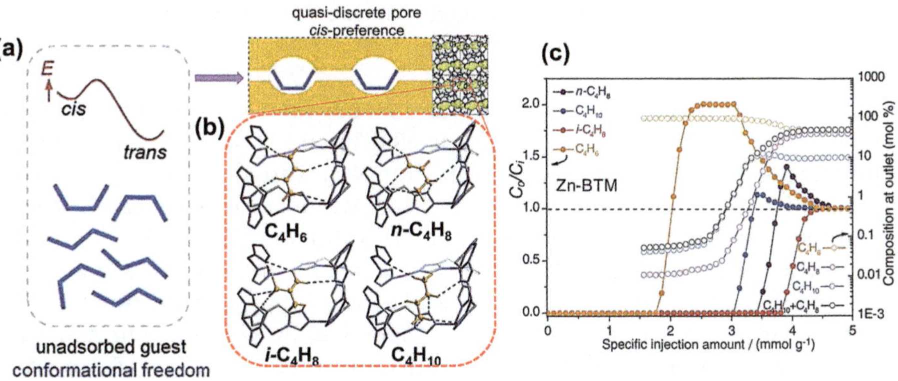

Figure 5.19 (a) Schematic diagrams of controlling the guest confirmation through the quasi-discrete pore in MAF-23. (b) Crystal structure of host–guest complexes of C_4H_6, $n\text{-}C_4H_8$, $i\text{-}C_4H_8$, and C_4H_{10}. (c) Breakthrough curves of MAF-23 for a 5 : 2 : 2 : 1 C_4H_6–$n\text{-}C_4H_8$–$i\text{-}C_4H_8$–C_4H_{10} mixture. Reproduced from ref. 66 with permission from the American Association for the Advancement of Science, Copyright 2017.

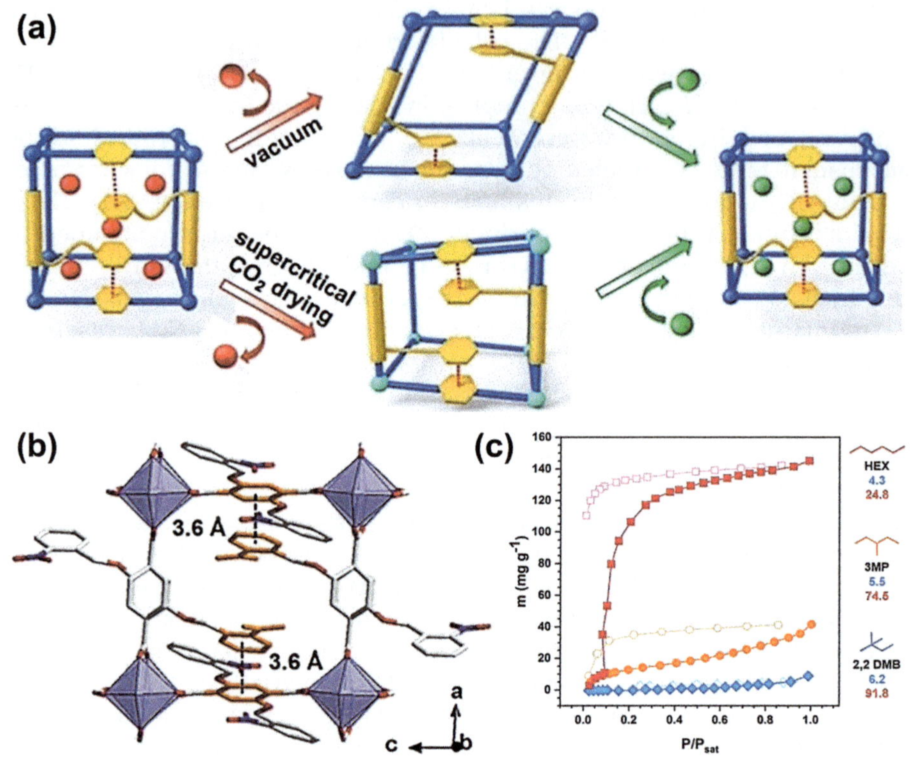

Figure 5.20 (a) MOF with local and global flexibility integrated by π–π interaction. (b) Crystal structure of as-synthesised $[Zn_2(bdc\text{-}NO_2)_2(bpy)]$. (c) Adsorption isotherms of $[Zn_2(bdc\text{-}R)_2(bpy)]$ for *n*-hexane (HEX), mono-branched 3-methylpentane (3MP), and di-branched 2,2-dimethylbutane (2,2-DMB). Adapted from ref. 70 with permission from John Wiley & Sons, Copyright 2021.

space of 10.8% featuring a π–π interaction between the nitro-substituted arenes and the central benzene of bdc-NO_2 ligand. The result of variable-temperature PXRD indicated that $[Zn_2(bdc\text{-}NO_2)_2(bpy)]$ could maintain its crystallinity after structural transformation while its isostructural $[Zn_2(bdc\text{-}H)_2(bpy)]$ lost long-range order in the crystallographic *a* and *b* directions due to the absence of intra-framework π–π interactions. Interestingly, two different guest-free structures of $[Zn_2(bdc\text{-}NO_2)_2(bpy)]$ can be obtained by vacuum and supercritical CO_2 drying, respectively, owing to their substituent-framework cooperative flexibility. Furthermore, sorption experimental results revealed that $[Zn_2(bdc\text{-}R)_2(bpy)]$ can preferentially adsorb HEX and 3MP whereas excluding 2,2-DMB at 303 K, which is more suitable for alkane isomer separation.

The separation of C_8 aromatics (xylene isomers and ethylbenzene) in industry is an extremely energy-intensive and challenging process. Flexible MOFs could achieve the separation of C_8 aromatics through energy-efficient adsorption separation technology.[71–76] A 1D coordination polymer [Mn(dhbq)(H_2O)$_2$] (H_2dhbq = 2,5-dihydroxy-1,4-benzoquinone) linked by intra-chain hydrogen bonds was recently reported to exhibit temperature-dependent and adsorbate-dependent adsorption behaviours for *p*-, *m*-, and *o*-xylene isomers in their both vapour and liquid phases (Figure 5.21).[72] The structure swelling of this coordination polymer accounts for its specific molecular recognition and separation for xylene isomers.

Temperature-dependent and adsorbate-dependent gate-opening of adsorbents for C_8 aromatics can also be found in flexible MOFs. A non-porous layered material [Co(4,4′-bipyridine)$_2$(NCS)$_2$]$_n$ (sql-1-Co-NCS) was reported to exhibit pore-opening upon exposure to C_8 aromatics and can adsorptive separation of *o*-xylene from C_8 aromatics with a capacity of >80 wt% and selectivity ($S_{OX/EB}$) of ~60 (Figure 5.22).[75] After the inclusion of C_8 aromatics, there was inter-layer swelling observed in the open-pore structure (with cavity sizes of 7.5 Å × 7.5 Å). Vapour sorption isotherms indicated that each C_8 aromatic isomer can induce gate-opening of sql-1-Co-NCS at different switching

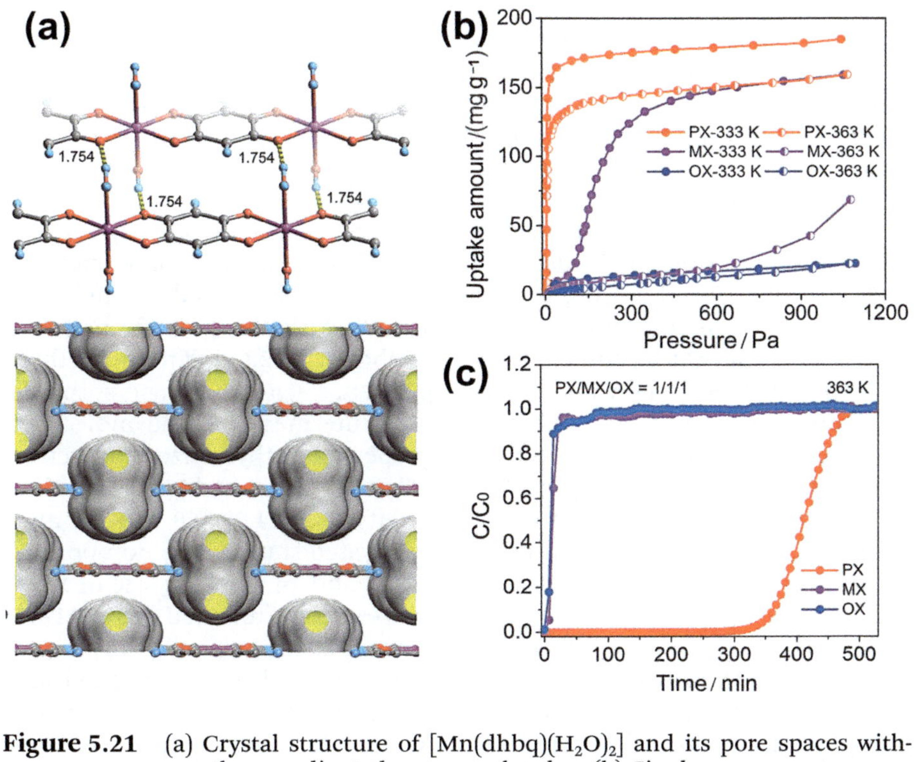

Figure 5.21 (a) Crystal structure of [Mn(dhbq)(H_2O)$_2$] and its pore spaces without the coordinated water molecules. (b) Single-component vapour adsorption isotherms of xylene isomers on Mn-dhbq at 333 and 363 K. (c) Breakthrough curves of an equimolar ternary mixture of xylene isomers at 363 K for Mn-dhbq pellet samples. Adapted from ref. 72 with permission from the American Association for the Advancement of Science, Copyright 2022.

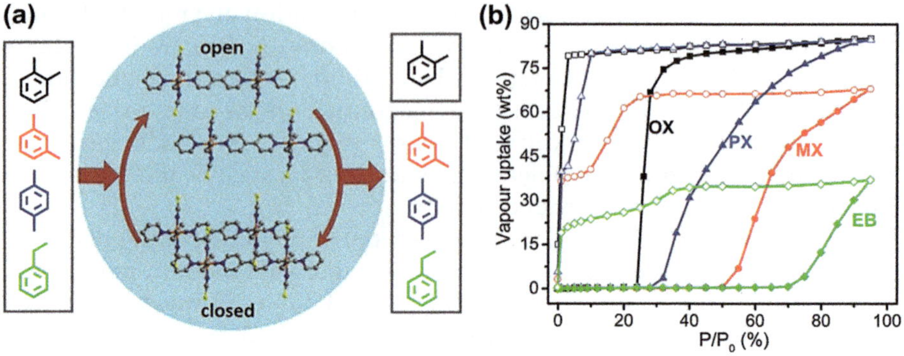

Figure 5.22 (a) Schematic diagram of sql-1-Co-NCS separation of *o*-xylene from C8 Aromatics. (b) Adsorption isotherms of sql-1-Co-NCS for *o*-xylene (OX) (298 K), *m*-xylene (MX) (298 K), *p*-xylene (PX) (298 K), ethylbenzene (EB) (298 K). Reproduced from ref. 75 with permission from John Wiley & Sons, Copyright 2019.

pressures (P_s), among which OX molecules showed the lowest switching pressure due to the strongest interactions with adjacent layers. Therefore, sql-1-Co-NCS exhibits OX preference with OX–PX, OX–MX and OX–EB selectivity values of 9.6, 7.5 and 60.1, respectively, as determined by liquid-phase binary tests.

Separation through size exclusion is a simple way to remove molecules of relatively large size. But co-adsorption of molecules of small size is difficult to avoid. Flexible MOFs usually exhibit guest-dependent pore-opening sorption behaviour, that differentiates guest molecules based on their different strength of intermolecular interactions with the host framework. A flexible MOF Cu_2(fbdim) (MAF-41, H_2fbdim = 2,6-ditrifluoromethylbenzodiimidazole) showing guest-dependent gate-opening sorption behaviours and size-sieving effect (so-called intermediate-sized molecular sieving effect) was applied to separate styrene from ethylbenzene–styrene–toluene–benzene mixture (Figure 5.23).[76] The as-synthesised MAF-41 with a 3D hinged-fence-like network possesses a cavity size of $4.2 \times 9.8 \times 10.1$ Å^3 and aperture sizes of 3.9×6.7 Å^2, whereas the activated states was a non-porous structure with a cavity size of $0.7 \times 6.0 \times 6.1$ Å^3 and an aperture size of 1.0×6.0 Å^2. Single-component vapour adsorption isotherms revealed that MAF-41 can adsorb styrene but exclude not only larger ethylbenzene *via* the sieving effect but also the smaller molecules (benzene and toluene) owing to their insufficient interactions for gate-opening. GCMC simulations confirmed a large phase transition energy (150.1 kJ mol^{-1}) hinders the adsorption of benzene and toluene. Mixed solvent adsorption experiments showed that MAF-41 has a high styrene selectivity of 1250 for an ethylbenzene–styrene mixture and 3300 for an ethylbenzene–styrene–toluene–benzene mixture. Liquid phase separation indicated that MAF-41 can make pure styrene (99.9%+) in a single adsorption–desorption cycle. These results indicate that combining

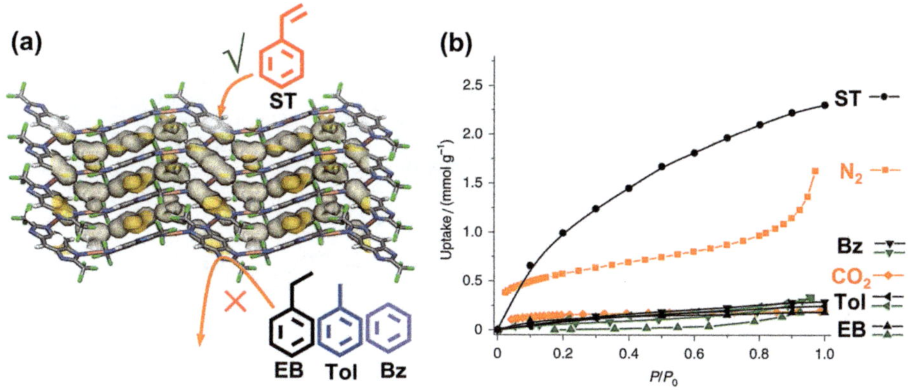

Figure 5.23 (a) Crystal structure of guest-free MAF-41 and showing the schematic diagram of the intermediate size molecule ST selectively adsorbed. (b) Adsorption isotherms of activated MAF-41 for ST (styrene, 298 K), EB (ethylbenzene, 298 K), Tol (toluene, 298 K), Bz (benzene, 298 K), N_2 (77 K) and CO_2 (195 K). Reproduced from ref. 76 with permission from Springer Nature, Copyright 2019.

mechanisms of kinetic and thermodynamic guest exclusion can realise the highly efficient separation of complex mixtures.

In contrast to the separation of small molecules such as alkanes and alkenes, the separation of high molecular weight polymers might be more complicated. There are flexible and rigid MOFs that emerged as promising candidates for the efficient separation and purification of polymer mixtures.[77] The dynamic character of flexible MOFs offers a distinct advantage over rigid MOFs in separating mixtures of polyethylene glycol (PEG) with closely similar terminal moieties, namely OH, OMe, and OEt because the subtle difference in polarity among these groups can greatly influence the pore-opening kinetics. The flexible MOF $[Co_2(ndc)_2(bpy)]$ (ndc = 2,6-napthalene dicarboxylate, bpy = 4,4-bipyridyl) exhibits a two-fold interpenetrated 3D structure with a large 1D channel of 4.7×4.7 Å^2, which becomes significantly constricted after removal of guest molecules. Interestingly, this study illustrated two different methods (solvent-evacuation method and neat insertion method) for separating polymers using a flexible MOF (Figure 5.24). The solvent-evacuation method facilitates the purification of polymers with termini exhibiting different steric hindrances *via* the sieving effect of opened-form MOF, whereas the neat insertion method enables precise discrimination of polymer termini as only PEGs with an appropriate terminal group can induce pore opening of closed-form MOF under molten PEG conditions. Time-resolved PXRD patterns studies revealed that PEG terminal groups with various polarities could lead to different insertion rates (Me > Et > Bu ≫ H). Therefore, the successful kinetic separation of Me-PEG-Me from H-PEG-H mixtures can be accomplished with high selectivity of 5.2. This work opens a new possibility for the purification of polymers.

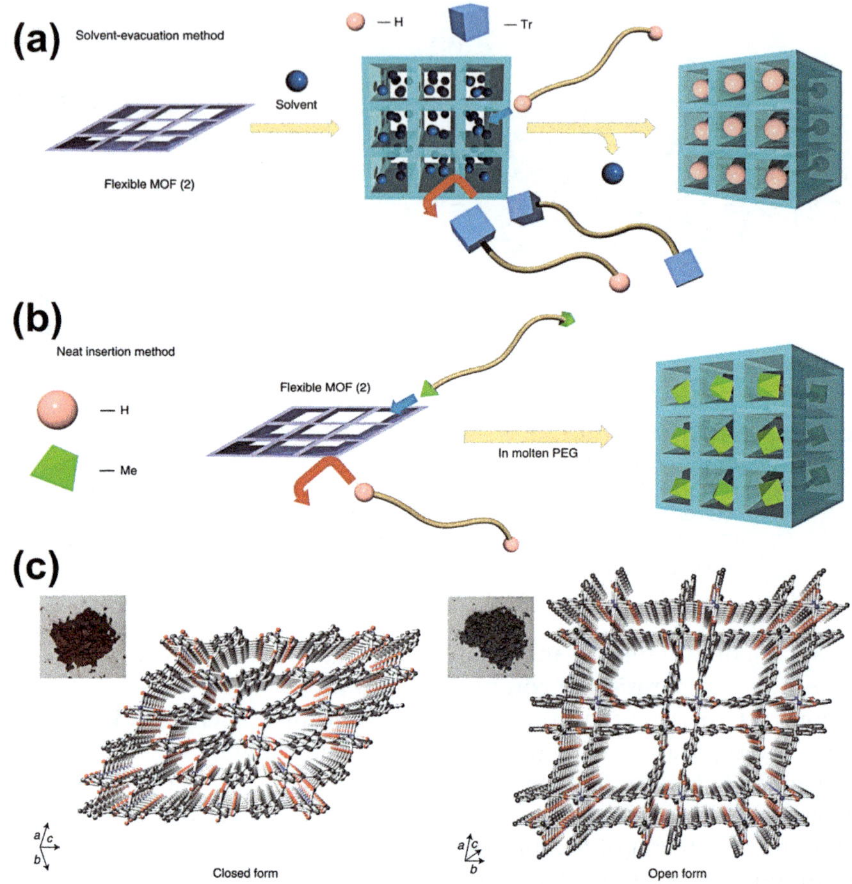

Figure 5.24 (a) Schematic diagram of a solvent evacuation method showing separation of polymers with different terminal groups using a flexible MOF. (b) Schematic diagram of neat insertion method showing separation of polymers with different terminal groups of similar sizes using a flexible MOF. (c) Crystal structure of two different forms of $[Co_2(ndc)_2(bpy)]$ and corresponding photograph. Adapted from ref. 77 with permission from Springer Nature, Copyright 2018.

5.4 Sensing

Flexible MOFs can generate multitudinous dynamic behaviours in response to the external environment.[11,26] Importantly, the coordination environment and ground-state structures can be fine-tuned by external stimuli such as guest molecules, light, temperature, mechanical force, and pressure, which can have great effects on the physicochemical properties.[78] These changes in properties can be applied as detection signals for rapid, sensitive, and specific detection. Therefore, flexible MOFs have emerged as a diversified platform for sensor detection.

Luminescence is one of the most common detection signals. Structural transformations of flexible MOFs can affect not only their luminescent properties but also those of guest species. A hybrid MOF, $[Zn_2(bdc)_2(dabco)]\supset DSB$ ($1\supset DSB$, H_2bdc = 1,4-benzenediarboxylic acid, dabco = 1,4-diazabicyclo-[2.2.2]octane, DSB = distyrylbenzene), was capable of selectively capturing and detecting gases by cooperative structural transformations of fluorescent molecules distyrylbenzene (DSB) and the flexible host framework (Figure 5.25).[79] In particular, the pore size and shape of MOF can be changed upon gas adsorption, resulting in the cooperative change in conformations of the DSB molecule. When CO_2 adsorption reaches a certain threshold pressure ($P = 30$ kPa) at 195 K, the configuration of DSB will change from a twisted state to almost planar, and the fluorescence spectrum (emission peak and position) remains unaffected by N_2, O_2 and Ar. Similarly, the solid-state fluorescence of the composite can also be changed at a lower pressure of C_2H_2 ($P = 7$ kPa). Therefore, solid-state fluorescence can be used to distinguish C_2H_2 and CO_2, despite their similar physicochemical properties. Such a collaborative strategy based on the fluorescence of guest molecules enables gas sensing and detection to be performed by general flexible MOFs.

Structural transformations of flexible MOFs during guest loading can enhance host–guest interactions, which could greatly impact on their luminescent properties. It was reported a 1,4,5,8-naphthalenediimide (NDI)-based entangled MOF showed displacement between two-fold interpenetrated frameworks after the introduction of guest molecules.[80] Such dynamic structural transformation makes the interaction between aromatic guests and scaffolds that contain NDI moieties stronger, leading to complexes with different charge-transfer (CT) properties. Aromatic species thus exhibited intense turn-on fluorescence emissions with different wavelengths. This study demonstrated a sensing approach for distinguishing

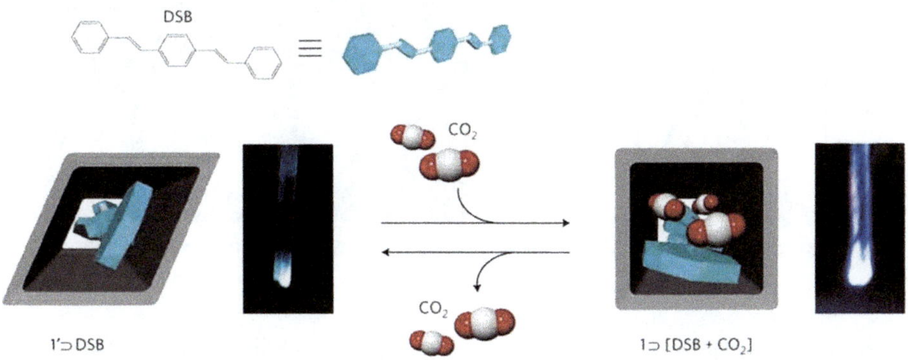

Figure 5.25 Schematic diagram of the coupled structural transformations of flexible MOFs and fluorescent reporter molecule (DSB), and fluorescence changes of the composite after co-adsorption of CO_2. Adapted from ref. 79 with permission from Springer Nature, Copyright 2011.

subtle structural differences in molecules through a complex transduction protocol and molecular decoding.

The fluorescence of a 3D flexible MOF {[Zn(ndc)(*o*-phen)]·Guest}$_n$ (*o*-phen = 1,10-phenanthroline, ndc = 2,6-napthalenedicarboxylate) with distorted rectangular-shaped 1D channels along crystallographic *a*-axis was reported can be modulated by removal or adsorption of guest solvent molecules (Figure 5.26).[81] The nanopore spaces can accommodate electron-donating aromatic amine guests of different molecular sizes, ranging from aniline (AN), *N*,*N*′-dimethyl-*p*-toluidine (DMPT), *N*,*N*′-dimethyl aniline (DMA), to *N*-methyl aniline (NMA), showing selective turn-on fluorescence emission. As indicated by time-resolved single-photon counting experiments, the selective turn-on emission is from the formation of charge-transfer complexes between *o*-phen ligands and aromatic amines as well as energy transfer of ndc chromophores in the framework.

To embed chromophores into flexible MOFs, an interesting 2D layered MOF (NUS-1) was developed using a tetraphenylethene (TPE)-based ligand, namely 4,4′-(2,2-diphenylethene-1,1-diyl)dibenzoic acid (DPEB), and Zn_4O-type building units.[82] The activated phase NUS-1a displays multistep gas adsorption behaviour, as well as turn-on fluorescence upon adsorption of various volatile organic compounds (Figure 5.27). This fluorescent MOF contains wide hexagonal pore channels in which two dangling phenyl rings from

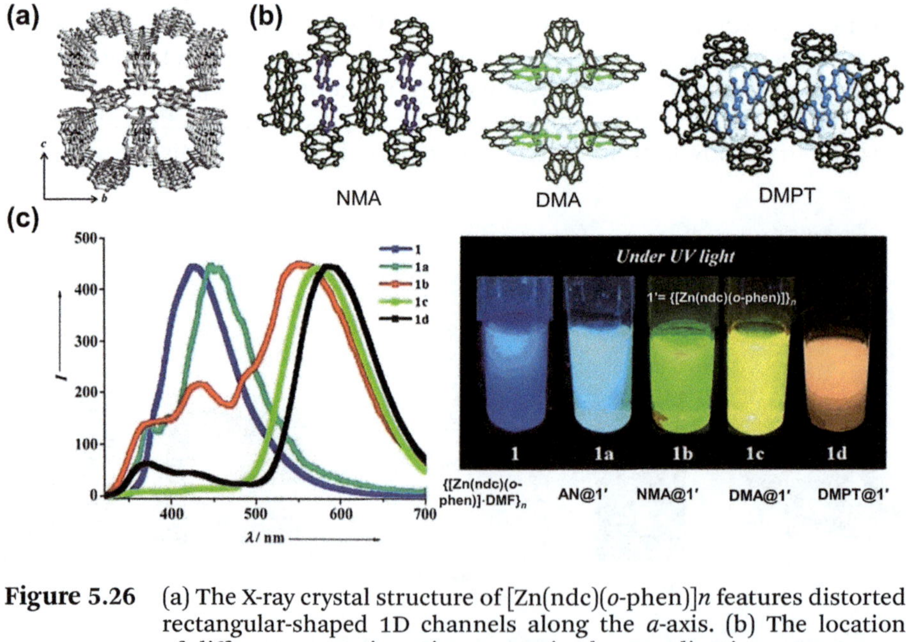

Figure 5.26 (a) The X-ray crystal structure of [Zn(ndc)(*o*-phen)]*n* features distorted rectangular-shaped 1D channels along the *a*-axis. (b) The location of different aromatic amine guests in the coordination nanospaces. (c) Emission spectra and photographs of [Zn(ndc)(*o*-phen)]*n* and inclusion compounds under irradiation with UV light. Adapted from ref. 81 with permission from John Wiley & Sons, Copyright 2014.

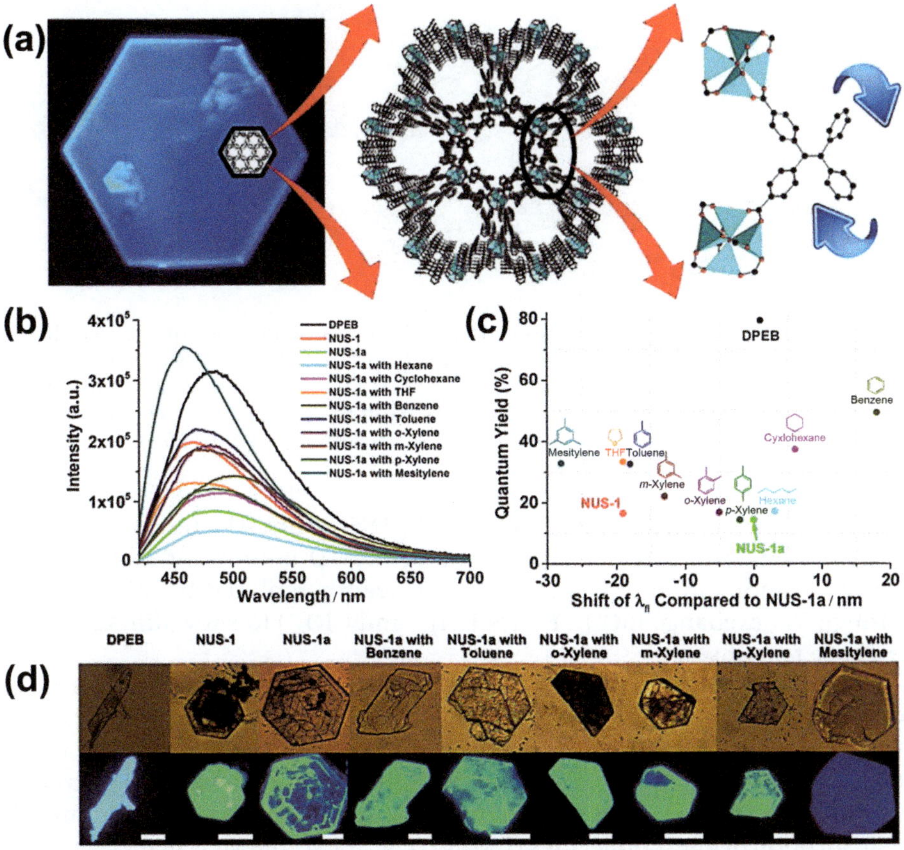

Figure 5.27 (a) Crystal structures of NUS-1. (b) Fluorescent emission spectra of corresponding samples. (c) Relationship of quantum yield *vs.* emission shift compared to NUS-1a. (d) Fluorescent microscopy images of NUS-1a in different guest inclusion states. Adapted from ref. 82 with permission from American Chemical Society, Copyright 2014.

DPEB ligands are aligned along the inner pore surface. The dangling phenyl rings can freely rotate/vibrate in the channel. The loading of analyte into this MOF would dimmish the rotation/vibration of benzene rings, resulting in a reduction in non-radiative transitions and thus fluorescence enhancement.

A porous cationic MOF consisting of 1D coordination chains in conjunction with free nitrate anions was reported to display responsive fluorescence towards various guests and anions (Figure 5.28).[83] The structure of $[\{Zn(edp)(MeOH)_2\}(NO_3)_2{\cdot}Guest]_n$ (where edp = (1*E*,1′*E*)-*N*,*N*′-(ethane-1,2-diylbis(4,1-phenylene))bis(1-(pyridin-2-yl)methanimine)) was assembled *via* packing of 1D chains that were connected with anion and solvent molecules through H-bonding interactions. The as-synthesised crystals transformed into another phase, namely $[\{Zn(edp)(H_2O)_2\}(NO_3)_2{\cdot}2H_2O]_n$, after 1 h of isolation from its mother liquor. This new phase showed different M–M–M angles

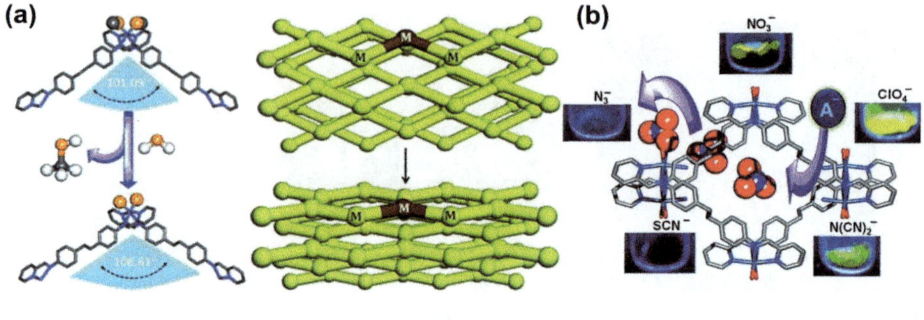

Figure 5.28 (a) Crystal structures of the cationic porous framework in two different coordination environments (top: $[\{Zn(edp)(MeOH)_2\}(NO_3)_2{\cdot}Guest]$, bottom: $[\{Zn(edp)(H_2O)_2\}(NO_3)_2{\cdot}2H_2O]$). (b) Photo of different anion-exchanged samples upon excitation at 394 nm. Adapted from ref. 83 with permission from John Wiley & Sons, Copyright 2013.

in contrast to those in $[\{Zn(edp)(MeOH)_2\}(NO_3)_2{\cdot}Guest]_n$, resulting in significant changes in the size and shape of the 1D pore channel. Moreover, the NO_3^- anions in pore channels can be readily replaced by other anions through ion exchange (SCN^-, N_3^-, $N(CN)_2^-$ and ClO_4^-) to show different fluorescence responses.

A flexible Mg-based MOF, $[Mg(H_2dhtp)(H_2O)_2]{\cdot}DMAc$ (where H_4dhtp = 2,5-dihydroxy-terepthalic acid; DMAc = *N*,*N*-dimethylacetamide), was reported to act as an effective sensor for detecting trace amounts of water (0.05–5%) in various organic solvents in real-time (Figure 5.29).[84] The framework of this MOF exhibits notable breathing behaviour upon the removal/re-adsorption of guest molecules (DMAc). The guest-free MOF features a weak yellowish emission while the hydrated MOF shows enhanced yellow-green emission. Further studies revealed that the mechanism of turn-on luminescence is different from that achieved by restricting non-radiative transitions.[82] Instead, it is dominated by the flexibility of the host framework, which induces changes in the separation between $H_2dhtp^{2-}{\cdots}H_2dhtp^{2-}$ and thereby alters their energy transfer. The long-distance between H_2dhtp^{2-} moieties in hydrated MOF hinders energy transfer, preventing self-quenching and thus leading to enhanced fluorescence.

In contrast to luminescent signals, visual colour change can also be applied as a sensing signal. A Co-based MOF, $[Co_{1.5}(tipb)(SO_4)(pta)_{0.5}](DMF)_{1.75}$ (BP⊃DMF, tipb = 1,3,5-tris(pimidazolylphenyl)benzene, H_2pta = terephthalic acid, DMF = *N*,*N*-dimethylformamide), has been reported to exhibit reversible single-crystal to single-crystal (SC–SC) structure transformation and corresponding colour changes upon the stimuli of coordinative molecules such as H_2O and NH_3 (Figure 5.30).[85] The as-synthesised MOF is a two-fold interpenetrated 3D network containing 1D micropore channels along crystallographic *c* axis. Upon exposure to air, the pta^{2-} ligands can be displaced by H_2O and lead to a change in the coordination configuration of the Co^{2+} ions and the colour of BP⊃DMF changes from blue to red. The isotherm at

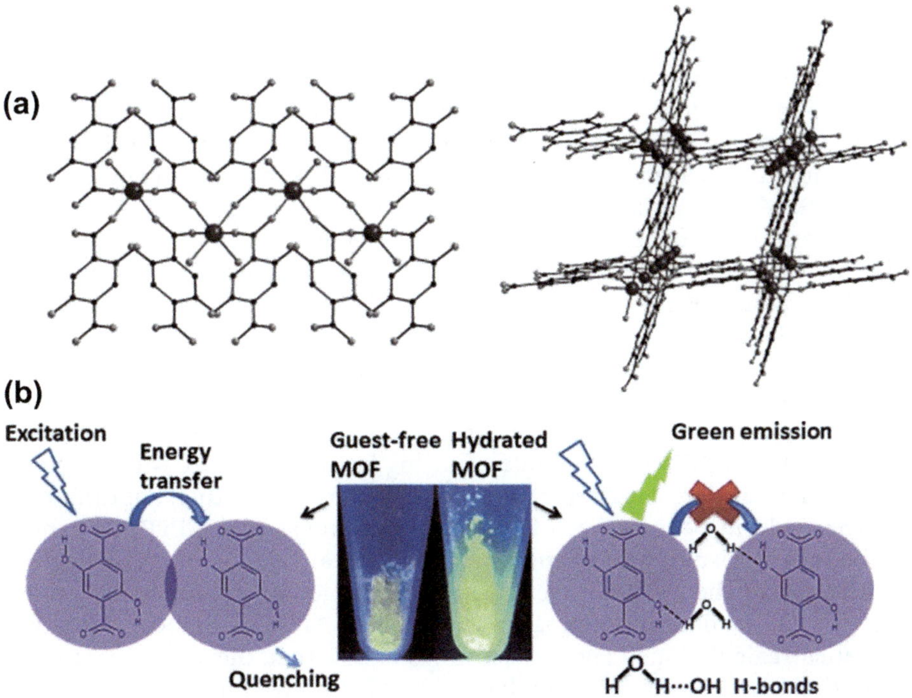

Figure 5.29 (a) Crystal structures of [$Mg(H_2dhtp)(H_2O)_2$]·DMAc. (b) Self-quenching of the emission of guest-free MOF and the intense turquoise fluorescence of hydrated MOF. Adapted from ref. 84 with permission from John Wiley & Sons, Copyright 2015.

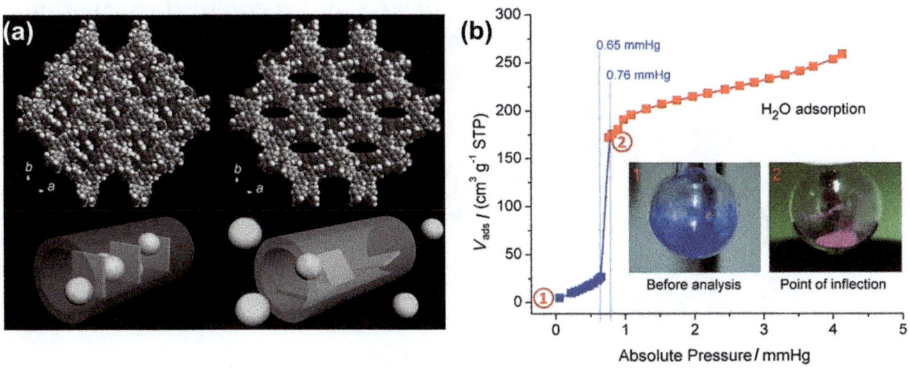

Figure 5.30 (a)The structural transformation of [$Co_{1.5}(tipb)(SO_4)(pta)_{0.5}$] during H_2O adsorption and corresponding schematic diagram. (b) Vapour adsorption isotherms of water for [$Co_{1.5}(tipb)(SO_4)(pta)_{0.5}$] at 273 K. Adapted from ref. 85 with permission from John Wiley & Sons, Copyright 2013.

273 K for H_2O showed that structural transformation can occur at pressures of 0.65–0.76 mmHg (P/P_0: 0.14–0.17). Interestingly, the coordinated pta^{2-} ligands can block the pore channels in BP⊃DMF while these pore channels can be opened when the pta^{2-} ligands are detached from the framework and displaced into the channels. Furthermore, other molecules such as NH_3 can also stimulate structural transformation making this material a candidate as a sensor for these coordinative molecules.

In general, electrochemical sensors can rapidly detect analytes by transforming the analyte–electrode interaction and chemical reactions into electrical signals. Flexible MOFs have been demonstrated to serve as electrochemical sensors. An interesting tetrathiafulvalene (TTF)-based MOF, namely MUV-2, was reported to show tunable electrochemical properties as response to its continuous breathing behaviour (Figure 5.31).[86] This flexible MOF was constructed by six-connected $[Fe_3(\mu_3O)(COO)_6]$ units and tetratopic TTF tetrabenzoic ($TTFTB^{4-}$) ligands, showing layered structure with 1D mesopore and micropore channels that mutually across each other. For different adsorbed solvents, the degree of curvature or flatness of the TTFTB ligands can be varied, resulting in changes in their crystal cell volume. The degree of curvature of the TTFTB linker directly affects its ionisation energy, thus changing the oxidation potential of (TTF)-based MOF, as demonstrated by theoretical calculations. These results indicated that the electrochemical properties of TTF-based MOF can be regulated by solvent-induced breathing behaviour, which allows this MOF to be applied for the detection of analytes.

An eight-fold interpenetrating 3D MOF, namely JUK-8 [Zn(oba)(pip)]*n* (where oba = 4′-oxybis(benzenedicarboxylate), pip = 4-pyridyl functionalised benzene-1,3-dicarbohydrazide), has been reported to show unique framework flexibility in response to temperature changes or guest molecule adsorption (H_2O and CO_2) (Figure 5.32).[87] The stepwise volume variation of JUK-8 upon water adsorption presents a promising opportunity for developing resistive humidity sensors. Structural analysis and theoretical calculations have revealed that the distinct breathing behaviour of JUK-8 is attributed to

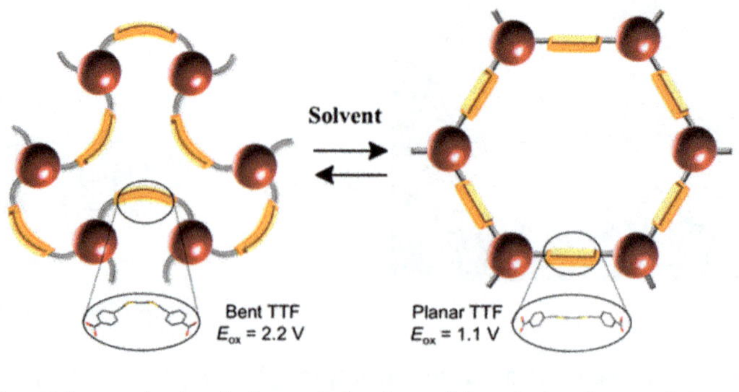

Figure 5.31 Schematic depiction of the breathing behaviour of TTF-based MOF upon solvent adsorption, and corresponding changes of electrochemical properties. Adapted from ref. 86 with permission from American Chemical Society, Copyright 2018.

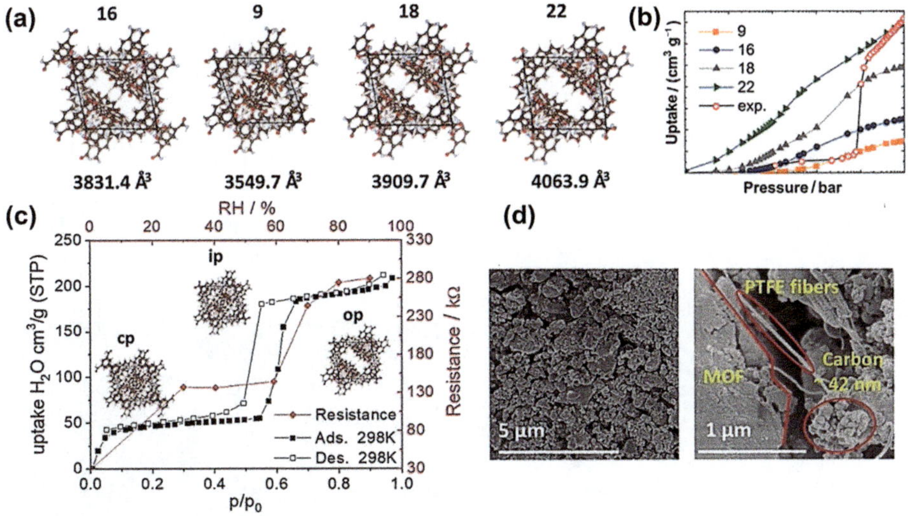

Figure 5.32 (a) Possible transition path for adsorption-induced transformation of JUK-8. Numbers indicate selected structures generated by DFT calculation; *V* denotes the volume of corresponding unit cells. (b) CO_2 adsorption isotherms of JUK-8 in different structures (simulated: closed symbols; experimental: open symbols). (c) Water vapour physisorption isotherm (298 K) for a composite (black), and the corresponding dependence of the composite resistance on relative humidity (red). (d) SEM images of the composite. The components are composed of MOF, conductive carbon black, and PTFE fibres. Adapted from ref. 87 with permission from John Wiley & Sons, Copyright 2020.

a combination of linker rotation and carboxylate hinge motion within the single network, which differs from the conventional subnetwork displacement mechanism commonly observed in interpenetrated MOFs. Interestingly, the unique breathing behaviour of JUK-8 gave three different phases, namely closed-pore (cp), intermediate-pore (ip) and open-pore (op) with void volumes of 0.8%, 5.0% and 27.2%, respectively. Below 298 K, the water vapour adsorption isotherm of JUK-8-cp exhibits a significant and reversible stepwise volume change upon water adsorption, enabling direct integration of JUK-8 into conductive composite membranes as an active component for threshold sensing of humidity. Resistive humidity sensing by JUK-8 was demonstrated with relative resistance responses of 833%, 333% and 115%, in the relative humidity range of 0–90%, 0–30% and 30–70%, respectively.

The contraction and expansion of flexible MOFs under stimulation can also be used as a visual signal. A flexible porous material $[Zn(L)_2(OH)_2]_n$·Guest (L = 4-(1*H*-naphtho[2,3-*d*]imidazol-1yl)benzoate, Guest = water or methanol) with a 1D channel was reported to show negative area compressibility (NAC), which is visible and can be modulated through the reversible exchange of guest molecules under static pressure (Figure 5.33).[88] This MOF consists

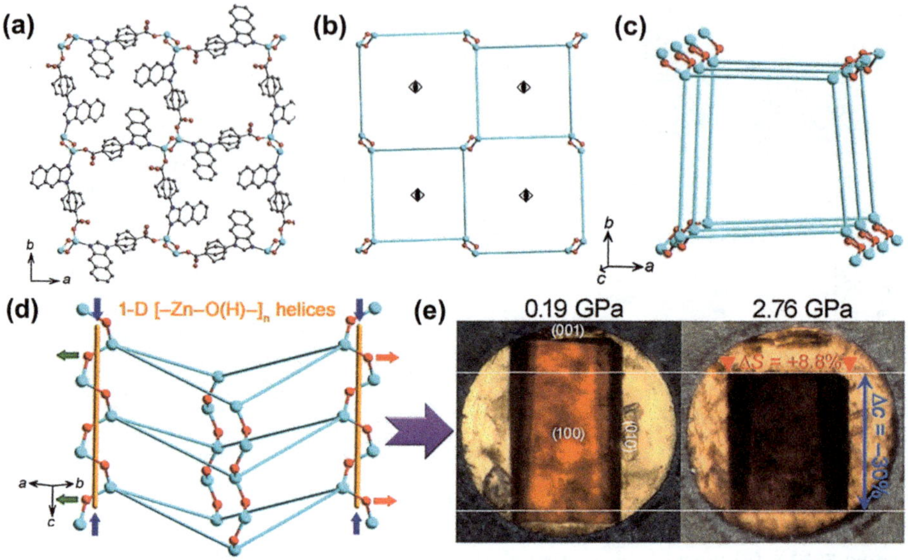

Figure 5.33 (a) Crystal structures of $[Zn(L)_2(OH)_2]_n$ viewed along the *c* axis. (b) and (c) Topological view of $[Zn(L)_2(OH)_2]_n$ showing a (4,4)-connected network. (d) A simplified diagram of 1D helical $[-Zn-O(H)-]_n$ chains along the [001] direction. (e) The distinct changes in crystal shape of $[Zn(L)_2(OH)_2]_n{\cdot}H_2O$ after hydrostatically compressed in isopropanol. Adapted from ref. 88 with permission from American Chemical Society, Copyright 2015.

of highly twisted quadrangular networks that are interconnected *via* flexible helical $[-Zn-O(H)-]_n$ chains running through the crystallographic [001] direction. The application of static pressure can lead to a compression of helical chains and an expansion of quadrilateral rings $[-Zn-L-]_4$, resulting in a substantial piezoelectric mechanical reaction. Moreover, incorporating various guest molecules into pore channels can result in different NAC in MOF.

Magnetism as one of the basic characteristics of matter can also be used as a sensing signal. Among numerous magnetic materials, one type that is well-explored are the spin crossover (SCO) materials. A coordination polymer $[Fe_2(azpy)_4(NCS)_4]{\cdot}$guest (azpy is 4,4′-azopyridine) featuring noticeable flexibility during guest uptake and release, was reported with regulation of its spin crossover behaviour (Figure 5.34).[89] $[Fe_2(azpy)_4(NCS)_4]{\cdot}$guest is constructed by interpenetration of 2D rhombic grids showing 1D channels along the *c*-axis. The structure shows guest-dependent pore structure changes. Crystallographic and magnetic analyses indicated that guest molecules (MeOH, EtOH, and *n*-PrOH) can form H-bonding interactions with the thiocyanate moiety on Fe centres, which affect the electron state of iron centres.

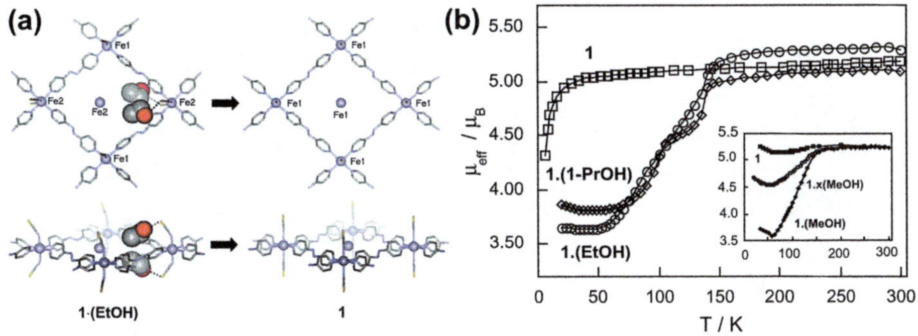

Figure 5.34 (a) Single net structure diagram of $Fe_2(azpy)_4(NCS)_4$·(EtOH) and $Fe_2(azpy)_4(NCS)_4$. (b) The temperature-dependent magnetic moment of $Fe_2(azpy)_4(NCS)_4$·x(guest). Adapted from ref. 89 with permission from American Association for the Advancement of Science, Copyright 2002.

5.5 Catalysis

The dynamic nature of flexible MOFs enables them to recognise and bind small molecules specifically with responsive structural transitions. Incorporated with certain catalytic centres for chemical conversion, flexible MOFs are a relatively less-explored but promising platform for selective catalysis and even biomimicking catalysis. A two-fold interpenetrated MOF Zn-DPA·$2H_2O$ (DPA = 4,4′,4″-tricarboxyltriphenylamine and (*E*)-1,2-di-(pyridin-4-yl)diazene) that features 1D rectangular channels of 13.2 × 9.7 Å^2 can undergo rotational rearrangement of the propeller-like DPA ligand during release and capture of guest molecules, leading to changes on pore channels (Figure 5.35).[90] Zn-DPA exhibits high catalytic activity for CO_2 cycloaddition owing to its high CO_2 uptake and Lewis acid sites. The reaction of 2-(phenoxymethyl)oxirane with CO_2 can be realised using this MOF as a catalyst with a high turnover number of ~40 000 per zinc cluster. The catalytic conversion was performed in a size-selective manner owing to the diffusion limit of the flexible pores in Zn-DPA.

Epoxidation of CO_2 was also reported in flexible PCN-700-Me_2 with Lewis acid sites, which shows closed pores after the rotation of organic moieties and twisting of metal clusters upon solvent removal.[91] The cycloaddition of CO_2 with epoxides with PCN-700-Me_2 as catalyst gave cyclic carbonates. The open-phase PCN-700-o and closed-phase PCN-700-c show dramatically different catalytic activities. PCN-700-o can drastically accelerate the cycloaddition reaction, whereas catalytic activity was diminished in the closed phase (Figure 5.36). A phase transition of PCN-700 occurs after exposure to trifluoroacetic acid–DMF. This study demonstrated the catalytic activity of flexible MOFs can be efficiently controlled by the dynamic host frameworks, affording a new dimension of developing smart catalysts for chemical conversions.

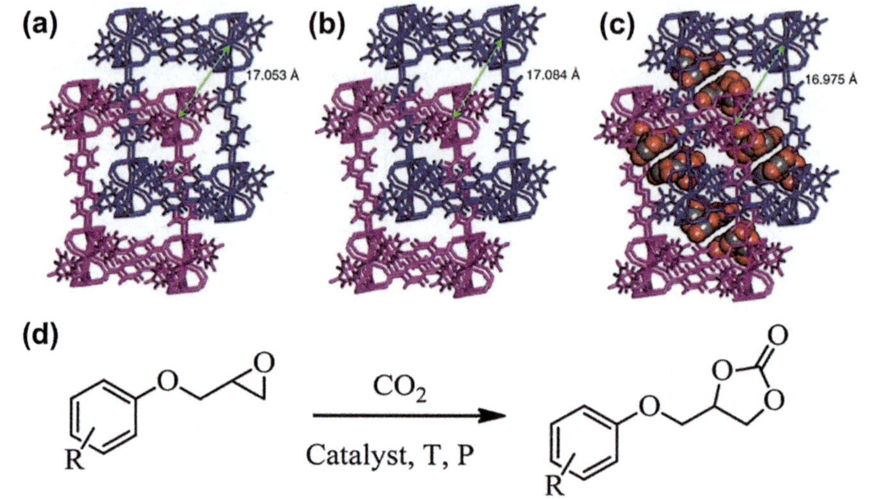

Figure 5.35 The two-fold interpenetrated frameworks in (a) Zn-DPA·$2H_2O$, (b) guest-free Zn-DPA, and (c) CO_2-containing Zn-DPA·$2CO_2$. (d) Zn-DPA-catalysed coupling of epoxides with CO_2. Adapted from ref. 90 with permission from Springer Nature, Copyright 2019.

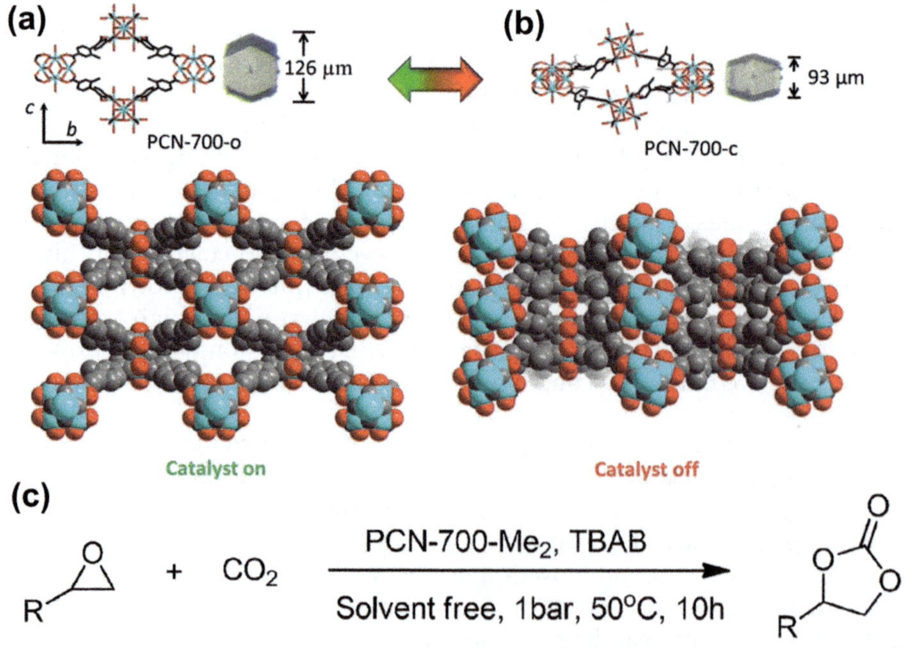

Figure 5.36 Structure of open (a) and closed (b) PCN-700-Me_2 with catalytic activity in on-state to off-state. (c) The catalytic reaction of CO_2 with epoxides. Adapted from ref. 91 with permission from John Wiley & Sons, Copyright 2016.

A series of flexible MOFs PCN-605, -606, and -700 serve as adsorbents for controlled delivery of bromine in bromination reactions.[92] After bromine adsorption, these MOF materials undergo significant structural transformations with a pore size expansion of 1.2–4.5% along crystallographic *a*-axis (Figure 5.37). The bromine-loaded flexible MOFs were then used as bromine sources for bromination reactions showing improved yields of up to 86% in contrast to only 46% by using liquid bromine.

A dynamic Zr-MOF NU-1400 with less-explored four-connected Zr_6 nodes shows guest- and temperature-dependent structural flexibility.[93] This flexible MOF exhibits hinge-like motion upon different solvent accommodations owing to the bending and twisting of coordinated carboxylate groups (Figure 5.38). Such flexibility leads to variable pore sizes in NU-1400 that are considered for shape-selective catalysis in the hydrolysis of a nerve agent simulant dimethyl 4-nitrophenyl phosphate. NU-1400 with different pore conformations showed drastically different catalytic activity. Open NU-1400 can swell to accommodate substrates and shows much higher catalytic activity, whereas the small-pore phase shows a catalytic rate one order of magnitude lower.

Flexible MOFs possess highly adjustable structures and pores making them suitable as reaction vessels for polymerisation of monomers at multi-level control, including control on stereoregularity, monomer sequence, molecular weight, helicity, networks *etc.*[94,95] Additionally, the flexibility of MOFs

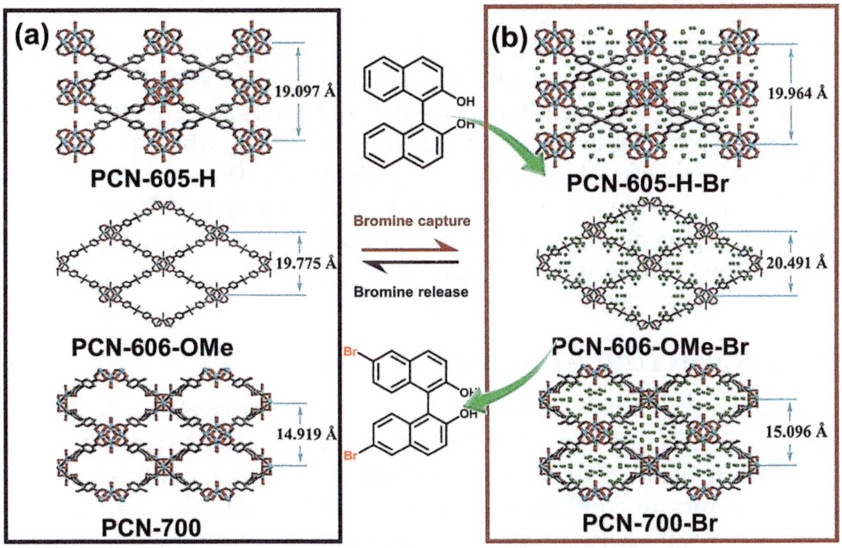

Figure 5.37 Structures of flexible MOFs PCN-605-H, PCN-606-OMe, and PCN-700 before (a) and after (b) bromine adsorption along with the schematic diagram of brominating bisaphthol utilizing the bromine pre-absorbed MOFs. Adapted from ref. 92 with permission from John Wiley & Sons, Copyright 2017.

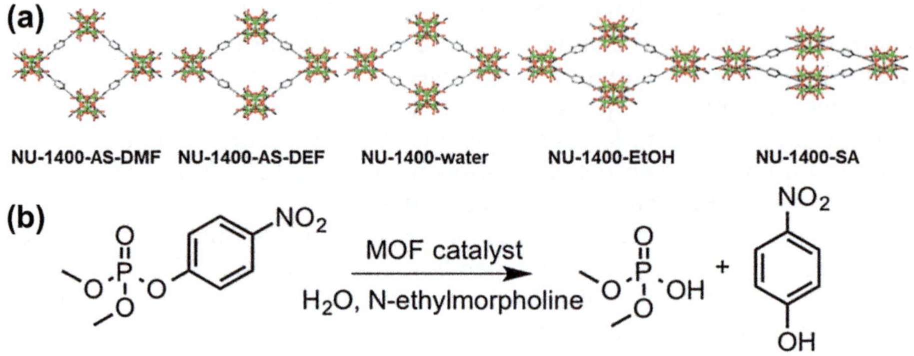

Figure 5.38 (a) Structures of NU-1400-ASDMF, NU-1400-AS-DEF, NU-1400-water, NU-1400-EtOH, and NU-1400-SA viewed along the *a*-axis. (b) Hydrolysis reaction of dimethyl 4-nitrophenyl phosphate. Adapted from ref. 93 with permission from American Chemical Society, Copyright 2018.

allows for highly selective recognition of target molecules and promotes tight packing of these molecules, which is important for interesting polymerisation reactions. For instance, $[M_2(bdc)_2(ted)]_n$ (where bdc = benzenedicarboxylate; ted = triethylenediamine; M = Zn^{2+} and Cu^{2+}) exhibits regular and uninterrupted 1D nanochannels along the *c*-axis, with a pore size of 7.8 × 7.8 Å^2.[96] Upon the adsorption of 1,4-divinylbenzene monomer (*p*-DVB; molecular dimensions = 8.5 × 4.4 Å^2), $[Zn_2(bdc)_2(ted)]_n$ displays a lattice expansion compared to the original host, while $[Cu_2(bdc)_2(ted)]_n$ does not exhibit such changes, as demonstrated by XRD measurements. The spectroscopic analysis of the resulting poly(*p*-DVB) in $[M_2(bdc)_2(ted)]_n$ has revealed that a soluble linear poly(*p*-DVB) can be achieved in $[Zn_2(bdc)_2(ted)]_n$, whereas no evidence of polymerisation was observed in $[Cu_2(bdc)_2(ted)]_n$ (Figure 5.39). Synchrotron PXRD has demonstrated that the flexible and dynamic properties of $[Zn_2(bdc)_2(ted)]_n$ facilitate the adsorption and close arrangement of *p*-DVB molecules, thus promoting polymerisation.

5.6 Conducting Materials

Proton conducting materials have been studied intensively as the key components in fuel cell devices.[97] It has been reported that crystalline MOFs with a certain flexibility and high porosity have great potential for high-performance proton-conductive materials.[98,99] To a certain extent, the proton conductivity of MOFs can be regulated by incorporating acidic functional groups or hybridising with charge-neutral proton carriers.[100] The flexibility inherent in MOFs offers a new avenue for enhancing the proton conductivity of relevant materials.

A stable Cr(III)-based flexible MOF $Cr_3(\mu_3\text{-}O)(H_2O)_3(NDC(SO_3H_{5/6})_2)_3$ (namely BUT-8(Cr)A, where $NDC(SO_3H)_2$ = 4,8-disulfonaphthalene-2,6-dicarboxylate)

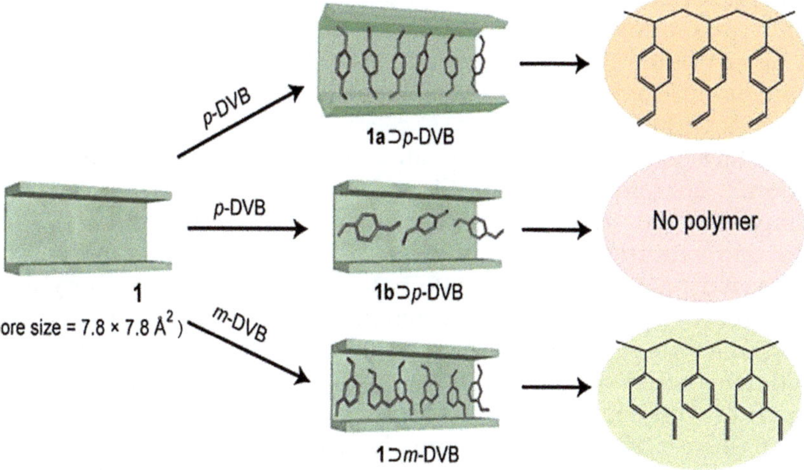

Figure 5.39 (a) Encapsulation and polymerisation of DVBs in the nanochannels of $[M_2(bdc)_2(ted)]_n$ ($M = Zn^{2+}$ and Cu^{2+}). Adapted from ref. 96 with permission from John Wiley & Sons, Copyright 2007.

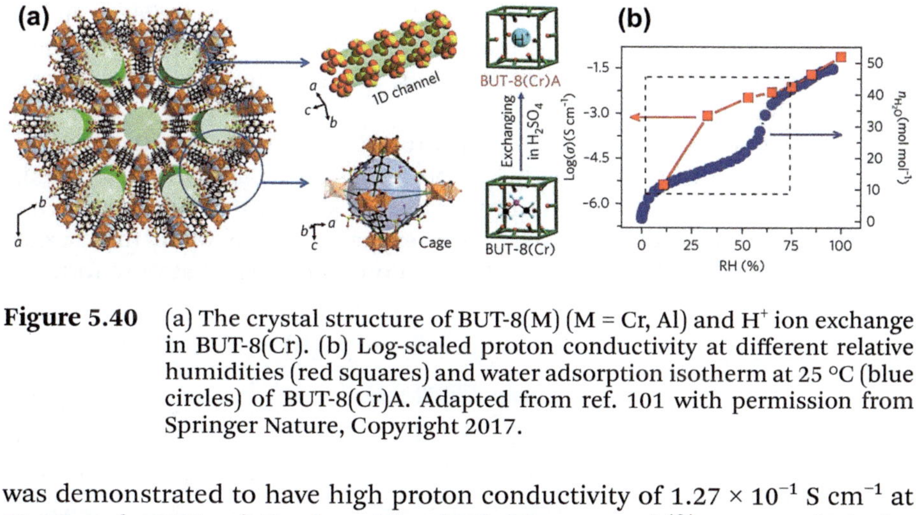

Figure 5.40 (a) The crystal structure of BUT-8(M) (M = Cr, Al) and H^+ ion exchange in BUT-8(Cr). (b) Log-scaled proton conductivity at different relative humidities (red squares) and water adsorption isotherm at 25 °C (blue circles) of BUT-8(Cr)A. Adapted from ref. 101 with permission from Springer Nature, Copyright 2017.

was demonstrated to have high proton conductivity of 1.27×10^{-1} S cm^{-1} at 80 °C and 100% relative humidity (RH) (Figure 5.40).[101] In BUT-8(Cr), the as-synthesised phase has a 3D structure with 1D pore channels with bi-pyramidal cages. The pore surfaces are extensively decorated with sulfonate groups. $NH_2(CH_3)^{2+}$ cation exists in the 1D channels that can be exchanged with H^+ to obtain $-SO_3H$-functionalised new phase (BUT-8(Cr)A). The study on the adsorption properties of BUT-8(Cr)A demonstrated that it could selectively adsorb H_2O over other gases, such as H_2, N_2, CO_2, C_2H_4, C_3H_6, C_2H_6 and C_3H_8. The structure of BUT-8(Cr)A has changed greatly following water absorption. In addition, the proton conductivities of BUT-8(Cr)A exhibit quasi-linear correlation with RH and still retain moderately high proton conductivities at a wide RH range and temperature range. These characteristics might be attributed to a certain degree of self-adaptation exhibited by the structure upon adsorption of water, allowing it to maintain a hydrogen-bonding network within the pores even at relatively low RH. Hydrogen-bonding networks have been shown to play an essential role in promoting proton migration.[98] Therefore, the flexibility of MOFs provides a new avenue to designing materials that exhibit high-proton conductivities over a broad range of temperatures and humidity.

A Ca-based MOF (Ca-PiPhtA-I, PiPhtA = 5-(dihydroxyphosphoryl)isophthalic acid) with a pillared layered structure and water-filled 1D channels was reported to exhibit guest-dependent proton conductivities (Figure 5.41).[102] The full dehydrated structure (Ca-PiPhtA-II) could be obtained by thermal treatment of Ca-PiPhtA-I. Structural analysis revealed that the removal of water can cause slight changes in pore dimensions. Moreover, the water molecules in the Ca-PiPhtA-I channels can reach an equilibrium state upon preheating at 40 °C for 2 h giving a different phase Ca-PiPhtA-III. The proton conductivity of Ca-PiPhtA-I, II, III are 5.7×10^{-4}, 3.6×10^{-4} and 1.3×10^{-3} S cm^{-1}, respectively, at 24 °C and 98% RH. Interestingly, a new isostructural (Ca-PiPhtA-NH_3) can be obtained when exposing Ca-PiPhtA-I to ammonia vapours and showing the highest proton conductivity (6.6×10^{-3} S cm^{-1}) among these materials. The enhancement in proton conductivity can be attributed to partial breaking and conversion of the pillared structure which could greatly increase the number of proton carriers and create new proton transfer pathways.

A highly stable Eu-based MOF proton conductor $(Me_2NH_2)[Eu(L)]$ (where H_4L = 5-(phosphonomethyl)isophthalic acid) has been developed, which is constructed by a 2D anionic framework $[Eu(L)]^-$ and counter cations $(Me_2NH_2)^+$ residing in the interlayer space.[103] In this MOF, there exist robust hydrogen-bonded chains (N–H···O) along the *c*-axis that were formed

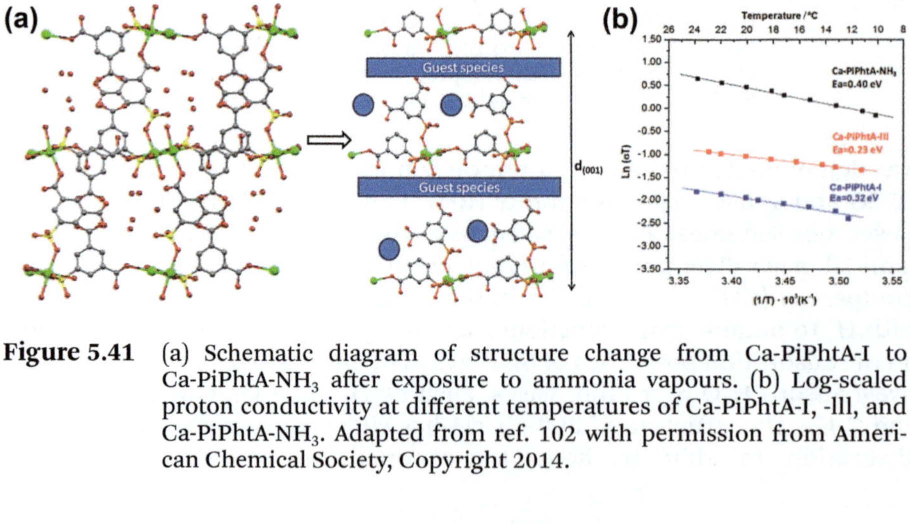

Figure 5.41 (a) Schematic diagram of structure change from Ca-PiPhtA-I to Ca-PiPhtA-NH_3 after exposure to ammonia vapours. (b) Log-scaled proton conductivity at different temperatures of Ca-PiPhtA-I, -lll, and Ca-PiPhtA-NH_3. Adapted from ref. 102 with permission from American Chemical Society, Copyright 2014.

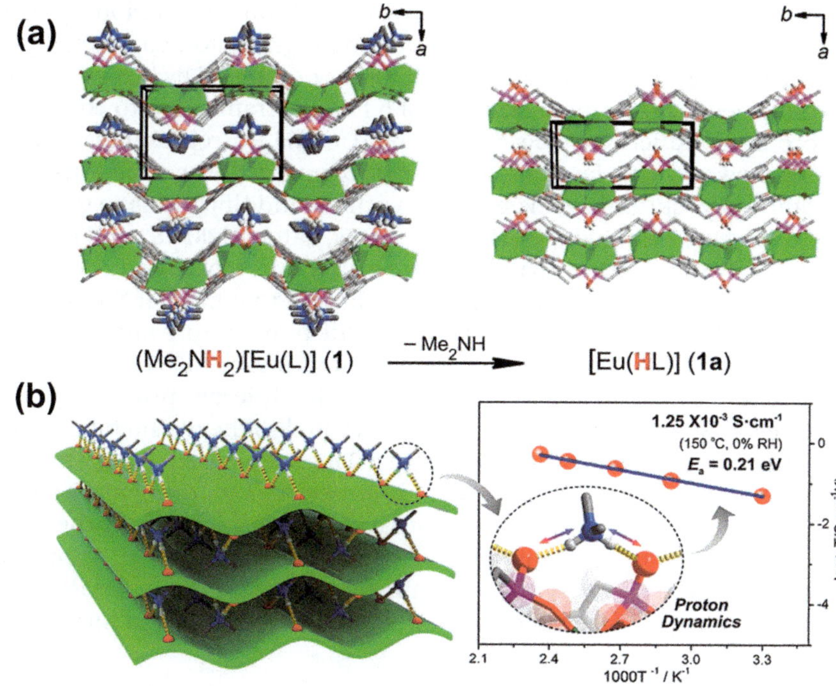

Figure 5.42 (a) The crystal structure transforms from an anionic framework to a neutral one. (b) Schematic diagram of strongly hydrogen-bonded (N–H···O) along the *c*-axis with the parallel arrangement and Arrhenius plot of proton conductivity of $(Me_2NH_2)[Eu(L)]$ under anhydrous atmospheres. Adapted from ref. 103 with permission from American Chemical Society, Copyright 2017.

between the NH_2Me^{2+} ions and uncoordinated O atom of phosphonate groups (Figure 5.42). Proton conductivity measurements indicated that this proton conductor shows a high proton conductivity not only in the presence of water (3.76×10^{-3} S cm^{-1} at 98% RH and 100 °C) but also in anhydrous conditions (1.25×10^{-3} S cm^{-1} at 150 °C). The high proton conductivity can be attributed to unique proton vibration and transfer in this MOF, which is achieved by strong hydrogen bonding and electrostatic interaction among acid–base pairs. The transport process of protons was further elucidated by diffuse reflectance infrared Fourier transform spectrum (DRIFTS), PXRD and single-crystal X-ray diffraction (SCXRD). It is noteworthy that the anionic host framework $(Me_2NH_2)[Eu(L)]$ can be converted to a neutral framework [Eu(HL)] by slow heat treatment, accompanied by a large shrinkage of the unit-cell volume and proton transfer. However, the proton conductivity of [Eu(HL)] drops dramatically (5.29×10^{-5} S cm^{-1} at 98% RH and 100 °C) compared with that of $(Me_2NH_2)[Eu(L)]$, which further demonstrates the crucial role of strong hydrogen bond chains in rapid proton transfer.

Flexible MOFs can promote efficient hydrogen-bond formation by self-adaption of their frameworks, thus maintaining high proton conductivity. A synergistic strategy of multivariate flexibility was developed by incorporating flexible linkages into the main skeleton of a flexible framework to construct highly proton-conductive MOFs.[104] MIL-88B(Cr)-NH_2, a representative of a flexible framework with large breathing behaviour,[19,105] was functionalised with various alkyl sulfonic acids by post-synthetic modifications to give two kinds of multivariate self-adaptive flexible MOFs, namely, PMNS1 and PMNS2 (Figure 5.43). Proton conductivity measurements on PMNS1 and PMNS2 at various relative humidity and temperatures showed that both of them could exhibit moderate proton conductivity over a wide temperature and humidity range, with a maximum conductivity value at 80 °C under 100% RH (PMNS1: 1.52×10^{-1} S cm^{-1}; PMNS2: 4.68×10^{-2} S cm^{-1}). The different proton conductivity between PMNS1 and PMNS2 can be attributed to the distinct polarities of alkyl sulfonic acids with varying lengths. The cooperative adaptation of dynamic torsion at alkyl sulfonic acid and framework breathing plays a crucial role in achieving high proton conductivity. In addition, given the high proton conductivity and stability of these flexible MOFs, a series of proton exchange membranes (PEMs) was then fabricated. Among them, the PMNS1-40 hybrid membrane (dispersed in poly(arylene ether ketone) matrix with

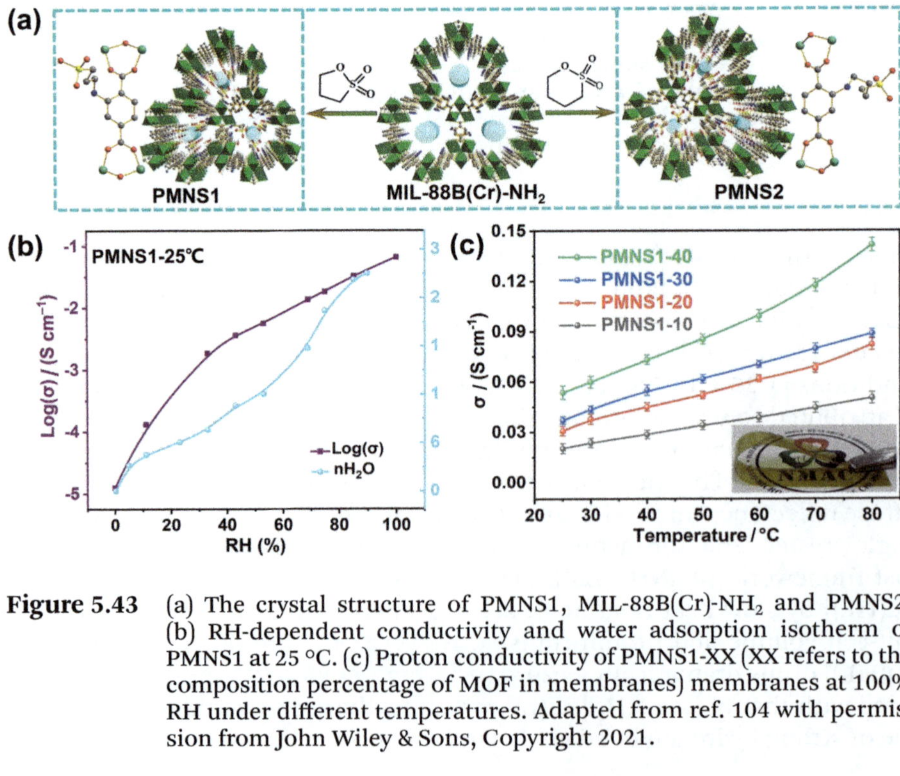

Figure 5.43 (a) The crystal structure of PMNS1, MIL-88B(Cr)-NH_2 and PMNS2. (b) RH-dependent conductivity and water adsorption isotherm of PMNS1 at 25 °C. (c) Proton conductivity of PMNS1-XX (XX refers to the composition percentage of MOF in membranes) membranes at 100% RH under different temperatures. Adapted from ref. 104 with permission from John Wiley & Sons, Copyright 2021.

40% PMNS1 powder content) was demonstrated with long-term durability and the highest power density of 34.76 mW cm^{-2}.

Generally, integrating photochromic molecules into conductive flexible MOFs can make materials with responsive and dynamic characteristics upon multiple synergistic stimuli.[106] For instance, photo-switchable MOF films were fabricated using layer-by-layer deposition of a pillared-layered MOF $Cu_2(F_2AzOBDC)_2$(dabco) (F_2AzoBDC = (*E*)-2-((2,6-difluorophenyl)diazenyl)terephthalic acid; dabco = 1,4-diazabicyclo[2.2.2]octane) on a quartz substrate (Figure 5.44).[107] In this structure, *o*-fluoroazobenzene groups were embedded into the backbone of MOF structures, which is capable of *trans–cis* isomerisation in the channels upon different UV radiation. The host–guest interactions between MOF and proton carriers (1,4-butanediol and 1,2,3-triazole) can then be regulated by *trans–cis* photoisomerisation, accompanying the switching of proton conduction between relatively high conducting in *trans* form and low conducting in *cis* form. Theoretical calculations revealed that proton-conducting mediators create more robust hydrogen bonding interactions with azo moieties in the *cis* state, which reduces the migration rate of a proton between carriers, resulting in a significant decrease in proton conductivity.

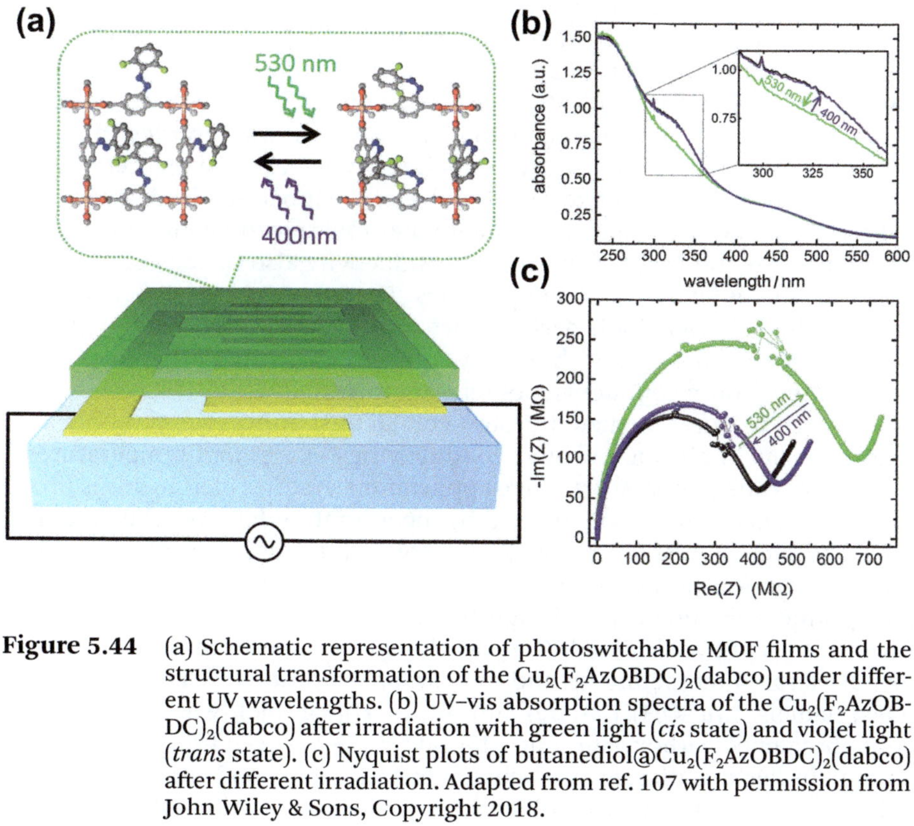

Figure 5.44 (a) Schematic representation of photoswitchable MOF films and the structural transformation of the $Cu_2(F_2AzOBDC)_2$(dabco) under different UV wavelengths. (b) UV–vis absorption spectra of the $Cu_2(F_2AzOBDC)_2$(dabco) after irradiation with green light (*cis* state) and violet light (*trans* state). (c) Nyquist plots of butanediol@$Cu_2(F_2AzOBDC)_2$(dabco) after different irradiation. Adapted from ref. 107 with permission from John Wiley & Sons, Copyright 2018.

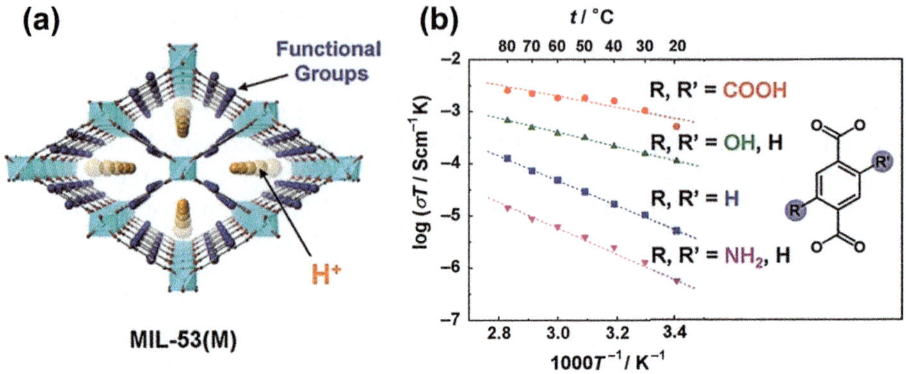

Figure 5.45 (a) The single-crystal structure of MIL-53 (M) derivatives and showing the functional group modification sites. (b) Temperature-dependent conductivity of MIL-53 (M) derivatives at 95% RH. Adapted from ref. 108 with permission from American Chemical Society, Copyright 2011.

Metal substitution and ligand functionalisation not only enable flexible control of MOFs but also alter the hydrophilicity or acidity of their pore surface, thereby increasing the proton migration rate as well as their loading capacity, which ultimately regulates the proton conductivity of the materials. A series of isostructural MIL-53 (M(OH)(bdc-R)(H_2O), M = Al^{3+} or Fe^{3+}; R = –OH, –H, –NH_2 or COOH; BDC = 1,4-benzenedicarboxylate) have been systematically studied as proton conductors (Figure 5.45).[108] It was reported that the magnitude of proton conductivity was not linearly correlated with their water adsorption but rather related to the pK_a value of the functionalised ligands in MOFs, *i.e.*, the lower the pK_a value, the higher their proton conductivity. As the temperature increases, their proton conductivity also increases, exhibiting pK_a-dependent activation energy values. Therefore, MIL-53(Fe)-$(COOH)_2$ demonstrates relatively high performance and low activation energy (2.0×10^{-6} at 298 K and 95% relative humidity, E_a = 0.21 eV). This work guides regulation of the proton conductivity of relevant materials.

In contrast to water-mediated proton conductors, anhydrous proton conductors can maintain high proton conductivities at elevated temperatures (>100 °C) for wide temperature range applications. A series of anhydrous proton conductors has been constructed by incorporating heterocyclic organic molecules such as histamine, benzimidazole, and imidazole into suitable nanochannels for various MOFs.[109–112] An early example was realised by loading imidazole molecules of high boiling points as proton carriers into pore channels of two flexible MOFs, developed for superior proton conductivity at elevated temperatures and anhydrous conditions.[109] These flexible MOFs, namely Al(μ_2-OH)(1,4-bdc) and Al(μ_2-OH)(1,4-ndc) (where 1,4-ndc = 1,4-naphthalenedicarboxylate and 1,4-bdc = benzene-1,4-dicarboxylate), contain similar one-dimensional channels and show good thermal stability, but differ in pore shapes and surfaces (Figure 5.46). Al(μ_2-OH)(1,4-ndc)

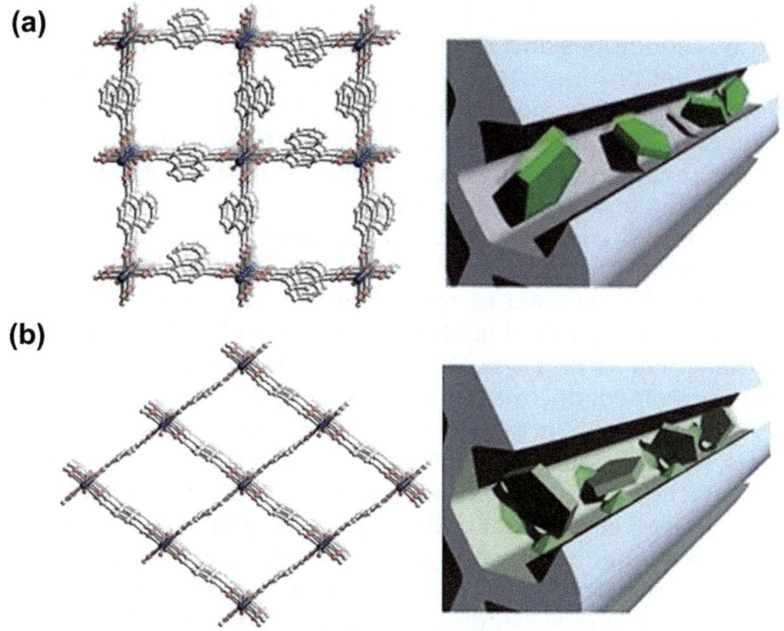

Figure 5.46 (a) The crystal structure of Al(μ_2-OH)(1,4-ndc) and schematic diagram of imidazole molecules incorporated in the 1D channel without strong interaction with the host framework. (b) The crystal structure of Al(μ_2-OH)(1,4-bdc) and the schematic diagram of imidazole molecules interact strongly with the host framework in a 1D channel. Adapted from ref. 109 with permission from Springer Nature, Copyright 2009.

exhibits hydrophobic pore surfaces as a result of a steric hindrance effect of naphthalene rings, whereas Al(μ_2-OH)(1,4-bdc) presents hydrophilic pore surfaces due to a large amount of μ_2-OH and carboxylate group exposed to the pore surface. Therefore, polar imidazole molecules can form relatively strong interactions with the main framework in Al(μ_2-OH)(1,4-bdc) in contrast to those in Al(μ_2-OH)(1,4-ndc). Conductivity measurements indicated that Al(μ_2-OH)(1,4-ndc) exhibits extremely low conductivity of 10^{-13} S cm^{-1}, that can be increased obviously at RT (to 5.5×10^{-8} S cm^{-1}) upon introducing Im into the framework of Al(μ_2-OH)(1,4-ndc). The hybrid material (Al(μ_2-OH)(1,4-bdc)⊃Im) has a proton conductivity of 10^{-10} S cm^{-1}. In addition, the proton conductivity of both hybrid materials (Al(μ_2-OH)(1,4-ndc)⊃Im and Al(μ_2-OH)(1,4-bdc)⊃Im) can be increased further to 2.2×10^{-5} S cm^{-1} and 1.0×10^{-7} S cm^{-1}, respectively, when the temperature is increased to 120 °C. Solid-state NMR studies revealed that strong interactions between guest molecules and host framework can weaken the mobility of guest imidazole molecules, resulting in low proton conductivity.

To further enhance the proton conductivity, another hybrid material (Al(μ_2-OH)(1,4-bdc)⊃His) (His = histamine) was obtained by incorporating histamine with more proton-hopping sites into the host framework of

Al(μ_2-OH)(1,4-ndc), giving a high proton conductivity of 1.7 × 10^{-3} S cm^{-1} at anhydrous conditions and 150 °C.[110] This proton conductivity was much higher than that of Al(μ_2-OH)(1,4-bdc)⊃Im at 120 °C in anhydrous conditions. In contrast to Im molecules embedded in Al(μ_2-OH)(1,4-bdc), histamine molecules showed a relatively high concentration of proton carriers, diverse structural conformation, and molecular isotropy, which plays a key role in improving proton conductivity.

Organic hydroxyl compounds such as butanol (Bu), hydroquinone (Hq), and cyclohexanol (Ch), and with lower pK_a values than *N*-containing aryl-molecules can also be used as proton carriers. By loading these organic hydroxyl compounds, a flexible MOF $[Zn_3(tz)_2(bdc)_2]_n$ (FJU-31, where H_2bdc = terephthalic acid, Htz = 1*H*-1,2,3-triazole) has been used to construct three anhydrous proton conductors (Figure 5.47).[113] FJU-31@Ch was reported to show a high proton conductivity of 2.65 × 10^{-4} S cm^{-1} at 125 °C in anhydrous conditions, low activation energy (E_a = 0.18 eV) and stable proton conduction over a broad temperature range (−40 to 125 °C). This work revealed that flexible MOFs have great potential in developing anhydrous proton conductors with high proton conductivity over a broad temperature range.

The study of the electrically conductive MOFs is another important subject.[114] Given the poorly conductive organic moieties, MOFs with high

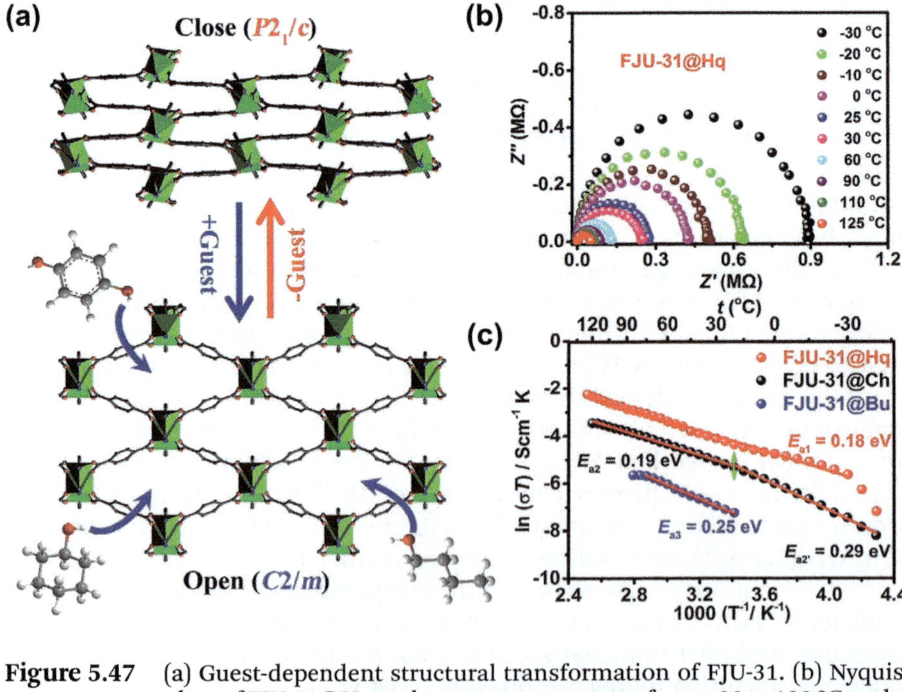

Figure 5.47 (a) Guest-dependent structural transformation of FJU-31. (b) Nyquist plots of FJU-31@Hq at the temperature range from −30 to 120 °C under anhydrous atmospheres. (c) Temperature-dependent conductivity of FJU-31@Hq, FJU-31@Ch, and FJU-31@Bu. Adapted from ref. 113 with permission from Royal Society of Chemistry.

conductivity have rarely been reported, except for some 2D MOFs with relatively compact packing between layered structures.[115,116] There is also little research on flexible MOFs with properties of electronic conductivity. But there are theoretical calculations that indicate that framework flexibility could be utilised to regulate conductivity. For instance, a flexible MOF M(NDIDP), modelled using a template framework of M(BDP)[18] (H_2BDP = 1,4-di(1*H*-pyrazol-4-yl)benzene), features redox-active 1,4,5,8-naphthalene diimide (NDI) moieties and exhibits highly dependent charge transport characteristics.[117] Therefore, its conductivity can be fine-tuned through external stimuli, making it a potential multifunctional electronic device material. The characteristic of flexible MOFs responding to external stimuli has been used for electrochemical sensors to detect analytes[86] or for constructing threshold humidity sensors.[87]

5.7 Biomedical Applications

As the structure of flexible MOFs is sensitive to external stimuli, there is great potential for application in controlled drug delivery. Early examples of drug delivery using MOFs were demonstrated in rigid MOFs.[118] Two rigid MOFs MIL-100 and MIL-101 showed a large drug storage capacity and controlled drug release for model drug ibuprofen. Subsequently, the adsorption and delivery of ibuprofen have been achieved in two flexible MOFs MIL-53(Cr) and MIL-53(Fe) (M(OH)[bdc], M = Cr^{3+} or Fe^{3+}) (Figure 5.48).[118] Both flexible MOFs showed a maximum ibuprofen loading capacity of 20 wt%, indicating that the type of metal in the MOFs has little impact on the loading capacity of drugs. These MOFs exhibited a very slow ibuprofen-release rate under physiological conditions with zero-order kinetics. It took three weeks to completely release ibuprofen from ibuprofen-loaded MIL-53 (Cr) and MIL-53 (Fe). Such a low drug release rate can be attributed to a flexible framework self-adaptive to drug molecules that optimises the interactions between drug molecules and host frameworks.

In addition to high drug loading and controlled drug release rates, biocompatibility, non-toxic and easily degradable are also crucial characteristics that need to be considered for nanocarriers. A series of non-toxic Fe(III)-based MOFs including four flexible frameworks (MIL-53, MIL-88B, MIL-88A, MIL-89)[119–121] and two rigid frameworks (MIL-100 and MIL-101_NH_2)[122–124] can be synthesised in water or ethanol solution (Figure 5.49).[125] These materials exhibit good degradability, biocompatibility, and good imaging properties, and can be used for delivery and capture of anticancer or antiviral drugs. The three microporous flexible frameworks namely MIL-88A, MIL-53 and MIL-89 can load up to 8.0, 14.3 and 9.8 wt% of busulfan, respectively. It is believed the sustained release of the delivery process was dominated by the interactions between drug molecules and the host, as well as drug–matrix interactions. The adaptive host framework in MOF materials would strengthen relevant host–guest interactions, making flexible MOF show great potential as drug-delivery nanocarriers for the controlled release of drugs.

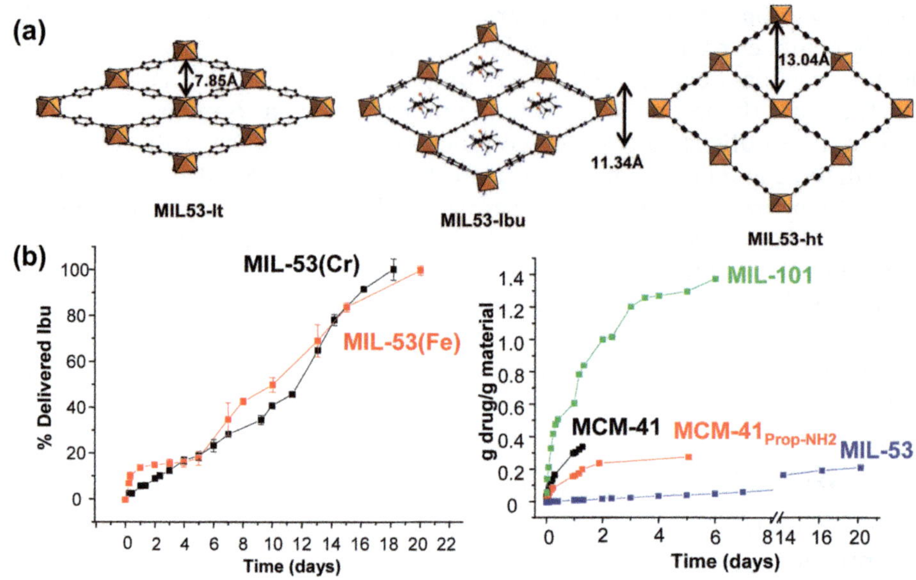

Figure 5.48 (a) Schematic diagram of the breathing effect of the MIL-53(Cr). (b) Ibuprofen release from MIL-53(Cr) and MIL-53(Fe) and (c) comparison with other materials. Adapted from ref. 118 with permission from American Chemical Society, Copyright 2008.

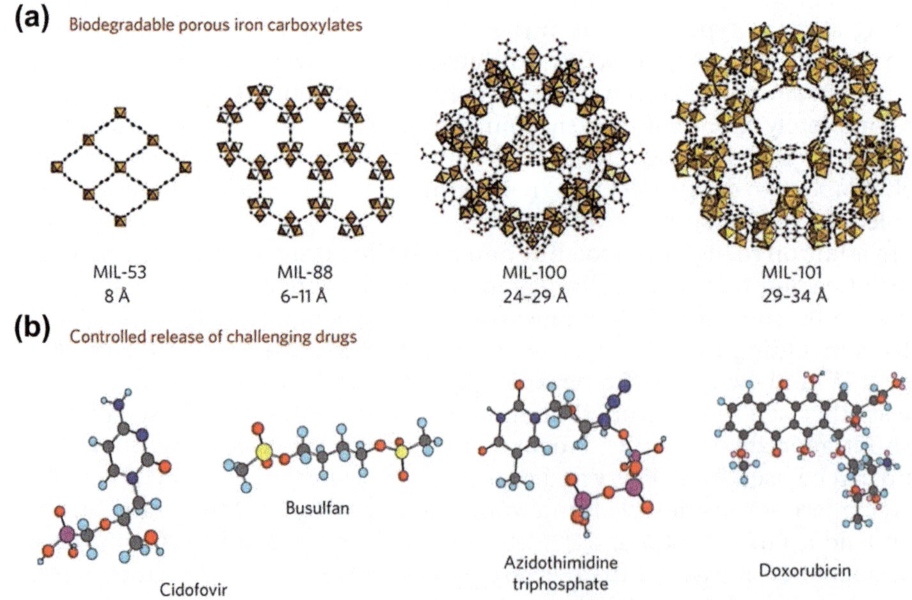

Figure 5.49 (a) Crystal structure of MIL-53, MIL-88, MIL-100 and MIL-101. (b) Molecular structure of anticancer or antiviral drugs. Adapted from ref. 125 with permission from Springer Nature, Copyright 2010.

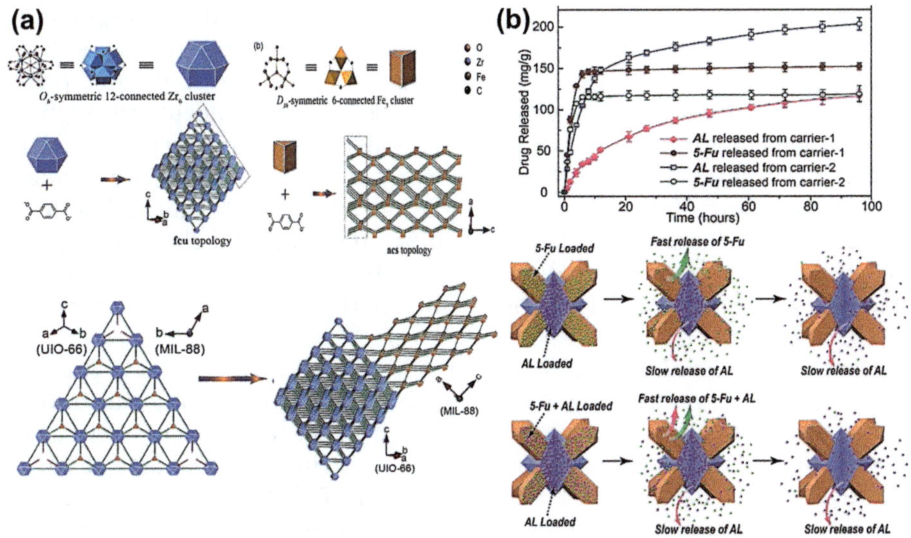

Figure 5.50 (a) Schematic diagram of MIL-88B-on-UiO-66 hybrid-phase MOF structure. (b) AL (alendronate) and 5-Fu (5-fluoracil) release from MIL-88B-on-UiO-66 with different AL loading values and corresponding schematic diagram of drug release. Adapted from ref. 126 with permission from John Wiley & Sons, Copyright 2020.

Interestingly, hybrid-phase MOF composites have been developed for dual-drug adsorption and controllable release.[126] These hybrid-phase MOFs were constructed from rigid UiO-66 (**fcu** topology) and flexible MIL-88B (**acs** topology). High-magnification field emission scanning electron microscopy images showed that hexagonal rod MIL-88B was grown vertically on each face of octahedral UiO-66 nanoparticles to form a rods-on-octahedron shape with MIL-88B-on-UiO-66 heterostructure (Figure 5.50). This structure is formed because the MOFs have a partially matched topology, and MIL-88B is also able to adapt to the UiO-66 lattice by structural adjustment. MIL-88B can still retain its flexibility in hybrid-phase MOFs. The investigation of the adsorption and release of drugs in pure-phase UiO-66 and MIL-88B demonstrated that they displayed different drug loading and releasing capacities, owing to their different pore sizes and distinct interactions with drugs. In MIL-88B-on-UiO-66 heterostructure, two drugs (5-fluoracil (5-Fu) and alendronate (AL)) can be co-loaded into the framework with a controllable loading ratio and various spatial distributions, realising the programmable release of drugs.

5.8 Molecular Machines

Introducing dynamic parts into MOFs to construct artificial molecular machines is an interesting research topic. The incorporation of motional units in MOF frameworks was initiated for their potential to construct spatially organised molecular motors and investigate their responsive dynamic behaviours in the solid state. A moto-MOF was obtained by replacing the DPNI (dipyridyl-naphthalenediimide) linkers in zinc pillared-paddlewheel BrYO-MOF with a dipyridyl molecular motor molecule through a post-synthetic method (Figure 5.51).[127] The alkene-based rotary molecular unit dangling on the pillars of moto-MOF can maintain large-amplitude rotations due to enough free volume existing in the moto-MOF. Polarised optical microscopy and Raman spectroscopy studies revealed that the majority of rotor units rotate in the same direction. Notably, the rotation speed (rate of thermal helix inversion) of these motor units in the solid state can be similar to that in solution.

Recently, an interesting elastic metal–organic crystal $[Co_2(^{CTN}L)\cdot 7H_2O\cdot 4DMF]_n$ (CTN = catenane), referred to as catenane-based MOF (^{CTN}MOF), was developed with MIM (mechanically interlocked molecule) backbones and showed guest- and temperature-dependent structural transformations (Figure 5.52).[128] The structural transformation of ^{CTN}MOF can occur even in a low temperature range showing a sensitive thermally adaptive feature. Nanoindentation revealed that the Young's modulus of ^{CTN}MOF was 6.31 ±

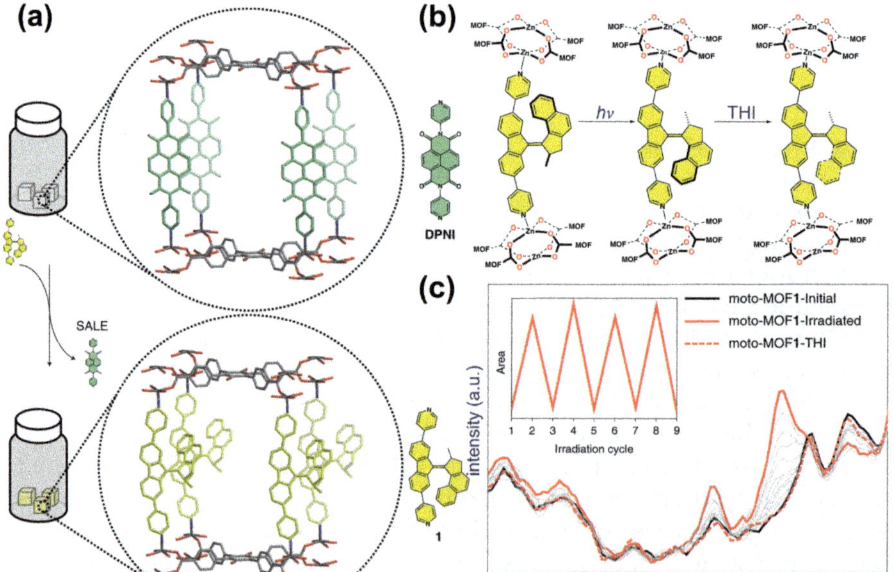

Figure 5.51 (a) Syntheses and structures of the moto-MOFs by replacing the DPNI pillars with bipyridine motor 1, 2-*E*, or 2-*Z* in the structure. (b) Schematic diagram of structural changes during the photochemical and subsequent thermal isomerisation of motor units incorporated in moto-MOF1. (c) Changes in the Raman spectrum of moto-MOF1. Adapted from ref. 127 with permission from Springer Nature, Copyright 2019.

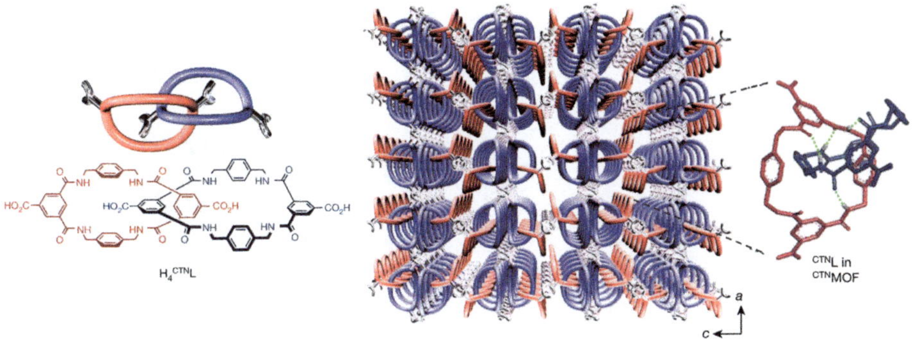

Figure 5.52 Crystal structure of ^{CTN}MOF. Reproduced from ref. 128 with permission from Springer Nature, Copyright 2021.

0.14 GPa in air, while lower values of 1.77 ± 0.16 GPa and 1.63 ± 0.13 GPa were found when tested in *N*,*N*-dimethylformamide and tetrahydrofuran, respectively. In addition, hydrostatic compression indicated the greatest deformation of DMF-containing ^{CTN}MOF along its *c*-axis while still maintaining structural integrity (5% contraction at 0.88 GPa), owing to the ability of interlocked macrocycles to slide against each other along the *c*-axis, as demonstrated by *in situ* high-pressure SCXRD. These results combined with *in situ* spectroscopy experiments suggest that the hydrogen bonding between the interlocking macrocycles can be regulated by solvent guests, allowing the material to switch between rigidity and elasticity. This work provides an idea for the synthesis of squeezable MOFs, which is expected to solve the dilemma of porous materials that are difficult to release high-affinity guests.

A series of dynamic MOFs (termed as UWDM-$1_{(22)}$, UWDM-$1_{(B24)}$, and UWDM-$1_{(24)}$) was synthesised from MIM components.[129,130] The first reported was UWDM-$1_{(24)}$ $[Cu_2(MIM)(H_2O)_2]\cdot 3H_2O$ (Figure 5.53(a)). Its MIM linker is composed of [24]-crown-6, one benzylaniline recognising site, and two 3,5-benzenedicarboxylic acid groups that served as a wheel, crossbar, and linker struts, respectively. Another two dynamic MOFs, UWDM-$1_{(22)}$, and UWDM-$1_{(B24)}$ were synthesised using crown ether macrocycles of different sizes and shapes ([22]crown-6 and benzo[24]crown-6). Each material exhibits varying degrees and rates of macrocycle rotation (Figure 5.53(a)).[130] The motional characteristic of the wheel can be demonstrated by variable-temperature ^{2}H solid-state NMR (SSNMR) after the removal of guest molecules. The result showed that crown ether can rotate freely and quickly at high temperatures, and exhibits different motional regimes at different temperatures, including two-position jumping of the CD_2 moiety, partial rotations, or full fast rotations of the macrocyclic ring. Moreover, the rotation of crown ether can be opened and closed by temperature control. Among these materials, UWDM-$1_{(24)}$ exhibits the strongest mobility with three different motional modes, while the other two materials only exhibit one or two modes.

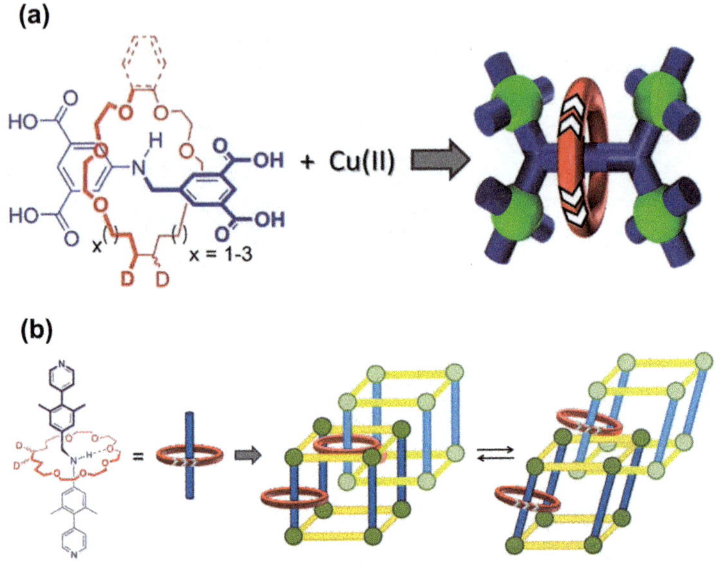

Figure 5.53 (a) Schematic diagram of the synthesis of UWDM-$1_{(24)}$, UWDM-$1_{(22)}$ and UWDM-$1_{(B24)}$. Adapted from ref. 130 with permission from American Chemical Society, Copyright 2015. (b) Schematic illustration of the phase transition between the α-UWDM-3 phase (left) and the β-UWDM-3 (right) phase. Adapted from ref. 131 with permission from American Chemical Society, Copyright 2014.

A novel class of MIM-pillared MOFs, UWDM-2 (1,4-diazophenyl-dicarboxylate) and UWDM-3 (1,4-biphenyl-dicarboxylate), were further constructed by MIM struts and periodic Zn-carboxylate layers with various linkers, where MIM linker is made up of a [24]crown-6 macrocyclic moiety appended with a pyridine-based receptor as a terminal group (Figure 5.53(b)).[131] Structural analysis showed that both UWDM-2 and UWDM-3 possess interpenetrating structures with three-fold and two-fold interpenetration, respectively. The interpenetration of the structure greatly hampers the movement of [24]crown-6 macrocycle, which was demonstrated by VT ^{2}H SSNMR spectra. Interestingly, the structure of UWDM-3 was capable of a dynamic response after removing the solvent and this gives rise to a new phase (β-UWDM-3). The motion of macrocyclic can thus be triggered and shows thermal-driven full rotation after the conversion of the α-UWDM-3 phase into the β-UWDM-3 phase.

In addition to rotation, macrocyclic molecules can also move in translation.[132] A [2]rotaxane molecular shuttle was used as a linker to fabricate another novel type of dynamic Zn-based MOF (UWDM-4) (Figure 5.54).[133] Variable-temperature ^{2}H SSNMR spectra revealed that the macrocyclic ring can perform large translational movements in the rigid framework of the MOF. A sequence of Zr-based MOFs containing an H-shaped [2]rotaxane molecular shuttle was further studied.[134,135] These materials incorporate diverse mobile constituents, including a rigid rotor (such as phenyl-d_4, tetrazine,

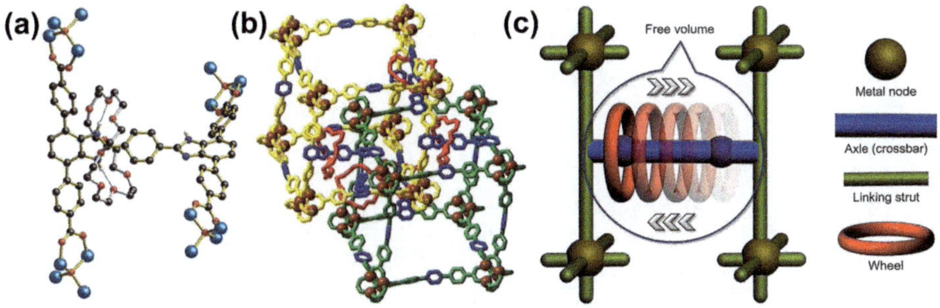

Figure 5.54 Crystal structures of (a) a MIM linker coordinated with four Zn_4O clusters and (b) UWDM-4. (c) The components of a mechanically interlocked molecule (MIM) and showing a schematic diagram of the molecular shuttle. Adapted from ref. 133 with permission from Springer Nature, Copyright 2015.

or tetramethylphenyl) and a rotaxane macrocycle ([24]crown-8). This study demonstrated that the dynamic movements of distinct mobile constituents can have a mutual influence. Specifically, the position of macrocyclic wheel translation is determined by the volume and rotational direction of the rotor, while rotor dynamics are impacted by the translation position of macrocycle.

5.9 Thermal Management

Generally, thermal management involves utilising heating or cooling methods to maintain a constant temperature within a system, thus ensuring the absence of issues such as reduced service life, damage, or decreased efficiency caused by excessively high or low temperatures.[136] For example, during gas storage and separation processes, exothermic and endothermic effects arising from adsorption and desorption, respectively, can result in adsorption to deviate from the ideal adsorption temperature of adsorbents, which would make a decline in storage capacity or separation performance.[18,30] Therefore, it is necessary to implement appropriate thermal management strategies to mitigate these effects. Flexible MOFs have been demonstrated to possess intrinsic thermal management, that the heat released/absorbed during the adsorption/desorption process can be neutralised by the heat absorbed/released during a phase transition.[18,46,137–140] This property makes flexible MOFs competitive in practical applications of gas storage and separation.

The thermal management of flexible MOFs was initially reported for methane storage. Two flexible MOFs, Co(bdp) and Fe(bdp) (where bdp^{2-} = 1,4-benzenedipyrazolate), showing reversible phase transitions with a sharp increase in their methane uptake at a certain threshold pressure, achieving a high CH_4 available capacity of up to 197 $cm^3\ cm^{-3}$ (5.8–65 bar).[18] The mutual agreement between the differential enthalpies calculated from variable-temperature adsorption isotherms and those measured by variable pressure microcalorimetry demonstrated that the phase transition enthalpy partially

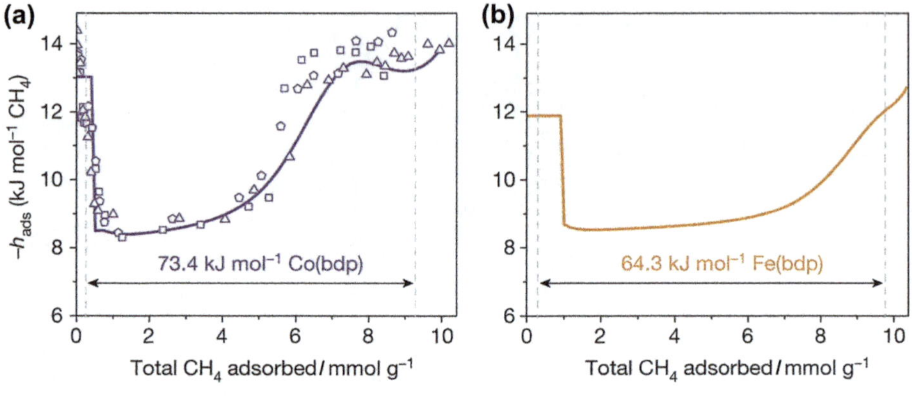

Figure 5.55 (a) Differential enthalpies of CH_4 adsorption for Co(bdp) (a) and Fe(bdp) (b). Dashed grey lines in (a) and (b) indicate the amount of CH_4 adsorbed at 5.8 bar and 35 bar. Adapted from ref. 18 with permission from Springer Nature, Copyright 2015.

offsets the adsorption and desorption enthalpies, resulting in relatively low differential enthalpies in the steepest region of the CH_4 adsorption isotherms (Figure 5.55). Moreover, due to the larger enthalpy change related to the phase transition of Fe(bdp) (8.1 kJ mol^{-1}) in comparison to that of Co(bdp) (7.0 kJ mol^{-1}), the heat of adsorption per litre of adsorbent at 35 bar CH_4 (64.3 kJ) is significantly lower, with a reduction of 12% compared to Co(bdp) and 41% when compared to the rigid MOF HKUST-1.

The thermal management capability of flexible MOFs also contributes to the advancement of efficient vacuum pressure swing systems for gas adsorption and separation. Elastic layer-structured ELM-11 [$Cu(BF_4)_2$(4,4′-bipyridine)$_2$] with a 2D grid framework was reported to show stepwise adsorption towards CO_2, that would facilitate its capability for storage and separation (Figure 5.56).[138] Recently, time-resolved *in situ* synchrotron XRPD was employed to investigate the rate of structural transition in ELM-11 after introduction of CO_2. The results demonstrated its ability to rapidly respond to gas feed, achieving a 95% structural transition at 298 K and CO_2 gas pressure of 250 kPa. Moreover, ELM-11 exhibits excellent thermal management abilities as its framework phase transition energy offsets ~40% of the adsorption heat over the temperature range of 248–335 K. The separation performance for equimolar CO_2–CH_4 mixture under adiabatic conditions indicates that ELM-11 surpasses the rigid MOF HKUST-1 by its relatively excellent CO_2 uptake, selectivity, renewability, and working capacity under adiabatic operating conditions. Given these characteristics, ELM-11 can be employed to establish a vacuum swing adsorption system that operates efficiently.

Other flexible MOFs such as the MIL-53 family,[141] exhibiting breathing effects where their pore size can switch between narrow-pore (np) and large-pore (lp) phases during adsorption and desorption of guest molecules, have also been shown to have good thermal management capabilities. Continuous

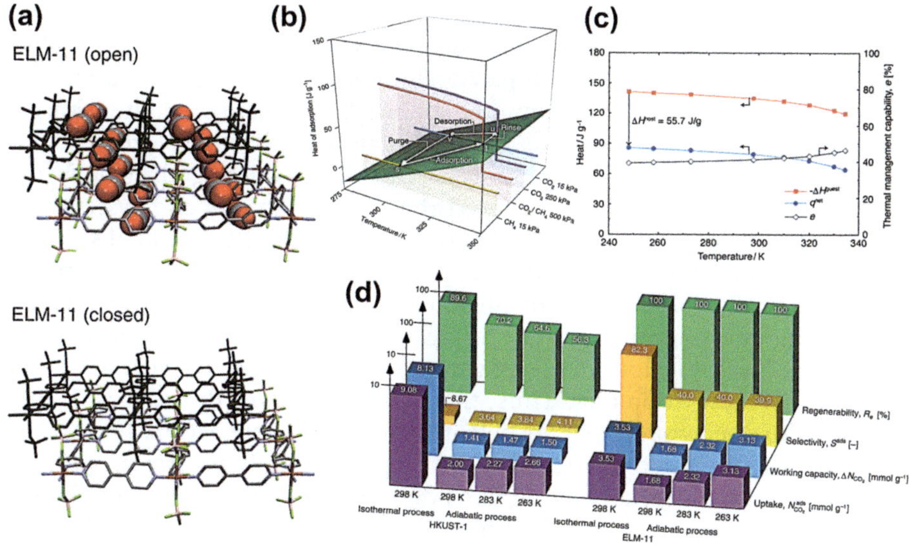

Figure 5.56 (a) Framework structures of ELM-11 in different phases. (b) Temperature dependence of heat of adsorption for ELM-11. (c) Temperature dependence of the thermal management capability of ELM-11 during the adsorption process for an equimolar CO_2–CH_4 mixture gas at 500 kPa ($-\Delta H_{guest}$ = exothermic heat, q_{net} = net heat, and e = thermal management capability). (d) CO_2 separation characteristics of HKUST-1 and ELM-11 under adiabatic conditions and those under isothermal conditions. Adapted from ref. 138 with permission from Nature Publishing Group, Copyright 2020.

tests of the energetic profiles of MIL-53(Al) for adsorption and desorption of CO_2 using pressure-ramped differential scanning calorimetry at various temperatures showed that its total enthalpy values for the entire sorption process were reduced by approximately 8–18% relative to rigid frameworks.[139] In addition to flexible MOFs with large crystal volume changes, flexible MOFs with gate-opening mechanisms, such as ZIF-7 (also known as MAF-3,[142] $Zn(bim)_2$, bim = benzimidazole) and its isoreticular structure (ZIF-9, $Co(bim)_2$; CdIF-13, $Cd(bim)_2$; SOD topology), have been applied for thermal management.[141] Taken together these results suggest that changes in the metal centre of flexible MOFs can significantly affect gas adsorption behaviour (mainly threshold pressure), which might inspire future regulation and prediction of phase transition pressure in flexible MOFs.

5.10 Conclusion and Outlook

Flexible MOFs represent an exceptional type of porous solid regarding their responsive functions towards various chemical and physical stimuli, affording new dimensions to develop functional smart materials. The diverse dynamism in flexible MOFs increases the complexity of rational structural design and property modulation in contrast to rigid and robust MOFs. In general, by stimuli such as different chemical species, temperature, pressure, light, electric or magnetic fields, structural transitions in flexible MOFs are accompanied by changes in pore size and shape, void volume, spin state, charge state, mass or heat balance. Thus, flexible MOFs have been explored for gas storage and separation, sensing, catalysis, proton conducting, drug delivery, and molecular machines. It is believed that new functions and systems of high complexity will be explored using the chemistry of flexible MOFs.

There are challenges for the future demonstration of the great competitive advantages of flexible MOFs. For instance, simultaneous structural transitions during guest accommodation in flexible MOFs can drastically change their adsorption behaviours including kinetic aspects, making such dynamic processes difficult to predict. Thus, there are significant difficulties in the prediction of the properties and performance of flexible MOFs. For MOFs with gate-opening behaviours, it is hard to predict their separation performance with static sorption results, considering the pore opening would also facilitate the co-adsorption of multiple components in the mixture. In this context, any attempts at semi-empirical or computational models would facilitate the understanding and control of dynamic phenomena and processes in flexible MOFs, whereas direct observation might be only possible when advanced analytical techniques are applied. The unique flexible nature of MOFs is expected to advance the field of flexible MOFs into new realms and it is believed that vast opportunities exist for the chemistry of flexible MOFs in other fields. For example, the understanding of molecular dynamics of rotaxane macrocycle results in interesting artificial molecular machines including molecular shuttles and rotary molecular motors, which would facilitate future research on controllable molecular delivery, directional mass transportation, or responsive pumps in microfluidic devices. Comprehensive analyses of energy barriers between different phases of flexible MOFs by massive computational calculations could help to well understand the mechanism of their structural transitions, which in turn will facilitate the rational design of new flexible MOFs. In this context, computational screening *via* machine learning would be an interesting field for the design of complex flexible MOF systems. This is a new materials space that is emerging with novel kinds of functions. The comprehensive study on stimulus-response behaviours of flexible MOFs would facilitate potential applications on smart devices like artificial muscles and adaptive actuators, or media for photodynamic therapy. More endeavours should also be dedicated to issues of composite materials, including interface mechanical properties of hybrid flexible MOFs with other functional matrices, macroscopic size or shape change in practical devices after structural transitions, and durability upon continuous phase transformations. Simultaneously, different flexibility mechanisms in MOFs need to be further studied. For example, flexible-robust MOFs have so far been overlooked especially for gas separation. The control by different stimuli such as temperature and pressure can lead to the variation of their pore spaces and consequent modulation of separation performance. There is great application potential in this type of material that has not been fully demonstrated. Overall, the chemistry of flexible MOFs is a comparatively less-explored topic and this chapter aims to initiate more extensive studies in the future.

Acknowledgements

R.-B.L. acknowledges supports from the National Natural Science Foundation of China (22090061, 22375221, 22101307), the Program for Guangdong Introducing Innovative and Entrepreneurial Teams (2017ZT07C069), the Hundred Talents Program of Sun Yat-Sen University and the Fundamental Research Funds for the Central Universities, (Sun Yat-Sen University, 22qntd2301). This work is also supported by the ERATO and ACCEL project of the Japan Science and Technology Agency (JST) and KAKENHI Grantin-Aid for Specially Promoted Research (No. 25000007) and Scientific Research (S) (JP18H05262) from the Japan Society of the Promotion of Science (JSPS).

References

1. S. Kitagawa, *Angew. Chem., Int. Ed.*, 2015, **54**, 10686.
2. S. Kitagawa and M. Kondo, *Bull. Chem. Soc. Jpn.*, 1998, **71**, 1739.
3. H. Furukawa, K. E. Cordova, M. O'Keeffe and O. M. Yaghi, *Science*, 2013, **341**, 1230444.
4. R.-B. Lin, Z. Zhang and B. Chen, *Acc. Chem. Res.*, 2021, **54**, 3362.
5. R. Kitaura, K. Fujimoto, S. i. Noro, M. Kondo and S. Kitagawa, *Angew. Chem.*, 2002, **114**, 141.
6. C. Serre, F. Millange, C. Thouvenot, M. Nogues, G. Marsolier, D. Louer and G. Ferey, *J. Am. Chem. Soc.*, 2002, **124**, 13519.
7. K. Uemura, R. Matsuda and S. Kitagawa, *J. Solid State Chem.*, 2005, **178**, 2420.
8. G. Férey and C. Serre, *Chem. Soc. Rev.*, 2009, **38**, 1380.
9. C. Gucuyener, J. van den Bergh, J. Gascon and F. Kapteijn, *J. Am. Chem. Soc.*, 2010, **132**, 17704.
10. A. Schneemann, V. Bon, I. Schwedler, I. Senkovska, S. Kaskel and R. A. Fischer, *Chem. Soc. Rev.*, 2014, **43**, 6062.
11. J.-P. Zhang, H.-L. Zhou, D.-D. Zhou, P.-Q. Liao and X.-M. Chen, *Natl. Sci. Rev.*, 2018, **5**, 907.
12. S. Krause, N. Hosono and S. Kitagawa, *Angew. Chem.*, 2020, **59**, 15325.
13. S.-Q. Wang, S. Mukherjee and M. J. Zaworotko, *Faraday Discuss.*, 2021, **231**, 9.
14. N. Behera, J. Duan, W. Jin and S. Kitagawa, *EnergyChem*, 2021, **3**, 100067.
15. J.-P. Zhang, H.-L. Zhou, D.-D. Zhou, P.-Q. Liao and S. Kitagawa, in *Comprehensive Supramolecular Chemistry II*, ed. J. L. Atwood, Elsevier, Oxford, 2017, pp. 425–474.
16. D. Li and K. Kaneko, *Chem. Phys. Lett.*, 2001, **335**, 50.

17. R. Kitaura, K. Seki, G. Akiyama and S. Kitagawa, *Angew. Chem., Int. Ed.*, 2003, **42**, 428.
18. J. A. Mason, J. Oktawiec, M. K. Taylor, M. R. Hudson, J. Rodriguez, J. E. Bachman, M. I. Gonzalez, A. Cervellino, A. Guagliardi, C. M. Brown, P. L. Llewellyn, N. Masciocchi and J. R. Long, *Nature*, 2015, **527**, 357.
19. C. Serre, C. Mellot-Draznieks, S. Surble, N. Audebrand, Y. Filinchuk and G. Férey, *Science*, 2007, **315**, 1828.
20. P. Horcajada, F. Salles, S. Wuttke, T. Devic, D. Heurtaux, G. Maurin, A. Vimont, M. Daturi, O. David, E. Magnier, N. Stock, Y. Filinchuk, D. Popov, C. Riekel, G. Férey and C. Serre, *J. Am. Chem. Soc.*, 2011, **133**, 17839.
21. Y.-S. Wei, K.-J. Chen, P.-Q. Liao, B.-Y. Zhu, R.-B. Lin, H.-L. Zhou, B.-Y. Wang, W. Xue, J.-P. Zhang and X.-M. Chen, *Chem. Sci.*, 2013, **4**, 1539.
22. B. Chen, C. Liang, J. Yang, D. S. Contreras, Y. L. Clancy, E. B. Lobkovsky, O. M. Yaghi and S. Dai, *Angew. Chem., Int. Ed.*, 2006, **45**, 1390.
23. J.-P. Zhang and X.-M. Chen, *J. Am. Chem. Soc.*, 2008, **130**, 6010.
24. C. A. Fernandez, J. Liu, P. K. Thallapally and D. M. Strachan, *J. Am. Chem. Soc.*, 2012, **134**, 9046.
25. E. J. Carrington, C. A. McAnally, A. J. Fletcher, S. P. Thompson, M. Warren and L. Brammer, *Nat. Chem.*, 2017, **9**, 882.
26. S. Horike, S. Shimomura and S. Kitagawa, *Nat. Chem.*, 2009, **1**, 695.
27. T. K. Maji, G. Mostafa, R. Matsuda and S. Kitagawa, *J. Am. Chem. Soc.*, 2005, **127**, 17152.
28. B.-Q. Song, Q.-Y. Yang, S.-Q. Wang, M. Vandichel, A. Kumar, C. Crowley, N. Kumar, C.-H. Deng, V. GasconPerez and M. Lusi, *J. Am. Chem. Soc.*, 2020, **142**, 6896.
29. V. I. Nikolayenko, D. C. Castell, D. Sensharma, M. Shivanna, L. Loots, K. A. Forrest, C. J. Solanilla-Salinas, K.-i. Otake, S. Kitagawa, L. J. Barbour, B. Space and M. J. Zaworotko, *Nat. Chem.*, 2023, **15**, 542.
30. H. J. Choi, M. Dinca and J. R. Long, *J. Am. Chem. Soc.*, 2008, **130**, 7848.
31. F. Salles, G. Maurin, C. Serre, P. L. Llewellyn, C. Knoefel, H. J. Choi, Y. Filinchuk, L. Oliviero, A. Vimont and J. R. Long, *J. Am. Chem. Soc.*, 2010, **132**, 13782.
32. R. Matsuda, R. Kitaura, S. Kitagawa, Y. Kubota, R. V. Belosludov, T. C. Kobayashi, H. Sakamoto, T. Chiba, M. Takata, Y. Kawazoe and Y. Mita, *Nature*, 2005, **436**, 238.
33. M. Bonneau, C. Lavenn, J.-J. Zheng, A. Legrand, T. Ogawa, K. Sugimoto, F.-X. Coudert, R. Reau, S. Sakaki, K.-i. Otake and S. Kitagawa, *Nat. Chem.*, 2022, **14**, 816.
34. Y. Gu, J. J. Zheng, K. I. Otake, K. Sugimoto, N. Hosono, S. Sakaki, F. Li and S. Kitagawa, *Angew. Chem., Int. Ed.*, 2020, **59**, 15517.
35. S. Krause, V. Bon, I. Senkovska, U. Stoeck, D. Wallacher, D. M. Tobbens, S. Zander, R. S. Pillai, G. Maurin, F. X. Coudert and S. Kaskel, *Nature*, 2016, **532**, 348.
36. S. Krause, V. Bon, I. Senkovska, D. M. Tobbens, D. Wallacher, R. S. Pillai, G. Maurin and S. Kaskel, *Nat. Commun.*, 2018, **9**, 1573.

37. B. Garai, V. Bon, S. Krause, F. Schwotzer, M. Gerlach, I. Senkovska and S. Kaskel, *Chem. Mater.*, 2020, **32**, 889.
38. S. Krause, J. D. Evans, V. Bon, I. Senkovska, S. Ehrling, P. Iacomi, D. M. Tobbens, D. Wallacher, M. S. Weiss, B. Zheng, P. G. Yot, G. Maurin, P. L. Llewellyn, F. X. Coudert and S. Kaskel, *Chem. Sci.*, 2020, **11**, 9468.
39. S. Krause, F. S. Reuter, S. Ehrling, V. Bon, I. Senkovska, S. Kaskel and E. Brunner, *Chem. Mater.*, 2020, **32**, 4641.
40. J. J. M. Beenakker, V. D. Borman and S. Y. Krylov, *Chem. Phys. Lett.*, 1995, **232**, 379.
41. J. Y. Kim, L. Zhang, R. Balderas-Xicohtencatl, J. Park, M. Hirscher, H. R. Moon and H. Oh, *J. Am. Chem. Soc.*, 2017, **139**, 17743.
42. J. Y. Kim, J. Park, J. Ha, M. Jung, D. Wallacher, A. Franz, R. Balderas-Xicohtencatl, M. Hirscher, S. G. Kang, J. T. Park, I. H. Oh, H. R. Moon and H. Oh, *J. Am. Chem. Soc.*, 2020, **142**, 13278.
43. M. Jung, J. Park, R. Muhammad, J. Y. Kim, V. Grzimek, M. Russina, H. R. Moon, J. T. Park and H. Oh, *Adv. Mater.*, 2021, **33**, e2007412.
44. L. Zhang, S. Jee, J. Park, M. Jung, D. Wallacher, A. Franz, W. Lee, M. Yoon, K. Choi, M. Hirscher and H. Oh, *J. Am. Chem. Soc.*, 2019, **141**, 19850.
45. H. Sato, W. Kosaka, R. Matsuda, A. Hori, Y. Hijikata, R. V. Belosludov, S. Sakaki, M. Takata and S. Kitagawa, *Science*, 2014, **343**, 167.
46. M. K. Taylor, T. e. Runčevski, J. Oktawiec, J. E. Bachman, R. L. Siegelman, H. Jiang, J. A. Mason, J. D. Tarver and J. R. Long, *J. Am. Chem. Soc.*, 2018, **140**, 10324.
47. H. Zeng, M. Xie, Y. L. Huang, Y. Zhao, X. J. Xie, J. P. Bai, M. Y. Wan, R. Krishna, W. Lu and D. Li, *Angew. Chem., Int. Ed.*, 2019, **58**, 8515.
48. X.-W. Zhang, D.-D. Zhou and J.-P. Zhang, *Chem*, 2021, **7**, 1006.
49. R.-B. Lin, L. Li, H. Wu, H. Arman, B. Li, R. G. Lin, W. Zhou and B. Chen, *J. Am. Chem. Soc.*, 2017, **139**, 8022.
50. J. Wang, Y. Zhang, P. Zhang, J. Hu, R.-B. Lin, Q. Deng, Z. Zeng, H. Xing, S. Deng and B. Chen, *J. Am. Chem. Soc.*, 2020, **142**, 9744.
51. J. Wang, Y. Zhang, Y. Su, X. Liu, P. Zhang, R.-B. Lin, S. Chen, Q. Deng, Z. Zeng, S. Deng and B. Chen, *Nat. Commun.*, 2022, **13**, 200.
52. S. Sen, N. Hosono, J. J. Zheng, S. Kusaka, R. Matsuda, S. Sakaki and S. Kitagawa, *J. Am. Chem. Soc.*, 2017, **139**, 18313.
53. C. Gu, N. Hosono, J.-J. Zheng, Y. Sato, S. Kusaka, S. Sakaki and S. Kitagawa, *Science*, 2019, **363**, 387.
54. R.-B. Lin, H. Wu, L. Li, X.-L. Tang, Z. Li, J. Gao, H. Cui, W. Zhou and B. Chen, *J. Am. Chem. Soc.*, 2018, **140**, 12940.
55. Y. Yang, L. Li, R.-B. Lin, Y. Ye, Z. Yao, L. Yang, F. Xiang, S. Chen, Z. Zhang, S. Xiang and B. Chen, *Nat. Chem.*, 2021, **13**, 933.
56. Y. Chen, Y. Yang, Y. Wang, Q. Xiong, J. Yang, S. Xiang, L. Li, J. Li, Z. Zhang and B. Chen, *J. Am. Chem. Soc.*, 2022, **144**, 17033.
57. Y. Yang, H. Zhang, Z. Yuan, J.-Q. Wang, F. Xiang, L. Chen, F. Wei, S. Xiang, B. Chen and Z. Zhang, *Angew. Chem.*, 2022, **61**, e202207579.
58. D. Wang and Y. Zhao, *Angew. Chem., Int. Ed.*, 2023, e202217903.

59. L. Li, R.-B. Lin, R. Krishna, X. Wang, B. Li, H. Wu, J. Li, W. Zhou and B. Chen, *J. Am. Chem. Soc.*, 2017, **139**, 7733.
60. B. Li, X. Cui, D. O'Nolan, H.-M. Wen, M. Jiang, R. Krishna, H. Wu, R.-B. Lin, Y.-S. Chen, D. Yuan, H. Xing, W. Zhou, Q. Ren, G. Qian, M. J. Zaworotko and B. Chen, *Adv. Mater.*, 2017, **29**, 1704210.
61. L. Li, H. M. Wen, C. He, R.-B. Lin, R. Krishna, H. Wu, W. Zhou, J. Li, B. Li and B. Chen, *Angew. Chem., Int. Ed.*, 2018, **57**, 15183.
62. M. R. Ryder, B. Civalleri, T. D. Bennett, S. Henke, S. Rudic, G. Cinque, F. Fernandez-Alonso and J. C. Tan, *Phys. Rev. Lett.*, 2014, **113**, 215502.
63. A. Knebel, B. Geppert, K. Volgmann, D. I. Kolokolov, A. G. Stepanov, J. Twiefel, P. Heitjans, D. Volkmer and J. Caro, *Science*, 2017, **358**, 347.
64. H. Zeng, M. Xie, T. Wang, R. J. Wei, X. J. Xie, Y. Zhao, W. Lu and D. Li, *Nature*, 2021, **595**, 542.
65. Q. Dong, X. Zhang, S. Liu, R.-B. Lin, Y. Guo, Y. Ma, A. Yonezu, R. Krishna, G. Liu, J. Duan, R. Matsuda, W. Jin and B. Chen, *Angew. Chem.*, 2020, **59**, 22756.
66. P.-Q. Liao, N.-Y. Huang, W.-X. Zhang, J.-P. Zhang and X.-M. Chen, *Science*, 2017, **356**, 1193.
67. D. Peralta, G. Chaplais, A. Simon-Masseron, K. Barthelet and G. D. Pirngruber, *Ind. Eng. Chem. Res.*, 2012, **51**, 4692.
68. H. Wang, X. Dong, E. Velasco, D. H. Olson, Y. Han and J. Li, *Energy Environ. Sci.*, 2018, **11**, 1226.
69. Q. Chen, S. Xian, X. Dong, Y. Liu, H. Wang, D. H. Olson, L. J. Williams, Y. Han, X. H. Bu and J. Li, *Angew. Chem.*, 2021, **60**, 10593.
70. P. Wang, K. I. Otake, N. Hosono and S. Kitagawa, *Angew. Chem.*, 2021, **60**, 7030.
71. L. Alaerts, C. E. Kirschhock, M. Maes, M. A. van der Veen, V. Finsy, A. Depla, J. A. Martens, G. V. Baron, P. A. Jacobs, J. F. Denayer and D. E. De Vos, *Angew. Chem., Int. Ed.*, 2007, **46**, 4293.
72. L. Li, L. Guo, D. H. Olson, S. Xian, Z. Zhang, Q. Yang, K. Wu, Y. Yang, Z. Bao, Q. Ren and J. Li, *Science*, 2022, **377**, 335.
73. J. E. Warren, C. G. Perkins, K. E. Jelfs, P. Boldrin, P. A. Chater, G. J. Miller, T. D. Manning, M. E. Briggs, K. C. Stylianou, J. B. Claridge and M. J. Rosseinsky, *Angew. Chem., Int. Ed.*, 2014, **53**, 4592.
74. X. Yang, H.-L. Zhou, C.-T. He, Z.-W. Mo, J.-W. Ye, X.-M. Chen and J.-P. Zhang, *Research*, 2019, **2019**, 9463719.
75. S. Q. Wang, S. Mukherjee, E. Patyk-Kazmierczak, S. Darwish, A. Bajpai, Q. Y. Yang and M. J. Zaworotko, *Angew. Chem., Int. Ed.*, 2019, **58**, 6630.
76. D.-D. Zhou, P. Chen, C. Wang, S.-S. Wang, Y. Du, H. Yan, Z.-M. Ye, C.-T. He, R.-K. Huang, Z.-W. Mo, N.-Y. Huang and J.-P. Zhang, *Nat. Mater.*, 2019, **18**, 994.
77. B. Le Ouay, C. Watanabe, S. Mochizuki, M. Takayanagi, M. Nagaoka, T. Kitao and T. Uemura, *Nat. Commun.*, 2018, **9**, 3635.
78. R.-B. Lin, S.-Y. Liu, J.-W. Ye, X.-Y. Li and J.-P. Zhang, *Adv. Sci.*, 2016, **3**, 1500434.

79. N. Yanai, K. Kitayama, Y. Hijikata, H. Sato, R. Matsuda, Y. Kubota, M. Takata, M. Mizuno, T. Uemura and S. Kitagawa, *Nat. Mater.*, 2011, **10**, 787.
80. Y. Takashima, V. M. Martinez, S. Furukawa, M. Kondo, S. Shimomura, H. Uehara, M. Nakahama, K. Sugimoto and S. Kitagawa, *Nat. Commun.*, 2011, **2**, 168.
81. R. Haldar, R. Matsuda, S. Kitagawa, S. J. George and T. K. Maji, *Angew. Chem., Int. Ed.*, 2014, **53**, 11772.
82. M. Zhang, G. Feng, Z. Song, Y. P. Zhou, H. Y. Chao, D. Yuan, T. T. Tan, Z. Guo, Z. Hu, B. Z. Tang, B. Liu and D. Zhao, *J. Am. Chem. Soc.*, 2014, **136**, 7241.
83. B. Manna, A. K. Chaudhari, B. Joarder, A. Karmakar and S. K. Ghosh, *Angew. Chem., Int. Ed.*, 2013, **52**, 998.
84. A. Douvali, A. C. Tsipis, S. V. Eliseeva, S. Petoud, G. S. Papaefstathiou, C. D. Malliakas, I. Papadas, G. S. Armatas, I. Margiolaki, M. G. Kanatzidis, T. Lazarides and M. J. Manos, *Angew. Chem., Int. Ed.*, 2015, **54**, 1651.
85. Q. Chen, Z. Chang, W. C. Song, H. Song, H. B. Song, T. L. Hu and X. H. Bu, *Angew. Chem., Int. Ed.*, 2013, **52**, 11550.
86. M. Souto, J. Romero, J. Calbo, I. J. Vitorica-Yrezabal, J. L. Zafra, J. Casado, E. Orti, A. Walsh and G. Minguez Espallargas, *J. Am. Chem. Soc.*, 2018, **140**, 10562.
87. K. Roztocki, F. Formalik, A. Krawczuk, I. Senkovska, B. Kuchta, S. Kaskel and D. Matoga, *Angew. Chem.*, 2020, **59**, 4491.
88. W. Cai, A. Gladysiak, M. Aniola, V. J. Smith, L. J. Barbour and A. Katrusiak, *J. Am. Chem. Soc.*, 2015, **137**, 9296.
89. G. J. Halder, C. J. Kepert, B. Moubaraki, K. S. Murray and J. D. Cashion, *Science*, 2002, **298**, 1762.
90. P. Wu, Y. Li, J. J. Zheng, N. Hosono, K. I. Otake, J. Wang, Y. Liu, L. Xia, M. Jiang, S. Sakaki and S. Kitagawa, *Nat. Commun.*, 2019, **10**, 4362.
91. S. Yuan, L. Zou, H. Li, Y. P. Chen, J. Qin, Q. Zhang, W. Lu, M. B. Hall and H. C. Zhou, *Angew. Chem.*, 2016, **128**, 10934.
92. J. Pang, S. Yuan, D. Du, C. Lollar, L. Zhang, M. Wu, D. Yuan, H. C. Zhou and M. Hong, *Angew. Chem., Int. Ed.*, 2017, **56**, 14622.
93. Y. Zhang, X. Zhang, J. Lyu, K.-i. Otake, X. Wang, L. R. Redfern, C. D. Malliakas, Z. Li, T. Islamoglu and B. Wang, *J. Am. Chem. Soc.*, 2018, **140**, 11179.
94. T. Uemura, N. Yanai and S. Kitagawa, *Chem. Soc. Rev.*, 2009, **38**, 1228.
95. N. Hosono and T. Uemura, *Acc. Chem. Res.*, 2021, **54**, 3593.
96. T. Uemura, D. Hiramatsu, Y. Kubota, M. Takata and S. Kitagawa, *Angew. Chem., Int. Ed.*, 2007, **46**, 4987.
97. K. D. Kreuer, S. J. Paddison, E. Spohr and M. Schuster, *Chem. Rev.*, 2004, **104**, 4637.
98. D. W. Lim and H. Kitagawa, *Chem. Rev.*, 2020, **120**, 8416.
99. Y. Ye, L. Gong, S. Xiang, Z. Zhang and B. Chen, *Adv. Mater.*, 2020, **32**, e1907090.
100. S. Horike, D. Umeyama and S. Kitagawa, *Acc. Chem. Res.*, 2013, **46**, 2376.

101. F. Yang, G. Xu, Y. Dou, B. Wang, H. Zhang, H. Wu, W. Zhou, J.-R. Li and B. Chen, *Nat. Energy*, 2017, **2**, 877.
102. M. Bazaga-Garcia, R. M. Colodrero, M. Papadaki, P. Garczarek, J. Zon, P. Olivera-Pastor, E. R. Losilla, L. Leon-Reina, M. A. Aranda, D. Choquesillo-Lazarte, K. D. Demadis and A. Cabeza, *J. Am. Chem. Soc.*, 2014, **136**, 5731.
103. Y. S. Wei, X. P. Hu, Z. Han, X. Y. Dong, S. Q. Zang and T. C. Mak, *J. Am. Chem. Soc.*, 2017, **139**, 3505.
104. Z.-H. Li, H. Zeng, G. Zeng, C. Ru, G. Li, W. Yan, Z. Shi and S. Feng, *Angew. Chem., Int. Ed.*, 2021, **60**, 26577.
105. Y.-H. Shih, S.-H. Lo, N.-S. Yang, B. Singco, Y.-J. Cheng, C.-Y. Wu, I. H. Chang, H.-Y. Huang and C.-H. Lin, *ChemPlusChem*, 2012, **77**, 982.
106. A. M. Rice, C. R. Martin, V. A. Galitskiy, A. A. Berseneva, G. A. Leith and N. B. Shustova, *Chem. Rev.*, 2020, **120**, 8790.
107. K. Muller, J. Helfferich, F. Zhao, R. Verma, A. B. Kanj, V. Meded, D. Bleger, W. Wenzel and L. Heinke, *Adv. Mater.*, 2018, **30**, 1706551.
108. A. Shigematsu, T. Yamada and H. Kitagawa, *J. Am. Chem. Soc.*, 2011, **133**, 2034.
109. S. Bureekaew, S. Horike, M. Higuchi, M. Mizuno, T. Kawamura, D. Tanaka, N. Yanai and S. Kitagawa, *Nat. Mater.*, 2009, **8**, 831.
110. D. Umeyama, S. Horike, M. Inukai, Y. Hijikata and S. Kitagawa, *Angew. Chem., Int. Ed.*, 2011, **50**, 11706.
111. D. Umeyama, S. Horike, M. Inukai, T. Itakura and S. Kitagawa, *J. Am. Chem. Soc.*, 2015, **137**, 864.
112. D. Umeyama, S. Horike, M. Inukai and S. Kitagawa, *J. Am. Chem. Soc.*, 2013, **135**, 11345.
113. Y. Ye, X. Wu, Z. Yao, L. Wu, Z. Cai, L. Wang, X. Ma, Q.-H. Chen, Z. Zhang and S. Xiang, *J. Mater. Chem. A*, 2016, **4**, 4062.
114. L. Sun, M. G. Campbell and M. Dincă, *Angew. Chem., Int. Ed.*, 2016, **55**, 3566.
115. S. Takaishi, M. Hosoda, T. Kajiwara, H. Miyasaka, M. Yamashita, Y. Nakanishi, Y. Kitagawa, K. Yamaguchi, A. Kobayashi and H. Kitagawa, *Inorg. Chem.*, 2009, **48**, 9048.
116. X. Huang, P. Sheng, Z. Tu, F. Zhang, J. Wang, H. Geng, Y. Zou, C.-a. Di, Y. Yi, Y. Sun, W. Xu and D. Zhu, *Nat. Commun.*, 2015, **6**, 7408.
117. S. Chong, S. M. J. Rogge and J. Kim, *Chem. Mater.*, 2022, **34**, 254.
118. P. Horcajada, C. Serre, G. Maurin, N. A. Ramsahye, F. Balas, M. Vallet-Regí, M. Sebban, F. Taulelle and G. Férey, *J. Am. Chem. Soc.*, 2008, **130**, 6774.
119. C. Serre, F. Millange, S. Surble and G. Férey, *Angew. Chem., Int. Ed.*, 2004, **43**, 6285.
120. S. Surble, C. Serre, C. Mellot-Draznieks, F. Millange and G. Férey, *Chem. Commun.*, 2006, 284.
121. T. R. Whitfield, X. Q. Wang, L. M. Liu and A. J. Jacobson, *Solid State Sci.*, 2005, **7**, 1096.

122. S. Bauer, C. Serre, T. Devic, P. Horcajada, J. Marrot, G. Férey and N. Stock, *Inorg. Chem.*, 2008, **47**, 7568.
123. P. Horcajada, S. Surble, C. Serre, D. Y. Hong, Y. K. Seo, J. S. Chang, J. M. Greneche, I. Margiolaki and G. Férey, *Chem. Commun.*, 2007, 2820.
124. P. Horcajada, C. Serre, M. Vallet-Regi, M. Sebban, F. Taulelle and G. Férey, *Angew. Chem., Int. Ed.*, 2006, **45**, 5974.
125. P. Horcajada, T. Chalati, C. Serre, B. Gillet, C. Sebrie, T. Baati, J. F. Eubank, D. Heurtaux, P. Clayette, C. Kreuz, J. S. Chang, Y. K. Hwang, V. Marsaud, P. N. Bories, L. Cynober, S. Gil, G. Férey, P. Couvreur and R. Gref, *Nat. Mater.*, 2010, **9**, 172.
126. X. G. Wang, L. Xu, M. J. Li and X. Z. Zhang, *Angew. Chem., Int. Ed.*, 2020, **59**, 18078.
127. W. Danowski, T. van Leeuwen, S. Abdolahzadeh, D. Roke, W. R. Browne, S. J. Wezenberg and B. L. Feringa, *Nat. Nanotechnol.*, 2019, **14**, 488.
128. W. Meng, S. Kondo, T. Ito, K. Komatsu, J. Pirillo, Y. Hijikata, Y. Ikuhara, T. Aida and H. Sato, *Nature*, 2021, **598**, 298.
129. V. N. Vukotic, K. J. Harris, K. Zhu, R. W. Schurko and S. J. Loeb, *Nat. Chem.*, 2012, **4**, 456.
130. V. N. Vukotic, C. A. O'Keefe, K. Zhu, K. J. Harris, C. To, R. W. Schurko and S. J. Loeb, *J. Am. Chem. Soc.*, 2015, **137**, 9643.
131. K. Zhu, V. N. Vukotic, C. A. O'Keefe, R. W. Schurko and S. J. Loeb, *J. Am. Chem. Soc.*, 2014, **136**, 7403.
132. P. L. Anelli, N. Spencer and J. F. Stoddart, *J. Am. Chem. Soc.*, 1991, **113**, 5131.
133. K. Zhu, C. A. O'Keefe, V. N. Vukotic, R. W. Schurko and S. J. Loeb, *Nat. Chem.*, 2015, **7**, 514.
134. G. Gholami, B. H. Wilson, K. Zhu, C. A. O'Keefe, R. W. Schurko and S. J. Loeb, *Faraday Discuss.*, 2021, **225**, 358.
135. B. H. Wilson, C. S. Vojvodin, G. Gholami, L. M. Abdulla, C. A. O'Keefe, R. W. Schurko and S. J. Loeb, *Chem*, 2021, 7, 202.
136. B. Li, H.-M. Wen, W. Zhou, J. Q. Xu and B. Chen, *Chem*, 2016, **1**, 557.
137. S. Hiraide, H. Tanaka, N. Ishikawa and M. T. Miyahara, *ACS Appl. Mater. Interfaces*, 2017, **9**, 41066.
138. S. Hiraide, Y. Sakanaka, H. Kajiro, S. Kawaguchi, M. T. Miyahara and H. Tanaka, *Nat. Commun.*, 2020, **11**, 3867.
139. W. K. Feldmann, C. Esterhuysen and L. J. Barbour, *ChemSusChem*, 2020, **13**, 5220.
140. C. M. McGuirk, T. Runčevski, J. Oktawiec, A. Turkiewicz, M. K. Taylor and J. R. Long, *J. Am. Chem. Soc.*, 2018, **140**, 15924.
141. F. Millange and R. I. Walton, *Isr. J. Chem.*, 2018, **58**, 1019.
142. X.-C. Huang, J.-P. Zhang and X.-M. Chen, *Chin. Sci. Bull.*, 2003, **48**, 1531.

Index